Cell Press Reviews:
Core Concepts in Cell Biology

Cell Press Reviews

Core Concepts in Cell Biology

Curated by

Rebecca Alvania; Scientific Editor, Cell Press

Original Articles Edited by the Following Cell Press Scientific Editors

Rebecca Alvania
Marie Bao
Karen Carniol
Kara Cerveny
Michaeleen Doucleff
Anne Knowlton
Robert Kruger
Florian Maderspacher
Cyrus Martin
Sri Devi Narasimhan
Deborah Taylor

AMSTERDAM • BOSTON • HEIDELBERG • LONDON
NEW YORK • OXFORD • PARIS • SAN DIEGO
SAN FRANCISCO • SINGAPORE • SYDNEY • TOKYO

AP Cell is an imprint of Elsevier

ELSEVIER

CellPress

Emilie Marcus, CEO, Editor-in-Chief
Joanne Tracy, Vice President of Business
Development
Keith Wollman, Vice President of Operations
Peter Lee, Publishing Director
Deborah Sweet, Publishing Director
Katja Brose, Editorial Director, Reviews
Strategy
Elena Porro, Editorial Director, Content
Development
Meredith Adinolfi, Director of Production
Jonathan Atkinson, Director of Marketing

Science and Technology Books

Suzanne BeDell, Managing Director
Laura Colantoni, Vice President & Publisher
Amorette Pedersen, Vice President, Channel
Management & Marketing Operations
Tommy Doyle, Senior Vice President,
Strategy, Business Development & Continuity
Publishing
Dave Cella, Publishing Director, Life Sciences
Janice Audet, Publisher
Elizabeth Gibson, Editorial Project Manager
Julia Haynes, Production Manager
Ofelia Chernock, Portfolio Marketing
Manager
Melissa Fulkerson, Senior Channel Manager
Cory Polonetsky, Director, Channel Strategy &
Pricing

AP Cell is an imprint of Elsevier
32 Jamestown Road, London NW1 7BY, UK
225 Wyman Street, Waltham, MA 02451, USA
525 B Street, Suite 1800, San Diego, CA 92101-4495, USA

British Library Cataloguing-in-Publication Data
A catalogue record for this book is available from the British Library

Library of Congress Cataloging-in-Publication Data
A catalog record for this book is available from the Library of Congress

ISBN: 978-0-12-420193-4

For information on all AP Cell publications
visit our website at www.store.elsevier.com

Typeset by TNQ Books and Journals Pvt. Ltd.

This book has been manufactured using Print On Demand technology.
Each copy is produced to order.
14 15 16 17 18   10 9 8 7 6 5 4 3 2 1

Contents

About Cell Press

Cell Press is a leading publisher in the biological sciences and is committed to improving scientific communication through the publication of exciting research and reviews. Cell Press publishes 30 journals, including the Trends reviews series, spanning the breadth of the biological sciences. Research titles published by Cell Press include *Cell*, *Cancer Cell*, *Cell Stem Cell*, *Cell Host & Microbe*, *Cell Metabolism*, *Chemistry & Biology*, *Current Biology*, *Developmental Cell*, *Immunity*, *Molecular Cell*, *Neuron*, *Structure*, and the open-access journal *Cell Reports*. In addition to publishing high-impact findings, Cell Press research journals publish a wide variety of peer-reviewed review and opinion articles, essays from leaders in the field, graphical Snap-Shots, science news articles, and much more. Cell Press is also the publisher of three society journals: *Biophysical Journal*, *American Journal of Human Genetics*, as well as the open-access journal *Stem Cell Reports*.

The *Trends* reviews journals are also part of the Cell Press family and consist of 14 monthly review titles that publish in a range of areas across the biological sciences. Peer-reviewed and thoroughly edited review and opinion articles cover the most recent developments in relevant fields in an authoritative, succinct and broadly accessible manner. Together with a range of additional shorter formats, *Trends* journals collectively provide a forum for hypothesis and debate.

As part of its mission to be a leader in scientific communication, Cell Press also organizes scientific meetings across a wide range of topics, hosts online webinars to bring leading scientists to the broadest international audience, and is committed to promoting innovation in science publishing.

Contributors

Corresponding authors and affiliations

Bruno Antonny
Institut de Pharmacologie Moléculaire et Cellulaire, Université de Nice Sophia Antipolis et CNRS, 06560 Valbonne, France

Vytas A. Bankaitis
Department of Cell and Developmental Biology, University of North Carolina School of Medicine, Chapel Hill, NC 27599-7090, USA; Lineberger Comprehensive Cancer Center, University of North Carolina School of Medicine, Chapel Hill, NC 27599-7090, USA

Clemens Cabernard
Biozentrum, University of Basel, CH-4056 Basel, Switzerland

David G. Drubin
Department of Molecular and Cell Biology, University of California, Berkeley, CA 94720-3202, USA

Zvulun Elazar
Department of Biological Chemistry, The Weizmann Institute of Science, Rehovot, 76100, Israel

Scott D. Emr
Weill Institute for Cell and Molecular Biology, Cornell University, Weill Hall, Ithaca, NY 14853, USA; Department of Molecular Biology and Genetics, Cornell University, Weill Hall, Ithaca, NY 14853, USA

Daniel A. Fletcher
Bioengineering Department and Biophysics Program, University of California, Berkeley, CA 94720, USA; Physical Biosciences Division, Lawrence Berkeley National Laboratory, Berkeley, CA 94720, USA

Nathan W. Goehring
Max Planck Institute of Molecular Cell Biology and Genetics (MPI-CBG), Pfotenhauerstraße 108, 01307 Dresden, Germany; Max Planck Institute for the Physics of Complex Systems (MPI-PKS), Nöthnitzer Straße 38, 01187 Dresden, Germany; Cancer Research UK London Research Institute, Lincoln's Inn Fields Laboratories, 44 Lincoln's Inn Fields, London WC2A 3LY, UK

Stephan W. Grill
Max Planck Institute of Molecular Cell Biology and Genetics (MPI-CBG), Pfotenhauerstraße 108, 01307 Dresden, Germany; Max Planck Institute for the Physics of Complex Systems (MPI-PKS), Nöthnitzer Straße 38, 01187 Dresden, Germany

Gregg G. Gundersen
Department of Pathology and Cell Biology, College of Physicians and Surgeons, Columbia University, 630 West 168th Street, New York, NY 10032, USA

Barry Honig
Department of Biochemistry and Molecular Biophysics, Columbia University, 1150 Saint Nicholas Avenue, New York, NY 10032, USA; Howard Hughes Medical Institute, Columbia University, 1130 Saint Nicholas Avenue, New York, NY 10032, USA; Center for Computational Biology and Bioinformatics, Columbia University, 1130 Saint Nicholas Avenue, New York, NY 10032, USA

Junjie Hu
Department of Genetics and Cell Biology, College of Life Sciences, and Tianjin Key Laboratory of Protein Sciences, Nankai University, Tianjin 300071, China

Geert J.P.L. Kops
Department of Medical Oncology, Department of Molecular Cancer Research and Cancer Genomics Centre, University Medical Center Utrecht, 3584 CG Utrecht, The Netherlands

Galit Lahav
Department of Systems Biology, Harvard Medical School, Boston, MA 02115, USA

Jean-Claude Martinou
Department of Cell Biology, University of Geneva, Faculty of Sciences, 30 quai Ernest-Ansermet, 1211 Geneva 4, Switzerland; Surgical Neurology Branch, NINDS, National Institute of Health, Bethesda, MD 20892, USA

Tom Misteli
National Cancer Institute, National Institutes of Health, Bethesda, MD 20892, USA

Ewa Paluch
Max Planck Institute of Molecular Cell Biology and Genetics, Dresden, 01307, Germany; International Institute of Molecular and Cell Biology, Warsaw, 02-109, Poland

Tom A. Rapoport
Howard Hughes Medical Institute, Department of Cell Biology, Harvard Medical School, Boston, MA 02115, USA

Anne J. Ridley
Randall Division of Cell and Molecular Biophysics, King's College London, New Hunt's House, Guy's Campus, London SE1 1UL, United Kingdom

David C. Rubinsztein
Department of Medical Genetics, Cambridge Institute for Medical Research, Wellcome/MRC Building, Addenbrooke's Hospital, Hills Road, Cambridge CB2 0XY, UK

Guillaume Salbreux
Max Planck Institute for the Physics of Complex Systems, Dresden, 01187, Germany

Lawrence Shapiro
Department of Biochemistry and Molecular Biophysics, Columbia University, 1150 Saint Nicholas Avenue, New York, NY 10032, USA

Berend Snel
Theoretical Biology and Bioinformatics, Department of Biology, Science Faculty, Utrecht University, 3584 CH Utrecht, The Netherlands

Manuel Théry
Laboratoire de Physiologie Cellulaire et Végétale, Institut de Recherche en Technologies et Sciences pour le Vivant, CNRS/UJF/INRA/CEA, 17 Rue des Martyrs, 38054, Grenoble, France

Frank Uhlmann
Chromosome Segregation Laboratory, Cancer Research UK London Research Institute, 44 Lincoln's Inn Fields, London WC2A 3LY, UK

Orion D. Weiner
Cardiovascular Research Institute and Department of Biochemistry, University of California San Francisco, San Francisco, CA 94143, USA

Howard J. Worman
Department of Pathology and Cell Biology, College of Physicians and Surgeons, Columbia University, 630 West 168th Street, New York, NY 10032, USA; Department of Medicine, College of Physicians and Surgeons, Columbia University, 630 West 168th Street, New York, NY 10032, USA

Richard J. Youle
Surgical Neurology Branch, NINDS, National Institute of Health, Bethesda, MD 20892, USA

Preface

We are very pleased to present *Cell Press Reviews: Core Concepts in Cell Biology*, which brings together review articles from Cell Press journals in order to offer readers a comprehensive and accessible entry point into some of the most important topics in cell biology today. Articles were selected by the editorial staff at Cell Press with an eye toward providing readers an introduction to timely and cutting-edge research written by leaders in the field. While *Cell Press Reviews: Core Concepts in Cell Biology* is not an exhaustive overview of current cell biological advances, our aim is to give readers insight into some of the most exciting recent developments and the challenges that remain. A wide range of topics are covered within this publication, including the cell biology of genomes, mechanochemical patterning in cell polarity, mechanisms of membrane curvature, and insights into processes such as organelle growth, cell motility, and morphogenesis.

We are pleased to be able to include contributions from Tom Misteli, National Cancer Institute; Galit Lahav, Harvard Medical School; Scott D. Emr, Cornell University; David G. Drubin, University of California, Berkeley; Tom Rapoport, Harvard Medical School; Anthony A. Hyman, Max Planck Institute of Molecular and Cell Biology, Dresden; and many other prominent researchers in the field. Their insights will offer readers, both experts and those new to the field, a fascinating perspective into this critically important and evolving area of research.

Cell Press Reviews: Core Concepts in Cell Biology is one in a series of books being published as part of an exciting new collaboration between Cell Press and Elsevier Science and Technology Books. Each book in this series is focused on a highly timely topic in the biological sciences. Editors at Cell Press carefully select recently published review articles in order to provide a comprehensive overview of the topic. With the wide range of journals within the Cell Press family, including research journals such as *Cell, Current Biology*, and *Developmental Cell* as well as review journals like *Trends in Cell Biology*, these compilations provide a diverse and accessible assortment of articles appropriate for a wide variety of readers. You can find additional titles in this series at http://www.store.elsevier.com/CellPressReviews. We are happy to be able to offer this series to such a wide audience via the collaboration with Elsevier Science and Technology Books, and we welcome all feedback from readers on how we might continue to improve the series.

CHAPTER 1

Cell

The Cell Biology of Genomes: Bringing the Double Helix to Life

Tom Misteli[1,*]

[1]National Cancer Institute, National Institutes of Health, Bethesda, MD 20892, USA

*Correspondence: mistelit@mail.nih.gov

Cell, Vol. 152, No. 6, March 14, 2013 © 2013 Elsevier Inc.

http://dx.doi.org/10.1016/j.cell.2013.02.048

SUMMARY

The recent ability to routinely probe genome function at a global scale has revolutionized our view of genomes. One of the most important realizations from these approaches is that the functional output of genomes is affected by the nuclear environment in which they exist. Integration of sequence information with molecular and cellular features of the genome promises a fuller understanding of genome function.

INTRODUCTION

It was a moment of scientific amazement in 1953 when Watson and Crick revealed the structure of DNA. The magnificence of the double helix and its elegant simplicity were awe inspiring. But more than just being beautiful, the double helix immediately paved the way forward; its structure implied fundamental biological processes such as semiconservative replication and the notion that chemical changes in its composition may alter heritable traits. The linear structure of DNA laid the foundation for the concept that a string of chemical entities could encode the information that determines the very essence of every living organism. The beauty of the double helix was the promise that, if the sequence of bases in the genome could be mapped and decoded, the genetic information that underlies all living organisms would be revealed and the secret of biological systems would be unlocked.

The idea of linearly encoded genetic information has been spectacularly successful, culminating in the recent development of powerful high-throughput sequencing methods that now allow the routine reading of entire genomes. The conceptual elegance of the genome is that the information contained

1

in the DNA sequence is absolute. The order of bases can be determined by sequencing, and the result is always unequivocal. The ability to decipher and accurately predict the behavior of genome sequences was appealing to the early molecular biologists, has given rise to the discipline of molecular genetics, and has catalyzed the reductionist thinking that has driven and dominated the field of molecular biology since its inception.

But the apparent simplicity and deterministic nature of genomes can be deceptive. One of the most important lessons learned from our ability to exhaustively sequence DNA and to probe genome behavior at a global scale by mapping chromatin properties and expression profiling is that the sequence is only the first step in genome function. In intact living cells and organisms, the functional output of genomes is modulated, and the hard-wired information contained in the sequence is often amplified or suppressed. While mutations are an extreme case of genome modulation, most commonly occurring changes in genome function are more subtle and consist of fluctuations in gene expression, temporary silencing, or temporary activation of genes. Although not caused by mutations, these genome activity changes are functionally important.

Several mechanisms modulate genome function (Figure 1). At the transcription level, the limited availability of components of the transcription machinery at specific sites in the genome influences the short-term behavior of genes and may make their expression stochastic. Epigenetic modifications are capable of overriding genetically encoded information via chemical modification of chromatin. Similarly, changes in higher-order chromatin organization and gene positioning within the nucleus alter functional properties of genome regions.

The existence of mechanisms that modulate the output of genomes makes it clear that a true understanding of genome function requires integration of what we have learned about genome sequence with what we are still discovering about how genomes are modified and how they are organized in vivo in the cell nucleus.

THE STOCHASTIC GENOME

The genome is what defines an organism and an individual cell. It is therefore tempting to assume that identical genomes behave identically in a population of cells. We now know that this is not the case. Individual, genetically identical cells can behave very differently even in the same physiological environment. It is rare to find a truly homogeneous population of cells even under controlled laboratory conditions, as anyone who has tried to make a cell line stably expressing a transgene knows. Much of the variability in

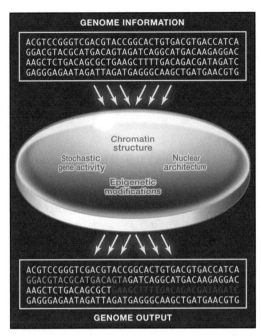

FIGURE 1 From Primary Sequence to Genome Output

The hard-wired primary information contained in the genome sequence is modulated at short or long timescales by several molecular and cellular events. Modulation may lead to activation (green) or silencing (red) of genome regions.

biological behavior between individual cells comes from stochastic activity of genes (Raj and van Oudenaarden, 2008).

Genes are by definition low-copy-number entities, as each typically only exists in two copies in the cell. Similarly, many transcription factors are present in relatively low numbers in the cell nucleus. The low copy number of genes and transcription factors makes gene expression inherently prone to stochastic effects (Raj and van Oudenaarden, 2008). Numerous observations make it clear that gene expression is stochastic in vivo. For example, dose-dependent increases in gene expression after treatment of cell populations with stimulating ligands, such as hormones, are often brought about by high expression of target genes in a relatively small number of cells in the population rather than by a uniform increase in the activity in all cells. Stochastic gene behavior is most evident in single-cell imaging approaches, and mapping by fluorescence in situ hybridization of multiple genes, which according to population-based PCR analysis are active in a given cell population, shows that only a few cells transcribe all "constitutively active" genes at any given time. Most cells only express a subset of genes, and the combinations vary considerably between individual cells. These observations suggest that many genes blink on and off and are expressed in bursts rather than in a continuous fashion (Larson et al., 2009).

The molecular basis for stochastic gene expression is unknown. There are several candidate mechanisms, all of which are related to genome or nuclear organization. Most genes require some degree of chromatin remodeling for activity, which is thought to make regulatory regions accessible to the transcription machinery. Several observations suggest that chromatin remodeling contributes to the stochastic bursting of gene expression. Maybe most compelling is the finding that genes located near each other on the same chromosome show correlated blinking behavior, indicating that a local chromosome property, such as chromatin structure, drives stochastic behavior (Becskei et al., 2005). Furthermore, altering chromatin, for example by deletion of chromatin remodeling machinery, affects stochastic variability in yeast. It can be envisioned that the stochastic behavior of genes is caused by the requirement for cyclical opening of chromatin regions. Open chromatin has a limited persistence time, and maintaining chromatin in an open state requires the cyclical action of chromatin remodelers. Whether an "active" gene is transcribed at any given time may thus depend on the transient condensation status of its chromatin at a particular moment.

A second mechanism to impose nonuniform stochastic genome activity may be the local availability of the transcription machinery at a gene. Although transcription factors are able to relatively freely diffuse through the nuclear space, and in this way effectively scan the genome for binding sites, their availability and functionality at a given local site may undergo significant temporal fluctuations (Misteli, 2001). The local availability of transcription complexes may affect transcription frequency positive or negatively. On the one hand, it is possible that relatively stable preinitiation complexes persist on a given gene, where they may support multiple rounds of transcription and in this way boost initiation frequency. On the other hand, assembly of the full polymerase is a stochastic and relatively inefficient event itself. In order for a functional polymerase complex to assemble, individual transcription machinery components associate with chromatin in a step-wise fashion, and formation of the mature polymerase complex involves multiple partially assembled intermediates, many of which are unstable and disintegrate before a functionally competent complex is formed (Misteli, 2001). The inefficiency of polymerase assembly may create stochasticity at an individual locus.

A further contributor to stochastic gene expression may be the organization of transcription events in transcription factories. These hubs of transcription consist of accumulations of transcription factors to which multiple genes, often located on distinct chromosomes, are recruited (Edelman and Fraser, 2012). Typically only a few hundred such transcription factories are observed in a mammalian cell nucleus. It is possible that some genes need to physically relocate from nucleoplasmic locations to transcription factories. A nominally "active" gene locus that is not associated with a transcription factory may thus be stochastically silent. The relatively low number of transcription sites makes them a limiting factor in the transcription process and thus a potential mediator of stochastic gene expression.

EPIGENETICS—AND WHEN EPIGENETICS IS NOT EPIGENETICS

Stochastic effects modulate genome output on short timescales. A mechanism to modulate the hardwired information of genomes on longer timescales is via epigenetics. The Greek-derived "Epi" means "over" or "above," and epigenetic effects are defined as heritable changes in genome activity caused by mechanisms other than changes in DNA sequence. Epigenetic events are mediated by chemical modifications of DNA or core histones in complex patterns by methylation, acetylation, ubiquitination, phosphorylation, etc. These modifications alter gene expression by changing the chromatin surface and in this way affect the binding of regulatory factors. Well-established examples of such effects include binding of the DNA-methylation-dependent binding of the MeCP2 protein or the binding of PHD-domain-containing proteins to trimethylated histone H3 tails. Prominent biological effects based on epigenetic regulation are phenotypic differences between homozygous twins or imprinted genes that are expressed from only one allele in a diploid organism.

A central tenet in the definition of epigenetic regulation is that its effects are heritable, i.e., transmittable over generations. In fact, the concept of epigenetics was inspired by epidemiological findings that nutrient availability in preadolescents during the 19th century Swedish famine determined life expectance of their grandchildren. The epidemiological studies have recently been complemented by controlled laboratory studies in mice (Rando, 2012), and they have been extended to the molecular level by the findings that loss of the histone H3K4-trimethylation prolongs lifespan in *C. elegans* in a heritable fashion for several generations (Greer et al., 2011).

A complicating aspect of epigenetics is that the same modifications that mediate heritable epigenetic regulation may also bring about nonheritable transient modulations of the genome. In fact, the term "epigenetic" is nowadays often used in a very cavalier manner to refer to any biological effect, heritable or not, that is affected by histone modifications. Even if they are not heritable, histone modifications are biologically relevant modulators of genome function. The system of histone modifications is in many ways akin to the mechanisms by which signal transduction pathways work

(Schreiber and Bernstein, 2002). Just as in signal transduction pathways, posttranslational modifications on histone tails create binding sites that are then recognized by adaptor or reader proteins, which in turn elicit downstream effects such as activation of kinases in the case of signaling cascades or recruitment of transcription factors in the case of histone modifications. In further analogy to the reversible events in signaling pathways, histone modifications can be altered or erased by modifying enzymes. Such transient and reversible modulatory effects of histone modifications have been implicated in every step of gene expression, starting from chromatin remodeling to recruitment of transcription machinery and even to downstream events that were thought to be chromatin independent, such as alternative pre-mRNA splicing (Luco et al., 2011). It is often difficult to determine heritability of these histone modification effects, and it therefore remains unclear how many of them are truly epigenetic. Regardless, DNA and histone modifications are an obvious source of modulation of the information contained in the genome sequence.

GENOME ORGANIZATION AS A MODULATOR OF GENOME FUNCTION

Genomes of course do not exist as linear, naked DNA in the cell nucleus but are organized into higher-order chromatin fibers, chromatin domains, and chromosomes. Many correlations between genome organization and activity have been made—most prominently, the findings that transcriptionally active genes are generally located in decondensed chromatin and that transcriptionally repressed genome regions are often found at the nuclear periphery. These observations point to the possibility that the spatial organization of the genome modulates its functional output.

But in considering the relationship of genome structure with its function, we are faced with a perpetual chicken-and-egg problem. Does structure drive function, or is structure merely a reflection of function? Much of the thinking on this topic has been guided by observations on individual genes. How representative these were for the genome as a whole has been a confounding concern. Recent unbiased genome-wide analysis of structure/function relationships has validated the tight link between structure and function. Large-scale analysis of chromatin structure, histone modifications, and expression profiles shows that genomes are portioned into well-defined domains that closely correlate with their activity status and the presence of active or repressive histone marks (Sexton et al., 2012). The domains are separated by sharp boundaries marked by particular histone modification patterns and binding sites for chromatin insulator proteins such as CTCF. Even stronger evidence comes from the analysis of physical interactions *between* chromatin domains.

At least in fruit flies, functionally equivalent domains tend to preferentially interact; that is, domains containing silent regions cluster in three-dimensional space, as do domains containing active regions (Sexton et al., 2012).

But can genome structure drive its function? The best example for structure-mediated gene expression effects is the silencing of genes when they become juxtaposed to heterochromatin domains, be it in the nuclear interior or at the nuclear periphery (Beisel and Paro, 2011). Gene activity has also frequently been linked to the position of a gene within the cell nucleus. The strongest evidence for such a relationship is experiments in which genes are transplanted from the nuclear interior to the lamina, leading to their repression or making them refractory to activation (Geyer et al., 2011). Based on these and similar experiments, it is often quite categorically stated that active genome regions are found in the interior of the nucleus and inactive ones at the periphery. This is a somewhat misleading oversimplification. Although lamina-associated genome regions are generally gene poor and are not transcribed, transcription labeling experiments reveal numerous active transcription sites at the periphery, and genes that are near the periphery, but not physically associated with it, are often active. On the other hand, inactive genes are frequently found in the interior. As far as we can tell, nuclear position per se does not determine activity, but association with repressive regions of the nucleus, be it at the periphery or the interior, does.

So, how then should we think about the chicken-and-egg problem of nuclear structure and function? How can it be that clear evidence exists for both "function-driving-structure" as well for "structure-driving-function"? The likely answer is that both effects are at play and are part of an overarching principle in which the mutual interplay of structure and function at multiple levels influences gene expression. The fact that there are very few known heterochromatic active genes suggests that a structural change in the form of chromatin decondensation is a crucial early step in gene activation. However, because chromatin states are generally unstable, mechanisms that reinforce a decondensed chromatin state must be in force for a gene to remain active. Such reinforcing mechanisms are dependent on gene activity and represent the "activity-drives-function" aspect of gene expression. Reinforcement mechanisms might be mediated by what we consider "active" histone modifications, some of which are known to be deposited during transcription as the polymerases traverse genes. On the flipside, a chromatin domain may also impose its effect on neighboring regions, either in *cis* on the same chromosome by spreading or in *trans* on distinct chromosomes. This effect represents the "structure-drives-function" aspect of genome function. Such a bidirectional, self-enforcing function-structure-function model accounts for most experimental observations on structure-function relationships in gene expression.

FACING THE COMPLEXITY

Since the discovery of the double helix, we have come to realize that understanding genomes requires more than reading their sequence and that the information contained in the sequence is modulated by the cellular environment. How then do we gain full knowledge of the functional information encoded in genomes?

To get a comprehensive picture of the functional output of genomes, the sequence information needs to be integrated with other information parameters such as epigenetic patterns, higher-order chromatin landscapes, and noncoding RNA profiles. The technology to do this is now available, and intense efforts are currently underway to comprehensively gather these data sets in various biological systems. The first examples of such multilevel mapping analyses are emerging, such as the recent flurry of reports from the ENCODE consortium, which has systematically mapped genome properties ranging from histone modification profiles to regulatory elements and chromatin structure (Ecker et al., 2012). Given the scale and complexity of the generated data, not to mention the technical difficulties in gathering it, this is a challenging undertaking that will require a series of progressively larger studies. Ideally, future studies should be designed to systematically map multiple genome properties for focused biological systems such as specific human diseases.

Large-scale mapping of genome-related parameters and their comparison is a logical and necessary next step in the exploration of genomes and their function. These efforts will create invaluable catalogs of genome properties, and the hope is that, by cross-comparing data sets, insight into the rules that govern genome regulation will be gleaned. One can go one step further and advocate for an even more comprehensive approach in which genome expression data are then compared to other cellular characteristics such as proteomic, metabolomic, morphological, and physiological data to systematically link genome activity to biological behavior. The ultimate version of such an approach was recently described in a report by the US National Academies of Sciences entitled "Toward Precision Medicine," which envisioned a fully minable biomedical data repository that would include information ranging from genomic and epigenetic parameters to physiological features and clinical symptoms.

The elegant simplicity of the DNA structure revealed by Watson and Crick is still stunning. True to its promise when it was first discovered, it opened up the floodgates to understanding heredity. But one of the most profound lessons from the ensuing decades of genome exploration must be that the linear arrangement of bases in the DNA is not an absolute set of instructions but is malleable by the cellular environment. We are just beginning to uncover some of the mechanisms that are responsible for these effects. As is the rule in biology, wherein the whole is often greater than the sum of its parts, we are realizing that the genome is far more complex than the sequence of its DNA.

ACKNOWLEDGMENTS

Due to space limitations, mostly review articles were cited. Work in the author's laboratory is supported by the Intramural Research Program of the National Institutes of Health (NIH), NCI, Center for Cancer Research.

REFERENCES

Becskei, A., Kaufmann, B.B., and van Oudenaarden, A. (2005). Contributions of low molecule number and chromosomal positioning to stochastic gene expression. Nat. Genet. *37*, 937–944.

Beisel, C., and Paro, R. (2011). Silencing chromatin: comparing modes and mechanisms. Nat. Rev. Genet. *12*, 123–135.

Ecker, J.R., Bickmore, W.A., Barroso, I., Pritchard, J.K., Gilad, Y., and Segal, E. (2012). Genomics: ENCODE explained. Nature *489*, 52–55.

Edelman, L.B., and Fraser, P. (2012). Transcription factories: genetic programming in three dimensions. Curr. Opin. Genet. Dev. *22*, 110–114.

Geyer, P.K., Vitalini, M.W., and Wallrath, L.L. (2011). Nuclear organization: taking a position on gene expression. Curr. Opin. Cell Biol. *23*, 354–359.

Greer, E.L., Maures, T.J., Ucar, D., Hauswirth, A.G., Mancini, E., Lim, J.P., Benayoun, B.A., Shi, Y., and Brunet, A. (2011). Transgenerational epigenetic inheritance of longevity in Caenorhabditis elegans. Nature *479*, 365–371.

Larson, D.R., Singer, R.H., and Zenklusen, D. (2009). A single molecule view of gene expression. Trends Cell Biol. *19*, 630–637.

Luco, R.F., Allo, M., Schor, I.E., Kornblihtt, A.R., and Misteli, T. (2011). Epigenetics in alternative pre-mRNA splicing. Cell *144*, 16–26.

Misteli, T. (2001). Protein dynamics: implications for nuclear architecture and gene expression. Science *291*, 843–847.

Raj, A., and van Oudenaarden, A. (2008). Nature, nurture, or chance: stochastic gene expression and its consequences. Cell *135*, 216–226.

Rando, O.J. (2012). Daddy issues: paternal effects on phenotype. Cell *151*, 702–708.

Schreiber, S.L., and Bernstein, B.E. (2002). Signaling network model of chromatin. Cell *111*, 771–778.

Sexton, T., Yaffe, E., Kenigsberg, E., Bantignies, F., Leblanc, B., Hoichman, M., Parrinello, H., Tanay, A., and Cavalli, G. (2012). Three-dimensional folding and functional organization principles of the Drosophila genome. Cell *148*, 458–472.

CHAPTER 2

Current Biology

Condensin, Chromatin Crossbarring and Chromosome Condensation

Rahul Thadani[1], Frank Uhlmann[1,*], Sebastian Heeger[1]

[1]*Chromosome Segregation Laboratory, Cancer Research UK London Research Institute, 44 Lincoln's Inn Fields, London WC2A 3LY, UK*
*Correspondence: frank.uhlmann@cancer.org.uk

Current Biology, Vol. 22, No. 23, R1012–R1021, December 4, 2012 © 2012 Elsevier Inc.
http://dx.doi.org/10.1016/j.cub.2012.10.023

SUMMARY

The processes underlying the large-scale reorganisation of chromatin in mitosis that form compact mitotic chromosomes and ensure the fidelity of chromosome segregation during cell division still remain obscure. The chromosomal condensin complex is a major molecular effector of chromosome condensation and segregation in diverse organisms ranging from bacteria to humans. Condensin is a large, evolutionarily conserved, multisubunit protein assembly composed of dimers of the structural maintenance of chromosomes (SMC) family of ATPases, clasped into topologically closed rings by accessory subunits. Condensin binds to DNA dynamically, in a poorly understood cycle of ATP-modulated conformational changes, and exhibits the ability to positively supercoil DNA. During mitosis, condensin is phosphorylated by the cyclin-dependent kinase (CDK), Polo and Aurora B kinases in a manner that correlates with changes in its localisation, dynamics and supercoiling activity. Here we review the reported architecture, biochemical activities and regulators of condensin. We compare models of bacterial and eukaryotic condensins in order to uncover conserved mechanistic principles of condensin action and to propose a model for mitotic chromosome condensation.

INTRODUCTION

The propagation of the blueprint of life, at a molecular level, can be described as the accurate transmission of replicated genetic material to daughter cells. To enable this, cells must compact centimetre-long DNA molecules, within

11

the confines of micrometre-sized nuclei, into stable chromosomes that can withstand the forces generated during segregation. This condensation of chromatin — into the thread-like chromosomes that give mitosis its name (from the Greek *mitos*, i.e., thread) — is one of the most striking morphological events of the cell cycle. Yet more than a century after Walther Flemming first observed mitotic chromosomes [1], and Theodor Boveri proposed they maintained their identity through interphase [2], mechanistic explanations of chromosome condensation remain elusive.

Early indications of a mitosis-specific condensation factor in cells began to emerge in the 1970s: classical cell fusion experiments showed that premature chromosome condensation could be induced in interphase HeLa cells that were fused to mitotic ones [3]. In a cell-free system derived from *Xenopus* eggs, metaphase chromosomes were assembled in interphase nuclei incubated with mitotic extracts [4,5]. Subsequent studies in budding and fission yeasts [6,7], *Xenopus* egg extracts [8] and chicken cells [9] led to the identification of Smc2 and Smc4, core components of the condensin complex, as proteins essential for chromosome condensation and segregation.

Accumulating lines of evidence over two decades indicate that the chromosomal condensin complex is the principal effector of condensation [10,11]. Condensin is a large, evolutionarily conserved multisubunit protein assembly that is found, with a broadly similar architecture, throughout the domains of life, including bacteria, archaea and eukarya (Figure 1). Along with cohesin and Smc5/6, it is one of three complexes built from dimers of SMC proteins, members of the structural maintenance of chromosomes family of ATPases, that are intimately involved in diverse aspects of higher order chromosome organisation. Indeed, condensin has been ascribed roles in several cellular processes apart from chromosome condensation; these have been extensively described elsewhere [12,13]. Conversely, the related SMC complexes might also contribute to chromosome condensation [14]. In this review, we focus on the role of condensin in mitotic chromosome condensation.

MOLECULAR ARCHITECTURE OF CONDENSIN

Eukaryotic condensin is a large pentameric complex that comprises a core catalytic Smc2–Smc4 heterodimer. As is characteristic of SMC proteins, Smc2 and Smc4 contain three globular parts — two terminal and one central — linked by long coiled coils. Each SMC protein folds back on itself through antiparallel coiled-coil arm interactions. This forms an SMC dimerisation hinge domain from the central part at one end, and an ATPase head domain from association of the terminal globular parts at the other (Figure 1). The catalytic head domain features canonical ATP-binding cassette motifs: the amino-terminal 'Walker A' motif, and carboxy-terminal 'Walker B' and

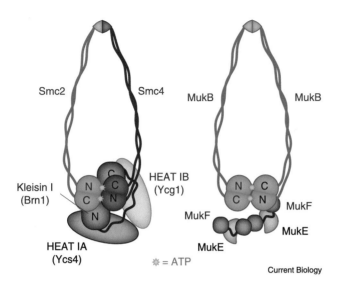

FIGURE 1 Molecular architecture of condensin.
Subunit composition of eukaryotic (left) and bacterial (right) condensins. Condensins are composed of a core dimer of SMC or SMC-like ATPases with a dimerisation hinge at one end and catalytic head domain at the other. The core dimers are closed into rings by kleisins, which are monomeric in eukaryotes and dimeric in prokaryotes. One or more additional regulatory subunits interact with the kleisin and/or SMC core.

'C/signature' motifs. The amino-terminal globular part of one SMC subunit engages in trans with a carboxy-terminal part from the other to form a bipartite ATP-binding pocket. Three accessory subunits bind to the SMC heterodimer and regulate its activity. Kleisin I/Brn1, a member of the kleisin family [15], interacts at its amino terminus with Smc2, and at its carboxyl terminus with Smc4 to form a topologically closed ring [10,16,17]. HEAT IA/Ycs4 [18,19] and HEAT IB/Ycg1 [20] contain HEAT repeats, and interact with the amino- and carboxy-terminal halves of Kleisin I/Brn1, respectively, and weakly with each other [21]. All three accessory subunits of condensin are required for its association with chromatin and function in chromosome condensation [22]. It is noteworthy that while the integrity of the ring-like structure of condensin is necessary for its function [17], details of its mechanistic significance remain to be determined. This is in contrast to the case of cohesin, where it has been shown that the complex topologically encircles sister chromatids until separase-driven cleavage of the kleisin Scc1 at anaphase onset enables them to segregate [23,24].

Most eukaryotes possess two isoforms of condensin, termed condensin I and II. These are built from identical core heterodimers of Smc2 and Smc4 but differing accessory subunits (Table 1), which may modulate the differential localisation patterns, dynamics, and functions of the two condensins. Condensin I is termed the canonical condensin due to its phylogenetic

Table 1 Condensin subunits involved in mitotic chromosome condensation

Subunits	*S. cerevisiae*	*S. pombe*	*C. elegans*	Others
Core (condensin I and II)				
SMC2	Smc2	Cut14	MIX-1	CAP-E
SMC4	Smc4	Cut3	SMC-4	CAP-C
Condensin I-specific				
Kleisin I (γ)	Brn1	Cnd2	DPY-26	CAP-H
HEAT IA	Ycs4	Cnd1	DPY-28	CAP-D2
HEAT IB	Ycg1	Cnd3	CAPG-1	CAP-G
Condensin II-specific				
Kleisin II (β)	–	–	KLE-2	CAP-H2
HEAT IIA	–	–	HCP-6	CAP-D3
HEAT IIB	–	–	CAPG-2	CAP-G2

ubiquity (Figure 2) and relative cellular abundance, although the ratio of condensin I and II varies substantially in different organisms, ranging from 1:1 in HeLa cells and 5:1 in *Xenopus* egg extracts [25] to 10:1 in chicken DT40 cells [26,27]. The presence of condensin I and II in diverse eukaryotic taxa suggests that their last common ancestor possessed both condensin isoforms [13]. This implies that condensin II was independently lost from the genomes of organisms such as fungi and ciliates that possess only a single known isoform of condensin.

Prokaryotes were believed to possess one of two SMC-related complexes: the SMC-ScpAB complex [28-30] widespread in bacteria and archaea, or the MukBEF complex [31,32] present in γ-proteobacteria such as *Escherichia coli*. These two complexes are divergent at the sequence level but share a common architecture. Inactivation of the two complexes produces defects reminiscent of condensin mutants, including decondensed nucleoids, chromosome segregation failure, anucleate cell formation and temperature-sensitive growth [28–32]. SMC-ScpAB is composed of a catalytic SMC homodimer, the kleisin ScpA and accessory protein ScpB, both of which are likely binary in the complex. Similarly, MukBEF comprises an SMC-like core MukB homodimer, while the kleisin MukF and accessory protein MukE again form a dimeric frame that interacts with the MukB heads [33,34] (Figure 1). A third family of MukBEF-like SMC protein, termed MksBEF, has recently been identified in diverse bacterial genomes [35]. MksBEF is often present alongside SMC-ScpAB, MukBEF, or even other MksBEFs, suggesting that prokaryotic genome organisation may be more complex than previously appreciated. This also raises the possibility of as yet undiscovered molecular drivers of chromosome condensation in the larger and

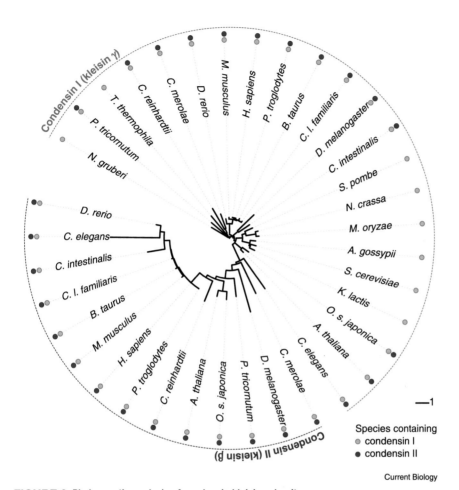

FIGURE 2 Phylogenetic analysis of condensin kleisin subunits.

Phylogram representing the evolutionary relationships among eukaryotic kleisin subunits. A maximum likelihood unrooted tree was constructed in PHYLIP [118] from a ClustalW multiple alignment [119] and rendered radially using iTOL [120]. The excavate *Naegleria gruberi* was used as an outgroup (top left), from which taxa fan out clockwise in order of increasing branch lengths. Note that kleisins from species with a single known condensin isoform, such as fungi and the ciliate *Tetrahymena thermophila*, cluster with the condensin I sequences. The scale bar represents one unit of evolutionary distance along branches, as computed by the Jones-Taylor-Thornton method [121].

incompletely annotated genomes of eukaryotes. We refer to the bacterial SMC-ScpAB and MukBEF complexes as prokaryotic condensins, due to their condensin-like null phenotypes. In the following sections, we draw mechanistic parallels with eukaryotic condensin, but note that the two prokaryotic complexes are not strictly phylogenetically closer to eukaryotic condensin than to other eukaryotic SMC complexes.

DIFFERENTIAL CONTRIBUTIONS OF CONDENSIN I AND II TO CHROMOSOME STRUCTURE

Condensin I and II exhibit distinct spatial staining patterns on chromosome axes, as well as differing temporal localisation patterns through the cell cycle [25,36], suggesting that they may have non-redundant roles in chromosome organisation. For instance, in HeLa cells, condensin I is excluded from the nucleus in interphase and binds to chromatin only on nuclear envelope breakdown (NEBD) in prometaphase. In contrast, condensin II is essentially nuclear in interphase, is stabilised on chromatin in early prophase, and remains associated with chromosomes throughout mitosis [36–38].

The differential contributions of the two condensin complexes to chromosome condensation are as yet poorly understood. In *Xenopus* egg extracts, the phenotypes following immunodepletion of condensin I- or II-specific subunits indicate that condensin I plays the major role in condensation [25]. By contrast, in HeLa cells, the siRNA-mediated knockdown of either condensin by RNAi leads to only minor abnormalities in chromosome morphology [25,37], with condensin I depletion producing swollen chromosomes, and condensin II depletion making them somewhat longer and curled. The varying severity of condensation phenotypes subsequent to condensin depletion in these two systems may be ascribed either to dose-dependence, given the differing ratios of the two condensin isoforms, or to incomplete silencing by RNAi. Consistent with the observed aberrant chromosome morphologies in HeLa cells, further studies in chicken DT40 cells [27] and *Xenopus* egg extracts [39] implicate condensin II in the early mitotic axial shortening of chromosome arms, and condensin I in their later lateral compaction. More work is needed to determine how two very similar complexes are able to bind to distinct chromosomal regions, and whether it is their differential localisation or intrinsic activity that is responsible for their separable contributions to condensation.

CELL-CYCLE REGULATION OF CHROMOSOME CONDENSATION

Numerous aspects of condensin biology are, reportedly, subject to control by the cell cycle machinery, making possible a multi-layered regulation of its function. Condensin activity may be regulated at the level of holocomplex formation, subcellular localisation, chromosomal loading, or chromatin binding dynamics, one or more of which are likely altered by post-translational modifications (Figure 3).

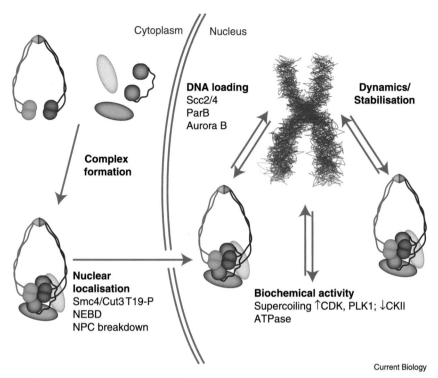

Cytoplasm / Nucleus

DNA loading
Scc2/4
ParB
Aurora B

**Dynamics/
Stabilisation**

**Complex
formation**

**Nuclear
localisation**
Smc4/Cut3 T19-P
NEBD
NPC breakdown

Biochemical activity
Supercoiling ↑CDK, PLK1; ↓CKII
ATPase

Current Biology

FIGURE 3 Regulation of condensin activity.
Condensin can be regulated at several different levels: complex formation, nuclear import, chromosomal localisation, binding dynamics, and ATPase activity. This regulation is likely performed by posttranslational modifications, which can modulate the biochemical activities of the complex.

Holocomplex Formation

In the test tube, condensin is often found in two major forms corresponding to the Smc2–Smc4 heterodimer and the holocomplex, as seen in early immunoaffinity purifications from *Xenopus* egg extracts [10], as well as reconstitutions of recombinant human condensin [21]. In addition, yeast Smc2 and Smc4 can form a stable heterodimer in cell extracts [40]. *In vitro* studies have ascribed differing activities to the SMC/MukB dimer and the holocomplex [10,41]. However, it should be noted that the uncomplexed SMC dimer has not been directly observed *in vivo*. Interestingly, the reported role of condensin in disassembly of the *Drosophila* nurse cell polytene chromosomes depends on a timely upregulation of only Kleisin II/CAP-H2 [42], indicating there may be a role for a limiting subunit in the regulation of complex assembly. Similarly, HEAT IA/CAP-D2 is a rate-limiting factor for the assembly of functional condensin I complexes in *Xenopus* oocytes [43]. This is reminiscent of cohesin, where levels of the kleisin Scc1 vary through the cell cycle, and determine the chromosomal association of the complex [44]. This principle can be applied more broadly, raising the possibility that varying expression levels of condensin I- and II-specific subunits determine the changing shapes of mitotic chromosomes in a developmental context [39]. Investigations of complex assembly and DNA binding dynamics of individual condensin subunits, for instance by fluorescence recovery after photobleaching (FRAP) assays, would prove instructive in further elucidating regulation at the level of condensin holocomplex formation.

Subcellular Localisation

In bacteria and archaea that lack a nuclear envelope, condensin is free to interact with chromatin at any time, limited only by the possible regulation of complex formation and DNA loading reactions. In eukaryotes, however, the nuclear envelope offers a potential barrier to chromatin access and consequently a possible mode of regulation. In organisms with a closed mitosis like the budding and fission yeasts, condensin has to be imported into the nucleus at or before chromatin compaction in mitosis. Intriguingly, in the budding yeast *Saccharomyces cerevisiae*, condensin localises to the nucleus throughout the cell cycle, a behaviour reminiscent of condensin II of higher eukaryotes, despite the complex being a homologue of condensin I at the sequence level (Figure 2). On the other hand, condensin in the fission yeast *Schizosaccharomyces pombe* is predominantly cytoplasmic in interphase and nuclear during mitosis, an enrichment that requires the CDK-dependent phosphorylation of the T19 site in Smc4/Cut3 [45]. In higher eukaryotes with an open mitosis, nuclear envelope breakdown (NEBD) at mitotic onset ensures that the chromatin association of condensin is not hindered. In *Drosophila melanogaster* [46], *Caenorhabditis elegans* [47], zebrafish [48] and HeLa [36–38] cells, condensin I is cytoplasmic in interphase and nuclear in mitosis, while condensin II is nuclear throughout the cell cycle, at least in HeLa cells. An interesting but as yet unresolved question is how this differential subcellular localisation of condensin I and II is achieved, either by modifications to their regulatory subunits or recognition by additional factors. Intriguingly, in *Drosophila* embryos, Kleisin I/Barren associates with chromatin several minutes before NEBD [46]. In addition, HeLa cells depleted of condensin II still initiate chromosome compaction just before NEBD [38]. These observations suggest that chromosome compaction in these organisms may be functionally coupled to disassembly of nuclear pore complexes (NPCs) rather than the nuclear membrane [38,49]. This is probably also the case in *C. elegans*, where NEBD is completed only in anaphase but chromosome condensation is initiated in prophase, accompanied by NPC breakdown [47].

Dynamic Binding to DNA

Chromosome condensation is a dynamic process, and condensin is required not only to assemble chromosomes, but also to maintain them in a condensed state throughout mitosis. Immunodepletion of condensin from mitotic chromosomes assembled in *Xenopus* egg extracts results in their rapid decondensation [8]. In mitotic yeast cells, the inactivation of condensin leads to dramatic defects in ribosomal DNA (rDNA) compaction [22], and decondenses chromosome arms to G1-like levels [50].

Interestingly, while condensin exchanges dynamically from chromosomes, condensin I and II exhibit different dynamic behaviour through the cell cycle. FRAP experiments on GFP-tagged kleisin subunits in HeLa cells have shown that condensin I, which is excluded from the nucleus in interphase and binds to chromatin on nuclear envelope breakdown, dynamically exchanges from chromosomes throughout mitosis. In contrast, condensin II, which is nuclear throughout the cell cycle, is stabilised on chromatin at the onset of condensation in prophase [38]. What brings about this change in kinetic turnover, whether it is related to cell-cycle-dependent posttranslational modifications, and what the consequences are on the chromosome condensation status are important questions to explore. Curiously, the budding yeast condensin is bound to chromatin even in interphase, and its levels and distribution along chromosome arms remain largely unaltered through the cell cycle [50,51]; the dynamic behaviour of this complex as a function of the cell cycle has not yet been investigated.

Condensin Loading

A critical but poorly understood aspect of condensin function is its loading onto chromosomes. The two — not mutually exclusive — possibilities are that condensin directly recognises DNA sequence elements or chromatin features, or that additional recruiting factors are responsible for loading condensin onto DNA. There is some evidence for the latter. In bacteria, the DNA-binding protein ParB recruits SMC to centromere-like parS sequences that cluster near the origin of replication [52,53]. In budding yeast, the cohesin loader Scc2–Scc4 promotes functional association of condensin with chromosomes, and may in turn be recruited at least in part by the transcription factor TFIIIC [50]. In the same model organism, replication fork barrier (RFB) sites at the 3' end of rDNA genes serve as *cis* elements for condensin recruitment in a manner dependent on the RFB-binding protein Fob1, the topoisomerase-I-interacting protein Tof2 and the monopolin subunits Csm1 and Lrs4 [54]. In addition, the fission yeast homologues of these monopolin subunits, Pcs1 and Mde4, act as recruitment factors for condensin enrichment at the kinetochore [55]. In the case of condensin II, it has been reported that protein phosphatase

2/PP2A acts in a noncatalytic fashion to recruit the complex to frog and human mitotic chromatin [56].

Budding and fission yeast remain the only eukaryotes in which condensin binding sites have been comprehensively characterised [50,51,57]. These studies show condensin to be enriched at centromeres, an enrichment that becomes particularly striking in mitosis compared to interphase. The molecular events governing this centromeric enrichment of condensin remain unclear. Condensin underlies, in part, the stiff elastic properties of centromeres during spindle attachment [36,38,58,59]. Along chromosome arms, condensin is found at TFIIIC-bound, RNA polymerase III-transcribed genes, notably tRNA genes, as well as a subset of other strongly transcribed genes, including ribosomal protein genes. A TFIIIC binding element is sufficient to recruit condensin to previously unoccupied sites in *S. cerevisiae* [50], suggesting that condensin binding along chromosomes is determined by certain *cis*-acting elements. What exactly is recognised by condensin, as well as the Scc2–Scc4 loading complex that is found at the same sites, remains to be elucidated. A chromosomal binding map of the *Bacillus subtilis* SMC complex has also been obtained, and shows striking similarities to yeast condensin, with an enrichment at the bacterial centromere-like partitioning locus as well as tRNA, ribosomal protein and other strongly expressed genes [60]. Little is as yet known about chromosomal condensin binding patterns in higher eukaryotes, an area of great interest for forthcoming studies.

Post-Translational Modifications

Mitotic CDK-dependent phosphorylation has been shown to stimulate the ATPase and supercoiling activities of *Xenopus* condensin I [61,62]. Subsequent studies using budding yeast condensin have established that CDK-mediated phosphorylation of Smc4 primes the complex for hyperphosphorylation of its three regulatory subunits by the Polo kinase PLK1/Cdc5. This hyperphosphorylation then further activates the DNA supercoiling activity of condensin [63]. Similarly, CDK has been shown to phosphorylate the T1415 residue of HEAT IIA/hCAPD3 in HeLa cells, priming the complex for hyperphosphorylation by PLK1, and ensuring the fidelity of chromosome assembly [64]. Taken together, these observations establish CDK and PLK1 as major molecular cell cycle regulators of condensin function. Large-scale phosphoproteomic studies [65–67] have described numerous additional cell-cycle-regulated condensin phosphorylation events (some of which are summarised in [68]) the role of which remains to be explored. A casein kinase II-mediated, interphase-enriched phosphorylation of condensin I in human cell cultures is the only known phosphorylation that negatively regulates the supercoiling activity of condensin [69].

The chromosomal passenger complex (CPC) Aurora B kinase has been implicated in chromosome condensation in a number of species. In the fission yeast *S. pombe*, the CPC member Cut17/Bir1 is essential for proper localisation of the Aurora B-like kinase Ark1, condensin recruitment and chromosome condensation [70]. In this system, Aurora B kinase activity is required for the mitotic chromatin association of condensin but not its nuclear localisation [71,72]. In *Drosophila* Schneider cells, RNAi against Aurora B is associated with a loss of chromatin-bound Kleisin I/Barren, leading to incomplete chromosome condensation, abnormal segregation, a failure of cytokinesis and polyploidy [73]. In budding yeast, the anaphase condensation of rDNA arrays requires Aurora B/Ipl1 [74]. Similarly, *C. elegans* Smc2/MIX1 fails to be recruited to chromatin after Aurora B RNAi [75], and chromosome condensation is consequently delayed [76]. In HeLa cells, RNAi, as well as treatment with the Aurora B inhibitor hesparadin, leads to a loss of chromatin association of condensin I but not condensin II [77]. Maximal anaphase chromosome compaction in rat kidney cells also depends on Aurora B [78], and condensin I association with chromatin is reduced after depletion of Aurora B from *Xenopus* egg extracts [79].

Two common themes emerge from these studies on the role of Aurora B in chromosome condensation. First, the depletion of Aurora B impairs the association of condensin I with chromatin, an observation consistent in *S. pombe* [71], *Drosophila* [73] and HeLa [77] cells, as well as *Xenopus* egg extracts [79]. Second, maximal chromosome compaction occurs in anaphase in a manner requiring Aurora B [74,78] and presumably its kinase activity [71,72]. However, a direct link between Aurora B kinase and condensin remained elusive until Aurora B/Ipl1 was shown to promote the mitosis-specific phosphorylation of the budding yeast condensin accessory subunits Kleisin I/Brn1, HEAT IA/Ycs4 and HEAT IB/Ycg1 in mitosis [63]. Mass spectrometry studies have since identified three Aurora B consensus sites in fission yeast Kleisin I/Cnd2 — S5, S41 and S52 — of which S52 phosphorylation was shown to depend on Aurora B kinase *in vivo*; substitutions of the three serine residues to nonphosphorylatable alanines lead to chromosome missegregation [80]. These observations have been confirmed and extended to mammals. A series of *in vitro* and *in vivo* experiments showed that mitotic phosphorylation of the condensin I kleisin subunit triggers its interaction with the basic amino-terminal tail of histones H2A and H2A.Z, which is required for chromatin association of the complex [55]. The conservation of phosphorylation-dependent condensin interactions with histone H2A variants between fission yeast and mammals, and the general requirement of Aurora B kinase for chromosome condensation in various species, suggests this may be a fundamental mechanism common to all eukaryotes.

In addition to the kinases CDK, Polo and Aurora B, the phosphatase Cdc14 also plays a role in chromosome condensation in budding yeast. Cdc14 is sequestered in the nucleolus until it is activated in anaphase by the Cdc-Fourteen Early Anaphase Release (FEAR) pathway and the mitotic exit network (MEN) [81]. In early anaphase, Cdc14 is essential for anaphase-specific condensation and segregation of the rDNA locus, in a manner dependent on condensin and Aurora B [82]. Cdc14 promotes condensin association with rDNA, which correlates with the sumoylation of HEAT IA/Ycs4 and phosphorylation of HEAT IB/Ycg1 [83]. The Cdc14 target(s) that promote rDNA condensation and segregation are not known. One possibility is that the Cdc14-dependent dephosphorylation of the CPC component Sli15 and the resultant anaphase-specific relocalisation of Aurora B to the spindle midzone play a role [84]. Aurora B at the spindle midzone has been proposed to promote hyper-condensation of trailing chromosome arms [55,85]. Potential roles of sumoylation or direct dephosphorylation of condensin subunits by Cdc14 remain to be explored, as is the role of Cdc14 in anaphase condensation in other species. In contrast to its early anaphase role in compaction, Cdc14 promotes chromosome decondensation subsequently in late anaphase. At this stage, CDK inhibition and Cdc14 activity impair the association of Brn1 with chromatin [86]. These results are consistent with the possibility that condensin dephosphorylation by Cdc14 promotes chromosome decondensation at mitotic exit. Since phosphorylation generally appears to stimulate the biochemical activity of the condensin complex, such as DNA binding and supercoiling, its dephosphorylation may reverse these effects to permit chromosome decondensation as cells return to interphase. Consistent with this idea, the protein phosphatase PP2A, which plays a role in mitotic exit in higher eukaryotes, has been implicated in the dephosphorylation of HEAT IIA/CAP-D3 [87].

TOWARDS A MECHANISTIC UNDERSTANDING OF CHROMOSOME CONDENSATION

Biochemical characterisation of condensin has uncovered a number of activities of the complex, including the ability to topologically encircle DNA, supercoil DNA and hydrolyse ATP. It is likely that models of mitosis-specific chromosome condensation by condensin will incorporate some or all of these activities, each of which could be modulated by post-translational modifications. Notably, postulating the topological entrapment of two DNA strands within a single circular condensin complex provides a succinct explanation of how condensin might mediate long-range chromosomal interactions [17]. Alternatively, a handcuff-like assembly of two tethered condensin rings, as is sometimes proposed in the case of cohesin [88], might bridge distant DNA regions (Figure 4A).

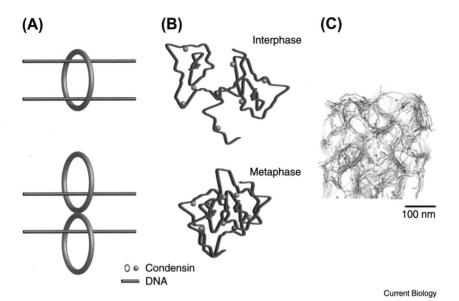

(A)

(B)

Interphase

Metaphase

(C)

100 nm

○ ● Condensin
━ DNA

Current Biology

FIGURE 4 Models of condensin action.
Proposed modes of condensin binding to DNA. (A) Top: A ring-shaped condensin complex topologically captures more than one strand of DNA in a sequential manner to bring about condensation. Bottom: A ring-shaped condensin complex binds to a single strand of DNA. The condensation reaction then involves dimerisation of more than one condensin complex in a handcuff-like assembly, or even their multimerisation. These models are not mutually exclusive, and it is easy to imagine the multimerisation of condensin rings that entrap more than one strand of DNA. (B) A meshwork of DNA interactions bridged by condensin. In this model, condensin constrains the expansion of a 'nucleosome melt' by bridging distant DNA segments. The condensation reaction involves a change in the dynamic binding of condensin complexes to DNA or each other. (C) Reconstruction of the DNA path and condensin in reconstituted *Xenopus* chromosomes by EM tomography, showing condensin enriched at sites of chromatin intersections. (Reprinted from König *et al.* [117] with kind permission from Springer Science and Business Media.)

Multimerisation

Evidence for the formation of multimeric condensin assemblies stems largely from *in vitro* studies of bacterial condensins. In electron micrographs, purified *E. coli* MukBEF has been seen as an oligomer, forming extended fibres and rosette-like configurations. In contrast, the MukB homodimer is rarely multimeric, suggesting that intermolecular MukE or MukF interactions mediate oligomerisation [89]. Protein volume measurements via atomic force microscopy show that *B. subtilis* SMC complexes form higher-order structures in the presence of ScpA and ScpB, further indicating that the accessory subunits may have a role in the organisation of SMC oligomers [90]. There are several indications that DNA compaction *in vitro* can proceed by the concerted action of several condensin complexes.

Direct force measurements in single molecule experiments demonstrate that MukBEF compacts DNA into stable, repetitive structures in a highly cooperative manner [91,92]. Similar cooperative behaviour of condensin I was observed during the ATP-dependent compaction of single nanomanipulated DNA fibres [93]. In contrast to these *in vitro* compaction reactions in the presence of excess amounts of condensin, the number of condensin complexes *in vivo* is relatively small. Thus, although MukB is found in clusters in living cells [94], the applicability of these results to the DNA condensation reaction at physiological concentrations of condensin remains undetermined. In both yeast and humans, chromosome condensation is achieved by one condensin complex per 5–10 kb of DNA [51,95]. Whether and how interactions between more than one condensin complex contribute to chromosome condensation *in vivo* is as yet unknown. Our initial attempts to detect signs of condensin interactions *in vivo* have been unsuccessful (S. Heeger, unpublished). Future experiments using superresolution microscopy and fluorescence resonance energy transfer (FRET) approaches will help to investigate this possibility.

DNA Supercoiling and Topological Selectivity

Eukaryotic and bacterial condensins have both been shown to possess the intrinsic ability to directly supercoil DNA *in vitro*, albeit with differing directionalities. Condensin purified from *Xenopus* egg extracts [10,61] and yeast cells [63] is able to introduce positive supercoils in circular plasmid DNA in the presence of topoisomerase I and ATP. The *Xenopus* condensin preparation also produces chiral knots in DNA in the presence of topoisomerase II, leading to the idea that condensin reconfigures DNA by the introduction of a global positive writhe [62]. Like its eukaryotic counterparts, the *E. coli* MukB introduces right-handed knots into DNA in the presence of phage topoisomerase II; the net supercoiling stabilised by MukB, however, is negative [96]. An interaction of condensin with type II topoisomerases has been demonstrated in *Drosophila* and *E. coli* [16,97,98]. However, this interaction has been implicated in the role of condensin in the decatenation of sister chromatids to facilitate chromosome segregation [16,99]. Any presumed role for topoisomerase II in chromosome condensation remains controversial. While supercoiling and the introduction of a writhe thus remain striking *in vitro* effects of condensin on DNA, their relevance to chromosome condensation *in vivo* in the face of abundant DNA topoisomerases that are adapted to relieve topological stress remains uncertain.

Condensin from *Xenopus* egg extracts was observed to preferentially bind cruciform DNA molecules over unstructured linear duplexes, and longer DNA molecules over shorter ones, in electromobility shift assays [100]. While such binding preferences to complex DNA structures *in vitro* remain difficult to interpret, they may in this case point to an interesting feature of condensin. In the case of *E. coli* MukB, single molecule recordings have shown the protein to stabilise interactions between two strands of DNA, with a marked preference for right-handed DNA crossings [92]. These observations suggest that condensin does not just capture randomly colliding DNA molecules but may recognise their topology. Such a topology distinction could contribute to a possible mechanism for distinguishing interactions within a chromatid from those between chromatids.

ATPase Activity

SMC proteins lack sequence or structural similarity to conventional motor proteins [101,102] and are thus unlikely to use the energy of ATP hydrolysis to move along chromosomes or physically reel in DNA. Instead the SMC ATPase cycle drives a series of conformational changes at the molecular level that likely influence the chromosomal association of the condensin complex. ATP binding to the SMC head domains leads to their engagement, while ATP hydrolysis allows the heads to move apart [102]. It remains unclear how precisely the catalytic cycle of ATP-dependent head engagement and disengagement is coupled to condensin's interaction with DNA, and whether this contributes to chromosome condensation. The SMC–kleisin interaction is a candidate for regulation by the ATPase cycle. A striking ATP-dependent conformational change has been observed in structural studies of the *E. coli* MukBEF complex, where ATP-binding-mediated engagement of the two MukB heads leads to the detachment of the dimeric MukF kleisin frame from one of the MukB heads [34]. This conformational change requires a transient loss of kleisin interaction with one of the MukB heads and thus might be coupled to topological DNA entry or exit from the MukB ring. Little is yet known about the consequences of ATP hydrolysis for condensin function *in vivo*, though in the case of cohesin it has been established that ATP hydrolysis is required for chromosome association of the complex [103,104]. A recent model for cohesin suggests two distinct entry and exit gates for DNA in the complex. DNA in this model enters the ring through the Smc1–Smc3 hinge dimerisation interface, and exits via opening of the Smc3–kleisin interaction [105]. How conformational changes at the ATPase heads could be translated to the hinge domain some 50 nm away to allow DNA entry remains to be investigated. Such a long range interaction has been demonstrated between the two parts of the cohesin complex [106]. In the case of condensin, a head–hinge interaction is apparent in atomic force microscopic images of the fission yeast complex [107]. The archaeal SMC hinge has also been implicated in its binding to DNA. Notably, an enzymatic crosstalk between DNA binding close to the hinge and ATP hydrolysis by the ATPase head domains has been observed [108,109]. At the same time, the hinges of the prokaryotic SMC and MukB proteins show substantial structural differences [108,110]. The impact of ATP binding and hydrolysis on SMC complexes therefore remains an important area of study that should shed light on their ability to dynamically associate with and condense chromosomes.

A MODEL FOR MITOTIC CHROMOSOME CONDENSATION

Although the precise mechanism by which condensin promotes chromosome condensation still remains to be worked out, a few facts about its action are evident. Condensin binds to specific sites along chromosomes, which likely involves the topological entrapment of DNA. Such condensin–DNA interactions could be translated into long-range chromosomal interactions that bring about condensation in two different ways. One scenario involves the sequential entrapment of two DNA strands by a single condensin ring. A second is the dimerisation — or even multimerisation — of condensin complexes that have captured one strand of DNA each (Figure 4A). These two possibilities need not, of course, be mutually exclusive. In order to understand chromosome condensation, it is necessary to not only address the mechanism by which condensin associates with DNA, but also determine which pairs of DNA sequences along a chromosome condensin brings together, and how this pairing pattern changes as a function of cell cycle progression. Techniques such as chromosome conformation capture and its variants [111] should be instructive in determining how condensin modulates intrachromosomal DNA interactions to drive mitotic chromosome condensation.

A model in which condensin promotes chromosome condensation by providing a meshwork of interactions between distant DNA sequences on the same chromosome is attractive for a number of reasons. Firstly, the biophysical properties of a mitotic vertebrate chromosome, as measured by mechanical micromanipulation studies, suggest that chromosomes are a composite network of DNA, crosslinked by protein interactions [112]. In contrast to popular models, no evidence for a contiguous protein scaffold has been found in native chromosomes. It should be emphasised that while a localised axis-like enrichment of condensin has been observed in fixed chromosome preparations [113,114], the imaging of fluorescent-tagged condensin in live cells does not support the notion of such a scaffold [37,38]. A scaffold would not be required if a broad network of condensin-mediated interactions between its binding sites compacts the chromosome. Recent structural studies of human mitotic chromosomes are also consistent with this mode of condensin action. Cryo-electron microscopy (cryo-EM) and X-ray scattering on close-to-native frozen chromosomes failed to find evidence for a hierarchical chromosome folding pattern of the kind portrayed in most molecular biology textbooks [115]. Instead of ordered structures, chromosomes appear to consist of a 'nucleosome melt', and we would like to argue that a network of DNA interactions between condensin binding sites would be ideally suited to constrain the expansion of such a melt. Evidence from three-dimensional imaging of lac operator arrays on metaphase chromosomes, in addition, shows that their folding pattern is irregular between cells and even between sister chromatids, as would be expected from a largely self-organising network of condensin binding site interactions [116]. The modulation, by mitotic regulators, of the on- or off-rates of condensin at its binding sites, or of interactions between condensin complexes, could shift the compaction equilibrium from a loose packing in interphase towards a more condensed state in mitosis (Figure 4B). A prediction from this model is that interphase chromosome architecture might similarly be governed by condensin. Pictures of condensin in striking agreement with this model have been obtained by EM tomography of reconstituted *Xenopus* chromosomes [117] (Figure 4C).

More work is required to determine the details of condensin action, and the exact nature of the interplay between its DNA binding dynamics on the one hand, and its various biochemical activities and posttranslational modifications on the other. Future investigations towards uncovering mechanisms of chromosome condensation will, no doubt, lead to a better understanding of the fascinating problem of how cells store, retrieve and transmit information using DNA molecules several orders of magnitude longer than the spatial confines of a nucleus.

ACKNOWLEDGEMENTS

We thank D. Agard for the chromosome reconstruction shown in Figure 4C. We are grateful to members of the Chromosome Segregation Laboratory for discussions and critical reading of the manuscript. This work was supported by Cancer Research UK and the European Research Council. R. Thadani and S. Heeger acknowledge support through a Boehringer Ingelheim Fonds PhD fellowship, and an EMBO long term fellowship, respectively.

REFERENCES

1 Flemming, W. (1882). Zellsubstanz, Kern und Zelltheilung. Leipzig: FCW Vogel.

2 Boveri, T. (1888). Zellen-Studien II: die Befruchtung und Teilung des Eies von Ascaris megalocephala, Volume 2 (Jena: Gustav Fischer).

3 Johnson, R.T., and Rao, P.N. (1970). Mammalian cell fusion: induction of premature chromosome condensation in interphase nuclei. Nature *226*, 717–722.

4 Lohka, M.J., and Maller, J.L. (1985). Induction of nuclear envelope breakdown, chromosome condensation, and spindle formation in cell-free extracts. J. Cell Biol. *101*, 518–523.

5 Newport, J., and Spann, T. (1987). Disassembly of the nucleus in mitotic extracts: membrane vesicularization, lamin disassembly, and chromosome condensation are independent processes. Cell *48*, 219–230.

6 Strunnikov, A.V., Hogan, E., and Koshland, D. (1995). SMC2, a Saccharomyces cerevisiae gene essential for chromosome segregation and condensation, defines a subgroup within the SMC family. Genes Dev. *9*, 587–599.

7 Saka, Y., Sutani, T., Yamashita, Y., Saitoh, S., Takeuchi, M., Nakaseko, Y., and Yanagida, M. (1994). Fission yeast cut3 and cut14, members of a ubiquitous protein family, are required for chromosome condensation and segregation in mitosis. EMBO J. *13*, 4938–4952.

8 Hirano, T., and Mitchison, T.J. (1994). A heterodimeric coiled-coil protein required for mitotic chromosome condensation in vitro. Cell *79*, 449–458.

9 Saitoh, N., Goldberg, I.G., Wood, E.R., and Earnshaw, W.C. (1994). ScII: an abundant chromosome scaffold protein is a member of a family of putative ATPases with an unusual predicted tertiary structure. J. Cell Biol. *127*, 303–318.

10 Hirano, T., Kobayashi, R., and Hirano, M. (1997). Condensins, chromosome condensation protein complexes containing XCAP-C, XCAP-E and a Xenopus homolog of the Drosophila Barren protein. Cell *89*, 511–521.

11 Freeman, L., Aragon-Alcaide, L., and Strunnikov, A. (2000). The condensin complex governs chromosome condensation and mitotic transmission of rDNA. J. Cell Biol. *149*, 811–824.

12 Wood, A.J., Severson, A.F., and Meyer, B.J. (2010). Condensin and cohesin complexity: the expanding repertoire of functions. Nat. Rev. Genet. *11*, 391–404.

13 Hirano, T. (2012). Condensins: universal organizers of chromosomes with diverse functions. Genes Dev. *26*, 1659–1678.

14 Heidinger-Pauli, J.M., Mert, O., Davenport, C., Guacci, V., and Koshland, D. (2010). Systematic reduction of cohesin differentially affects chromosome segregation, condensation, and DNA repair. Curr. Biol. *20*, 957–963.

15 Schleiffer, A., Kaitna, S., Maurer-Stroh, S., Glotzer, M., Nasmyth, K., and Eisenhaber, F. (2003). Kleisins: a superfamily of bacterial and eukaryotic SMC protein partners. Mol. Cell *11*, 571–575.

16 Bhat, M.A., Philp, A.V., Glover, D.M., and Bellen, H.J. (1996). Chromatid segregation at anaphase requires the barren product, a novel chromosome-associated protein that interacts with Topoisomerase II. Cell *87*, 1103–1114.

17 Cuylen, S., Metz, J., and Haering, C.H. (2011). Condensin structures chromosomal DNA through topological links. Nat. Struct. Mol. Biol. *18*, 894–901.

18 Biggins, S., Bhalla, N., Chang, A., Smith, D.L., and Murray, A.W. (2001). Genes involved in sister chromatid separation and segregation in the budding yeast Saccharomyces cerevisiae. Genetics *159*, 453–470.

19 Bhalla, N., Biggins, S., and Murray, A.W. (2002). Mutation of YCS4, a budding yeast condensin subunit, affects mitotic and nonmitotic chromosome behavior. Mol. Biol. Cell *13*, 632–645.

20 Ouspenski, II, Cabello, O.A., and Brinkley, B.R. (2000). Chromosome condensation factor Brn1p is required for chromatid separation in mitosis. Mol. Biol. Cell *11*, 1305–1313.

21 Onn, I., Aono, N., Hirano, M., and Hirano, T. (2007). Reconstitution and subunit geometry of human condensin complexes. EMBO J. *26*, 1024–1034.

22 Lavoie, B.D., Hogan, E., and Koshland, D. (2002). In vivo dissection of the chromosome condensation machinery: reversibility of condensation distinguishes contributions of condensin and cohesin. J. Cell Biol. *156*, 805–815.

23 Uhlmann, F., Lottspeich, F., and Nasmyth, K. (1999). Sister-chromatid separation at anaphase onset is promoted by cleavage of the cohesin subunit Scc1. Nature *400*, 37–42.

24 Gruber, S., Haering, C.H., and Nasmyth, K. (2003). Chromosomal cohesin forms a ring. Cell *112*, 765–777.

25 Ono, T., Losada, A., Hirano, M., Myers, M.P., Neuwald, A.F., and Hirano, T. (2003). Differential contributions of condensin I and condensin II to mitotic chromosome architecture in vertebrate cells. Cell *115*, 109–121.

26 Ohta, S., Bukowski-Wills, J.C., Sanchez-Pulido, L., Alves Fde, L., Wood, L., Chen, Z.A., Platani, M., Fischer, L., Hudson, D.F., Ponting, C.P., *et al*. (2010). The protein composition of mitotic chromosomes determined using multiclassifier combinatorial proteomics. Cell *142*, 810–821.

27 Green, L.C., Kalitsis, P., Chang, T.M., Cipetic, M., Kim, J.H., Marshall, O., Turnbull, L., Whitchurch, C.B., Vagnarelli, P., Samejima, K., *et al*. (2012). Contrasting roles of condensin I and condensin II in mitotic chromosome formation. J. Cell Sci. *125*, 1591–1604.

28 Britton, R.A., Lin, D.C., and Grossman, A.D. (1998). Characterization of a prokaryotic SMC protein involved in chromosome partitioning. Genes Dev. *12*, 1254–1259.

29 Mascarenhas, J., Soppa, J., Strunnikov, A.V., and Graumann, P.L. (2002). Cell cycle-dependent localization of two novel prokaryotic chromosome segregation and condensation proteins in Bacillus subtilis that interact with SMC protein. EMBO J. *21*, 3108–3118.

30 Soppa, J., Kobayashi, K., Noirot-Gros, M.F., Oesterhelt, D., Ehrlich, S.D., Dervyn, E., Ogasawara, N., and Moriya, S. (2002). Discovery of two novel families of proteins that are proposed to interact with prokaryotic SMC proteins, and characterization of the Bacillus subtilis family members ScpA and ScpB. Mol. Microbiol. *45*, 59–71.

31 Niki, H., Jaffe, A., Imamura, R., Ogura, T., and Hiraga, S. (1991). The new gene mukB codes for a 177 kd protein with coiled-coil domains involved in chromosome partitioning of E. coli. EMBO J. *10*, 183–193.

32 Yamazoe, M., Onogi, T., Sunako, Y., Niki, H., Yamanaka, K., Ichimura, T., and Hiraga, S. (1999). Complex formation of MukB, MukE and MukF proteins involved in chromosome partitioning in Escherichia coli. EMBO J. *18*, 5873–5884.

33 Fennell-Fezzie, R., Gradia, S.D., Akey, D., and Berger, J.M. (2005). The MukF subunit of Escherichia coli condensin: architecture and functional relationship to kleisins. EMBO J. *24*, 1921–1930.

34 Woo, J.S., Lim, J.H., Shin, H.C., Suh, M.K., Ku, B., Lee, K.H., Joo, K., Robinson, H., Lee, J., Park, S.Y., *et al*. (2009). Structural studies of a bacterial condensin complex reveal ATP-dependent disruption of intersubunit interactions. Cell *136*, 85–96.

35 Petrushenko, Z.M., She, W., and Rybenkov, V.V. (2011). A new family of bacterial condensins. Mol. Microbiol. *81*, 881–896.

36 Ono, T., Fang, Y., Spector, D.L., and Hirano, T. (2004). Spatial and temporal regulation of Condensins I and II in mitotic chromosome assembly in human cells. Mol. Biol. Cell *15*, 3296–3308.

37 Hirota, T., Gerlich, D., Koch, B., Ellenberg, J., and Peters, J.M. (2004). Distinct functions of condensin I and II in mitotic chromosome assembly. J. Cell Sci. *117*, 6435–6445.

38 Gerlich, D., Hirota, T., Koch, B., Peters, J.M., and Ellenberg, J. (2006). Condensin I stabilizes chromosomes mechanically through a dynamic interaction in live cells. Curr. Biol. *16*, 333–344.

39 Shintomi, K., and Hirano, T. (2011). The relative ratio of condensin I to II determines chromosome shapes. Genes Dev. *25*, 1464–1469.

40 Stray, J.E., and Lindsley, J.E. (2003). Biochemical analysis of the yeast condensin Smc2/4 complex: an ATPase that promotes knotting of circular DNA. J. Biol. Chem. *278*, 26238–26248.

41 Petrushenko, Z.M., Lai, C.H., and Rybenkov, V.V. (2006). Antagonistic interactions of kleis-ins and DNA with bacterial Condensin MukB. J. Biol. Chem. *281*, 34208–34217.

42 Hartl, T.A., Smith, H.F., and Bosco, G. (2008). Chromosome alignment and transvection are antagonized by condensin II. Science *322*, 1384–1387.

43 Watrin, E., Cubizolles, F., Osborne, H.B., Le Guellec, K., and Legagneux, V. (2003). Expression and functional dynamics of the XCAP-D2 condensin subunit in Xenopus laevis oocytes. J. Biol. Chem. *278*, 25708–25715.

44 Uhlmann, F., and Nasmyth, K. (1998). Cohesion between sister chromatids must be estab-lished during DNA replication. Curr. Biol. *8*, 1095–1101.

45 Sutani, T., Yuasa, T., Tomonaga, T., Dohmae, N., Takio, K., and Yanagida, M. (1999). Fission yeast condensin complex: essential roles of non-SMC subunits for condensation and Cdc2 phosphorylation of Cut3/SMC4. Genes Dev. *13*, 2271–2283.

46 Oliveira, R.A., Heidmann, S., and Sunkel, C.E. (2007). Condensin I binds chromatin early in prophase and displays a highly dynamic association with Drosophila mitotic chromo-somes. Chromosoma *116*, 259–274.

47 Collette, K.S., Petty, E.L., Golenberg, N., Bembenek, J.N., and Csankovszki, G. (2011). Different roles for Aurora B in condensin targeting during mitosis and meiosis. J. Cell Sci. *124*, 3684–3694.

48 Seipold, S., Priller, F.C., Goldsmith, P., Harris, W.A., Baier, H., and Abdelilah-Seyfried, S. (2009). Non-SMC condensin I complex proteins control chromosome segregation and survival of proliferating cells in the zebrafish neural retina. BMC Dev. Biol. *9*, 40.

49 Lenart, P., Rabut, G., Daigle, N., Hand, A.R., Terasaki, M., and Ellenberg, J. (2003). Nuclear envelope breakdown in starfish oocytes proceeds by partial NPC disassembly followed by a rapidly spreading fenestration of nuclear membranes. J. Cell Biol. *160*, 1055–1068.

50 D'Ambrosio, C., Schmidt, C.K., Katou, Y., Kelly, G., Itoh, T., Shirahige, K., and Uhlmann, F. (2008). Identification of cis-acting sites for condensin loading onto budding yeast chromo-somes. Genes Dev. *22*, 2215–2227.

51 Wang, B.D., Eyre, D., Basrai, M., Lichten, M., and Strunnikov, A. (2005). Condensin bind-ing at distinct and specific chromosomal sites in the Saccharomyces cerevisiae genome. Mol. Cell Biol. *25*, 7216–7225.

52 Sullivan, N.L., Marquis, K.A., and Rudner, D.Z. (2009). Recruitment of SMC by ParB-parS organizes the origin region and promotes efficient chromosome segregation. Cell *137*, 697–707.

53 Minnen, A., Attaiech, L., Thon, M., Gruber, S., and Veening, J.W. (2011). SMC is recruited to oriC by ParB and promotes chromosome segregation in Streptococcus pneumoniae. Mol. Microbiol. *81*, 676–688.

54 Johzuka, K., and Horiuchi, T. (2009). The cis element and factors required for condensin recruitment to chromosomes. Mol. Cell *34*, 26–35.

55 Tada, K., Susumu, H., Sakuno, T., and Watanabe, Y. (2011). Condensin association with histone H2A shapes mitotic chromosomes. Nature *474*, 477–483.

56 Takemoto, A., Maeshima, K., Ikehara, T., Yamaguchi, K., Murayama, A., Imamura, S., Imamoto, N., Yokoyama, S., Hirano, T., Watanabe, Y., et al. (2009). The chromosomal association of condensin II is regulated by a noncatalytic function of PP2A. Nat. Struct. Mol. Biol. *16*, 1302–1308.

57 Schmidt, C.K., Brookes, N., and Uhlmann, F. (2009). Conserved features of cohesin bind-ing along fission yeast chromosomes. Genome Biol. *10*, R52.

58 Ribeiro, S.A., Gatlin, J.C., Dong, Y., Joglekar, A., Cameron, L., Hudson, D.F., Farr, C.J., McEwen, B.F., Salmon, E.D., Earnshaw, W.C., et al. (2009). Condensin regulates the stiff-ness of vertebrate centromeres. Mol. Biol. Cell *20*, 2371–2380.

59 Stephens, A.D., Haase, J., Vicci, L., Taylor, R.M., 2nd, and Bloom, K. (2011). Cohesin, condensin, and the intramolecular centromere loop together generate the mitotic chromatin spring. J. Cell Biol. *193*, 1167–1180.

60 Gruber, S., and Errington, J. (2009). Recruitment of condensin to replication origin regions by ParB/SpoOJ promotes chromosome segregation in B. subtilis. Cell *137*, 685–696.

61 Kimura, K., Hirano, M., Kobayashi, R., and Hirano, T. (1998). Phosphorylation and activation of 13S condensin by Cdc2 in vitro. Science *282*, 487–490.

62 Kimura, K., Rybenkov, V.V., Crisona, N.J., Hirano, T., and Cozzarelli, N.R. (1999). 13S condensin actively reconfigures DNA by introducing global positive writhe: implications for chromosome condensation. Cell *98*, 239–248.

63 St-Pierre, J., Douziech, M., Bazile, F., Pascariu, M., Bonneil, E., Sauve, V., Ratsima, H., and D'Amours, D. (2009). Polo kinase regulates mitotic chromosome condensation by hyperactivation of condensin DNA supercoiling activity. Mol. Cell *34*, 416–426.

64 Abe, S., Nagasaka, K., Hirayama, Y., Kozuka-Hata, H., Oyama, M., Aoyagi, Y., Obuse, C., and Hirota, T. (2011). The initial phase of chromosome condensation requires Cdk1-mediated phosphorylation of the CAP-D3 subunit of condensin II. Genes Dev. *25*, 863–874.

65 Nousiainen, M., Sillje, H.H., Sauer, G., Nigg, E.A., and Korner, R. (2006). Phosphoproteome analysis of the human mitotic spindle. Proc. Natl. Acad. Sci. USA *103*, 5391–5396.

66 Hegemann, B., Hutchins, J.R., Hudecz, O., Novatchkova, M., Rameseder, J., Sykora, M.M., Liu, S., Mazanek, M., Lenart, P., Heriche, J.K., *et al.* (2011). Systematic phosphorylation analysis of human mitotic protein complexes. Sci. Signal. *4*, rs12.

67 Pagliuca, F.W., Collins, M.O., Lichawska, A., Zegerman, P., Choudhary, J.S., and Pines, J. (2011). Quantitative proteomics reveals the basis for the biochemical specificity of the cell-cycle machinery. Mol. Cell *43*, 406–417.

68 Bazile, F., St-Pierre, J., and D'Amours, D. (2010). Three-step model for condensin activation during mitotic chromosome condensation. Cell Cycle *9*, 3243–3255.

69 Takemoto, A., Kimura, K., Yanagisawa, J., Yokoyama, S., and Hanaoka, F. (2006). Negative regulation of condensin I by CK2-mediated phosphorylation. EMBO J. *25*, 5339–5348.

70 Morishita, J., Matsusaka, T., Goshima, G., Nakamura, T., Tatebe, H., and Yanagida, M. (2001). Bir1/Cut17 moving from chromosome to spindle upon the loss of cohesion is required for condensation, spindle elongation and repair. Genes Cells *6*, 743–763.

71 Petersen, J., and Hagan, I.M. (2003). S. pombe aurora kinase/survivin is required for chromosome condensation and the spindle checkpoint attachment response. Curr. Biol. *13*, 590–597.

72 Nakazawa, N., Nakamura, T., Kokubu, A., Ebe, M., Nagao, K., and Yanagida, M. (2008). Dissection of the essential steps for condensin accumulation at kinetochores and rDNAs during fission yeast mitosis. J. Cell Biol. *180*, 1115–1131.

73 Giet, R., and Glover, D.M. (2001). Drosophila aurora B kinase is required for histone H3 phosphorylation and condensin recruitment during chromosome condensation and to organize the central spindle during cytokinesis. J. Cell Biol. *152*, 669–682.

74 Lavoie, B.D., Hogan, E., and Koshland, D. (2004). In vivo requirements for rDNA chromosome condensation reveal two cell-cycle-regulated pathways for mitotic chromosome folding. Genes Dev. *18*, 76–87.

75 Kaitna, S., Pasierbek, P., Jantsch, M., Loidl, J., and Glotzer, M. (2002). The aurora B kinase AIR-2 regulates kinetochores during mitosis and is required for separation of homologous Chromosomes during meiosis. Curr. Biol. *12*, 798–812.

76 Maddox, P.S., Portier, N., Desai, A., and Oegema, K. (2006). Molecular analysis of mitotic chromosome condensation using a quantitative time-resolved fluorescence microscopy assay. Proc. Natl. Acad. Sci. USA *103*, 15097–15102.

77 Lipp, J.J., Hirota, T., Poser, I., and Peters, J.M. (2007). Aurora B controls the association of condensin I but not condensin II with mitotic chromosomes. J. Cell Sci. 120, 1245–1255.

78 Mora-Bermudez, F., Gerlich, D., and Ellenberg, J. (2007). Maximal chromosome compaction occurs by axial shortening in anaphase and depends on Aurora kinase. Nat. Cell Biol. 9, 822–831.

79 Takemoto, A., Murayama, A., Katano, M., Urano, T., Furukawa, K., Yokoyama, S., Yanagisawa, J., Hanaoka, F., and Kimura, K. (2007). Analysis of the role of Aurora B on the chromosomal targeting of condensin I. Nucleic Acids Res. 35, 2403–2412.

80 Nakazawa, N., Mehrotra, R., Ebe, M., and Yanagida, M. (2011). Condensin phosphorylated by the Aurora-B-like kinase Ark1 is continuously required until telophase in a mode distinct from Top2. J. Cell Sci. 124, 1795–1807.

81 Queralt, E., and Uhlmann, F. (2008). Cdk-counteracting phosphatases unlock mitotic exit. Curr. Opin. Cell Biol. 20, 661–668.

82 Sullivan, M., Higuchi, T., Katis, V.L., and Uhlmann, F. (2004). Cdc14 phosphatase induces rDNA condensation and resolves cohesin-independent cohesion during budding yeast anaphase. Cell 117, 471–482.

83 D'Amours, D., Stegmeier, F., and Amon, A. (2004). Cdc14 and condensin control the dissolution of cohesin-independent chromosome linkages at repeated DNA. Cell 117, 455–469.

84 Pereira, G., and Schiebel, E. (2003). Separase regulates INCENP-Aurora B anaphase spindle function through Cdc14. Science 302, 2120–2124.

85 Neurohr, G., Naegeli, A., Titos, I., Theler, D., Greber, B., Diez, J., Gabaldon, T., Mendoza, M., and Barral, Y. (2011). A midzone-based ruler adjusts chromosome compaction to anaphase spindle length. Science 332, 465–468.

86 Varela, E., Shimada, K., Laroche, T., Leroy, D., and Gasser, S.M. (2009). Lte1, Cdc14 and MEN-controlled Cdk inactivation in yeast coordinate rDNA decompaction with late telophase progression. EMBO J. 28, 1562–1575.

87 Yeong, F.M., Hombauer, H., Wendt, K.S., Hirota, T., Mudrak, I., Mechtler, K., Loregger, T., Marchler-Bauer, A., Tanaka, K., Peters, J.M., et al. (2003). Identification of a subunit of a novel Kleisin-beta/SMC complex as a potential substrate of protein phosphatase 2A. Curr. Biol. 13, 2058–2064.

88 Zhang, N., Kuznetsov, S.G., Sharan, S.K., Li, K., Rao, P.H., and Pati, D. (2008). A handcuff model for the cohesin complex. J. Cell Biol. 183, 1019–1031.

89 Matoba, K., Yamazoe, M., Mayanagi, K., Morikawa, K., and Hiraga, S. (2005). Comparison of MukB homodimer versus MukBEF complex molecular architectures by electron microscopy reveals a higher-order multimerization. Biochem. Biophys. Res. Commun. 333, 694–702.

90 Fuentes-Perez, M.E., Gwynn, E.J., Dillingham, M.S., and Moreno-Herrero, F. (2012). Using DNA as a fiducial marker to study SMC complex interactions with the atomic force microscope. Biophys. J. 102, 839–848.

91 Cui, Y., Petrushenko, Z.M., and Rybenkov, V.V. (2008). MukB acts as a macromolecular clamp in DNA condensation. Nat. Struct. Mol. Biol. 15, 411–418.

92 Petrushenko, Z.M., Cui, Y., She, W., and Rybenkov, V.V. (2010). Mechanics of DNA bridging by bacterial condensin MukBEF in vitro and in singulo. EMBO J. 29, 1126–1135.

93 Strick, T.R., Kawaguchi, T., and Hirano, T. (2004). Real-time detection of single-molecule DNA compaction by condensin I. Curr. Biol. 14, 874–880.

94 Ohsumi, K., Yamazoe, M., and Hiraga, S. (2001). Different localization of SeqA-bound nascent DNA clusters and MukF-MukE-MukB complex in Escherichia coli cells. Mol. Microbiol. 40, 835–845.

95 MacCallum, D.E., Losada, A., Kobayashi, R., and Hirano, T. (2002). ISWI remodeling complexes in Xenopus egg extracts: identification as major chromosomal components that are regulated by INCENP-aurora B. Mol. Biol. Cell *13*, 25–39.

96 Petrushenko, Z.M., Lai, C.H., Rai, R., and Rybenkov, V.V. (2006). DNA reshaping by MukB. Right-handed knotting, left-handed supercoiling. J. Biol. Chem. *281*, 4606–4615.

97 Hayama, R., and Marians, K.J. (2010). Physical and functional interaction between the condensin MukB and the decatenase topoisomerase IV in Escherichia coli. Proc. Natl. Acad. Sci. USA *107*, 18826–18831.

98 Li, Y., Stewart, N.K., Berger, A.J., Vos, S., Schoeffler, A.J., Berger, J.M., Chait, B.T., and Oakley, M.G. (2010). Escherichia coli condensin MukB stimulates topoisomerase IV activity by a direct physical interaction. Proc. Natl. Acad. Sci. USA *107*, 18832–18837.

99 D'Ambrosio, C., Kelly, G., Shirahige, K., and Uhlmann, F. (2008). Condensin-dependent rDNA decatenation introduces a temporal pattern to chromosome segregation. Curr. Biol. *18*, 1084–1089.

100 Kimura, K., and Hirano, T. (1997). ATP-dependent positive supercoiling of DNA by 13S condensin: a biochemical implication for chromosome condensation. Cell *90*, 625–634.

101 van den Ent, F., Lockhart, A., Kendrick-Jones, J., and Lowe, J. (1999). Crystal structure of the N-terminal domain of MukB: a protein involved in chromosome partitioning. Structure *7*, 1181–1187.

102 Lammens, A., Schele, A., and Hopfner, K.P. (2004). Structural biochemistry of ATP-driven dimerization and DNA-stimulated activation of SMC ATPases. Curr. Biol. *14*, 1778–1782.

103 Arumugam, P., Gruber, S., Tanaka, K., Haering, C.H., Mechtler, K., and Nasmyth, K. (2003). ATP hydrolysis is required for cohesin's association with chromosomes. Curr. Biol. *13*, 1941–1953.

104 Weitzer, S., Lehane, C., and Uhlmann, F. (2003). A model for ATP hydrolysis-dependent binding of cohesin to DNA. Curr. Biol. *13*, 1930–1940.

105 Chan, K.L., Roig, M.B., Hu, B., Beckouet, F., Metson, J., and Nasmyth, K. (2012). Cohesin's DNA exit gate is distinct from its entrance gate and is regulated by acetylation. Cell *150*, 961–974.

106 Mc Intyre, J., Muller, E.G., Weitzer, S., Snydsman, B.E., Davis, T.N., and Uhlmann, F. (2007). In vivo analysis of cohesin architecture using FRET in the budding yeast Saccharomyces cerevisiae. EMBO J. *26*, 3783–3793.

107 Yoshimura, S.H., Hizume, K., Murakami, A., Sutani, T., Takeyasu, K., and Yanagida, M. (2002). Condensin architecture and interaction with DNA: regulatory non-SMC subunits bind to the head of SMC heterodimer. Curr. Biol. *12*, 508–513.

108 Griese, J.J., and Hopfner, K.P. (2010). Structure and DNA-binding activity of the Pyrococcus furiosus SMC protein hinge domain. Proteins *79*, 558–568.

109 Hirano, M., and Hirano, T. (2006). Opening closed arms: long-distance activation of SMC ATPase by hinge-DNA interactions. Mol. Cell *21*, 175–186.

110 Ku, B., Lim, J.H., Shin, H.C., Shin, S.Y., and Oh, B.H. (2010). Crystal structure of the MukB hinge domain with coiled-coil stretches and its functional implications. Proteins *78*, 1483–1490.

111 de Wit, E., and de Laat, W. (2012). A decade of 3C technologies: insights into nuclear organization. Genes Dev. *26*, 11–24.

112 Poirier, M.G., and Marko, J.F. (2002). Mitotic chromosomes are chromatin networks without a mechanically contiguous protein scaffold. Proc. Natl. Acad. Sci. USA *99*, 15393–15397.

113 Maeshima, K., and Laemmli, U.K. (2003). A two-step scaffolding model for mitotic chromosome assembly. Dev. Cell *4*, 467–480.

114 Hudson, D.F., Ohta, S., Freisinger, T., Macisaac, F., Sennels, L., Alves, F., Lai, F., Kerr, A., Rappsilber, J., and Earnshaw, W.C. (2008). Molecular and genetic analysis of condensin function in vertebrate cells. Mol. Biol. Cell *19*, 3070–3079.

115 Nishino, Y., Eltsov, M., Joti, Y., Ito, K., Takata, H., Takahashi, Y., Hihara, S., Frangakis, A.S., Imamoto, N., Ishikawa, T., *et al*. (2012). Human mitotic chromosomes consist predominantly of irregularly folded nucleosome fibres without a 30-nm chromatin structure. EMBO J. *31*, 1644–1653.

116 Strukov, Y.G., and Belmont, A.S. (2009). Mitotic chromosome structure: reproducibility of folding and symmetry between sister chromatids. Biophys. J. *96*, 1617–1628.

117 Konig, P., Braunfeld, M.B., Sedat, J.W., and Agard, D.A. (2007). The three-dimensional structure of in vitro reconstituted Xenopus laevis chromosomes by EM tomography. Chromosoma *116*, 349–372.

118 Felsenstein, J. (1989). PHYLIP - Phylogeny Inference Package (Version 3.2). Cladistics *5*, 164–166.

119 Larkin, M.A., Blackshields, G., Brown, N.P., Chenna, R., McGettigan, P.A., McWilliam, H., Valentin, F., Wallace, I.M., Wilm, A., Lopez, R., *et al*. (2007). Clustal W and Clustal X version 2.0. Bioinformatics *23*, 2947–2948.

120 Letunic, I., and Bork, P. (2011). Interactive Tree Of Life v2: online annotation and display of phylogenetic trees made easy. Nucleic Acids Res. *39*, W475–W478.

121 Jones, D.T., Taylor, W.R., and Thornton, J.M. (1992). The rapid generation of mutation data matrices from protein sequences. Comput. Appl. Biosci. *8*, 275–282.

ell

Nuclear Positioning

Gregg G. Gundersen[1,*], Howard J. Worman[1,2,*]

[1]Department of Pathology and Cell Biology, College of Physicians and Surgeons,
Columbia University, 630 West 168th Street, New York, NY 10032, USA
[2]Department of Medicine, College of Physicians and Surgeons, Columbia University,
630 West 168th Street, New York, NY 10032, USA
*Correspondence: ggg1@columbia.edu (G.G.G.), hjw14@columbia.edu (H.J.W.)

Cell, Vol. 152, No. 6, March 14, 2013 © 2013 Elsevier Inc.
http://dx.doi.org/10.1016/j.cell.2013.02.031

SUMMARY

The nucleus is the largest organelle and is commonly depicted in the center of the cell. Yet during cell division, migration, and differentiation, it frequently moves to an asymmetric position aligned with cell function. We consider the toolbox of proteins that move and anchor the nucleus within the cell and how forces generated by the cytoskeleton are coupled to the nucleus to move it. The significance of proper nuclear positioning is underscored by numerous diseases resulting from genetic alterations in the toolbox proteins. Finally, we discuss how nuclear position may influence cellular organization and signaling pathways.

INTRODUCTION

Diagrams in biology textbooks usually depict the nucleus as a spheroid in the center of the cell. However, the position of nuclei varies dramatically from this simple view. Nuclei are frequently positioned asymmetrically depending on cell type, stage of the cell cycle, migratory state, and differentiation status. For example, during cell division in budding yeast, nuclei are moved into the bud neck so that each daughter cell receives one (Figure 1A). Nuclei are actively positioned in the middle of the fission yeast *S. pombe*, ensuring that the division plane produces two equal daughter cells. In fertilized mammalian and invertebrate eggs, male and female pronuclei move toward each other and fuse near the middle of the zygote, ensuring that the ensuing cell division creates two equal daughter blastomeres. Asymmetric

35

CellPress

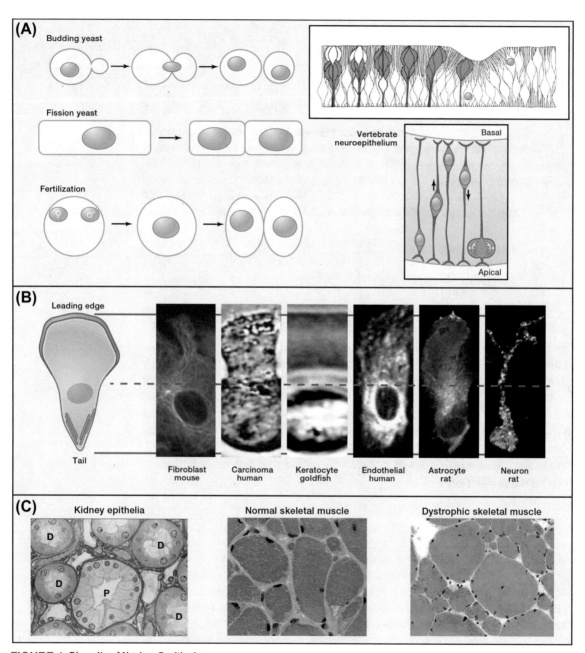

FIGURE 1 Diversity of Nuclear Positioning

(A) Schematics of nuclear positioning in dividing cells and developing epithelium. Arrows indicate movements of nuclei (blue). The nucleus is positioned relative to the plane of division in yeast and fertilized eggs. The diagram of insect optic epithelium (adapted from Patterson et al., 2004; Tomlinson and Ready, 1986) represents a longitudinal section of a larval eye disc; two nuclei are shown. Nuclei that are anterior (A) to the morphogenetic furrow (mf), which moves anteriorly, move basally. Nuclei that are posterior (P) to the furrow move divisions—typical of early embryos and stem cells—frequently reflect a prepositioning of the nucleus.

Though nuclear positioning to affect the cell division plane makes intuitive sense, asymmetric positioning occurs in nondividing cells, where the purpose is less obvious. For example, in the developing optic epithelium in *Drosophila*, nuclei move basally and then apically to establish the characteristic arrangement of cells in the ommatidium (Figure 1A). An analogous movement of nuclei occurs over the cell cycle in the developing vertebrate neuroepithelium. In most migrating cells, the nucleus is positioned in the rear, well removed from the protruding front (Figure 1B). Nuclei in numerous differentiated animal tissues, such as skeletal muscle, many epithelia, and neurons, are also asymmetrically positioned (Figure 1C and Table 1). These examples suggest that nuclei are positioned for specialized cellular functions and that abnormal positioning could lead to dysfunction and disease.

Position of nuclei can be modified secondarily to changes in cytoplasmic organization. For example, when macrovesicular fat accumulates in hepatocytes in alcoholic or nonalcoholic steatosis, nuclei are forced to the cell's periphery. Similar changes in nuclear position may occur in cells with abundant secretory granules. However, recent research has discovered regulated, cytoplasmic mechanical systems that function primarily to exert forces on the nucleus via connections to the nuclear envelope. These systems maintain the position of the nucleus or move it during processes such as cell migration and differentiation. Though their role in homeostatic nuclear positioning is poorly understood, mechanistic details are being deciphered in cases where nuclei move.

We review systems in which progress is being made in understanding nuclear movement and positioning, and we identify the molecular toolbox that cells use for these processes. This toolbox includes specific nuclear envelope connections to cytoskeletal force-generating systems. We then

apically as cells are recruited into clusters comprising ommatidium (white cells, cones; gray cells, R cells). The diagram of vertebrate neuroepithelium represents a longitudinal section of the developing cerebral cortex. Nuclei move basally during G1 and apically during G2. Mitosis (M) occurs near the apical surface. Adapted from Buchman and Tsai (2008) with permission.

(B) Rearward nuclear position is typical of migrating cells. (Left) Schematic of a migrating cell with protruding leading edge and contracting tail. (Red) Actin filaments. (Right) Montage of migrating cells with front-back dimensions normalized. Dotted line represents the midpoint between the front and back. Nuclei are positioned along the front-back axis but always rearward of the cell center. Images reproduced with permission from: fibroblast (Gomes et al., 2005), breast carcinoma (McNiven, 2013), keratocyte (Barnhart et al., 2010), endothelial cell (Tsai and Meyer, 2012), astrocyte (Osmani et al., 2006), and neuron (Godin et al., 2012).

(C) Nuclear positioning in mammalian tissues. Cross-sections of kidney cortex and skeletal muscle stained with hematoxylin and eosin. Nuclei are positioned centrally in the distal (D) convoluted tubules and basally in proximal (P) convoluted tubules. Nuclei are positioned at the periphery of normal skeletal muscle fibers but are found centrally in dystrophic tissue.

Table 1 Nuclear Positions in Mammalian Cells and Tissues

Cell Tissue	Nuclear Position	Axis Alignment[a]	Comments
Proliferating Cells			
Somatic cells	central	NA	
Stem cells	usually asymmetric	various; niche related	
Germ cells, oocytes	asymmetric	NA	moves centrally after fertilization
Migrating Cells			
1D (cultured fibroblast)	asymmetric	front-*rear*	
2D (cultured; many types)	asymmetric	front-*rear*	see Figure 1
3D (cultured fibroblast)	asymmetric	front-*rear*	
3D (dermal sarcoma cells)	asymmetric	front-*rear*	
3D (neurons in cortex)	asymmetric	front-*rear*	
Macrophages, neutrophils	asymmetric	front-*rear*	
Tissues			
Muscle, skeletal	asymmetric, complex	*peripheral*-central	clustered at neuromuscular junction
Muscle, cardiac	central	NA	
Muscle, smooth	central	NA	
Epithelia, squamous	central	NA	
Epithelia, cuboidal	central	NA	
Epithelia, columnar	asymmetric	apical-*basal*	
Epithelia, pseudostratified	asymmetric	apical-*basal*	cell-cycle dependent
Epithelia, secretory	asymmetric	apical-*basal*	aligned with secretory axis
Neurons	asymmetric	*proximal*-distal	
Astrocytes/oligodendricytes	central	NA	
Connective Tissue			
Osteoblasts/osteocytes	central	NA	
Osteocytes, actively secreting	asymmetric	front-*rear*	relative to secretory axis
Osteoclasts	asymmetric	front-*rear*	
Chondroblasts/chondrocytes	central	NA	
Chondrocytes, actively secreting	asymmetric	front-*rear*	relative to secretory axis
Fibrocytes, resting	central	NA	
Adipocytes	asymmetric	NA	
Hematopoetic			
Macrophages	asymmetric	front-*rear*	
T cells, migrating or contacting target cell	asymmetric	front-*rear*	
B cells, plasma cells	asymmetric	front-*rear*	

[a]Position in italics.

evaluate how this toolbox is employed and identify conserved mechanisms that use microtubules (MTs) and actin filaments as force generators. Genes encoding toolbox proteins are targets of mutations that cause disease, raising the possibility that inappropriate nuclear positioning contributes to pathogenesis. As active nuclear movement suggests that its relative position may influence other cellular systems, we consider the significance of nuclear positioning for cytoskeletal organization, signaling, and transcriptional control.

THE NUCLEAR POSITIONING TOOLBOX

The molecular toolbox for nuclear positioning contains: (1) elements of the cytoskeleton and (2) protein complexes of the nuclear envelope. The cytoskeletal elements generate forces to move the nucleus. The protein complexes spanning the nuclear membranes mediate attachment of cytoskeletal elements to the nucleoskeleton (Figure 2).

Cytoskeletal Elements

Actin filaments, MTs, and associated motor proteins are the principal cytoskeletal elements of the nuclear positioning toolbox. Cytoplasmic intermediate filaments may also play a role, but this is currently poorly defined. In some cases, a single cytoskeletal element drives nuclear movement, as in MT-dependent movement of male and female pronuclei after fertilization and actin-dependent rearward movement of nuclei in fibroblasts polarizing for migration. In other cases, MTs and actin filaments collaborate to move nuclei, as in migrating neuronal cells. The role of these cytoplasmic elements in different systems is discussed in detail below.

Protein Complexes in the Nuclear Envelope

An exciting advance in the past few years has been the identification of the *linker* of *nucleoskeleton* and *cytoskeleton* (LINC) complex in the nuclear envelope that mediates connections to both MTs and actin filaments (Crisp et al., 2006). LINC complexes are composed of outer nuclear membrane KASH (*klarsicht*, *Anc1*, and *Syne homology*) proteins and inner nuclear membrane SUN (*Sad1* and *Unc-83*) proteins, both of which are type II membrane proteins with a single transmembrane segment (Starr and Fridolfsson, 2010) (Figure 2A). KASH and SUN proteins have been described in metazoan, fungi, and recently plants (Razafsky and Hodzic, 2009; Zhou et al., 2012a). KASH proteins are characterized by a conserved ~60 residue KASH domain at their C terminus, which includes a transmembrane segment and up to 30 residues that project into the perinuclear space between inner and outer nuclear membranes. KASH

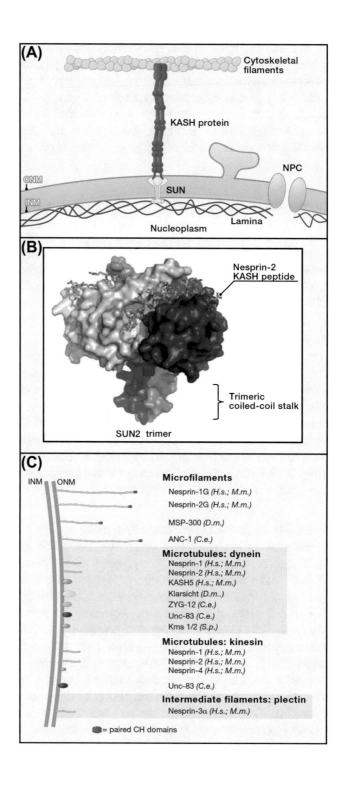

(A)

Cytoskeletal filaments

KASH protein

NPC

ONM

INM

SUN

Lamina

Nucleoplasm

(B)

Nesprin-2 KASH peptide

Trimeric coiled-coil stalk

SUN2 trimer

(C)

INM   ONM

Microfilaments
Nesprin-1G *(H.s.; M.m.)*
Nesprin-2G *(H.s.; M.m.)*
MSP-300 *(D.m.)*
ANC-1 *(C.e.)*

Microtubules: dynein
Nesprin-1 *(H.s.; M.m.)*
Nesprin-2 *(H.s.; M.m.)*
KASH5 *(H.s.; M.m.)*
Klarsicht *(D.m..)*
ZYG-12 *(C.e.)*
Unc-83 *(C.e.)*
Kms 1/2 *(S.p.)*

Microtubules: kinesin
Nesprin-1 *(H.s.; M.m.)*
Nesprin-2 *(H.s.; M.m.)*
Nesprin-4 *(H.s.; M.m.)*
Unc-83 *(C.e.)*

Intermediate filaments: plectin
Nesprin-3α *(H.s.; M.m.)*

= paired CH domains

domains in fungi and plants are less conserved than those in metazoans. SUN proteins contain a conserved SUN domain located within the perinuclear space. Five genes encode SUN proteins in mammals, although only two of these (SUN1 and SUN2) are widely expressed; lower eukaryotes have one or two SUN proteins (Starr and Fridolfsson, 2010).

The crystal structure of SUN2 reveals an interesting mushroom-like trimer with a "cap" composed of SUN domains and a triple coiled-coil stalk, which is required for trimer formation (Figure 2B) (Sosa et al., 2012; Zhou et al., 2012b). Predictions of the length of this stalk suggest that the SUN protein could span the nearly 50 nm between inner and outer nuclear membranes (Sosa et al., 2012). Each SUN protein binds three KASH peptides in deep grooves between adjacent SUN domains in the trimer (Figure 2B). A KASH-SUN disulfide bond may further stabilize the complex.

The trimeric SUN-KASH structure raises intriguing questions about higher-ordered KASH-SUN protein assemblies, particularly if KASH proteins are indeed dimeric molecules as predicted. The binding pocket between SUN2 subunits suggests that it will accommodate related KASH domains and that SUN1 and SUN2 bind KASH proteins promiscuously (Starr and Fridolfsson, 2010). Yet there is an example in cells in which a specific KASH-SUN pair assembles to move the nucleus (Luxton et al., 2011). The apparent tight packing within the SUN-KASH complex also raises questions about its assembly and regulation. KASH and SUN proteins have diffusional mobilities similar to other nuclear membrane proteins, indicating that they are likely in dynamic complexes (Östlund et al., 2009). TorsinA is a

←

FIGURE 2 Molecular Toolbox for Nuclear Movement/Positioning

(A) Schematic of an idealized LINC complex in nuclear envelope. The inner nuclear membrane (INM) SUNs bind within the perinuclear space to outer nuclear membrane (ONM) KASH proteins. KASH proteins bind directly or indirectly to cytoskeletal filaments, including MTs, actin microfilaments, and cytoplasmic intermediate filaments. In metazoans, SUNs bind to the nuclear lamina; in yeast and plants, other intranuclear proteins bind SUNs. A nuclear pore complex (NPC) is shown for reference.

(B) Side view of the structure of the SUN2-nesprin2 KASH complex. Trimeric SUN2 domains are represented by different shades of blue, and the KASH peptide is in orange. The structure illustrates the orientation of the KASH peptide between adjacent SUN domains. Modified from Sosa et al. (2012) with permission.

(C) Schematic diagrams of KASH proteins from representative organisms and the cytoskeletal filaments to which they bind. Binding to actin filaments is mediated by CH domains and binding to cytoplasmic intermediate filaments by plectin. Binding to MTs is mediated by dynein and kinesins; direct binding to MTs has not been reported. The specific splice variants of nesprin-1 and nesprin-2 that interact with MT motors are unknown; for simplicity, a short variant of each is depicted. *H.s.*, *Homo sapiens*; *M.m.*, *Mus musculus*; *D.m.*, *Drosophila melanogaster*; *C.e.*, *Caenorhabditis elegans*; *S.p.*, *Schizosaccharomyces pombe*.

potential regulator of the LINC complex, as it localizes to the ER lumen and perinuclear space and shows affinity for KASH domains (Nery et al., 2008; Tanabe et al., 2009). TorsinA's homology to AAA ATPases suggests that it may chaperone assembly or disassembly of LINC complexes (Tanabe et al., 2009).

Specificity of LINC complexes is determined by the N termini of KASH proteins, which are variable in size and ability to bind cytoskeletal elements (Figure 2C). In mammals, KASH proteins (termed nesprins) are encoded by five genes, some of which generate multiple isoforms by alternative RNA splicing. The "giant" isoforms nesprin-1G and nesprin-2G (>800 kDa) encoded by *SYNE1* and *SYNE2*, respectively, bind actin through calponin homology (CH) domains near their N termini (Luxton et al., 2011; Zhang et al., 2001). Much of their large cytoplasmic region is predicted to be composed of spectrin repeats, suggesting a structure reminiscent of dystrophin with an extended but flexible core and the potential for dimerization. Nesprin-1 and nesprin-2 isoforms also interact with the MT motors kinesin-1 and dynein, although whether binding is direct is unknown (Yu et al., 2011; Zhang et al., 2009). In *C. elegans*, the KASH protein Unc-83 interacts directly with kinesin-1, dynein, and dynein regulators, including BicaudalD and NudE homologs (Fridolfsson et al., 2010; Fridolfsson and Starr, 2010). Nesprin-3α, an isoform encoded by *SYNE3*, binds the crosslinking protein plectin, which binds cytoplasmic intermediate filaments (Wilhelmsen et al., 2005). Nesprin-4 encoded by *SYNE4* has a short N terminus that associates with MTs through kinesin-1 and is restricted in expression to highly secretory cells and hair cells of the cochlea (Horn et al., 2013; Roux et al., 2009). Aside from spectrin repeats, there are no other recognizable domains in nesprins 1–4. A meiosis-specific "nesprin" termed KASH5 binds the dynein regulator dynactin (Morimoto et al., 2012). Lower eukaryotes express actin- and MT motor-binding KASH proteins, although there is less genetic complexity in these organisms. For example, there are two KASH proteins in *Drosophila* and four in *C. elegans* (Figure 2C) (Starr and Fridolfsson, 2010).

At the intranuclear side of the LINC complex, SUN proteins bind to nuclear lamins (Crisp et al., 2006; Haque et al., 2006). Lamins are intermediate filament proteins that polymerize to form the nuclear lamina, a meshwork underlying the inner nuclear membrane. Lamins A and C (A-type lamins), which are alternative splice isoforms of the same gene, and lamins B1 and B2 are the predominant lamins expressed in differentiated mammalian somatic cells. N termini of SUN1 and SUN2 bind to lamin A, mediating their interaction with the lamina. Hence, the LINC complex, via KASH protein interactions with cytoskeletal proteins and SUN protein interactions with lamins, connects the nucleoskeleton to the cytoskeleton.

In mammalian cells lacking A-type lamins, SUN proteins still localize to the nucleus (Crisp et al., 2006; Haque et al., 2006), although they and their nesprin partners have increased membrane diffusional mobility (Östlund et al., 2009). This suggests that other factors contribute to LINC complex anchoring. Indeed, yeast lack lamins but still employ KASH and SUN proteins to attach the nucleus to the cytoskeleton. In *S. pombe*, the heterochromatin-binding protein Ima1 anchors the SUN protein Sad1, a component of the spindle pole body (King et al., 2008). SAMP1, the mammalian Ima1 ortholog, localizes to LINC complex assemblies that attach actin to the nucleus (Borrego-Pinto et al., 2012). Emerin, which is an integral protein predominantly localized to the inner nuclear membrane, binds to lamins and nesprins (Mislow et al., 2002; Zhang et al., 2005). Additionally, SUN1 associates with nuclear pore complexes (Liu et al., 2007).

LINC complex components constitute the major tools for connecting the nucleus to the cytoskeleton, yet they may not be the only ones. Dynein interacts with Bicaudal2, which in turn binds to RANBP2 at the cytoplasmic face of the nuclear pore complex (Splinter et al., 2010). This association targets dynein to the nucleus during G2 and may contribute to nuclear envelope breakdown. However, it could be an alternative means to target dynein for nuclear movement. Certain muscle-specific nuclear membrane proteins accumulate along MTs, suggesting that the nuclear positioning toolbox may also contain tissue-specific tools (Wilkie et al., 2011).

INITIATION OF NUCLEAR MOVEMENT

Specific sets of tools become activated to move the nucleus in response to stimuli. In pronuclear migrations in fertilized eggs, formation of MTs by the sperm centrosome initiates movement of both male and female pronuclei. Activation of the Rho GTPase Cdc42 by the serum factor lysophosphatidic acid (LPA) initiates nuclear movement in migrating fibroblasts by activating actin retrograde flow (Gomes et al., 2005; Palazzo et al., 2001). Cdc42 is also essential for nuclear movements in neuronal migration (Solecki et al., 2004) and neuronal precursors in the neuroepithelium (Cappello et al., 2006). Nuclear movement in the neuroepithelium is under cell-cycle control, and interference with cell-cycle progression prevents it (Taverna and Huttner, 2010). These examples indicate that initiating nuclear movements involves the de novo assembly of cytoskeletal components of the toolbox. However, this is a fledgling area of inquiry, and other processes such as activation of motors or relaxation of nuclear anchoring may contribute to initiating nuclear movement. Almost nothing is known about factors terminating nuclear movement.

CHARACTERISTICS OF NUCLEAR MOVEMENTS

Nuclear movements occur in different cellular contexts and are powered by different cytoskeletal elements. It is therefore not surprising that they have different characteristics (Table 2). Velocities vary between 0.1 and 1.0 μm/min, although peak rates can be >10 μm/min for sperm pronuclei in *Xenopus* eggs. Distances transversed during single episodes are generally one nuclear diameter (5–10 μm) or less, although they are longer in fertilized

Table 2 Physical Characteristics of Typical Nuclear Movements

System	Rate (μm min⁻¹)	Distance (μm)	Mode	Dependence	References
Fertilized Egg					
Male pronucleus, *Xenopus*	16	100–300	?	MT polymerization	Reinsch and Gönczy, 1998
Female pronucleus, *Xenopus*	0.2–1.5	100–300	?	dynein	Reinsch and Gönczy, 1998
Migrating Neurons					
Cortical brain slice	0.33	1–5	saltatory	MT and myosin II	Tsai et al., 2007
SVZ explants, matrigel	1.2–5	2–5	saltatory		Schaar and McConnell, 2005
Granular neurons on radial glia	1.0	1.3	saltatory	MT and myosin II	Solecki et al., 2004; Solecki et al., 2009
Radial Glia INM, Cortical Brain Slice					
Basal directed	0.14	30–50	intermittent with long pauses	kinesin3	Tsai et al., 2010
Apical directed	0.06	30–50	continuous	dynein	Tsai et al., 2007
Other Systems					
Fibroblasts polarizing for migration	0.28–0.35	5–10	continuous	actomyosin flow	Gomes et al., 2005; Luxton et al., 2010
Astrocytes polarizing for migration	0.05	~10	continuous	actomyosin flow	Dupin et al., 2011
Oocyte (*D.m.*)	0.07	5–10	continuous	MT polymerization	Zhao et al., 2012
Hypodermal cell (*C.e.*)	0.23	3.3	continuous	kinesin1	Fridolfsson and Starr, 2010
Budding yeast	1.18	1–2	continuous	dynein (and MT depolymerization)	Adames and Cooper, 2000

D.m., Drosophila melanogaster; C.e., Caenorhabditis elegans.

eggs and in the neuroepithelium. Nuclear movements are usually continuous and unidirectional. However, high-temporal-resolution imaging of nuclei in *C. elegans* hypodermal cells revealed short pauses and bidirectional movements, suggesting additional complexity (Fridolfsson and Starr, 2010). During basal movement in the rat neuroepithelium, nuclei pause for hour-long intervals before continuing in the same direction, suggesting complex regulation. This diversity of nuclear movements provided an early clue that there is more than one mechanism responsible.

MT-MEDIATED NUCLEAR MOVEMENT

Pioneering studies on invertebrate and vertebrate eggs revealed that there are distinct mechanisms by which MTs connect to the nucleus to move it (reviewed in Reinsch and Gönczy, 1998). The male pronucleus, which forms after sperm entry into the egg, nucleates MTs from its centrosome and moves toward the middle of the cell. The female pronucleus laterally engages MTs emanating from the male pronuclear-centrosome complex and moves along them to join the male nucleus near the cell center. Male pronuclear movement is generated by MT growth and pushing along cortical sites and/or sites within the cytoplasm (Reinsch and Gönczy, 1998). Force is transmitted to the nucleus through its intimate association with the centrosome and centrosomal MTs. Female pronuclear movement is generated by attached cytoplasmic dynein motors that move it toward MT minus ends at the sperm centrosome. Research on nuclear movement has progressed from fertilized eggs to more molecularly tractable systems, yet the idea that distinct MT-dependent processes move the nucleus has persisted and has been strengthened by newer studies.

Nuclear Movement by MT Pushing and Pulling Forces

In the male pronuclear form of nuclear movement, an MT organizing center (MTOC) connects the nucleus to MTs, and MT dynamics power movement (Figure 3A). This form of nuclear movement occurs before cell division in the budding yeast *S. cerevisiae* (Adames and Cooper, 2000), the fission yeast *S. pombe* (Tran et al., 2001), early *C. elegans* embryos (Gönczy et al., 1999), *Drosophila* oocytes (Zhao et al., 2012), and cultured mammalian cells (Levy and Holzbaur, 2008). The MTOC is either embedded in the nuclear envelope (yeast spindle pole body) or is tightly associated with it (other systems). In *C. elegans*, the centrosome connects to the nuclear envelope through the LINC complex proteins Zyg-12, a KASH protein, and SUN1 (Malone et al., 2003). Outer nuclear membrane Zyg-12 binds to dynein, moving the centrosome close to the nucleus and promoting association between Zyg-12 and a centrosomal splice variant lacking the transmembrane domain. Zyg-12 is not

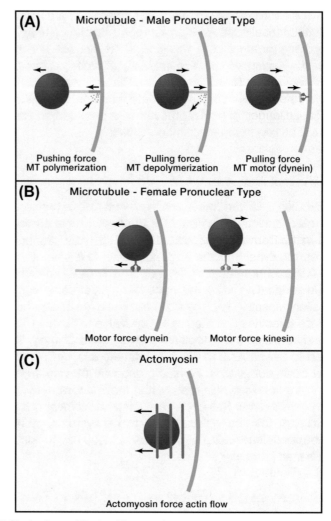

FIGURE 3 Mechanisms of Nuclear Movement
(A) Schematic of male pronuclear-type nuclear movement mediated by MTs (green). Forces (arrows) can be generated by MT polymerization, depolymerization, or dynein motors (red) anchored in the cortex or cytoplasmic sites.
(B) Schematic of female pronuclear-type nuclear movement mediated by MT dynein (red) and kinesin (orange) motors. Forces (arrows) are generated by motors that laterally connect nuclei to MTs.
(C) Schematic of actomyosin-type nuclear movement. Force (arrows) is generated by the actomyosin-dependent flow of dorsal actin cables (red).

conserved, so whether a similar mechanism is present in other organisms is unclear. Defects in A-type lamins and emerin increase spacing between the nucleus and centrosome in mammalian cells (Lee et al., 2007; Salpingidou et al., 2007); however, it is not clear that these proteins directly link them.

For male pronuclear type of nuclear movement, forces are generated by MTs interacting with cortical or cytoplasmic sites (Figure 3A). The interaction can be simply physical or mediated by anchored dynein. In *S. pombe*, interaction of growing MTs with the periphery generates pushing forces that maintain the nucleus in the middle of the cell (Tran et al., 2001). Pushing forces are restricted to systems in which relatively short distances (~10 μm) are involved because longer MTs cannot withstand compressive forces. Thus, in larger cells, MT pulling forces contribute to centrosome movements. In most cases, pulling forces are generated by cortically anchored dynein (Grill et al., 2003; Schmoranzer et al., 2009), as originally described in budding yeast, where dynein immobilized in the bud pulls on spindle-pole-body-associated MTs, moving the nucleus toward the bud neck (Adames and Cooper, 2000).

In syncytial cells with multiple nuclei, a more complex MT pulling mechanism exists. In the filamentous fungus *Aspergillus*, in which genetic screens revealed roles for dynein and its regulators in nuclear positioning (Morris et al., 1998), MT anchoring at cortical sites appears to evenly space nuclei in the syncytial hyphae (Gladfelter and Berman, 2009). In differentiating insect and mammalian muscle cells, which lack active centrosomes, MT minus ends associate directly with the nuclear envelope through uncharacterized factors. These cells also use dynein pulling and MT sliding by kinesin-1 and MT-associated proteins to cluster nuclei near the center of syncytial myotubes (Folker et al., 2012; Metzger et al., 2012).

Nuclear Movement by Attached MT Motor Forces

In the female pronuclear form of nuclear movement, nuclei associate laterally with MTs and move along them, powered by nuclear-envelope-associated motors (Figure 3B). This is typical of nuclear movements that occur during developmental events. Genetic screens that identified KASH (Unc83) and SUN (Unc84) proteins in *C. elegans* revealed that these proteins were required for nuclear movement in various cell types (Starr and Fridolfsson, 2010). Unc83 recruits both dynein and kinesin-1 motors to the nuclear envelope, where kinesin-1 is responsible for moving the nucleus while dynein contributes to directionality (Fridolfsson and Starr, 2010).

Female pronuclear-type nuclear movements are pronounced in the developing nervous system. Early genetic screens in *Drosophila* identified Klarsicht, or Klar, as a founding member of the KASH protein family, and it is required for apical movements of nuclei that establish the proper arrangement of cells in the ommatidium (Mosley-Bishop et al., 1999). Klar function has been linked to kinesin and dynein (Welte, 2004), suggesting that it may recruit these MT motors to the nuclear envelope. Mutants in the dynein regulators dynactin and DLis1 have similar nuclear migration defects as *Klar* mutants (Fan and Ready, 1997; Swan et al., 1999). Mutants in lamin Dm(0) and the SUN protein klaroid disrupt Klar localization to the nuclear envelope and apical movement of the nucleus, generating the same *Klar* phenotype (Kracklauer et al., 2007; Patterson et al., 2004). This result was the first to suggest that the nuclear lamina anchored the LINC complex.

Female pronuclear-type nuclear movements occur during two stages of vertebrate central nervous system development. In neuroepithelial radial glial cells, which serve as neuronal precursors, nuclear movement occurs along the apical-basal axis in a cell-cycle-dependent fashion. This has been termed interkinetic nuclear migration (INM). During INM, the nucleus moves basally during G1 and returns during G2 to an apical location where mitosis occurs (Taverna and Huttner, 2010). As the centrosome remains apical, basal and apical movements occur in MT plus and minus end directions, respectively. MT motors have been implicated in these movements. The kinesin-3 family member, Kif1a, has been implicated for plus-end-directed movement and dynein for minus-end-directed movement (Tsai et al., 2007; Tsai et al., 2010). Nesprin-1 and nesprin-2 may serve as recruitment factors for MT motors in vertebrate INM. Knockout of their genes in mice and zebrafish leads to defective INM in the neocortex and retina, and mouse nesprin-2 coimmunoprecipitates with dynein and kinesin-1 (Tsujikawa et al., 2007; Yu et al., 2011; Zhang et al., 2009). Interfering with the dynactin and Lis1 gives similar phenotypes (Tsai et al., 2005; Tsujikawa et al., 2007). Nuclear movements in vertebrate INM may be more complex than in the *Drosophila* eye, as myosin II and actomyosin contractility may also play a role (Norden et al., 2009; Schenk et al., 2009).

The second stage of vertebrate central nervous system development involving female pronuclear-type movements is neuron migration. After their "birth" in the neuroepithelium, neurons migrate significant distances to their final locations. Most migrating neurons exhibit a characteristic two-stroke form of migration in which the narrow leading process extends; the centrosome then moves forward into a swelling in the leading process followed by the nucleus and the rest of the soma (Tsai and Gleeson, 2005). Nuclear movement toward MT minus ends at the centrosome is dependent on dynein and its regulators Lis1 and NudE (Shu et al., 2004; Tsai et al., 2007). The centrosome also moves in a dynein- and Lis1-dependent fashion. Lis1 binds to a specific nucleotide state of dynein and enhances force generation, which may be necessary for moving the nucleus (McKenney et al., 2010). Nesprin-2 and SUN1/SUN2, which are also required for the forward movement of the nucleus, may recruit dynein to the nucleus (Zhang et al., 2009). Doublecortin, a MT-associated protein, is also required for nuclear movement during neuron migration (Koizumi et al., 2006). Importantly, nuclear movement during neuronal migration is also dependent on actomyosin contraction (see below), so this is not a pure form of female pronuclear-type movement.

The two-stroke mode of migration with a large separation (5–18 μm) between the centrosome and nucleus is thought to be a particular feature of neurons and is not typically observed in other migrating cells. Nonetheless, the same anterior localization of the centrosome relative to the nucleus, albeit in closer proximity, occurs in many migrating cell types, and dynein has been implicated in nuclear movements in migrating nonneuronal cells (Luxton and Gundersen, 2011).

ACTIN-MEDIATED NUCLEAR MOVEMENT

A groundbreaking study in *C. elegans* identified an outer nuclear membrane protein, termed Anc-1, which bound to actin and was essential for anchoring nuclei in the syncytial hypodermal and intestinal cells (Starr and Han, 2002). Anc-1 is one of the founding members of the KASH protein family and requires the SUN protein Unc84 for its outer nuclear membrane localization. While the discovery of Anc-1 showed that nuclear connections to the actin cytoskeleton anchor nuclei, we now know that nuclei are also actively moved through actin-dependent processes, typically in cells polarizing for migration.

Nuclear Movement by Tethering to Moving Actin Cables

The rearward positioning of the nucleus in migrating cells (Figure 1B) may result, at least in part, from an extension of the leading edge. Yet, studies in a number of cultured cell types have revealed that rearward nuclear positioning is an active process independent of cell protrusion (Desai et al., 2009; Dupin et al., 2011; Gomes et al., 2005; Luxton et al., 2010).

A direct mechanism for moving the nucleus has been established in experiments utilizing wounded monolayers of serum-starved fibroblasts treated with LPA, which stimulates cell polarization, but not protrusion or migration (Luxton et al., 2010). In this system, rearward-moving dorsal actin cables induced by Cdc42 provide force to move the nucleus (Figure 3C). Movement of dorsal actin cables is likely powered by myosin II, as its inhibition prevents actin flow and nuclear movement (Gomes et al., 2005). These cables are directly coupled to the nucleus by nesprin-2G and SUN2, which accumulate along them to form linear assemblies termed transmembrane actin-associated nuclear (TAN) lines. Actin-binding CH domains of nesprin-2G are required for TAN line formation and nuclear movement. A-type lamins anchor TAN lines to the nucleoskeleton, and in their absence, TAN lines slip over an immobile nucleus (Folker et al., 2011). This anchorage is presumably mediated through SUN2 binding to A-type lamins. Additional anchorage may be mediated by SUN2 binding to SAMP1, which also localizes to TAN lines and is necessary for nuclear movement (Borrego-Pinto et al., 2012).

Nuclear Movement by Actomyosin Contraction

Nuclear movement appears to be rate limiting for cells migrating through narrow extracellular spaces in which nuclei become deformed (Friedl et al., 2011). In at least some of these cases, passage through a narrow opening specifically requires myosin II (Beadle et al., 2008), suggesting that actomyosin-mediated nuclear movement is necessary. Myosin II is also necessary for the forward movement of the nucleus in migrating neurons (Solecki et al., 2009; Tsai et al., 2007), localizing behind it where it may provide contractile forces that help to move it into the leading process. This may reflect the difficulty of moving the nucleus into the narrow leading process, which requires nuclei to become elongated.

NUCLEAR POSITIONING AND DISEASE

We have provided several examples of nuclear positioning events that are required for specific cellular processes. Given this requirement, one could imagine that defects in the molecular toolbox for nuclear positioning could lead to cellular dysfunction. Indeed, results from human subjects with inherited diseases and mouse models have shown that alterations in proteins involved in nuclear positioning are associated with pathology. Mutations in genes encoding proteins involved in MT function, LINC complex components, and the nuclear lamina all cause human diseases (Table 3).

Lissencephaly is characterized by mislocalization of cortical neurons, resulting in decreased cortical complexity and a smooth brain surface. Affected children have severe psychomotor retardation, seizures, muscle spasticity, and failure to thrive. At the cellular level, neuronal migration required for brain development is blocked. Most cases of "classic" lissencephaly are caused by deletion or truncating mutations in *LIS1* (Reiner et al., 1993). The Lis1 protein is required for INM and nuclear and centrosomal movement during two-stroke neuronal migration (Shu et al., 2004; Tsai et al., 2007). Similarly, mutations in *DCX* encoding doublecortin cause X-linked lissencephaly and defective nuclear movement in neurons (Gleeson et al., 1998; Koizumi et al., 2006). De novo mutations in *TUBA3* encoding α-1 tubulin also cause lissencephaly and defective nuclear movement in neurons (Keays et al., 2007).

Intriguingly, depletion of lamin B1, lamin B2, or both in mice causes lissencephaly-like phenotypes (Coffinier et al., 2010, 2011). These phenotypes result from neuronal migration defects, which likely have accompanying abnormalities in nuclear movement, although this has not been assessed directly. Nuclei spin in mouse fibroblasts lacking lamin B1, suggesting that B-type lamins function in nuclear anchoring (Ji et al., 2007). B-type lamins may therefore anchor LINC complexes. Mutations in genes encoding B-type

Table 3 Genes Encoding Proteins Functioning in Nuclear Positioning Linked to Human Disease

Human Gene	Protein	Function	Human Disease(s)	Disease Phenotypes
DCX	doublecortin	stabilizes microtubules	lissencephaly	mislocalization of cortical neurons, "smooth brain"
LIS1	Lis1	dynein regulation	lissencephaly	mislocalization of cortical neurons, "smooth brain"
TUBA3	α-tubulin	MT component	lissencephaly	mislocalization of cortical neurons, "smooth brain"
LMNB1	lamin B1	lamina component	adult onset leukodystrophy results from gene duplication	demyelination
LMNB2	lamin B2	lamina component	susceptibility to acquired partial lipodystrophy	regional fat loss
SUN1	Sun1	LINC complex	none to date	
SUN2	Sun2	LINC complex	none to date	
SYNE1	nesprin-1	LINC complex	(1) cerebellar ataxia; (2) myopathies; (3) arthrogryposis	(1) coordination defects; 2) cardiomyopathy and muscular dystrophy; 3) congenital joint contractures and muscle weakness
SYNE2	nesprin-2	LINC complex	myopathies	cardiomyopathy and skeletal muscular dystrophy
SYNE4	nesprin-4	LINC complex	high-frequency hearing loss	progressive high-frequency hearing loss
LMNA	A-type lamins	lamina components	(1) myopathy; (2) partial lipodystrophy; (3) peripheral neuropathy; (4) progeria	(1) cardiomyopathy with variable skeletal muscular dystrophy; (2) fat loss from extremities; (3) peripheral nerve defects; (4) accelerated aging phenotypes

lamins have not yet been linked to human developmental brain disorders, but duplications in *LMNB1* cause overexpression of lamin B1 and an adult-onset demyelinating disease (Padiath et al., 2006).

Experiments in knockout mice implicate SUN1, SUN2, nesprin-1, and nesprin-2 in nuclear migration during neurogenesis and migration (Zhang et al., 2009).

However, mutations in genes encoding nesprins have been linked to diseases other than lissencephaly. Mutations in *SYNE1* encoding nesprin-1 cause adult-onset autosomal-recessive cerebellar ataxia characterized by diffuse cerebellar atrophy and impaired walking, dysarthria, and poor coordination (Gros-Louis et al., 2007). This could potentially result from neuronal migration defects in a specific region of the brain. Mutations in *SYNE1* have also been reported to cause an autosomal-recessive form of arthrogryposis multiplex congenita characterized by congenital joint contractures, muscle weakness, and progressive motor decline (Attali et al., 2009). Mutations in *SYNE1* and *SYNE2* have further been reported to cause Emery-Dreifuss muscular dystrophy (EDMD)-like phenotypes (Zhang et al., 2007a). Mutations in the gene encoding the LINC-complex-associated protein emerin were first reported to cause X-linked EDMD (Bione et al., 1994), and mutations in *LMNA*-encoding A-type lamins are responsible for most autosomally inherited cases (Bonne et al., 1999). This suggests an association between LINC complex function and EDMD-like phenotypes, which generally share a dilated cardiomyopathy with variable skeletal muscle involvement. More recently, mutation in *SYNE4* encoding nesprin-4 has been shown to cause autosomal-recessive, progressive high-frequency hearing loss (Horn et al., 2013).

Nuclear positioning defects caused by *SYNE1* and *SYNE2* mutations have been described. One patient with a *SYNE1* mutation and cerebellar ataxia was reported to have fewer muscle nuclei under neuromuscular junctions (Gros-Louis et al., 2007). Similarly, deletion of the KASH domain from nesprin-1 in mice abolishes synaptic nuclei clustering and disrupts spacing of nonsynaptic nuclei in skeletal muscle; deletion of the nesprin-2 KASH domain has no effect but exacerbates the defect in mice lacking nesprin-1 (Zhang et al., 2007b). Nesprin-2 deletion in mice disrupts nuclear movement in cells of the neocortex and retina, causing reduced thickness of the cortex and the outer nuclear layer into which newly formed photoreceptor cells migrate (Yu et al., 2011; Zhang et al., 2009). Mice lacking nesprin-4 suffer from deafness, mimicking the human mutation phenotype, and have abnormal positioning of nuclei in cochlear outer hair cells (Horn et al., 2013). Although no disease-causing mutations in *SUN1* or *SUN2* have been described in humans, depletion of both proteins from mice cause nuclear positioning defects in muscle, retina, and developing brain, similar to those in mice lacking nesprin-1 and nesprin-2 (Lei et al., 2009; Yu et al., 2011; Zhang et al., 2009). Mice without SUN1 also have hearing loss and abnormal nuclear positioning in cochlear outer hair cells (Horn et al., 2013).

The tissue-selective human diseases and pathology in mice that occur in response to alterations in different SUNs and nesprins may result because only certain isoforms are necessary in different tissues. Data from mice demonstrate tissue-selective differences in the expression of nesprins and

SUNs, yet there is no comprehensive analysis of the expression patterns and tissue-type functionality of all of the different nesprins and SUNs. Results from knockout mice also suggest redundancy in the function of SUN1 and SUN2 and different tissue effects of nesprin-1 and nesprin-2.

Mutations in *LMNA* encoding the A-type lamins cause a broad range of human diseases often referred to as "laminopathies" (Dauer and Worman, 2009). *LMNA* mutations that cause EDMD and related myopathies are mostly missense or small in-frame deletions, which lead to expression of variant proteins, splice site truncations, or promoter mutations. Depletion of A-type lamins from mice leads primarily to cardiac and skeletal muscle phenotypes, suggesting that *LMNA* mutations, even dominant ones leading to variant protein expression, somehow cause loss of function (Sullivan et al., 1999). Skeletal muscles from humans with autosomal dominant EDMD and *Lmna* null mice both have nuclei in the center of myofibers rather than at their normal peripheral localization. However, this also occurs in other myopathies not associated with defects in proteins directly implicated in nuclear positioning. For more on laminopathies, please see the Review by Schreiber and Kennedy on page 1365 of this issue (Schreiber and Kennedy, 2013).

In migrating fibroblasts depleted of A-type lamins or expressing variants associated with myopathy, actin-dependent rearward nuclear movement fails to occur (Folker et al., 2011). In these cells, nesprin-2G assembles into TAN lines that slip over the nucleus rather than moving with it, indicating an anchorage defect. Amino acid substitutions within an immunoglobulin-like motif in the tail of A-type lamins cause partial lipodystrophy, which is characterized by fat loss from the extremities. In contrast to those causing myopathy, expression of lamin A variants that cause lipodystrophy inhibit MT-dependent centrosome positioning, but not actin-dependent nuclear movement in migrating fibroblasts (Folker et al., 2011).

Except for cases in which nuclear positioning defects associate with abnormal neuronal migration, the relationship of the positioning defects observed in model systems to pathogenic mechanisms remains uncertain. It is not known why alterations in the nuclear positioning proteins affect only cells in certain tissues when the proteins are widely expressed. In some instances, observed nuclear positioning defects may not directly connect to the disease, such as mispositioning of nuclei at the neuromuscular junction in cerebellar ataxia. Overall, alterations in the nuclear positioning toolbox most often affect tissues, such as the nervous system and striated muscle, in which cell migration plays an important role in organ development or homeostasis. Abnormal force transmission between the nucleus and cytoplasm may also render cells more susceptible to damage by mechanical stress, leading to activation of stress response or apoptotic pathways, resulting, respectively, in cell dysfunction or death.

CELLULAR SIGNIFICANCE OF NUCLEAR POSITIONING: HYPOTHESES AND PERSPECTIVES

Our understanding of why cells move and position their nuclei is still rudimentary. Yet, interfering with proteins involved in nuclear movement inhibits many cell functions. Defects in the nuclear positioning toolbox also cause disease. Thus, nuclear positioning itself may influence other cellular activities. Here, we put into perspective evidence supporting the hypotheses that nuclear positioning influences the organization and mechanical properties of the surrounding cytoplasm, cytoplasmic signaling, and accessibility of the nucleus to signaling pathways.

The Nuclear Envelope as a Cytoskeletal Integrator

Identification of the LINC complex and other proteins mediating nucleo-cytoskeletal connections raises the possibility that the nucleus not only attaches to the cytoskeleton, but also organizes it. Even before the identification of specific nucleocytoskeletal connectors, a classical experiment by Ingber and colleagues revealed that the nucleus was physically connected to integrins in the plasma membrane (Maniotis et al., 1997). These investigators showed that applying force to fibronectin beads attached to integrins moved the nucleus tens of microns away. Although the nature of the connection was not identified, this observation clearly reflects linkages that exist between the nucleus and the plasma membrane.

The nucleus influences the MT cytoskeleton through its association with MTOCs, which determine where MT minus ends are anchored. A more direct influence of the nucleus on MT distribution occurs in cells with noncentrosomal MTs. In multinucleated myotubes, which lack functional centrosomes, MTs minus ends are attached to nuclei by unidentified linkers, contributing to an overall bipolar array of MTs with mixed polarity (Tassin et al., 1985). The nucleus may also affect organization of the actin cytoskeleton. CH-domain-containing nesprins tether the nucleus to actin filaments, but whether they organize actin arrays around it is less certain. In fibroblasts polarizing for migration, depleting nesprin-2G or A-type lamins does not alter the overall distribution of actin filaments or the formation and movement of dorsal actin cables (Folker et al., 2011; Luxton et al., 2010). However, alterations in actin filaments and focal adhesions have been reported when LINC complex components are perturbed (Hale et al., 2008; Khatau et al., 2009). This may reflect lack of direct connection of the actin arrays to the nuclear envelope or indirect effects. These findings suggest that, at least under some circumstances, the nucleus actively participates in organizing certain actin structures.

Additional evidence that the nucleus organizes the cytoplasm comes from biophysical measurements. Cytoplasmic stiffness adjacent and distal to the nucleus is altered in cells depleted of A-type lamins (Broers et al., 2004; Lammerding et al., 2004). Whether this result solely reflects direct physical links between the nucleus and cytoskeleton or indirect effects of signaling pathways that are also modified by alterations in the nuclear envelope (see below) is presently unclear.

The Nuclear Envelope as a Regulator of Signaling Pathways

As the largest and most compression-resistant membrane-bound organelle in the cell, the nucleus has been likened to a "molecular shock absorber" (Dahl et al., 2004). Theoretically, movement of such a large, non-deformable organelle through the cytoplasm will result in tensile and/or compressive forces. Mediated by nuclear connections to the cytoskeleton, these forces could be transmitted to distal sites that are mechanical transducers, such as integrin-based focal adhesions or cadherin-based cell-cell adhesions (Leckband et al., 2011; Parsons et al., 2010). In a sense, the nucleus would act like the bead in Ingber's experiment, except that force would originate inside rather than outside of the cell. Given that adhesions respond to mechanical stimuli by regulating Rho GTPase and mitogen-activated protein (MAP) kinase signaling, the prediction is that nuclear movement may affect the activity of these pathways.

The idea that nuclear movement may regulate cellular signaling pathways has not been directly tested. Yet there is evidence that alterations in the nuclear movement toolbox alter signaling pathways. Lamin A variants that cause myopathies increase MAP kinase signaling, as does knockdown of A-type lamins or emerin (Muchir et al., 2007, 2009). Similar results have been obtained for Rho signaling (Hale et al., 2008). Given that alterations in A-type lamins interfere with actin-dependent nuclear movement (Folker et al., 2011), it is possible that changes in signaling result from altered nuclear positioning. A-type lamins may also affect signaling by interacting with proteins in the pathway, for example, by binding the MAP kinase ERK1/2 (González et al., 2008). KASH proteins may recruit signaling molecules to the nuclear envelope and regulate their activities, as nesprin-2 binds active ERK1/2, and its knockdown results in prolonged ERK1/2 activity (Warren et al., 2010). As other actin-dependent membrane structures such as focal adhesions regulate signaling, TAN lines assembled on the surface of the nuclear envelope may also.

Nuclear Position as a Response Regulator of Signaling Pathways

The position of the nucleus may also alter its responsiveness to pathways that regulate transcription and mRNA transport and localization. It is generally assumed that latent cytoplasmic transcription factors and second messengers activated by plasma membrane receptors reach the nucleus in an unabated fashion. However, the distance that they travel may depend on encounters with costimulatory and inhibitory factors in the cytoplasm (Calvo et al., 2010). Thus, the nucleus's position relative to the origin of an external signal may modulate its response. This could be particularly important for asymmetrically encountered signals, for example, on the apical or basal aspects of epithelia or in gradients of external factors during development. The spatial relationship between the nucleus and the primary cilium changes in many developing epithelia, such as the neuroepithelium, and may affect the output of signaling pathways, such as the Sonic hedgehog pathway that requires the cilium (Goetz and Anderson, 2010). Signaling from intracellular sites, such as the signaling endosome, may enhance responsiveness by bringing the signal in close proximity to the nucleus.

Only one study has directly examined the relationship between nuclear position and asymmetrical signaling (Del Bene et al., 2008). A gradient of Notch signaling, highest at the apical surface, exists in the retinal neuroepithelium, as in other epithelia (Murciano et al., 2002). INM moves the nucleus basally during G1, exposing it to lower Notch activity. A mutation in the zebrafish *mok* gene encoding the dynactin p150^{glued} subunit causes longer and faster basal nuclear excursions, resulting in increased basal mitoses and the formation of early differentiating neurons at the expense of later ones (Del Bene et al., 2008). Notch overexpression rescues the *mok* phenotype, showing that it results from inadequate exposure of the nucleus to Notch due to defective nuclear movement. Alterations in *Syne-2* lead to similar changes in INM and cell fate in zebrafish retina (Tsujikawa et al., 2007). Deficiencies in Cep120 and TACC, proteins that affect the centrosome-MT connection, or in nesprin-2 or SUN1/2 also affect INM in developing mouse cerebral cortex and lead to early depletion of neural progenitors (Xie et al., 2007; Zhang et al., 2009). Although altered cell fate has not yet been demonstrated in these studies, they are consistent with altered response to Notch or other apical signals.

CONCLUSIONS

Rather than being a passive or random phenomenon, active mechanisms exist to position nuclei in cells. We have reviewed the molecular tools and mechanisms that move and position nuclei, most of which are conserved among eukaryotes. Human diseases result from genetic abnormalities in nuclear movement toolbox proteins and, in some cases, are linked to altered nuclear movement. We have highlighted potential mechanisms by which nuclear position may influence cellular processes and disease pathogenesis. Additional investigation is needed to understand how the nucleus affects these processes and to separate direct from indirect effects of its positioning. Future basic research on nuclear positioning and how it affects cellular processes is likely to significantly impact public health.

ACKNOWLEDGMENTS

We thank Susumu Antoku, Wakam Chang, Edgar Gomes, Gant Luxton, and Alex Palazzo for their comments and Wakam Chang for Figures 1B and 2A. The authors are supported by NIH grants R01GM099481, R01NS059352, R01HD070713, and R01AR048997.

REFERENCES

Adames, N.R., and Cooper, J.A. (2000). Microtubule interactions with the cell cortex causing nuclear movements in Saccharomyces cerevisiae. J. Cell Biol. *149*, 863–874.

Attali, R., Warwar, N., Israel, A., Gurt, I., McNally, E., Puckelwartz, M., Glick, B., Nevo, Y., Ben-Neriah, Z., and Melki, J. (2009). Mutation of SYNE-1, encoding an essential component of the nuclear lamina, is responsible for autosomal recessive arthrogryposis. Hum. Mol. Genet. *18*, 3462–3469.

Barnhart, E.L., Allen, G.M., Jülicher, F., and Theriot, J.A. (2010). Bipedal locomotion in crawling cells. Biophys. J. *98*, 933–942.

Beadle, C., Assanah, M.C., Monzo, P., Vallee, R., Rosenfeld, S.S., and Canoll, P. (2008). The role of myosin II in glioma invasion of the brain. Mol. Biol. Cell *19*, 3357–3368.

Bione, S., Maestrini, E., Rivella, S., Mancini, M., Regis, S., Romeo, G., and Toniolo, D. (1994). Identification of a novel X-linked gene responsible for Emery-Dreifuss muscular dystrophy. Nat. Genet. *8*, 323–327.

Bonne, G., Di Barletta, M.R., Varnous, S., Bécane, H.M., Hammouda, E.H., Merlini, L., Muntoni, F., Greenberg, C.R., Gary, F., Urtizberea, J.A., et al. (1999). Mutations in the gene encoding lamin A/C cause autosomal dominant Emery-Dreifuss muscular dystrophy. Nat. Genet. *21*, 285–288.

Borrego-Pinto, J., Jegou, T., Osorio, D.S., Auradé, F., Gorjánácz, M., Koch, B., Mattaj, I.W., and Gomes, E.R. (2012). Samp1 is a component of TAN lines and is required for nuclear movement. J. Cell Sci. *125*, 1099–1105.

Broers, J.L., Peeters, E.A., Kuijpers, H.J., Endert, J., Bouten, C.V., Oomens, C.W., Baaijens, F.P., and Ramaekers, F.C. (2004). Decreased mechanical stiffness in LMNA-/- cells is caused by defective nucleo-cytoskeletal integrity: implications for the development of laminopathies. Hum. Mol. Genet. *13*, 2567–2580.

Buchman, J.J., and Tsai, L.H. (2008). Putting a notch in our understanding of nuclear migration. Cell *134*, 912–914.

Calvo, F., Agudo-Ibáñez, L., and Crespo, P. (2010). The Ras-ERK pathway: understanding site-specific signaling provides hope of new anti-tumor therapies. Bioessays *32*, 412–421.

Cappello, S., Attardo, A., Wu, X., Iwasato, T., Itohara, S., Wilsch-Bräuninger, M., Eilken, H.M., Rieger, M.A., Schroeder, T.T., Huttner, W.B., et al. (2006). The Rho-GTPase cdc42 regulates neural progenitor fate at the apical surface. Nat. Neurosci. *9*, 1099–1107.

Coffinier, C., Chang, S.Y., Nobumori, C., Tu, Y., Farber, E.A., Toth, J.I., Fong, L.G., and Young, S.G. (2010). Abnormal development of the cerebral cortex and cerebellum in the setting of lamin B2 deficiency. Proc. Natl. Acad. Sci. USA *107*, 5076–5081.

Coffinier, C., Jung, H.J., Nobumori, C., Chang, S., Tu, Y., Barnes, R.H., 2nd, Yoshinaga, Y., de Jong, P.J., Vergnes, L., Reue, K., et al. (2011). Deficiencies in lamin B1 and lamin B2 cause neurodevelopmental defects and distinct nuclear shape abnormalities in neurons. Mol. Biol. Cell *22*, 4683–4693.

Crisp, M., Liu, Q., Roux, K., Rattner, J.B., Shanahan, C., Burke, B., Stahl, P.D., and Hodzic, D. (2006). Coupling of the nucleus and cytoplasm: role of the LINC complex. J. Cell Biol. *172*, 41–53.

Dahl, K.N., Kahn, S.M., Wilson, K.L., and Discher, D.E. (2004). The nuclear envelope lamina network has elasticity and a compressibility limit suggestive of a molecular shock absorber. J. Cell Sci. *117*, 4779–4786.

Dauer, W.T., and Worman, H.J. (2009). The nuclear envelope as a signaling node in development and disease. Dev. Cell *17*, 626–638.

Del Bene, F., Wehman, A.M., Link, B.A., and Baier, H. (2008). Regulation of neurogenesis by interkinetic nuclear migration through an apical-basal notch gradient. Cell *134*, 1055–1065.

Desai, R.A., Gao, L., Raghavan, S., Liu, W.F., and Chen, C.S. (2009). Cell polarity triggered by cell-cell adhesion via E-cadherin. J. Cell Sci. *122*, 905–911.

Dupin, I., Sakamoto, Y., and Etienne-Manneville, S. (2011). Cytoplasmic intermediate filaments mediate actin-driven positioning of the nucleus. J. Cell Sci. *124*, 865–872.

Fan, S.S., and Ready, D.F. (1997). Glued participates in distinct microtubule-based activities in Drosophila eye development. Development *124*, 1497–1507.

Folker, E.S., Östlund, C., Luxton, G.W., Worman, H.J., and Gundersen, G.G. (2011). Lamin A variants that cause striated muscle disease are defective in anchoring transmembrane actin-associated nuclear lines for nuclear movement. Proc. Natl. Acad. Sci. USA *108*, 131–136.

Folker, E.S., Schulman, V.K., and Baylies, M.K. (2012). Muscle length and myonuclear position are independently regulated by distinct Dynein pathways. Development *139*, 3827–3837.

Fridolfsson, H.N., and Starr, D.A. (2010). Kinesin-1 and dynein at the nuclear envelope mediate the bidirectional migrations of nuclei. J. Cell Biol. *191*, 115–128.

Fridolfsson, H.N., Ly, N., Meyerzon, M., and Starr, D.A. (2010). UNC-83 coordinates kinesin-1 and dynein activities at the nuclear envelope during nuclear migration. Dev. Biol. *338*, 237–250.

Friedl, P., Wolf, K., and Lammerding, J. (2011). Nuclear mechanics during cell migration. Curr. Opin. Cell Biol. *23*, 55–64.

Gladfelter, A., and Berman, J. (2009). Dancing genomes: fungal nuclear positioning. Nat. Rev. Microbiol. *7*, 875–886.

Gleeson, J.G., Allen, K.M., Fox, J.W., Lamperti, E.D., Berkovic, S., Scheffer, I., Cooper, E.C., Dobyns, W.B., Minnerath, S.R., Ross, M.E., and Walsh, C.A. (1998). Doublecortin, a brain-specific gene mutated in human X-linked lissencephaly and double cortex syndrome, encodes a putative signaling protein. Cell *92*, 63–72.

Godin, J.D., Thomas, N., Laguesse, S., Malinouskaya, L., Close, P., Malaise, O., Purnelle, A., Raineteau, O., Campbell, K., Fero, M., et al. (2012). p27(Kip1) is a microtubule-associated protein that promotes microtubule polymerization during neuron migration. Dev. Cell *23*, 729–744.

Goetz, S.C., and Anderson, K.V. (2010). The primary cilium: a signalling centre during vertebrate development. Nat. Rev. Genet. *11*, 331–344.

Gomes, E.R., Jani, S., and Gundersen, G.G. (2005). Nuclear movement regulated by Cdc42, MRCK, myosin, and actin flow establishes MTOC polarization in migrating cells. Cell *121*, 451–463.

Gönczy, P., Pichler, S., Kirkham, M., and Hyman, A.A. (1999). Cytoplasmic dynein is required for distinct aspects of MTOC positioning, including centrosome separation, in the one cell stage Caenorhabditis elegans embryo. J. Cell Biol. *147*, 135–150.

González, J.M., Navarro-Puche, A., Casar, B., Crespo, P., and Andrés, V. (2008). Fast regulation of AP-1 activity through interaction of lamin A/C, ERK1/2, and c-Fos at the nuclear envelope. J. Cell Biol. *183*, 653–666.

Grill, S.W., Howard, J., Schäffer, E., Stelzer, E.H., and Hyman, A.A. (2003). The distribution of active force generators controls mitotic spindle position. Science *301*, 518–521.

Gros-Louis, F., Dupré, N., Dion, P., Fox, M.A., Laurent, S., Verreault, S., Sanes, J.R., Bouchard, J.P., and Rouleau, G.A. (2007). Mutations in SYNE1 lead to a newly discovered form of autosomal recessive cerebellar ataxia. Nat. Genet. *39*, 80–85.

Hale, C.M., Shrestha, A.L., Khatau, S.B., Stewart-Hutchinson, P.J., Hernandez, L., Stewart, C.L., Hodzic, D., and Wirtz, D. (2008). Dysfunctional connections between the nucleus and the actin and microtubule networks in laminopathic models. Biophys. J. *95*, 5462–5475.

Haque, F., Lloyd, D.J., Smallwood, D.T., Dent, C.L., Shanahan, C.M., Fry, A.M., Trembath, R.C., and Shackleton, S. (2006). SUN1 interacts with nuclear lamin A and cytoplasmic nesprins to provide a physical connection between the nuclear lamina and the cytoskeleton. Mol. Cell. Biol. *26*, 3738–3751.

Horn, H.F., Brownstein, Z., Lenz, D.R., Shivatzki, S., Dror, A.A., Dagan-Rosenfeld, O., Friedman, L.M., Roux, K.J., Kozlov, S., Jeang, K.-T., et al. (2013). The LINC complex is essential for hearing. J. Clin. Invest. *123*, 740–750.

Ji, J.Y., Lee, R.T., Vergnes, L., Fong, L.G., Stewart, C.L., Reue, K., Young, S.G., Zhang, Q., Shanahan, C.M., and Lammerding, J. (2007). Cell nuclei spin in the absence of lamin B1. J. Biol. Chem. *282*, 20015–20026.

Keays, D.A., Tian, G., Poirier, K., Huang, G.J., Siebold, C., Cleak, J., Oliver, P.L., Fray, M., Harvey, R.J., Molnár, Z., et al. (2007). Mutations in alpha-tubulin cause abnormal neuronal migration in mice and lissencephaly in humans. Cell *128*, 45–57.

Khatau, S.B., Hale, C.M., Stewart-Hutchinson, P.J., Patel, M.S., Stewart, C.L., Searson, P.C., Hodzic, D., and Wirtz, D. (2009). A perinuclear actin cap regulates nuclear shape. Proc. Natl. Acad. Sci. USA *106*, 19017–19022.

King, M.C., Drivas, T.G., and Blobel, G. (2008). A network of nuclear envelope membrane proteins linking centromeres to microtubules. Cell *134*, 427–438.

Koizumi, H., Higginbotham, H., Poon, T., Tanaka, T., Brinkman, B.C., and Gleeson, J.G. (2006). Doublecortin maintains bipolar shape and nuclear translocation during migration in the adult forebrain. Nat. Neurosci. *9*, 779–786.

Kracklauer, M.P., Banks, S.M., Xie, X., Wu, Y., and Fischer, J.A. (2007). Drosophila klaroid encodes a SUN domain protein required for Klarsicht localization to the nuclear envelope and nuclear migration in the eye. Fly (Austin) *1*, 75–85.

Lammerding, J., Schulze, P.C., Takahashi, T., Kozlov, S., Sullivan, T., Kamm, R.D., Stewart, C.L., and Lee, R.T. (2004). Lamin A/C deficiency causes defective nuclear mechanics and mechanotransduction. J. Clin. Invest. *113*, 370–378.

Leckband, D.E., le Duc, Q., Wang, N., and de Rooij, J. (2011). Mechanotransduction at cadherin-mediated adhesions. Curr. Opin. Cell Biol. *23*, 523–530.

Lee, J.S., Hale, C.M., Panorchan, P., Khatau, S.B., George, J.P., Tseng, Y., Stewart, C.L., Hodzic, D., and Wirtz, D. (2007). Nuclear lamin A/C deficiency induces defects in cell mechanics, polarization, and migration. Biophys. J. *93*, 2542–2552.

Lei, K., Zhang, X., Ding, X., Guo, X., Chen, M., Zhu, B., Xu, T., Zhuang, Y., Xu, R., and Han, M. (2009). SUN1 and SUN2 play critical but partially redundant roles in anchoring nuclei in skeletal muscle cells in mice. Proc. Natl. Acad. Sci. USA *106*, 10207–10212.

Levy, J.R., and Holzbaur, E.L. (2008). Dynein drives nuclear rotation during forward progression of motile fibroblasts. J. Cell Sci. *121*, 3187–3195.

Liu, Q., Pante, N., Misteli, T., Elsagga, M., Crisp, M., Hodzic, D., Burke, B., and Roux, K.J. (2007). Functional association of Sun1 with nuclear pore complexes. J. Cell Biol. *178*, 785–798.

Luxton, G.W., and Gundersen, G.G. (2011). Orientation and function of the nuclear-centrosomal axis during cell migration. Curr. Opin. Cell Biol. 23, 579–588.

Luxton, G.W., Gomes, E.R., Folker, E.S., Vintinner, E., and Gundersen, G.G. (2010). Linear arrays of nuclear envelope proteins harness retrograde actin flow for nuclear movement. Science 329, 956–959.

Luxton, G.W., Gomes, E.R., Folker, E.S., Worman, H.J., and Gundersen, G.G. (2011). TAN lines: a novel nuclear envelope structure involved in nuclear positioning. Nucleus 2, 173–181.

Malone, C.J., Misner, L., Le Bot, N., Tsai, M.C., Campbell, J.M., Ahringer, J., and White, J.G. (2003). The C. elegans hook protein, ZYG-12, mediates the essential attachment between the centrosome and nucleus. Cell 115, 825–836.

Maniotis, A.J., Chen, C.S., and Ingber, D.E. (1997). Demonstration of mechanical connections between integrins, cytoskeletal filaments, and nucleoplasm that stabilize nuclear structure. Proc. Natl. Acad. Sci. USA 94, 849–854.

McKenney, R.J., Vershinin, M., Kunwar, A., Vallee, R.B., and Gross, S.P. (2010). LIS1 and NudE induce a persistent dynein force-producing state. Cell 141, 304–314.

McNiven, M.A. (2013). Breaking away: matrix remodeling from the leading edge. Trends Cell Biol. 23, 16–21.

Metzger, T., Gache, V., Xu, M., Cadot, B., Folker, E.S., Richardson, B.E., Gomes, E.R., and Baylies, M.K. (2012). MAP and kinesin-dependent nuclear positioning is required for skeletal muscle function. Nature 484, 120–124.

Mislow, J.M., Holaska, J.M., Kim, M.S., Lee, K.K., Segura-Totten, M., Wilson, K.L., and McNally, E.M. (2002). Nesprin-1alpha self-associates and binds directly to emerin and lamin A in vitro. FEBS Lett. 525, 135–140.

Morimoto, A., Shibuya, H., Zhu, X., Kim, J., Ishiguro, K., Han, M., and Watanabe, Y. (2012). A conserved KASH domain protein associates with telomeres, SUN1, and dynactin during mammalian meiosis. J. Cell Biol. 198, 165–172.

Morris, N.R., Efimov, V.P., and Xiang, X. (1998). Nuclear migration, nucleokinesis and lissencephaly. Trends Cell Biol. 8, 467–470.

Mosley-Bishop, K.L., Li, Q., Patterson, L., and Fischer, J.A. (1999). Molecular analysis of the klarsicht gene and its role in nuclear migration within differentiating cells of the Drosophila eye. Curr. Biol. 9, 1211–1220.

Muchir, A., Pavlidis, P., Decostre, V., Herron, A.J., Arimura, T., Bonne, G., and Worman, H.J. (2007). Activation of MAPK pathways links LMNA mutations to cardiomyopathy in Emery-Dreifuss muscular dystrophy. J. Clin. Invest. 117, 1282–1293.

Muchir, A., Wu, W., and Worman, H.J. (2009). Reduced expression of A-type lamins and emerin activates extracellular signal-regulated kinase in cultured cells. Biochim. Biophys. Acta 1792, 75–81.

Murciano, A., Zamora, J., López-Sánchez, J., and Frade, J.M. (2002). Interkinetic nuclear movement may provide spatial clues to the regulation of neurogenesis. Mol. Cell. Neurosci. 21, 285–300.

Nery, F.C., Zeng, J., Niland, B.P., Hewett, J., Farley, J., Irimia, D., Li, Y., Wiche, G., Sonnenberg, A., and Breakefield, X.O. (2008). TorsinA binds the KASH domain of nesprins and participates in linkage between nuclear envelope and cytoskeleton. J. Cell Sci. 121, 3476–3486.

Norden, C., Young, S., Link, B.A., and Harris, W.A. (2009). Actomyosin is the main driver of interkinetic nuclear migration in the retina. Cell 138, 1195–1208.

Osmani, N., Vitale, N., Borg, J.P., and Etienne-Manneville, S. (2006). Scrib controls Cdc42 localization and activity to promote cell polarization during astrocyte migration. Curr. Biol. 16, 2395–2405.

Östlund, C., Folker, E.S., Choi, J.C., Gomes, E.R., Gundersen, G.G., and Worman, H.J. (2009). Dynamics and molecular interactions of linker of nucleoskeleton and cytoskeleton (LINC) complex proteins. J. Cell Sci. *122*, 4099–4108.

Padiath, Q.S., Saigoh, K., Schiffmann, R., Asahara, H., Yamada, T., Koeppen, A., Hogan, K., Ptácek, L.J., and Fu, Y.H. (2006). Lamin B1 duplications cause autosomal dominant leukodystrophy. Nat. Genet. *38*, 1114–1123.

Palazzo, A.F., Joseph, H.L., Chen, Y.J., Dujardin, D.L., Alberts, A.S., Pfister, K.K., Vallee, R.B., and Gundersen, G.G. (2001). Cdc42, dynein, and dynactin regulate MTOC reorientation independent of Rho-regulated microtubule stabilization. Curr. Biol. *11*, 1536–1541.

Parsons, J.T., Horwitz, A.R., and Schwartz, M.A. (2010). Cell adhesion: integrating cytoskeletal dynamics and cellular tension. Nat. Rev. Mol. Cell Biol. *11*, 633–643.

Patterson, K., Molofsky, A.B., Robinson, C., Acosta, S., Cater, C., and Fischer, J.A. (2004). The functions of Klarsicht and nuclear lamin in developmentally regulated nuclear migrations of photoreceptor cells in the Drosophila eye. Mol. Biol. Cell *15*, 600–610.

Razafsky, D., and Hodzic, D. (2009). Bringing KASH under the SUN: the many faces of nucleo-cytoskeletal connections. J. Cell Biol. *186*, 461–472.

Reiner, O., Carrozzo, R., Shen, Y., Wehnert, M., Faustinella, F., Dobyns, W.B., Caskey, C.T., and Ledbetter, D.H. (1993). Isolation of a Miller-Dieker lissencephaly gene containing G protein beta-subunit-like repeats. Nature *364*, 717–721.

Reinsch, S., and Gönczy, P. (1998). Mechanisms of nuclear positioning. J. Cell Sci. *111*, 2283–2295.

Roux, K.J., Crisp, M.L., Liu, Q., Kim, D., Kozlov, S., Stewart, C.L., and Burke, B. (2009). Nesprin 4 is an outer nuclear membrane protein that can induce kinesin-mediated cell polarization. Proc. Natl. Acad. Sci. USA *106*, 2194–2199.

Salpingidou, G., Smertenko, A., Hausmanowa-Petrucewicz, I., Hussey, P.J., and Hutchison, C.J. (2007). A novel role for the nuclear membrane protein emerin in association of the centrosome to the outer nuclear membrane. J. Cell Biol. *178*, 897–904.

Schaar, B.T., and McConnell, S.K. (2005). Cytoskeletal coordination during neuronal migration. Proc. Natl. Acad. Sci. USA *102*, 13652–13657.

Schenk, J., Wilsch-Bräuninger, M., Calegari, F., and Huttner, W.B. (2009). Myosin II is required for interkinetic nuclear migration of neural progenitors. Proc. Natl. Acad. Sci. USA *106*, 16487–16492.

Schmoranzer, J., Fawcett, J.P., Segura, M., Tan, S., Vallee, R.B., Pawson, T., and Gundersen, G.G. (2009). Par3 and dynein associate to regulate local microtubule dynamics and centrosome orientation during migration. Curr. Biol. *19*, 1065–1074.

Schreiber, K.H., and Kennedy, B.K. (2013). When lamins go bad: Nucleur structure and disease. Cell *152*, 1365–1375 this issue.

Shu, T., Ayala, R., Nguyen, M.D., Xie, Z., Gleeson, J.G., and Tsai, L.H. (2004). Ndel1 operates in a common pathway with LIS1 and cytoplasmic dynein to regulate cortical neuronal positioning. Neuron *44*, 263–277.

Solecki, D.J., Model, L., Gaetz, J., Kapoor, T.M., and Hatten, M.E. (2004). Par6alpha signaling controls glial-guided neuronal migration. Nat. Neurosci. *7*, 1195–1203.

Solecki, D.J., Trivedi, N., Govek, E.E., Kerekes, R.A., Gleason, S.S., and Hatten, M.E. (2009). Myosin II motors and F-actin dynamics drive the coordinated movement of the centrosome and soma during CNS glial-guided neuronal migration. Neuron *63*, 63–80.

Sosa, B.A., Rothballer, A., Kutay, U., and Schwartz, T.U. (2012). LINC complexes form by binding of three KASH peptides to domain interfaces of trimeric SUN proteins. Cell *149*, 1035–1047.

Splinter, D., Tanenbaum, M.E., Lindqvist, A., Jaarsma, D., Flotho, A., Yu, K.L., Grigoriev, I., Engelsma, D., Haasdijk, E.D., Keijzer, N., et al. (2010). Bicaudal D2, dynein, and kinesin-1 associate with nuclear pore complexes and regulate centrosome and nuclear positioning during mitotic entry. PLoS Biol. *8*, e1000350.

Starr, D.A., and Han, M. (2002). Role of ANC-1 in tethering nuclei to the actin cytoskeleton. Science *298*, 406–409.

Starr, D.A., and Fridolfsson, H.N. (2010). Interactions between nuclei and the cytoskeleton are mediated by SUN-KASH nuclear-envelope bridges. Annu. Rev. Cell Dev. Biol. *26*, 421–444.

Sullivan, T., Escalante-Alcalde, D., Bhatt, H., Anver, M., Bhat, N., Nagashima, K., Stewart, C.L., and Burke, B. (1999). Loss of A-type lamin expression compromises nuclear envelope integrity leading to muscular dystrophy. J. Cell Biol. *147*, 913–920.

Swan, A., Nguyen, T., and Suter, B. (1999). Drosophila Lissencephaly-1 functions with Bic-D and dynein in oocyte determination and nuclear positioning. Nat. Cell Biol. *1*, 444–449.

Tanabe, L.M., Kim, C.E., Alagem, N., and Dauer, W.T. (2009). Primary dystonia: molecules and mechanisms. Nat. Rev. Neurol. *5*, 598–609.

Tassin, A.M., Maro, B., and Bornens, M. (1985). Fate of microtubule-organizing centers during myogenesis in vitro. J. Cell Biol. *100*, 35–46.

Taverna, E., and Huttner, W.B. (2010). Neural progenitor nuclei IN motion. Neuron *67*, 906–914.

Tomlinson, A., and Ready, D.F. (1986). Sevenless: a cell-specific homeotic mutation of the Drosophila eye. Science *231*, 400–402.

Tran, P.T., Marsh, L., Doye, V., Inoué, S., and Chang, F. (2001). A mechanism for nuclear positioning in fission yeast based on microtubule pushing. J. Cell Biol. *153*, 397–411.

Tsai, L.H., and Gleeson, J.G. (2005). Nucleokinesis in neuronal migration. Neuron *46*, 383–388.

Tsai, F.C., and Meyer, T. (2012). Ca2+ pulses control local cycles of lamellipodia retraction and adhesion along the front of migrating cells. Curr. Biol. *22*, 837–842.

Tsai, J.W., Chen, Y., Kriegstein, A.R., and Vallee, R.B. (2005). LIS1 RNA interference blocks neural stem cell division, morphogenesis, and motility at multiple stages. J. Cell Biol. *170*, 935–945.

Tsai, J.W., Bremner, K.H., and Vallee, R.B. (2007). Dual subcellular roles for LIS1 and dynein in radial neuronal migration in live brain tissue. Nat. Neurosci. *10*, 970–979.

Tsai, J.W., Lian, W.N., Kemal, S., Kriegstein, A.R., and Vallee, R.B. (2010). Kinesin 3 and cytoplasmic dynein mediate interkinetic nuclear migration in neural stem cells. Nat. Neurosci. *13*, 1463–1471.

Tsujikawa, M., Omori, Y., Biyanwila, J., and Malicki, J. (2007). Mechanism of positioning the cell nucleus in vertebrate photoreceptors. Proc. Natl. Acad. Sci. USA *104*, 14819–14824.

Warren, D.T., Tajsic, T., Mellad, J.A., Searles, R., Zhang, Q., and Shanahan, C.M. (2010). Novel nuclear nesprin-2 variants tether active extracellular signal-regulated MAPK1 and MAPK2 at promyelocytic leukemia protein nuclear bodies and act to regulate smooth muscle cell proliferation. J. Biol. Chem. *285*, 1311–1320.

Welte, M.A. (2004). Bidirectional transport along microtubules. Curr. Biol. *14*, R525–R537.

Wilhelmsen, K., Litjens, S.H., Kuikman, I., Tshimbalanga, N., Janssen, H., van den Bout, I., Raymond, K., and Sonnenberg, A. (2005). Nesprin-3, a novel outer nuclear membrane protein, associates with the cytoskeletal linker protein plectin. J. Cell Biol. *171*, 799–810.

Wilkie, G.S., Korfali, N., Swanson, S.K., Malik, P., Srsen, V., Batrakou, D.G., de las Heras, J., Zuleger, N., Kerr, A.R., Florens, L., et al. (2011). Several novel nuclear envelope transmembrane proteins identified in skeletal muscle have cytoskeletal associations. Mol. Cell. Proteomics 10, M110.003129.

Xie, Z., Moy, L.Y., Sanada, K., Zhou, Y., Buchman, J.J., and Tsai, L.H. (2007). Cep120 and TACCs control interkinetic nuclear migration and the neural progenitor pool. Neuron *56*, 79–93.

Yu, J., Lei, K., Zhou, M., Craft, C.M., Xu, G., Xu, T., Zhuang, Y., Xu, R., and Han, M. (2011). KASH protein Syne-2/Nesprin-2 and SUN proteins SUN1/2 mediate nuclear migration during mammalian retinal development. Hum. Mol. Genet. *20*, 1061–1073.

Zhang, Q., Skepper, J.N., Yang, F., Davies, J.D., Hegyi, L., Roberts, R.G., Weissberg, P.L., Ellis, J.A., and Shanahan, C.M. (2001). Nesprins: a novel family of spectrin-repeat-containing proteins that localize to the nuclear membrane in multiple tissues. J. Cell Sci. *114*, 4485–4498.

Zhang, Q., Ragnauth, C.D., Skepper, J.N., Worth, N.F., Warren, D.T., Roberts, R.G., Weissberg, P.L., Ellis, J.A., and Shanahan, C.M. (2005). Nesprin-2 is a multi-isomeric protein that binds lamin and emerin at the nuclear envelope and forms a subcellular network in skeletal muscle. J. Cell Sci. *118*, 673–687.

Zhang, Q., Bethmann, C., Worth, N.F., Davies, J.D., Wasner, C., Feuer, A., Ragnauth, C.D., Yi, Q., Mellad, J.A., Warren, D.T., et al. (2007). Nesprin-1 and -2 are involved in the pathogenesis of Emery Dreifuss muscular dystrophy and are critical for nuclear envelope integrity. Hum. Mol. Genet. *16*, 2816–2833.

Zhang, X., Xu, R., Zhu, B., Yang, X., Ding, X., Duan, S., Xu, T., Zhuang, Y., and Han, M. (2007b). Syne-1 and Syne-2 play crucial roles in myonuclear anchorage and motor neuron innervation. Development *134*, 901–908.

Zhang, X., Lei, K., Yuan, X., Wu, X., Zhuang, Y., Xu, T., Xu, R., and Han, M. (2009). SUN1/2 and Syne/Nesprin-1/2 complexes connect centrosome to the nucleus during neurogenesis and neuronal migration in mice. Neuron *64*, 173–187.

Zhao, T., Graham, O.S., Raposo, A., and St Johnston, D. (2012). Growing microtubules push the oocyte nucleus to polarize the Drosophila dorsal-ventral axis. Science *336*, 999–1003.

Zhou, X., Graumann, K., Evans, D.E., and Meier, I. (2012). Novel plant SUN-KASH bridges are involved in RanGAP anchoring and nuclear shape determination. J. Cell Biol. *196*, 203–211.

Zhou, Z., Du, X., Cai, Z., Song, X., Zhang, H., Mizuno, T., Suzuki, E., Yee, M.R., Berezov, A., Murali, R., et al. (2012). Structure of Sad1-UNC84 homology (SUN) domain defines features of molecular bridge in nuclear envelope. J. Biol. Chem. *287*, 5317–5326.

evelopmental Cell

Evolution and Function of the Mitotic Checkpoint

Mathijs Vleugel[1], Erik Hoogendoorn[2], Berend Snel[2,*], Geert J.P.L. Kops[1,*]

[1]Department of Medical Oncology, Department of Molecular Cancer Research and Cancer Genomics Centre, University Medical Center Utrecht, 3584 CG Utrecht, The Netherlands, [2]Theoretical Biology and Bioinformatics, Department of Biology, Science Faculty, Utrecht University, 3584 CH Utrecht, The Netherlands
*Correspondence: b.snel@uu.nl (B.S.), g.j.p.l.kops@umcutrecht.nl (G.J.P.L.K.)

Developmental Cell, Vol. 23, No. 2, August 14, 2012 © 2012 Elsevier Inc.
http://dx.doi.org/10.1016/j.devcel.2012.06.013

SUMMARY

The mitotic checkpoint evolved to prevent cell division when chromosomes have not established connections with the chromosome segregation machinery. Many of the fundamental molecular principles that underlie the checkpoint, its spatiotemporal activation, and its timely inactivation have been uncovered. Most of these are conserved in eukaryotes, but important differences between species exist. Here we review current concepts of mitotic checkpoint activation and silencing. Guided by studies in model organisms and our phylogenomics analysis of checkpoint constituents and their functional domains and motifs, we highlight ancient and taxa-specific aspects of the core checkpoint modules in the context of mitotic checkpoint function.

MITOSIS, KINETOCHORES, AND THE MITOTIC CHECKPOINT

Accurate distribution of the replicated genome during mitosis is essential for the formation of genetically identical daughter cells. Errors in this process lead to genomic instability by causing aneuploidy and structural chromosome aberrations, both hallmarks of cancer (Gordon et al., 2012). Error-free chromosome segregation relies on dynamic linkages between chromosomes and the plus ends of spindle microtubules in a manner that connects sister

65

chromatids to opposite spindle poles. Such bioriented attachments are provided by large multiprotein complexes called kinetochores that are assembled on centromeric DNA (Cheeseman and Desai, 2008). Kinetochores attach to microtubules predominantly via the KMN network, a complex of eleven proteins that contains at least two microtubule-binding activities, provided by the Ndc80 complex and Knl1 (Cheeseman and Desai, 2008; Lampert and Westermann, 2011). Different evolutionary taxa have distinct additional factors that act in concert with the KMN network (Lampert and Westermann, 2011).

The mitotic checkpoint (MC, also called the spindle assembly checkpoint [SAC]) is a molecular safeguard mechanism that prevents premature chromosome segregation until all kinetochores have obtained connections to spindle microtubules (Musacchio and Salmon, 2007). There is some degree of debate about whether the checkpoint can distinguish unattached kinetochores from non-bioriented chromosomes, and we refer interested readers to some recent reviews on this matter (Khodjakov and Pines, 2010; Nezi and Musacchio, 2009). Kinetochores respond to lack of attachment by catalyzing the production of a molecular inhibitor of the anaphase promoting complex/cyclosome (APC/C), an E3 ubiquitin ligase that drives sister chromatid separation and mitotic exit by directing Securin and Cyclin B, respectively, for proteasomal degradation (Pines, 2011) (Figure 1A). As long as unattached kinetochores persist, the APC/C remains inactive toward these substrates and cells are stuck in a mitotic state with connected sister chromatids. The core machinery of the checkpoint comprises the APC/C inhibitor, also known as the mitotic checkpoint complex (MCC), as well as the proximal proteins that ensure its assembly by unattached kinetochores (Figure 1A). The MCC is a complex of Mad2, BubR1/Mad3, and Bub3 that is directly bound to the essential APC/C cofactor Cdc20. In addition, Bub1, Mps1, and Mad1 promote Cdc20 inhibition either directly through phosphorylation (Bub1) or indirectly through stimulating MCC assembly (Mps1 and Mad1) (Musacchio and Salmon, 2007). Several additional, sometimes taxa-specific, kinetochore proteins have been included in the group of checkpoint proteins and may aid in fine-tuning or amplifying checkpoint signals (see below).

Once the checkpoint is satisfied by attachment of the final kinetochore, the block on APC/C-Cdc20 by the MCC is quickly released, a process known as checkpoint silencing. This involves disassembly of the MCC, an active process that requires ubiquitination by the APC/C and a protein known as p31^{comet} (Hardwick and Shah, 2010). In addition, checkpoint proteins are removed from kinetochores by the dynein motor with the aid of kinetochore dynein recruiters such as Spindly and the Rod-Zwilch-ZW10 (RZZ) complex. Furthermore, phosphorylation events critical for MC function are reversed by kinetochore-localized protein phosphatases such as PP1 (Hardwick and Shah, 2010) (Figure 1B).

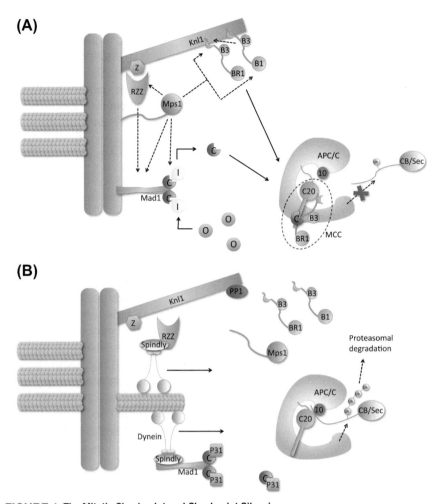

FIGURE 1 The Mitotic Checkpoint and Checkpoint Silencing

(A) Unattached kinetochores recruit Mad1, Bub1 (B1), BubR1/Mad3 (BR1), Bub3 (B3), and the RZZ complex (RZZ) either directly or indirectly via the MC scaffold Knl1/Zwint-1 (Z). The combined actions of these proteins and protein complexes promotes conversion of O-Mad2 (O) into C-Mad2 (C) through an intermediate state (I) after its dimerization with Mad1-bound C-Mad2. Soluble C-Mad2 and BubR1/ Mad3 then bind the APC/C coactivator Cdc20 (C20), blocking its substrate binding sites and repositioning Cdc20 away from the APC/C subunit Apc10 (10). As a result, APC/C-mediated ubiquitinations (Ub) of Cyclin B (CB) and Securin (Sec) are inhibited, maintaining sister chromatid cohesion and a mitotic state. Various steps in these processes are under control of Mps1.

(B) Attachment of vertebrate kinetochores causes dynein-dependent poleward stripping of MC proteins such as Mad1/Mad2, Spindly, and the RZZ complex. Mps1, Bub1, and BubR1/Mad3 are additionally dislodged from attached kinetochores. After satisfaction of the MC, when all kinetochores have achieved stable attachments, the MCC is disassembled by the action of p31comet (P31), resulting in APC/C-Cdc20 activity toward Cyclin B and Securin, followed by their proteasomal degradation. Mitotic exit further requires reversal of MC phosphorylations by PP1-like phosphatases that bind to the N terminus of Knl1.

With this review, we aim to provide an overview of the molecular workings of the MC and distill its core principles. To this end, we complement insights from experiments in various model organisms with our phylogenomics analysis of the MC machinery. This evolutionary perspective aids in distinguishing ancient from modern mechanisms and helps to uncover previously underappreciated concepts of the MC signaling pathway.

EVOLUTION OF THE MC AND ITS AUXILIARY PROTEINS

We used the publicly available genomes of 60 eukaryotes from all supergroups except rhizaria (Supplemental Experimental Procedures), to search for homologs of proteins from the core and auxiliary MC modules, including the MCC (Mad2, BubR1/Mad3, Bub3), kinetochore MC scaffolds (Mad1, Knl1), and kinases (Bub1, Mps1), as well as the contributing protein complex RZZ, the primary MC target Cdc20, and components of MC silencing mechanisms (Spindly, p31^{comet}) (Figure 2; see also Supplemental Experimental Procedures, Table S1, Figures S1–S3, and Supplemental Sequence File available online). We complemented our data with recent phylogenomic analysis of the APC/C by showing presence or absence of Apc1 (scaffold), Apc2 (cullin-domain), and Apc11 (RING-finger) homologs (Eme et al., 2011). For more in-depth analysis of evolution of functional domains within the identified homologs, we focused on a selection of species from different classes (indicated in bold in Table S1), representing the best-characterized species in the supergroups, as well as most of the common model organisms (Figure 2).

In general, our analyses indicate that most checkpoint components are ancient and were likely present in the last eukaryotic common ancestor (LECA). The exception is Spindly, with recognizable homologs only in most ophistokonta except for dikaryan fungi. Please note that we cannot formally exclude the possibility that poor genome annotation is an occasional reason for our inability to identify homologs in certain species. Although the core MC components can be found in at least one species in every supergroup, some may have been specifically lost in distinct supergroups or in major subbranches: Knl1 in chromalveolata and excavata, p31^{comet} in primitive fungi, and Zwilch in most but not all species that lack Spindly. In addition, some single-celled eukaryotes appear to lack one or more of the essential APC/C subunits, and most or all of the core MC components could not be found in the genomes of such species (e.g., *Encephalitozoon cuniculi*, *Plasmodium falciparum*, and *Cryptosporidium parvum*). Some organisms that contain APC/C subunits and Cdc20 are devoid of core MC components (*Paramecium tetraurelia*, *Tetrahymena thermophila*, *Leishmania major*,

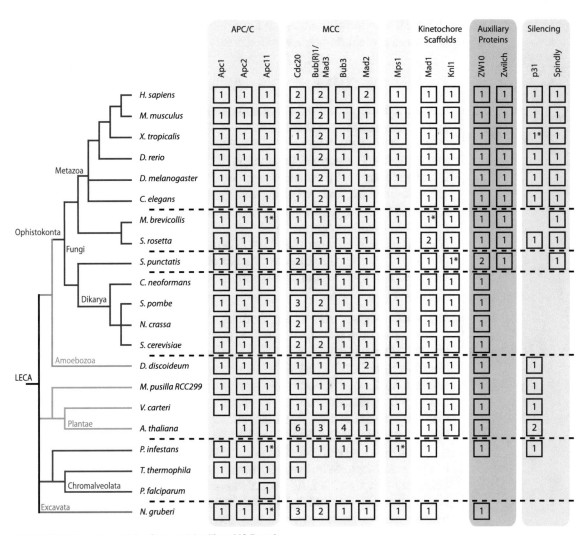

FIGURE 2 Homologs of the Core and Auxiliary MC Proteins

Schematic representation of eukaryotic tree of life in which a selection of eukaryotic species from the five different supergroups is indicated on the left. Checkpoint proteins are grouped in different functional groups (MCC, Mps1, kinetochore scaffolds, auxiliary proteins, silencing), and, whenever present, the number of homologs is indicated in black boxes (for gene IDs, see Table S1; for protein sequences, see Supplemental Sequence Files). Data on APC/C subunit homologs are adapted from (Eme et al., 2011); asterisks indicate potential homologs of MC subunits in genomic DNA from nonannotated genes (see Supplemental Experimental Procedures and Figure S3).

and *Trypanosoma brucei*). Certain eukaryotes may therefore do without a surveillance mechanism for chromosome segregation or may have evolved alternative ways of delaying cell division in the presence of unattached kinetochores. We will discuss these and other interesting evolutionary patterns in relation to established protein function in the following sections and expand it with insights obtained from detailed inspection of the evolution of functional protein domains within a subset of proteins.

THE INHIBITOR AND ITS TARGET: MCC AND CDC20

Polyubiquitination of Cyclin B and Securin by the APC/C requires destruction signals including a D(estruction) box (RxxLxxxx[EDNQ]) and/or KEN box (KEN) that are recognized by Cdc20. Recent structural insights have shown that the related cofactor Cdh1 and the APC/C subunit Apc10 form a bipartite D-box receptor that positions the substrate for catalysis by the Apc11/Apc2 catalytic core (Buschhorn et al., 2011; da Fonseca et al., 2011; Schreiber et al., 2011). Recognition of D or KEN boxes is provided by distinct surfaces on the WD40 repeat domain in Cdc20 and Cdh1 (Chao et al., 2012). An additional IR tail and an amino-terminal C box anchor the cofactor to the APC/C (Yu, 2007). Finally, Cdc20 itself has either a D or KEN box sequence in its amino-terminal region that is required for its degradation during later stages of mitosis (Yu, 2007).

The MCC inhibits substrate recognition by the APC/C by repositioning Cdc20 away from the Apc10 subunit, by blocking the KEN-box binding site, and by partially blocking the D-box binding site in Cdc20 (Chao et al., 2012; Herzog et al., 2009). This is achieved by a concerted effort of Mad2 and BubR1/Mad3. Mad2 directly interacts with Cdc20 through a motif preceding the WD40 repeat domain (Chao et al., 2012). Binding of Mad2 to Cdc20 disturbs interactions between Cdc20 and the APC/C (Yu, 2007) but, more importantly, allows BubR1 to bind Cdc20 (Kulukian et al., 2009). BubR1 has an amino-terminal KEN box that engages the KEN-box binding site in Cdc20 in a pseudosubstrate manner (Burton and Solomon, 2007; Chao et al., 2012; Sczaniecka et al., 2008). Additional interactions of the BubR1 tetratricopeptide repeat (TPR) domain with the WD40 repeat domain in Cdc20 sterically hinders access of substrate D-box sequences to the D-box binding site in Cdc20 (Chao et al., 2012). Finally, it has been proposed that a second KEN box in BubR1, carboxy-terminal to the TPR domain, directly engages the APC/C and may thus contribute to the inhibitory activity of MCC (Lara-Gonzalez et al., 2011).

EVOLUTION, FUNCTION, AND REGULATION OF CDC20

Cdc20 is found in one or multiple copies in virtually all genomes that we analyzed (Figure 2; Table S1). Most essential domains in Cdc20 have been strongly conserved during evolution, including the Mad2-binding motif, C box, WD40 repeats, IR tail, and, to a lesser extent, the degradation motifs (D and KEN box) (Figure 3; Figure S1). Interestingly, in many species with multiple Cdc20 paralogs, only one contains all the domains that in animals and fungi are required for the function and regulation of Cdc20. In budding and fission yeast, some of the other Cdc20-like proteins have meiosis-specific functions (Kimata et al., 2011; Tsuchiya et al., 2011). These paralogs have no Mad2-binding motif (Figure 3), raising the question of whether they are regulated by the state of kinetochore attachment. The Cdc20B paralog in humans (*H. sapiens* 2 in Figure 3) is highly degenerated. Besides a recent report that an intronic region in the gene encodes a miRNA that regulates proliferation (Lizé et al., 2010), it is unknown whether human Cdc20B or similarly degenerate Cdc20 proteins in other organisms have a cellular function.

FEEDBACK CONTROL OF THE MC: UBIQUITINATION OF CDC20 BY THE APC/C

Cdc20 expression is restricted to late S phase, G2, and early mitosis. This restriction is imposed by Cdh1, which recognizes Cdc20 as an APC/C substrate in anaphase, leading to persistent low Cdc20 protein levels in G1 and early S phase (Yu, 2007). Besides ensuring the absence of Cdc20 postanaphase, ubiquitination of Cdc20 has also been implicated in regulating MC function. Multiubiquitination (monoubiquitination on multiple residues) of Cdc20 by the APC/C was proposed to cause MCC dissociation and MC silencing (Reddy et al., 2007). A nonubiquitinatable mutant of Cdc20, however, still allows MCC dissociation upon MC satisfaction, challenging this notion of feedback inhibition (Mansfeld et al., 2011). Rather than multiubiquitination, Cdc20 seems to undergo polyubiquitination and subsequent degradation continuously, a process that is balanced by Cdc20 protein synthesis (Nilsson et al., 2008; Varetti et al., 2011; Zeng et al., 2010). This turnover could assist the MC in maintaining mitotic delays by keeping APC/C activity toward its relevant substrates low (Nilsson et al., 2008; Pan and Chen, 2004), or it could promote a certain rate of formation and disassembly of MCC-APC/C complexes to allow timely mitotic exit as soon as MCC production at kinetochores stops. The latter hypothesis is supported by evidence that Cdc20 turnover is aided by p31[comet], a structural Mad2 mimic that opposes MC function (Varetti et al., 2011). These two proposed models are difficult to reconcile, and further detailed studies will be required to clarify the role of Cdc20 degradation in mitosis. Regardless of the exact consequences of Cdc20 ubiquitination, it will be informative to examine whether the destruction motifs in Cdc20 contribute to this: p31[comet] does not necessarily co-occur in evolution with Cdc20 homologs containing such destruction motifs (e.g., *Neurospora crassa*, *Volvox carteri*, and *Phytophtora infestans*) (Figures 2 and 3). If destruction motifs are critical for Cdc20 turnover, this may suggest that p31[comet] has other functionalities in addition

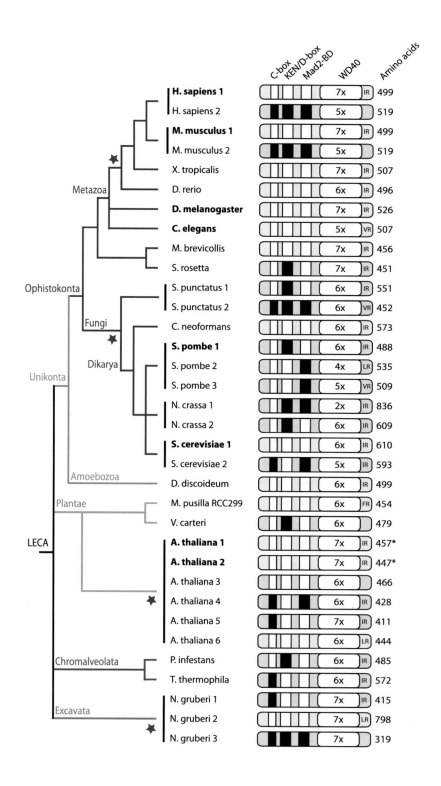

to promoting Cdc20 turnover. Conversely, in budding yeast, where Cdc20 turnover was described initially (Pan and Chen, 2004), we could not identify a p31^{comet} homolog, indicating that Cdc20 turnover can occur by p31^{comet}-independent mechanisms.

CATALYZING MCC PRODUCTION

An essential feature of the MC is the ability of Mad2 to bind Cdc20. Mad2 interacts with Cdc20 only when in a "closed" conformation (C-Mad2), production of which is catalyzed by unattached kinetochores through the action of Mad1. Mad1 and Mad2 localize to unattached kinetochores in mitosis (Musacchio and Salmon, 2007), and a significant pool of free Mad2 is present in the cytoplasm of mitotic cells (Chung and Chen, 2002). Cytoplasmic Mad2 is in an "open" conformation (O-Mad2) that has low affinity for Cdc20 but can be converted to C-Mad2 by virtue of dimerizing with Mad1-bound C-Mad2 at unattached kinetochores (De Antoni et al., 2005; Nezi et al., 2006). Structural conversion of O- to C-Mad2 then allows it to bind Cdc20 and ensures efficient MCC formation. Although this conversion and subsequent MCC formation can be strikingly recapitulated in vitro (Kulukian et al., 2009; Vink et al., 2006), efficient MCC formation in cells seems to require additional inputs from kinetochores. Mitotic delays in cells that express an artificial Mad1 protein that is maintained on attached kinetochores depends on kinetochore kinases, and targeting Mad1 to nonkinetochore chromosomal regions is not sufficient to delay mitosis (Maldonado and Kapoor, 2011). One possible explanation for this is that the kinetochore-localized MC kinase Mps1 aids Mad2 conversion by promoting Mad2 dimerization (Hewitt et al., 2010). In normal conditions, Mps1 further impacts Mad1-Mad2 function by promoting Mad1 localization to kinetochores (Lan and Cleveland, 2010). Clarifying the mechanism for this will require identification of the Mad1 receptor at kinetochores. Interestingly, the amino-terminal region of Mad1 that is required for its kinetochore binding was allowed to diverge during evolution (Figure S2). It has been suggested that this region determines checkpoint sensitivity, because the less-robust checkpoint in rodent cells can

FIGURE 3 Cdc20 Homologs in the Eukaryotic Tree of Life

Schematic representation of eukaryotic tree of life with Cdc20 homologs from species listed in Figure 2. Indicated for every homolog are the presence (white box) or absence (black box) of the C box (DR[YF]IP), KEN/D box (KEN/RxxLxxxx[EDNQ]), Mad2-binding domain ([KR][IV]LxxxP), the number of predicted WD40 repeats (using SMART-EMBL), the presence of an IR tail ([IVLF]R), and the ORF length in amino acids. Species in bold indicate experimentally confirmed Cdc20 homologs, blue protein bodies indicate homologs containing all essential domains, and the *A. thaliana* Cdc20 genes that are expressed are indicated by an asterisk. Red star shapes indicate probable gene duplication events based on phylogenetic alignments.

be made more stringent by ectopic expression of human Mad1 or a hybrid of murine Mad1 with a human amino-terminal domain (Haller et al., 2006).

The conversion of O-Mad2 to C-Mad2 relies on several features within the Mad1-Mad2 complex, including a Cdc20-like Mad2-binding motif in Mad1, Mad1 homodimerization, and a HORMA domain in Mad2 that is required for both Cdc20 and Mad1 binding in a mutually exclusive manner (Musacchio and Salmon, 2007). The Mad2 HORMA domains are highly similar between species in all supergroups analyzed, suggesting strict conservation of the Mad2-Cdc20 interface. Much like the Mad2-binding motif in Cdc20 and the HORMA domain in Mad2, the Mad2-binding motif in Mad1, when present, is highly conserved (Figure S2). Mutation of this motif abrogates MC activity (Maldonado and Kapoor, 2011). Interestingly, the Mad2-binding motif in Mad1 is absent from Mad1 homologs in *Salpingoeca rosetta*, *Micromonas pusilla*, and *Naegleria gruberi* (Figure S2). The related motif can be found in their Cdc20 homologs, suggesting that fundamentals of Mad2 binding have in principle not been altered in these species. If their Mad1 is nevertheless capable of binding Mad2, examining how may provide additional insight into molecular aspects of this interaction. Potentially important in this regard is the recent identification of S187 phosphorylation in fission yeast Mad2 that affects the Mad1-Mad2 interaction (Zich et al., 2012). Given the high conservation of the position of this serine in Mad2 homologs, such a regulatory mechanism for Mad2 function may be ancient.

THE APC/C PSEUDOSUBSTRATE INHIBITOR WITHIN THE MCC

Human BubR1 was identified as a Bub1-like gene mutated in chromosomally unstable colon cancer cell lines but was later recognized as the functional equivalent of the budding yeast spindle checkpoint protein Mad3p (Elowe, 2011). Mad3/BubR1 and Bub1 share extensive sequence homology and domain architecture. Both contain a TPR domain followed by a Gle2-binding sequence (GLEBS) motif, and in vertebrates and *Drosophila* both contain an unusual carboxy-terminal Ser/Thr kinase domain (Bolanos-Garcia and Blundell, 2011). This similarity stems from the fact that LECA contained a single protein, to which we refer as Madbub, that possessed the shared domains as well as the amino-terminal KEN box characteristic of Mad3/BubR1-like proteins (Suijkerbuijk et al., 2012) (Figure 4). Madbub subsequently took distinct paths of evolution: it either remained a Madbub and diverged little or it underwent a gene duplication event on multiple (probably nine) independent occasions. Duplication was followed either by loss of one of the copies, as in the case of some relatives of *Saccharomyces cerevisiae* (Murray, 2012), or by a striking example of parallel subfunctionalization, during which retainment

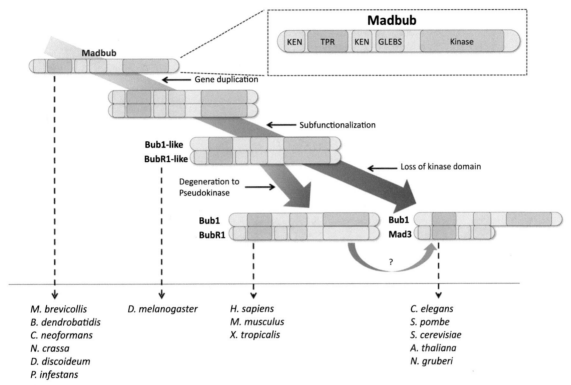

FIGURE 4 Proposed Model for Evolution of the Madbub Family

LECA possessed a Madbub protein containing the predominant functional domains (see inset: two KEN boxes, TPR domain, GLEBS motif, and a kinase domain). Madbubs are still present in numerous organisms, including those indicated at the bottom. At least nine independent gene duplications led to subfunctionalization of a Bub1-like (kinase) and a BubR1-like (KEN box) gene. After the loss of kinase requirement in the KEN-box protein, the kinase either degenerated into a pseudokinase (vertebrates) or was shed altogether. One notable exception is the Drosophilids, in which the KEN box was lost from one paralog, but the kinase was maintained with almost identical sequence in both.

of the KEN box or kinase domain was mutually exclusive in the vast majority (seven of nine) of cases. These parallel subfunctionalization events gave rise to present-day Bub1-BubR1/Mad3 paralogs (Figure 4) (Suijkerbuijk et al., 2012). Insightful exceptions to this rule are insects and vertebrates. The KEN-box-containing protein retained a kinase domain in vertebrates, but this domain was allowed to degenerate to a pseudokinase that is highly sensitive to destabilization by amino acid substitutions in various regions of the domain (Suijkerbuijk et al., 2012). Because destabilization is propagated to the whole protein, this liability may have contributed to selection for truncating mutations in so many nonvertebrate species (Figure 4). In *D. melanogaster*, however, the KEN-box-containing protein retained a proper Madbub-like, and therefore Bub1-like, kinase domain. *D. melanogaster* BubR1 may have kinase function, which is unique in eukaryotes, or may alternatively be in a transition state, one that is predicted to have occurred between a gene duplication event and evolution toward either a pseudokinase or shedding of the kinase domain (Figure 4).

THE FATE OF A PARALOG: EVOLUTION AND FUNCTION OF BUB1

Whereas the role of the KEN-box-containing Mad3/BubR1-like proteins in the MC is well defined, it is less so for the paralogs that retained the kinase domain. These Bub1-like kinases can be found in at least one copy in most eukaryotes examined, either as part of Madbub proteins or of the KEN-box-lacking paralog that originated after evolution from Madbub gene duplications (Figure 2) (Suijkerbuijk et al., 2012). Given the evidence from gene disruptions in mice, *Drosophila*, and both model fungi, Bub1 appears to be essential for MC function (Musacchio and Salmon, 2007). Whether this is mediated by kinase activity is unclear. Studies in human cells, *S. pombe*, and *X. laevis* extracts show that Bub1 kinase activity promotes but is not absolutely required for a robust MC response (Chen, 2004; Klebig et al., 2009; Yamaguchi et al., 2003), while the MC in *S. cerevisiae* responds properly when the Bub1 kinase domain is removed altogether (Fernius and Hardwick, 2007; Warren et al., 2002). Human Bub1 was found to modify Cdc20 on multiple residues in vitro, causing reduced APC/C activity (Tang et al., 2004). Some of these are relatively well conserved, but functional analysis of phosphomimetic substitutions in the background of inactive Bub1 in various organisms will be needed to clarify whether Cdc20 phosphorylation by Bub1 is conserved and part of the core MC. Bub1 kinase activity does have a conserved role in non-MC processes, such as centromere localization of Shugoshin via phosphorylation of T121 on the histone H2A (Kawashima et al., 2010). A recent study of Bub1 function in human cells pinpointed a short sequence, dubbed conserved domain I (CDI), as crucial for the MC (Klebig et al., 2009). Although it is unknown how CD1 has impact on MCC formation, it may have been part of LECA Madbub, as we can recognize CD1 in some Madbub homologs (Suijkerbuijk et al., 2012).

KINETOCHORE SCAFFOLDS FOR THE MITOTIC CHECKPOINT

Both Mad3/BubR1 and Bub1, as well as the Madbub proteins, have a highly similar TPR domain that interacts with the KMN network member Knl1. This interaction was mapped to the convex surface of the TPR domains and to two "KI" motifs in Knl1, which we and others can recognize only in vertebrate Knl1 homologs (Figure 5) (Bolanos-Garcia and Blundell, 2011; Bolanos-Garcia et al., 2011; Kiyomitsu et al., 2011; Krenn et al., 2012). The mode of Knl1-Bub interactions may be quite flexible and may rely on other motifs in nonvertebrates, because *D. melanogaster* Bub1 interacts with Knl1/Spc105, which is devoid of a clear KI1 motif (Schittenhelm et al., 2009). Knl1 depletion in human and fungal cells prevents Bub1 and BubR1/Mad3 kinetochore binding and checkpoint activation (Kiyomitsu et al., 2007; London et al.,

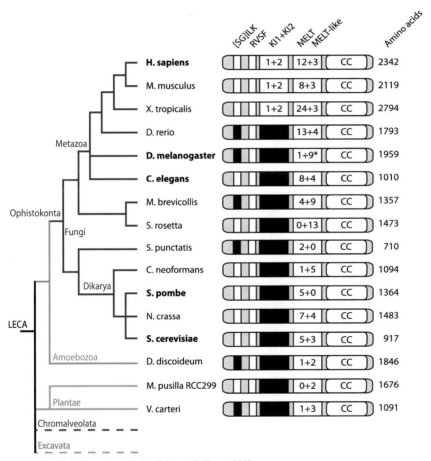

FIGURE 5 Knl1 Homologs in the Eukaryotic Tree of Life
Schematic representation of eukaryotic tree of life with Knl1 homologs from species listed in Figure 2. Indicated for every homolog are the presence (white box) or absence (black box) of the [SG]ILK and RVSF motifs, the KI motifs, the number of MELT (M[ED][ILVM][ST]) and MELT-like Mxxx (x = 2 out of 3 amino acids are D or E) motifs, the presence of a coiled coil (using the EMBnet coils server), and the ORF length in amino acids. Species in bold indicate experimentally confirmed homologs, and the asterisk indicates the presence of *Drosophila*-specific motifs.

2012; Shepperd et al., 2012). This is, however, independent of the interaction between the TPR domains and Knl1. Mutating this interface does not prevent localization of the Bubs and has only minor effects on the MC response (Bolanos-Garcia et al., 2011; Krenn et al., 2012). The functional relevance of the Bub-Knl1 interaction is unclear but may involve, at least for Bub1, an allosteric mechanism for kinase activation (Krenn et al., 2012). The essential role of Knl1 in localizing the Bubs to kinetochores is likely mediated by Bub3, a small globular protein that interacts with the Bub1/BubR1 GLEBS motifs. Like Knl1 depletion, mutating the GLEBS motif in either Bub1 or BubR1 prevents Bub3 binding, abrogates kinetochore localization of both Bubs, and disrupts their various functions in mitosis (Bolanos-Garcia and Blundell, 2011). Oddly, however, Bub3 depletion in human cells inhibits kinetochore localization of BubR1 but not Bub1 (Logarinho et al., 2008). Whether this reflects greater sensitivity of BubR1 to reductions in Bub3, possibly because of differences in kinetochore residence time (Howell et al., 2004), or whether this reflects a possible Bub3-independent role of the Bub1 GLEBS motif is presently unknown. Because the interaction of the Bub1 GLEBS motif to Bub3 leaves limited space for other interaction partners (Larsen et al., 2007), we favor the former possibility. To make matters more complicated, Bub1 is required for BubR1 localization, but not vice versa (Johnson et al., 2004; Klebig et al., 2009). Unraveling the intricate relationship between BubR1, Bub1, Knl1, and Bub3 will be of great interest. One recent insight may be a significant step forward in this regard: in fungi, Bub3 and Bub1 kinetochore localization depends on intact MELT motifs of Knl1 (London et al., 2012; Shepperd et al., 2012).

Knl1 may act as a molecular MC scaffold on more levels than localizing the three Bubs. Through its C-terminal region, Knl1 binds the kinetochore protein Zwint-1 that, in turn, localizes the RZZ complex to kinetochores (Kiyomitsu et al., 2011). Because RZZ promotes Mad1 kinetochore binding (Karess, 2005), Knl1 likely affects the ability of kinetochores to efficiently catalyze MCC formation. Additionally, Knl1 binds PP1 phosphatase through a SILK-RVSF motif near its N terminus. While this interaction is required to stabilize kinetochore-microtubule interactions in human cells (Liu et al., 2010), it silences MC signaling from attached kinetochores in fungi and *C. elegans* (Espeut et al., 2012; Meadows et al., 2011; Rosenberg et al., 2011).

Knl1 displays poor overall sequence conservation, which may explain its propensity, more than other MC components, to escape identification in our homolog searches (Table S1). Despite this, the SILK and RVSF motifs, as well as MELT motifs (defined as M[ED][ILVM][ST]), are well conserved in most identifiable Knl1 homologs, as is a defining C-terminal coiled coil (Figure 5). A striking observation is that the MELT motifs diverge highly in numbers, ranging from 0 (*S. rosetta* and *M. pusilla*) to 24 (*Xenopus tropicalis*). Their functional relevance likely goes beyond Bub recruitment, because a MELT-mutated Spc7/Knl1 in *Schizosaccharomyces pombe* results in profoundly more chromosome segregation problems than deletion of Bub3 (Shepperd et al., 2012). Interestingly, Spc105/Knl1 in Drosophilids have repeats of a slightly distinct motif, which seem largely dispensable for Spc105/Knl1 function in *D. melanogaster* (Schittenhelm et al., 2009). The MELT motifs are thus quite enigmatic, and uncovering their role in mitosis and the reasons for their highly variable numbers in different species will be a great value in our understanding of the connections between the KMN network, microtubule attachments, and the MC.

REGULATING THE MCC: ESSENTIAL AND CONSERVED CONTRIBUTIONS OF MITOTIC KINASES

Efficient formation of MCC in cells depends both directly and indirectly on kinase activities. Mps1 and Bub1 were among the original genes found to control the MC in *S. cerevisiae* (Musacchio and Salmon, 2007). In contrast with Bub1, inhibition of Mps1 ablates MC activity in all organisms tested (Lan and Cleveland, 2010; Musacchio and Salmon, 2007) and is therefore the only undisputed MC kinase. Nevertheless, activity of several other kinases, such as Aurora B, Cdk1, and PRP4, also affects MC function, but current evidence supports the notion that they do so predominantly by regulating Mps1 (Montembault et al., 2007; Morin et al., 2012; Saurin et al., 2011).

Mps1 orchestrates many events that contribute to APC/C inhibition, including localization of Bub1, BubR1, RZZ, and Mad1 to unattached kinetochores (in various organisms) (Lan and Cleveland, 2010), Mad2 phosphorylation (in fission yeast) (Zich et al., 2012), and Mad2 dimerization and MCC stabilization (in human cells) (Hewitt et al., 2010; Maciejowski et al., 2010). Recent studies have shown that Mps1 promotes Bub1 recruitment and subsequent MC activity in human cells and in budding and fission yeast by phosphorylating Knl1 on multiple of its MELT motifs (London et al., 2012; Shepperd et al., 2012; Yamagishi et al., 2012). The Bubs, however, interact with KI rather than MELT motifs. Because MELT phosphorylations also recruit Bub3, a likely scenario is that Mps1 controls Bub1 (and BubR1/Mad3) localization by promoting the interaction of Bub3 with Knl1, possibly by ensuring Bub3-pMELT binding. This could also contribute to the role of Mps1 in localization of Mad1 to unattached kinetochores, because Bub1 depletion prevents Mad1 kinetochore binding (Musacchio and Salmon, 2007). A MELT-phosphomimetic Knl1 retains Bub1 on kinetochores in the absence of Mps1. In contrast, this mutant Knl1 cannot force kinetochore recruitment of Mad1 under those conditions, suggesting that Mps1 has Bub1-independent mechanisms for recruiting Mad1 (Shepperd et al., 2012). This is supported by *D. melanogaster*, in which Spc105/Knl1 possesses species-specific MELT-like motifs that lack the phosphorylatable threonine (Schittenhelm et al., 2009), while Mad1 localization remains Mps1 dependent (Althoff et al., 2012) (Figure 5). Interestingly, these motifs contain an excess of negative charges (D or E), possibly bypassing phosphodependency of Bub recruitment, but excluding MELT-dependent control of this by Mps1. Our analysis of Knl1 homologs has revealed additional MELT-like methionine-based motifs (methionine followed by two or three acidic residues) in other organisms as well (Figure 5). Given the conservation of Mps1 function and MELT motifs in Knl1, we hypothesize that MELT phosphorylation by Mps1 is a fundamental MC regulatory principle. Whether the widely differing number of MELT and MELT-like motifs in species has any relation to this (for instance by adding levels of control or variable distance between PP1 and Bub [N-terminal] and the RZZ [C-terminal] binding sites) and, if so, how, are intriguing questions. Answers will require detailed insight into which MELT motifs are truly essential for mediating the impact of Mps1 on Bub localization and MC function. Based on species like *Spizellomyces punctatus* and *Dictyostelium discoideum*, we predict one or two will suffice.

Strikingly, although Mps1 is well conserved and essential for error-free chromosome segregation in all organisms tested, no sequence homolog in *C. elegans* can be detected (Figure 2). It is possible that the Mps1 homolog exists but diverged so much as to escape our detection. Alternatively, perhaps the fast-evolving nematode has bypassed a requirement for Mps1 in regulating MCC formation/function, for instance by modifying the mechanism of (Knl1-dependent) Bub3 localization. Finally, a distinct kinase may have replaced Mps1 in nematodes. In this respect it is of interest to note that Mps1 shares significant overlap in consensus phosphorylation sequence with the kinetochore-localized kinase Plk1 (Dou et al., 2011), which is expressed in *C. elegans* (Chase et al., 2000).

AUXILIARY MC PROTEINS: THE RZZ COMPLEX

The heterotrimeric RZZ complex plays an essential part in recruitment of Mad1/Mad2 to unattached kinetochores in human and *Drosophila* cells (Karess, 2005). In contrast to Mad1 and Mad2, however, the RZZ subunit Zwilch does not seem to have been retained in many species besides ophistokonta, indicating that RZZ function is a fairly recent add-on to the core MC (Figure 2). This may point to evolution in more complex eukaryotes toward a multiprotein kinetochore interface for Mad1 binding that includes RZZ (Kim et al., 2012). Whereas Zwilch is never found without co-occurrence of an identifiable ortholog of ZW10, the opposite is frequently observed, suggesting a non-RZZ function of ZW10. In support of this, ZW10 is involved in vesicle trafficking in interphase, during which it is part of the conserved NRZ complex that contains Nag and Rint1 in addition to ZW10 (Civril et al., 2010). Homology between Rod and Nag and lack thereof between Zwilch and Rint1 (Civril et al., 2010) indicates that the RZZ may have arisen from NRZ by initially replacing Rint1 with Zwilch, causing it to be retained in those organisms that utilized RZZ for distinct functions. RZZ is coupled to kineto-chores via an interaction between ZW10 and Zwint-1 that in turn binds the C terminus of Knl1 (Karess, 2005; Kiyomitsu et al., 2011). In contrast to Zwilch participation, the ZW10-Zwint-1 interaction is likely ancient, as Zwint-1, like NRZ, is suggested to regulate interphasic vesicular trafficking (van Vlijmen et al., 2008). Functions of ZW10-containing complexes furthermore involve the minus-end-directed microtubule motor dynein. ZW10 directly binds the dynactin subunit p50/dynamitin, and, as a result, RZZ ensures kineto-chore localization of dynein, required for both chromosome movements and checkpoint silencing (Karess, 2005). RZZ therefore promotes checkpoint activation while simultaneously setting the stage for checkpoint silencing. Because both ancient interphasic and more recent mitotic functions of ZW10 depend on dynein, recruitment of dynein to kinetochores may have provided an important selective force driving ZW10 toward RZZ evolution. Perhaps more complex kinetochore-microtubule interactions benefit from more ways to ensure inhibition of MCC production.

RELEASING THE BRAKE: MC SILENCING

APC/C activation upon attachment of the final kinetochore is very rapid, sug-gesting a switch-like release from the MC-inhibited state (Clute and Pines, 1999). MC silencing occurs on two levels: local shutdown of MCC produc-tion upon kinetochore attachment and global reversion of APC/C inhibition upon stable attachment of the final kinetochore (Figure 1B). Below, we will briefly outline different checkpoint silencing mechanisms, their mode of action, and to what extent they have been conserved throughout evolution.

Inhibiting MCC Production upon Kinetochore-Microtubule Interaction

Microtubule attachment depletes essential checkpoint components, including Mad1/Mad2, from kinetochores in a dynein-dependent man-ner. This is a critical step in MC silencing, because kinetochore-tethered Mad1 is sufficient to delay mitotic exit after full chromosome biorientation is achieved (Maldonado and Kapoor, 2011). Essential to this is the Spin-dly protein, which depends on RZZ for kinetochore localization and which localizes dynein to kinetochores via its so-called Spindly-box motif [GNSx-FxEVxD] (Barisic et al., 2010; Gassmann et al., 2010). Besides a receptor for dynein, Spindly, like RZZ, is also cargo, and it was recently suggested that in fact removal of Spindly from kinetochores is a primary function of dynein in MC silencing (Gassmann et al., 2010). It was proposed that Spindly prevents an undefined dynein-independent pathway for Mad1/Mad2 removal from attached kinetochores and that dynein-dependent removal of Spindly allowed this unknown pathway to operate. Spindly appears to be an ophistokont invention and shows a strong correlation with the presence of Zwilch homologs being absent from, for example, dikaryan fungi (Figure 2; Table S1). This observation is likely related to a role for dynein at mitotic kinetochores, which is nonexistent in either *S. cerevisiae* or *S. pombe*. Because such organisms nevertheless presumably also deplete Mad1/Mad2 from attached kinetochores, it has been speculated that the unknown dynein-independent Mad1/Mad2 removal pathway that Spindly normally prevents is an ancient one (Gassmann et al., 2010). A major challenge for the future will be to examine whether such a dynein-independent pathway for clearing MC proteins from kinetochores exits, what its molecular identity is, and how RZZ and Spindly affect its function. Given the conserved nature of Knl1, it may involve a recently defined MC silencing mechanism that relies on direct interaction of Knl1 with microtubules (Espeut et al., 2012).

Undoing the Actions of MC Kinases

Besides physically removing MC proteins from kinetochores as soon as they engage a microtubule, MC silencing requires dephosphorylation of essential MC targets. A PP1-like phosphatase is needed for the ability of yeast cells to exit from an MC-induced cell-cycle delay (Hardwick and Shah, 2010). Specific PP1 isoforms localize to bioriented kinetochores via the N-terminal SILK and RVSF motifs in Knl1 (Espeut et al., 2012; Liu et al., 2010; Meadows et al., 2011; Rosenberg et al., 2011). This specific interaction contributes to MC silencing in budding and fission yeast, as well as in *C. elegans* (Espeut et al., 2012; Meadows et al., 2011; Rosenberg et al., 2011), but it is unknown whether this is also true for vertebrates, in which a dynein-dependent MC silencing mechanism has evolved. Phosphatases are nevertheless likely required for exit from an MC arrest in human cells, as persistent kinetochore Mps1 maintains MC signaling from attached and bioriented kinetochores in metaphase (Jelluma et al., 2010). Similar reversal of Mps1-mediated phosphorylations also contributes to PP1's role in MC silencing in budding yeast (Pinsky et al., 2009), which could involve dephosphorylation of the Knl1/Spc105 MELT motifs (London et al., 2012).

Freeing the APC/C: Disassembly of MCC-APC/C Complexes

Once all kinetochores have achieved stable attachments to spindle microtubules, what remains for cells to initiate anaphase is releasing APC/C inhibition by MCC. As outlined in our discussions on Cdc20, this process requires

APC/C-dependent ubiquitination and the actions of the Mad2-mimetic p31[comet]. The same surface on Mad2 interacts with both p31[comet] and Mad3/BubR1, suggesting that p31[comet] actively disrupts MCC stability by competing out Mad2 (Chao et al., 2012; Westhorpe et al., 2011). This may simply be achieved by the observed high affinity of p31[comet] for C-Mad2 (Vink et al., 2006), but it somehow also involves Cdk1-dependent phosphorylation of Cdc20 (Miniowitz-Shemtov et al., 2012). How Cdk1, the APC/C, and p31[comet] collaborate to ensure efficient MCC disassembly is presently unclear but may involve, for example, Cdk1- and APC/C-mediated relaxation of structural constraints to allow more efficient p31[comet]-dependent exclusion of Mad2 from MCC. It will be of additional interest to examine how rapid disassembly by this pathway is regulated by kinetochore attachment. p31[comet] is located exclusively on unattached kinetochores with a residence time identical to Mad2, leading to a proposed model in which p31[comet] is modified by unattached kinetochores in order to prevent its premature action on MCC disassembly (Hagan et al., 2011). As postulated before (Yang et al., 2007), our analysis shows that Mad2 and p31[comet] are probably paralogs that have arisen by a pre-LECA gene duplication (Figure 2; Supplemental Experimental Procedures). In contrast to the widespread maintenance of MCC throughout evolution, p31[comet] was apparently lost in many species, which is particularly apparent in fungi (Figure 2; Table S1). Unlike most other fungi, the higher basidiomycete fungi *Ustilago maydis* contains p31[comet] and has a metazoa-like open mitosis and anaphase B-like spindle elongation (Steinberg and Perez-Martin, 2008). Examining the p31[comet] homolog in *U. maydis* cell division may provide intriguing insights into its mitotic functions and may help to reveal why p31[comet] was allowed to disappear from the genomes of some organisms while it was retained by others.

CONCLUDING THOUGHTS AND FUTURE DIRECTIONS

In this review, we have attempted to integrate current knowledge on the molecular workings of the MC with our evolutionary analysis of key MC (silencing) proteins and their functional domains and motifs. Inspired by this, we propose an outline of the ancient MC and its functional modules (Figure 6), a core that is conserved in species that utilize the MC and that was likely present in LECA. The various species-specific additions, deletions, and/or modifications to this core may be related to fundamental differences between mitoses in these organisms. Such differences include but are by no means limited to: open versus closed mitosis, holocentric versus point centromeres, the size of kinetochores and the amount of microtubules connecting these to the mitotic spindle, the number of chromosomes to be segregated, the size of the cells, and the amount of cells that make up the organism. Future studies on the relation between such differences and MC function will be of interest not only from an evolutionary perspective but also from the perspective of understanding the MC and its adaptability. Many additional outstanding questions remain in relation to the conserved MC activation and silencing mechanisms. How and where is the MCC formed? How is the signal amplified from individual kinetochores, and, possibly in relation to this, how are MC kinases activated and what are their critical substrates? How is MCC action reverted upon MC satisfaction, especially considering the poor conservation of p31^{comet}? How does the state of attachment of kinetochores translate to recruitment or removal of MC proteins? The lack of kinetochore dynein and Spindly/RZZ in most species points to another, more ancient, mechanism that might or might not be retained in all eukaryotes. Binding of MC proteins like Bub1,

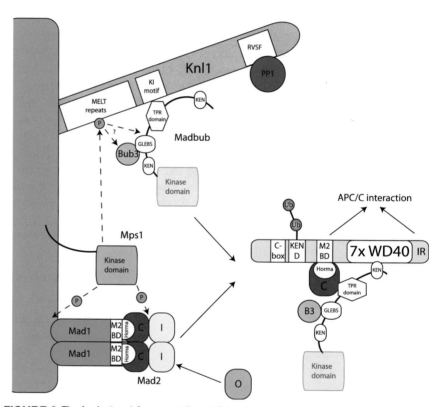

FIGURE 6 The Ancient and Conserved Core MC Proteins and Their Domains/Motifs
Phosphorylation of one or more MELT motifs on Knl1 by Mps1 recruits Bub3 and the Madbub protein. Mps1, together with the Madbub protein, further ensures kinetochore binding of Mad1 that interacts with the Horma domain of Mad2, allowing subsequent structural conversion of Mad2 into a closed form. C-Mad2 and Madbub/Bub3 assemble onto Cdc20 via various indicated domain/motif interactions to inhibit Cdc20 activity. PP1-mediated checkpoint silencing occurs through its interaction with the RVSF motif in Knl1. See text for further details.

Mad3/BubR1, and Mps1 with KMN network components are intriguing interactions on which silencing mechanisms could act to affect MC activity, but it is unknown whether such interactions are directly sensitive to microtubules. Uncovering which principles are ancient will require significant efforts in establishing sensitive real-time and biochemical assays for measuring MC activity, kinetochore changes upon microtubule binding, and MCC assembly and disassembly in a variety of model organisms. Because evolution has done most of the experiments for us, it may further be worthwhile to adopt less widely used and possibly even novel model organisms into this exciting field of research.

SUPPLEMENTAL INFORMATION

Supplemental Information includes three figures, one table, one sequence file, and Supplemental Experimental Procedures and can be found with this article online at http://www.sciencedirect.com/science/MiamiMultiMediaURL/1-s2.0-S1534580712002870/1-s2.0-S1534580712002870-mmc1.pdf/272236/FULL/S1534580712002870/9da40071e5180903134f-345c5478a84b/mmc1.pdf;
http://www.sciencedirect.com/science/MiamiMultiMediaURL/1-s2.0-S1534580712002870/1-s2.0-S1534580712002870-mmc2.xlsx/272236/FULL/S1534580712002870/8018ea0520648733d21c8b334a139524/mmc2.xlsx;
http://www.sciencedirect.com/science/MiamiMultiMediaURL/1-s2.0-S1534580712002870/1-s2.0-S1534580712002870-mmc3.zip/272236/FULL/S1534580712002870/532303384998861bf001e59b791ffaca/mmc3.zip.

ACKNOWLEDGMENTS

We apologize to all colleagues we were unable to cite due to space limitations. We are grateful to Jagesh Shah, Susanne Lens, and members of the Kops and Snel laboratories for critical reading of the manuscript and useful discussions. Work in the Kops laboratory is supported by the Dutch Cancer Society, by the Netherlands Organization for Scientific Research (NWO), and by the European Research Council (ERC-StG KINSIGN).

REFERENCES

Althoff, F., Karess, R.E., and Lehner, C.F. (2012). Spindle checkpoint-independent inhibition of mitotic chromosome segregation by Drosophila Mps1. Mol. Biol. Cell 23, 2275–2291.

Barisic, M., Sohm, B., Mikolcevic, P., Wandke, C., Rauch, V., Ringer, T., Hess, M., Bonn, G., and Geley, S. (2010). Spindly/CCDC99 is required for efficient chromosome congression and mitotic checkpoint regulation. Mol. Biol. Cell 21, 1968–1981.

Bolanos-Garcia, V.M., and Blundell, T.L. (2011). BUB1 and BUBR1: multifaceted kinases of the cell cycle. Trends Biochem. Sci. 36, 141–150.

Bolanos-Garcia, V.M., Lischetti, T., Matak-Vinković, D., Cota, E., Simpson, P.J., Chirgadze, D.Y., Spring, D.R., Robinson, C.V., Nilsson, J., and Blundell, T.L. (2011). Structure of a Blinkin-BUBR1 complex reveals an interaction crucial for kinetochore-mitotic checkpoint regulation via an unanticipated binding Site. Structure 19, 1691–1700.

Burton, J.L., and Solomon, M.J. (2007). Mad3p, a pseudosubstrate inhibitor of APCCdc20 in the spindle assembly checkpoint. Genes Dev. 21, 655–667.

Buschhorn, B.A., Petzold, G., Galova, M., Dube, P., Kraft, C., Herzog, F., Stark, H., and Peters, J.M. (2011). Substrate binding on the APC/C occurs between the coactivator Cdh1 and the processivity factor Doc1. Nat. Struct. Mol. Biol. 18, 6–13.

Chao, W.C., Kulkarni, K., Zhang, Z., Kong, E.H., and Barford, D. (2012). Structure of the mitotic checkpoint complex. Nature 484, 208–213.

Chase, D., Serafinas, C., Ashcroft, N., Kosinski, M., Longo, D., Ferris, D.K., and Golden, A. (2000). The polo-like kinase PLK-1 is required for nuclear envelope breakdown and the completion of meiosis in Caenorhabditis elegans. Genesis 26, 26–41.

Cheeseman, I.M., and Desai, A. (2008). Molecular architecture of the kinetochore-microtubule interface. Nat. Rev. Mol. Cell Biol. 9, 33–46.

Chen, R.H. (2004). Phosphorylation and activation of Bub1 on unattached chromosomes facilitate the spindle checkpoint. EMBO J. 23, 3113–3121.

Chung, E., and Chen, R.H. (2002). Spindle checkpoint requires Mad1-bound and Mad1-free Mad2. Mol. Biol. Cell 13, 1501–1511.

Civril, F., Wehenkel, A., Giorgi, F.M., Santaguida, S., Di Fonzo, A., Grigorean, G., Ciccarelli, F.D., and Musacchio, A. (2010). Structural analysis of the RZZ complex reveals common ancestry with multisubunit vesicle tethering machinery. Structure 18, 616–626.

Clute, P., and Pines, J. (1999). Temporal and spatial control of cyclin B1 destruction in metaphase. Nat. Cell Biol. 1, 82–87.

da Fonseca, P.C., Kong, E.H., Zhang, Z., Schreiber, A., Williams, M.A., Morris, E.P., and Barford, D. (2011). Structures of APC/C(Cdh1) with substrates identify Cdh1 and Apc10 as the D-box co-receptor. Nature 470, 274–278.

De Antoni, A., Pearson, C.G., Cimini, D., Canman, J.C., Sala, V., Nezi, L., Mapelli, M., Sironi, L., Faretta, M., Salmon, E.D., and Musacchio, A. (2005). The Mad1/Mad2 complex as a template for Mad2 activation in the spindle assembly checkpoint. Curr. Biol. 15, 214–225.

Dou, Z., von Schubert, C., Körner, R., Santamaria, A., Elowe, S., and Nigg, E.A. (2011). Quantitative mass spectrometry analysis reveals similar substrate consensus motif for human Mps1 kinase and Plk1. PLoS ONE 6, e18793.

Elowe, S. (2011). Bub1 and BubR1: at the interface between chromosome attachment and the spindle checkpoint. Mol. Cell. Biol. 31, 3085–3093.

Eme, L., Trilles, A., Moreira, D., and Brochier-Armanet, C. (2011). The phylogenomic analysis of the anaphase promoting complex and its targets points to complex and modern-like control of the cell cycle in the last common ancestor of eukaryotes. BMC Evol. Biol. 11, 265.

Espeut, J., Cheerambathur, D.K., Krenning, L., Oegema, K., and Desai, A. (2012). Microtubule binding by KNL-1 contributes to spindle checkpoint silencing at the kinetochore. J. Cell Biol. 196, 469–482.

Fernius, J., and Hardwick, K.G. (2007). Bub1 kinase targets Sgo1 to ensure efficient chromosome biorientation in budding yeast mitosis. PLoS Genet. 3, e213.

Gassmann, R., Holland, A.J., Varma, D., Wan, X., Civril, F., Cleveland, D.W., Oegema, K., Salmon, E.D., and Desai, A. (2010). Removal of Spindly from microtubule-attached kinetochores controls spindle checkpoint silencing in human cells. Genes Dev. 24, 957–971.

Gordon, D.J., Resio, B., and Pellman, D. (2012). Causes and consequences of aneuploidy in cancer. Nat. Rev. Genet. *13*, 189–203.

Hagan, R.S., Manak, M.S., Buch, H.K., Meier, M.G., Meraldi, P., Shah, J.V., and Sorger, P.K. (2011). p31(comet) acts to ensure timely spindle checkpoint silencing subsequent to kinetochore attachment. Mol. Biol. Cell *22*, 4236–4246.

Haller, K., Kibler, K.V., Kasai, T., Chi, Y.H., Peloponese, J.M., Yedavalli, V.S., and Jeang, K.T. (2006). The N-terminus of rodent and human MAD1 confers species-specific stringency to spindle assembly checkpoint. Oncogene *25*, 2137–2147.

Hardwick, K.G., and Shah, J.V. (2010). Spindle checkpoint silencing: ensuring rapid and concerted anaphase onset. F1000 Biol. Rep. *2*, 55.

Herzog, F., Primorac, I., Dube, P., Lenart, P., Sander, B., Mechtler, K., Stark, H., and Peters, J.M. (2009). Structure of the anaphase-promoting complex/cyclosome interacting with a mitotic checkpoint complex. Science *323*, 1477–1481.

Hewitt, L., Tighe, A., Santaguida, S., White, A.M., Jones, C.D., Musacchio, A., Green, S., and Taylor, S.S. (2010). Sustained Mps1 activity is required in mitosis to recruit O-Mad2 to the Mad1-C-Mad2 core complex. J. Cell Biol. *190*, 25–34.

Howell, B.J., Moree, B., Farrar, E.M., Stewart, S., Fang, G., and Salmon, E.D. (2004). Spindle checkpoint protein dynamics at kinetochores in living cells. Curr. Biol. *14*, 953–964.

Jelluma, N., Dansen, T.B., Sliedrecht, T., Kwiatkowski, N.P., and Kops, G.J. (2010). Release of Mps1 from kinetochores is crucial for timely anaphase onset. J. Cell Biol. *191*, 281–290.

Johnson, V.L., Scott, M.I., Holt, S.V., Hussein, D., and Taylor, S.S. (2004). Bub1 is required for kinetochore localization of BubR1, Cenp-E, Cenp-F and Mad2, and chromosome congression. J. Cell Sci. *117*, 1577–1589.

Karess, R. (2005). Rod-Zw10-Zwilch: a key player in the spindle checkpoint. Trends Cell Biol. *15*, 386–392.

Kawashima, S.A., Yamagishi, Y., Honda, T., Ishiguro, K.I., and Watanabe, Y. (2010). Phosphorylation of H2A by Bub1 prevents chromosomal instability through localizing shugoshin. Science *327*, 172–177.

Khodjakov, A., and Pines, J. (2010). Centromere tension: a divisive issue. Nat. Cell Biol. *12*, 919–923.

Kim, S., Sun, H., Tomchick, D.R., Yu, H., and Luo, X. (2012). Structure of human Mad1 C-terminal domain reveals its involvement in kinetochore targeting. Proc. Natl. Acad. Sci. USA *109*, 6549–6554.

Kimata, Y., Kitamura, K., Fenner, N., and Yamano, H. (2011). Mes1 controls the meiosis I to meiosis II transition by distinctly regulating the anaphase-promoting complex/cyclosome coactivators Fzr1/Mfr1 and Slp1 in fission yeast. Mol. Biol. Cell *22*, 1486–1494.

Kiyomitsu, T., Obuse, C., and Yanagida, M. (2007). Human Blinkin/AF15q14 is required for chromosome alignment and the mitotic checkpoint through direct interaction with Bub1 and BubR1. Dev. Cell *13*, 663–676.

Kiyomitsu, T., Murakami, H., and Yanagida, M. (2011). Protein interaction domain mapping of human kinetochore protein Blinkin reveals a consensus motif for binding of spindle assembly checkpoint proteins Bub1 and BubR1. Mol. Cell. Biol. *31*, 998–1011.

Klebig, C., Korinth, D., and Meraldi, P. (2009). Bub1 regulates chromosome segregation in a kinetochore-independent manner. J. Cell Biol. *185*, 841–858.

Krenn, V., Wehenkel, A., Li, X., Santaguida, S., and Musacchio, A. (2012). Structural analysis reveals features of the spindle checkpoint kinase Bub1-kinetochore subunit Knl1 interaction. J. Cell Biol. *196*, 451–467.

Kulukian, A., Han, J.S., and Cleveland, D.W. (2009). Unattached kinetochores catalyze production of an anaphase inhibitor that requires a Mad2 template to prime Cdc20 for BubR1 binding. Dev. Cell *16*, 105–117.

Lampert, F., and Westermann, S. (2011). A blueprint for kinetochores - new insights into the molecular mechanics of cell division. Nat. Rev. Mol. Cell Biol. *12*, 407–412.

Lan, W., and Cleveland, D.W. (2010). A chemical tool box defines mitotic and interphase roles for Mps1 kinase. J. Cell Biol. *190*, 21–24.

Lara-Gonzalez, P., Scott, M.I., Diez, M., Sen, O., and Taylor, S.S. (2011). BubR1 blocks substrate recruitment to the APC/C in a KEN-box-dependent manner. J. Cell Sci. *124*, 4332–4345.

Larsen, N.A., Al-Bassam, J., Wei, R.R., and Harrison, S.C. (2007). Structural analysis of Bub3 interactions in the mitotic spindle checkpoint. Proc. Natl. Acad. Sci. USA *104*, 1201–1206.

Liu, D., Vleugel, M., Backer, C.B., Hori, T., Fukagawa, T., Cheeseman, I.M., and Lampson, M.A. (2010). Regulated targeting of protein phosphatase 1 to the outer kinetochore by KNL1 opposes Aurora B kinase. J. Cell Biol. *188*, 809–820.

Lizé, M., Pilarski, S., and Dobbelstein, M. (2010). E2F1-inducible microRNA 449a/b suppresses cell proliferation and promotes apoptosis. Cell Death Differ. *17*, 452–458.

Logarinho, E., Resende, T., Torres, C., and Bousbaa, H. (2008). The human spindle assembly checkpoint protein Bub3 is required for the establishment of efficient kinetochore-microtubule attachments. Mol. Biol. Cell *19*, 1798–1813.

London, N., Ceto, S., Ranish, J.A., and Biggins, S. (2012). Phosphoregulation of Spc105 by Mps1 and PP1 regulates Bub1 localization to kinetochores. Curr. Biol. *22*, 900–906.

Maciejowski, J., George, K.A., Terret, M.E., Zhang, C., Shokat, K.M., and Jallepalli, P.V. (2010). Mps1 directs the assembly of Cdc20 inhibitory complexes during interphase and mitosis to control M phase timing and spindle checkpoint signaling. J. Cell Biol. *190*, 89–100.

Maldonado, M., and Kapoor, T.M. (2011). Constitutive Mad1 targeting to kinetochores uncouples checkpoint signalling from chromosome biorientation. Nat. Cell Biol. *13*, 475–482.

Mansfeld, J., Collin, P., Collins, M.O., Choudhary, J.S., and Pines, J. (2011). APC15 drives the turnover of MCC-CDC20 to make the spindle assembly checkpoint responsive to kinetochore attachment. Nat. Cell Biol. *13*, 1234–1243.

Meadows, J.C., Shepperd, L.A., Vanoosthuyse, V., Lancaster, T.C., Sochaj, A.M., Buttrick, G.J., Hardwick, K.G., and Millar, J.B. (2011). Spindle checkpoint silencing requires association of PP1 to both Spc7 and kinesin-8 motors. Dev. Cell *20*, 739–750.

Miniowitz-Shemtov, S., Eytan, E., Ganoth, D., Sitry-Shevah, D., Dumin, E., and Hershko, A. (2012). Role of phosphorylation of Cdc20 in p31(comet)-stimulated disassembly of the mitotic checkpoint complex. Proc. Natl. Acad. Sci. USA *109*, 8056–8060.

Montembault, E., Dutertre, S., Prigent, C., and Giet, R. (2007). PRP4 is a spindle assembly checkpoint protein required for MPS1, MAD1, and MAD2 localization to the kinetochores. J. Cell Biol. *179*, 601–609.

Morin, V., Prieto, S., Melines, S., Hem, S., Rossignol, M., Lorca, T., Espeut, J., Morin, N., and Abrieu, A. (2012). CDK-dependent potentiation of MPS1 kinase activity is essential to the mitotic checkpoint. Curr. Biol. *22*, 289–295.

Murray, A.W. (2012). Don'T make me mad, bub!. Dev. Cell *22*, 1123–1125.

Musacchio, A., and Salmon, E.D. (2007). The spindle-assembly checkpoint in space and time. Nat. Rev. Mol. Cell Biol. *8*, 379–393.

Nezi, L., and Musacchio, A. (2009). Sister chromatid tension and the spindle assembly checkpoint. Curr. Opin. Cell Biol. *21*, 785–795.

Nezi, L., Rancati, G., De Antoni, A., Pasqualato, S., Piatti, S., and Musacchio, A. (2006). Accumulation of Mad2-Cdc20 complex during spindle checkpoint activation requires binding of open and closed conformers of Mad2 in Saccharomyces cerevisiae. J. Cell Biol. *174*, 39–51.

Nilsson, J., Yekezare, M., Minshull, J., and Pines, J. (2008). The APC/C maintains the spindle assembly checkpoint by targeting Cdc20 for destruction. Nat. Cell Biol. *10*, 1411–1420.

Pan, J., and Chen, R.H. (2004). Spindle checkpoint regulates Cdc20p stability in Saccharomyces cerevisiae. Genes Dev. *18*, 1439–1451.

Pines, J. (2011). Cubism and the cell cycle: the many faces of the APC/C. Nat. Rev. Mol. Cell Biol. *12*, 427–438.

Pinsky, B.A., Nelson, C.R., and Biggins, S. (2009). Protein phosphatase 1 regulates exit from the spindle checkpoint in budding yeast. Curr. Biol. *19*, 1182–1187.

Reddy, S.K., Rape, M., Margansky, W.A., and Kirschner, M.W. (2007). Ubiquitination by the anaphase-promoting complex drives spindle checkpoint inactivation. Nature *446*, 921–925.

Rosenberg, J.S., Cross, F.R., and Funabiki, H. (2011). KNL1/Spc105 recruits PP1 to silence the spindle assembly checkpoint. Curr. Biol. *21*, 942–947.

Saurin, A.T., van der Waal, M.S., Medema, R.H., Lens, S.M., and Kops, G.J. (2011). Aurora B potentiates Mps1 activation to ensure rapid checkpoint establishment at the onset of mitosis. Nat. Commun. *2*, 316.

Schittenhelm, R.B., Chaleckis, R., and Lehner, C.F. (2009). Intrakinetochore localization and essential functional domains of Drosophila Spc105. EMBO J. *28*, 2374–2386.

Schreiber, A., Stengel, F., Zhang, Z., Enchev, R.I., Kong, E.H., Morris, E.P., Robinson, C.V., da Fonseca, P.C., and Barford, D. (2011). Structural basis for the subunit assembly of the anaphase-promoting complex. Nature *470*, 227–232.

Sczaniecka, M., Feoktistova, A., May, K.M., Chen, J.S., Blyth, J., Gould, K.L., and Hardwick, K.G. (2008). The spindle checkpoint functions of Mad3 and Mad2 depend on a Mad3 KEN box-mediated interaction with Cdc20-anaphase-promoting complex (APC/C). J. Biol. Chem. *283*, 23039–23047.

Shepperd, L.A., Meadows, J.C., Sochaj, A.M., Lancaster, T.C., Zou, J., Buttrick, G.J., Rappsilber, J., Hardwick, K.G., and Millar, J.B. (2012). Phosphodependent recruitment of Bub1 and Bub3 to Spc7/KNL1 by Mph1 kinase maintains the spindle checkpoint. Curr. Biol. *22*, 891–899.

Steinberg, G., and Perez-Martin, J. (2008). Ustilago maydis, a new fungal model system for cell biology. Trends Cell Biol. *18*, 61–67.

Suijkerbuijk, S.J., van Dam, T.J., Karagöz, G.E., von Castelmur, E., Hubner, N.C., Duarte, A.M., Vleugel, M., Perrakis, A., Rüdiger, S.G., Snel, B., and Kops, G.J. (2012). The Vertebrate Mitotic Checkpoint Protein BUBR1 Is an Unusual Pseudokinase. Dev. Cell *22*, 1321–1329.

Tang, Z., Shu, H., Oncel, D., Chen, S., and Yu, H. (2004). Phosphorylation of Cdc20 by Bub1 provides a catalytic mechanism for APC/C inhibition by the spindle checkpoint. Mol. Cell *16*, 387–397.

Tsuchiya, D., Gonzalez, C., and Lacefield, S. (2011). The spindle checkpoint protein Mad2 regulates APC/C activity during prometaphase and metaphase of meiosis I in Saccharomyces cerevisiae. Mol. Biol. Cell *22*, 2848–2861.

van Vlijmen, T., Vleugel, M., Evers, M., Mohammed, S., Wulf, P.S., Heck, A.J., Hoogenraad, C.C., and van der Sluijs, P. (2008). A unique residue in rab3c determines the interaction with novel binding protein Zwint-1. FEBS Lett. *582*, 2838–2842.

Varetti, G., Guida, C., Santaguida, S., Chiroli, E., and Musacchio, A. (2011). Homeostatic control of mitotic arrest. Mol. Cell *44*, 710–720.

Vink, M., Simonetta, M., Transidico, P., Ferrari, K., Mapelli, M., De Antoni, A., Massimiliano, L., Ciliberto, A., Faretta, M., Salmon, E.D., and Musacchio, A. (2006). In vitro FRAP identifies the minimal requirements for Mad2 kinetochore dynamics. Curr. Biol. *16*, 755–766.

Warren, C.D., Brady, D.M., Johnston, R.C., Hanna, J.S., Hardwick, K.G., and Spencer, F.A. (2002). Distinct chromosome segregation roles for spindle checkpoint proteins. Mol. Biol. Cell *13*, 3029–3041.

Westhorpe, F.G., Tighe, A., Lara-Gonzalez, P., and Taylor, S.S. (2011). p31comet-mediated extraction of Mad2 from the MCC promotes efficient mitotic exit. J. Cell Sci. *124*, 3905–3916.

Yamagishi, Y., Yang, C.H., Tanno, Y., and Watanabe, Y. (2012). MPS1/Mph1 phosphorylates the kinetochore protein KNL1/Spc7 to recruit SAC components. Nat. Cell Biol. *14*, 746–752.

Yamaguchi, S., Decottignies, A., and Nurse, P. (2003). Function of Cdc2p-dependent Bub1p phosphorylation and Bub1p kinase activity in the mitotic and meiotic spindle checkpoint. EMBO J. *22*, 1075–1087.

Yang, M., Li, B., Tomchick, D.R., Machius, M., Rizo, J., Yu, H., and Luo, X. (2007). p31comet blocks Mad2 activation through structural mimicry. Cell *131*, 744–755.

Yu, H. (2007). Cdc20: a WD40 activator for a cell cycle degradation machine. Mol. Cell *27*, 3–16.

Zeng, X., Sigoillot, F., Gaur, S., Choi, S., Pfaff, K.L., Oh, D.C., Hathaway, N., Dimova, N., Cuny, G.D., and King, R.W. (2010). Pharmacologic inhibition of the anaphase-promoting complex induces a spindle checkpoint-dependent mitotic arrest in the absence of spindle damage. Cancer Cell *18*, 382–395.

Zich, J., Sochaj, A.M., Syred, H.M., Milne, L., Cook, A.G., Ohkura, H., Rappsilber, J., and Hardwick, K.G. (2012). Kinase activity of fission yeast Mph1 is required for Mad2 and Mad3 to stably bind the anaphase promoting complex. Curr. Biol. *22*, 296–301.

urrent Biology

Cell Division Orientation in Animals

Taryn E. Gillies[1], Clemens Cabernard[2,*]

[1]Institute of Neuroscience, University of Oregon 1254, Eugene OR 97403, USA,
[2]Biozentrum, University of Basel, CH-4056 Basel, Switzerland
*Correspondence: clemens.cabernard@unibas.ch

Current Biology, Vol. 21, No. 15, R599–R609, August 9, 2011 © 2011 Elsevier Inc.
http://dx.doi.org/10.1016/j.cub.2011.06.055

SUMMARY

Cell division orientation during animal development can serve to correctly organize and shape tissues, create cellular diversity or both. The underlying cellular mechanism is regulated spindle orientation. Depending on the developmental context, extrinsic signals or intrinsic cues control the correct orientation of the mitotic spindle. Cell geometry has been known to be another determinant of spindle orientation and recent results have shed new light on the link between cellular shape and cell division orientation. The importance of controlling spindle orientation is manifested in neurodevelopmental defects such as microcephaly, tumor initiation as well as defects in tissue architecture and cell fate misspecification. Here, we summarize the role of oriented cell division during animal development and also outline the cellular and molecular mechanisms in selected invertebrate and vertebrate systems.

INTRODUCTION

The orientation of the division axis is a basic regulator of metazoan development. Oriented cell division serves two purposes: first, to elongate cell sheets and shape tissues and, second, to generate cellular diversity. In order to achieve these two mutually non-exclusive functions, the orientation of the mitotic spindle has to be controlled. Over 120 years ago, Oscar Hertwig [1] recognized that cell shape is a determinant of spindle orientation and cell division orientation. He was among the first to discover that cells divide along their long cell axis, an observation known as the 'long axis rule'. Cell shape has been considered to be a default mechanism of spindle

91

CellPress

orientation [2,3]. However, several cell types override the cell shape pathway and control spindle orientation through external or internal polarity cues [4,5].

In this review, we first describe oriented cell division in different developmental contexts by outlining some of the classic and emerging model systems. We further summarize the underlying molecular and cellular mechanisms and highlight recent reports showing how external cues affect cellular shape (and thus oriented cell division) and how external or internal cues are linked to the mitotic spindle.

SHAPING TISSUES AND ORGANS

Studies performed in zebrafish (*Danio rerio*) have made significant contributions to our understanding of oriented cell division during animal development. Analyzing the patterns of cell division within the surface layer of the epiblast during zebrafish gastrulation revealed that in the dorsal region of the midline, and later in the ventral region, cell divisions are highly oriented along the animal-vegetal axis [6–8]. Similarly, the development of the immature epithelium of the zebrafish neural keel, which forms a lumenized neuroepithelium, depends on stereotyped oriented cell divisions, generating two bilaterally distributed neural progenitors. As these progenitors divide, one cell is placed on the ipsilateral side of the neural keel and its daughter intercalates across the midline integrating into the contralateral neuroepithelium (Figure 1A) [6,7,9–13]. Cell division is required for neural keel development, and blocking cell division will prevent the majority of cells from crossing the midline [12]. The correct cell division orientation is achieved through a 90 degree rotation of the mitotic spindle (Figure 1A) [9,11]. Disruption of division orientation in the neuroepithelium results in mostly unilateral placement of progenitors in the neural tube [11].

FIGURE 1 Cell division orientation in model systems.

(A) Zebrafish neural keel development. Neural progenitors divide perpendicular to the midline (brown line) and position sibling cells on both sides of the midline (four representative cell pairs are highlighted in blue). The stereotypic division axis is indicated by grey arrows. The mitotic spindle undergoes a 90 degree rotation, positioning the cell division orientation perpendicularly to the midline. (A, anterior; P, posterior). (B) *Drosophila* wing imaginal disc. Mitotic twin spot clones close to the anterior-posterior (dashed purple line) and dorsal-ventral (D-V) axis (dashed blue line) are indicated in green and red, respectively. Clones show elongated shape based on cell division orientation respecting the proximal-distal (P-D) axis. The stereotypic division axis in boxed regions is indicated with grey arrows. Dachsous (Ds) is expressed in a gradient along the P-D axis. Dachs (D) localization depends on the orientation of the Ds gradient. (C) Neuroblasts in the larval *Drosophila* CNS (dark and light blue dots). Snapshot showing dividing neuroblasts in the larval brain. Several division axes are highlighted with a yellow arrow. Note that the division axis is random in relation to the A-P axis (yellow arrows and grey arrows on top), but is fixed with respect to the time axis (grey arrows below). Schematic neuroblast (Nb) showing the apical (green) and basal (red) polarity complexes. Neuroblasts divide asymmetrically, resulting in an apical self-renewed neuroblast and a basal ganglion mother cell (GMC). (D) Mouse epidermis. At embryonic day 12, basal cells divide preferentially within the plane of the epithelium. From embryonic day 15.5 onwards, perpendicular asymmetric cell divisions occur, producing differentiating siblings (yellow). LGN protein (green) is diffusely localized (light green throughout the cell) in symmetrically dividing basal cells but becomes asymmetrically localized (green crescent) from embryonic day 15.5 onwards, inducing asymmetric cell division.

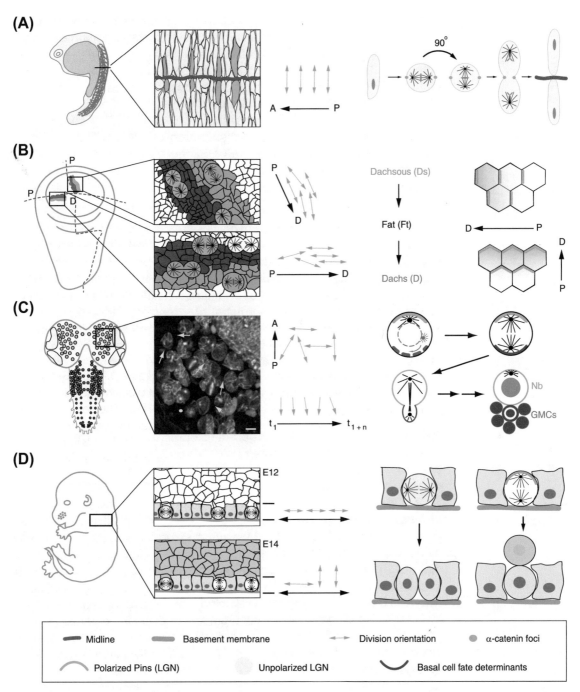

(A)

A ← P

90°

(B)

P → D

Dachsous (Ds)
↓
Fat (Ft)
↓
Dachs (D)

D ← P

D ↑ P

(C)

A ↑ P

t₁ → t₁₊ₙ

Nb

GMCs

(D)

E12

E14

| Midline | Basement membrane | ↔ Division orientation | ● α-catenin foci |
| Polarized Pins (LGN) | Unpolarized LGN | Basal cell fate determinants |

Similarly, in embryos of the frog *Xenopus laevis* it was shown that cells divide in three different manners in relation to the embryonic surface: parallel, oblique and perpendicular. The majority of divisions take place parallel to the surface. However, from the 32-cell stage onward, perpendicular cell divisions occur, resulting in the generation of deeper cells. An isolated 64-cell stage blastomere dividing in culture can generate a superficial cell that will express the bHLH gene *ESR6e* and a sibling cell that does not. Although the distribution of perpendicular divisions appears to differ between embryos and the division angle is not fixed between successive divisions, there is a strong correlation between cell division orientation and the generation of molecularly distinct deep and superficial cell layers [14].

Oriented cell division has also been observed in invertebrate species. During early *Drosophila melanogaster* embryogenesis the germband extends and elongates. During the fast phase of the elongation process cells divide preferentially along the anterior-posterior axis, corresponding to the long axis of the extending tissue. Blocking cell division does not completely prevent tissue elongation but reduces the amount of extension. Furthermore, mutant embryos lacking segmental patterning show randomized spindle orientation and isotropic increase in tissue size [15]. Although the above experiments suggest that oriented cell division is involved in germband extension, the overlap between cell division and morphogenetic movements makes it difficult to discern the individual contributions of these two processes to *Drosophila* embryogenesis.

Recently, it was shown that organ development can also be controlled, at least in part, through oriented cell division. *Drosophila* imaginal discs are epithelial structures originating from the embryonic ectoderm developing into adult organs such as the wing, legs and compound eyes. In wing and eye imaginal discs, a striking correlation between the shape of labeled clones of cells and the orientation of cell division has been described [16,17] (Figure 1B). Measuring spindle orientation in the wing blade revealed that the majority of divisions are oriented along the proximal-distal axis [16]. The correlation between oriented cell divisions and the shape of clones is maintained throughout wing development. Analysis of two- and four-cell clone clusters revealed that cell relocation plays a minor role in defining clonal shape. Measurements of division orientation outside the wing blade or in the eye disc showed similar results. Thus, oriented cell division appears to be a general mechanism to shape organs during fly organogenesis [16,17].

These examples illustrate how oriented cell division acts as a nearly ubiquitous morphogenetic force in multiple species. As we will see in the next paragraph, a different form of oriented cell division, asymmetric cell division, is iteratively used during development and across species.

GENERATING CELLULAR DIVERSITY

Asymmetric cell division generates cellular diversity [18]. This is achieved through the asymmetric partitioning of cell fate determinants, resulting in the generation of molecularly distinct sibling cells. An important aspect of asymmetric cell division is the correct alignment of the mitotic spindle in relation to an axis of internal or external polarity, ensuring asymmetric segregation of cell fate determinants. Asymmetric cell division has been studied in great detail and to great effect in model organisms such as the early *Caenorhabditis elegans* embryo [19–21], *Drosophila* neuroblasts [18] and *Drosophila* sensory organ precursor (SOP) cells [22].

Shortly after fertilization, the early *C. elegans* embryo becomes polarized — a prerequisite for the correct orientation of the mitotic spindle. Spindle orientation consists of two phases: first, during prophase, the nucleus–centrosome complex moves to the cell center and undergoes a 90 degree rotation; second, during metaphase and anaphase, the spindle is pulled towards the posterior of the cell. Spindle rotation and displacement require interactions between the mitotic spindle and the cortex and are dependent on intrinsic polarity cues [5,23]. Proper spindle orientation also ensures the correct segregation of cell fate determinants. While the anterior cell inherits cell fate determinants, the zinc-finger proteins MEX-5/MEX-6, another set of zinc-finger proteins, MEX-1, PIE-1 and POS-1, are partitioned asymmetrically into the posterior cell [20,24].

Similarly, *Drosophila* neural stem cell-like cells, called neuroblasts, orient their mitotic spindle along an established internal polarity axis. Stereotypic spindle orientation ensures that cell fate determinants, such as the coiled-coil protein Miranda (Mira), the transcriptional repressor Prospero (Pros; Prox1 in vertebrates), Numb (Numbl in vertebrates), Partner of Numb (Pon) and Brain tumor (Brat), are asymmetrically segregated into a small differentiating ganglion mother cell (GMC), while retaining a self-renewed apical neuroblast [18,25] (Figure 1C). Controlling the orientation of the mitotic spindle is required for the correct segregation of cell fate determinants [26]. As in *C. elegans*, delaminated embryonic neuroblasts rotate their mitotic spindle by 90 degrees and divide perpendicular to the neuroectoderm, which places the newly born GMCs into deeper tissue layers [27,28]. However, spindle rotation occurs only during the first neuroblast cell cycle after delamination [28]. From the second cell cycle onwards, one centrosome remains attached to the apical cortex, predetermining the future orientation of the mitotic spindle [28–30]. Elegant cell dissociation experiments showed that embryonic neuroblasts associated with neuroepithelial cells maintain their division axis over successive rounds of divisions; however, unassociated neuroblasts divide along random division axes [31]. Thus, in addition to neuroblast intrinsic polarity cues, some unknown extrinsic factors are required to maintain neuroblast division orientation in the fly embryo [31]. Surprisingly, live imaging experiments performed in intact larval brains or dissociated larval neuroblasts did not suggest a requirement for an extracellular signal to orient the mitotic spindle along a 'global' tissue axis [29]. Nevertheless, individual larval neuroblasts repetitively divide along the same axis with minor deviations, ensuring that sibling cells are positioned on the basal side of the neuroblast exclusively [26,32] (Figure 1C).

SWITCHING THE DIVISION AXIS TO GENERATE CELLULAR DIVERSITY

Cell division orientation can be switched during development. The switch from symmetric, proliferative divisions towards asymmetric, diversifying ones occurs in several different cell types.

Mammalian skin epidermis is a stratified epithelium. Stratification occurs through a change in the division axis. Before that, proliferative basal cells predominantly divide within the plane of the epithelium. In mice, around embryonic day 15 basal cells change their division orientation and start dividing perpendicular to the underlying basement membrane. This orientation will place one sibling, the suprabasal cell, into deeper layers of the epithelium and away from the underlying basement membrane (Figure 1D). Perpendicular divisions are also associated with a change in cell fate as the suprabasal cells express differentiation markers not seen in proliferative basal cells. Thus, the stratification of the mammalian epidermis consists of two phases: a proliferative, amplification phase in which symmetric divisions increase the surface area of the epithelium, followed by an asymmetric division phase generating distinct molecular identities [33–35]. Elegant lineage tracing experiments in mice further revealed that epidermal cells are not committed to one type of division but can change between symmetric and asymmetric divisions [35] (Figure 1D).

A similar situation is found in the vertebrate neuroepithelium. These columnar epithelial cells function as neural progenitors, extending from the apical to the basal surface of the cortex. In metaphase, they round up and divide at the apical side. Neuroepithelial cells initially expand their population during cortical development through symmetric divisions before they switch to an asymmetric division mode, giving rise to a self-renewing neural progenitor and a differentiating neuron [36–38]. Recent live imaging studies in mice revealed that the majority of divisions occur in the plane of the epithelium. However, randomization of the division plane in LGN mutant mice results in the loss of apically located neural progenitors [39]. In the chick neuroepithelium, it was found that randomizing spindle orientation does not affect cell fate but leads to a premature loss of neuroepithelial cells from the apical cortex; they become localized to the mantle zone where they proliferate aberrantly [40]. Interestingly, studies in mice further revealed that Lissencepaly1 (Lis1), its upstream regulator Magoh (a component of the exon junction complex (EJC)) and abnormal spindle-like microcephaly associated protein (ASPM) are all required to control spindle orientation in neuroepithelial stem cells [41–43].

The consequence of disrupted spindle orientation on cellular fate is controversial, as experimentally randomizing spindle orientation in neural progenitors yielded varying results. However, it is clear that neural progenitors tightly control the switch from proliferative to differentiating divisions.

MECHANISM OF CELL DIVISION ORIENTATION: EXTRACELLULAR CUES

Based on the diversity of cell types undergoing oriented cell division, the following immediate questions arise: what are the cellular and molecular determinants regulating correct cell division orientation, and how do these determinants differ between cell types and species?

Oriented cell division relies on the orientation of the mitotic spindle. Spindle orientation can be controlled through extrinsic factors, cell intrinsic molecules or physical constraints such as cell shape and cellular environment. During zebrafish gastrulation, cells of the epiblast give rise to the neural ectoderm on the dorsal side of the epidermis and the epidermis on the ventral side [10,44]. Blocking of the DEP-domain protein Dishevelled (Dsh) results in randomization of cell division orientation in all layers of the epiblast. In addition, the polarity of elongation of these cells is also affected [8]. The effect of Dsh knock-down is independent of convergent extension and acts cell-autonomously based on mosaic experiments. Dsh functions in multiple Wnt pathways such as the canonical Wnt/β-catenin pathway and the core planar cell polarity (PCP) pathway. Blocking canonical Wnt signaling did not affect oriented cell division. Conversely, inhibiting the two PCP ligands Wnt11 or Wnt5 partially affected oriented cell division, suggesting that they both act in parallel [8]. Furthermore, zebrafish lacking Fz7, the receptor for Wnt11 [45], and Stbm show disrupted oriented cell division [6,8]. Although these experiments demonstrated that PCP signaling is instrumental for oriented cell division, PCP-controlled oriented cell division has neither an instructive nor permissive role in body-axis elongation [6].

PCP signaling seems to be a general pathway to orient cell divisions within tissues and there are two separate but interconnected PCP pathways: the 'core PCP' pathway and the Fat/Dachsous (Ft/Ds) system, an overlapping local alignment pathway [46] (Figure 2A,B). The Ft/Ds system has been shown to be required for oriented cell division in the developing fly wing [16] (Figure 1B). Clones of cells lacking Dachsous or Fat protein show a rounded morphology as opposed to elongated wild-type clones; similar results have also been obtained if Dachsous or Fat are expressed ectopically. The cellular basis for this phenotype is a loss of oriented cell division [16,47]. Dachsous is expressed in a gradient along the proximal-distal axis in response to cues emanating from the compartment boundaries [48,49]. Dachsous is the ligand for Fat [50] and transduces the signal to the atypical myosin Dachs (D) (Figure 2B). Dachs is localized to the distal side of each cell's apical surface axis in response to the Dachsous gradient, corresponding to division

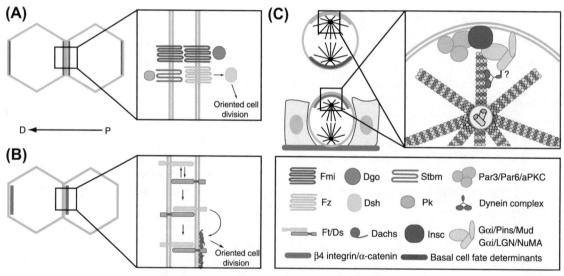

FIGURE 2 **Planar cell polarity pathways and cell intrinsic polarity pathway in oriented cell division.**
(A) The core planar cell polarity (PCP) pathway. The seven-pass transmembrane protein Frizzled (Fz) and the cytoplasmic proteins Dishevelled (Dsh) and Diego (Dgo) are localized to distal cell junctions. Strabismus (Stbm), another transmembrane protein, and the cytoplasmic protein Prickle (Pk) are localized to proximal cell junctions, while Flamingo (Fmi) is localized distally and proximally. The core PCP signaling pathway can be used for oriented cell division. (B) Fat/Dachsous system. Fat (Ft) and Dachsous (Ds) are large, atypical cadherins that interact heterophilically at cell junctions. A downstream effector of the Ft/Ds pathway, the atypical myosin Dachs, localizes to distal cell junctions and directs oriented cell division by influencing cell shape. (C) Pins/Mud/Gαi (in *Drosophila*), or LGN/NuMA/Gαi (in vertebrates) pathway. Asymmetric localization of Pins/LGN, Mud/NuMA, and Gαi directs spindle orientation in *Drosophila* neuroblasts and mammalian basal cells. The dynein complex, a minus-end directed microtubule motor protein complex, has been proposed to act as a force generator in both basal cells and neuroblasts. The Pins–Mud–Gαi, and LGN–NuMA–Gαi complexes, respectively, are connected to the Par3–Par6–aPKC apical polarity complex via Insc.

orientation in the developing fly wing [17,48] (Figures 1 and 2). Mutant *dachs* clones showed rounded morphology and randomized spindle orientation similar to *fat* or *dachsous* mutant clones. Furthermore, Dachsous is sufficient to change oriented cell division: reorienting Dachsous localization through genetically altering the direction of the Dachsous gradient changed Dachsous localization within cells and, as a consequence, altered the direction of oriented cell division [17]. These experiments strongly suggest that *Drosophila* organ shape is dependent on oriented cell division, regulated through the Ft/Ds PCP signaling system.

How does PCP signaling control the orientation of the mitotic spindle? Before we review literature addressing this question, we will summarize how cell intrinsic polarity is linked to spindle orientation and oriented cell division.

MECHANISM OF CELL DIVISION ORIENTATION: INTRINSIC CUES

Cell intrinsic polarity is a prominent feature of asymmetrically dividing cells (Figure 2C). The link between cell intrinsic polarity and spindle orientation, and thus oriented cell division, is best understood in invertebrate model systems. As this topic was the subject of a recent review [5], we will just provide a brief synopsis and then highlight similarities to vertebrate systems.

Genetic and molecular analyses in *C. elegans* zygotes revealed that cortical polarity is required for spindle positioning. In addition, laser severing experiments demonstrated that pulling forces on the posterior pole are larger than on the anterior, resulting in a displacement of the spindle towards the posterior pole [51–53]. This anaphase spindle positioning, as well as centration and rotation of the nucleus–centrosome complex, is controlled by the minus-end microtubule motor dynein (together with its components dynactin/Lis1). This molecular motor provides the pulling force and binds to cortically localized Lin-5 (nuclear mitotic apparatus (NuMA) in vertebrates). Lin-5 is found in a protein complex with the cortical proteins GPR1/2 (LGN/AGS3) and G-protein α-subunit (Gα) [54–57]. As the GPR1/2/Lin-5/Gα complex is asymmetrically localized, pulling forces differ between the anterior and posterior ends of the worm embryo, resulting in a displacement of the mitotic spindle [5,58,59].

Similarly, in *Drosophila* neuroblasts, lack of the NuMA orthologue *mushroom body defect* (*mud*) results in misaligned spindles in relation to the internal polarity axis; *mud* mutant neuroblasts are properly polarized. It was further shown that Mud is localized to the apical and basal cortex and to both centrosomes [60–62]. However, as Mud only co-localizes with Pins (LGN/AGS3 in vertebrates) on the apical cortex, the prevailing models propose that the apically localized Pins/Mud/Gαi complex is involved in neuroblast spindle orientation through interactions with dynein [60–62]. A physical connection between dynein and Mud has not been reported in *Drosophila* neuroblasts but genetic analysis revealed that mutations in the two dynein-complex proteins Lis1 and Glued result in spindle orientation defects similar to *mud* mutants [63]. Furthermore, dynein has been shown to physically interact with NuMA in *Xenopus* [64]. These results suggest that the *Drosophila* and *C. elegans* NuMA orthologues (Mud and Lin-5, respectively) are key effector proteins controlling the alignment of the mitotic spindle in relation to cell intrinsic polarity cues (Figure 2C) [5].

Recently, it was shown that a similar pathway is also used for spindle orientation in mammalian cells. In mitotic basal cells, NuMA has been found to be localized to spindle poles but also forms a polarized cortical crescent co-localizing with LGN, Inscuteable and Par3. This complex is localized opposite of integrins and the basement membrane. Furthermore, NuMA co-localizes with the p150glued subunit of the dynactin complex [34]. Neither NuMA nor LGN require microtubules for their localization to the cortex, since depolymerizing microtubules does not alter NuMA's cortical localization. Instead, the basement membrane component β1 integrin and α-catenin are required for LGN and NuMA localization [34]. LGN is also required for cortical NuMA localization but LGN does not depend on NuMA; the same relationship has been reported for the *Drosophila* orthologues Mud and Pins in neuroblasts (Figure 2C) [5,33,60,61,65].

How is the switch between symmetric and asymmetric spindle orientation controlled? Basal epidermal cells do not display predetermined spindle orientation before metaphase. During prometaphase, spindle orientation is still random but by metaphase the mitotic spindle aligns either perpendicularly or in parallel to the basement membrane, indicating that the spindle rotates into its final position [35]. Furthermore, in mice, LGN protein shows diffuse localization in interphase or symmetrically dividing basal cells but changes towards apical localization from embryonic day 15.5 onwards. Metaphase spindles are further aligned perpendicular to the LGN crescent [34]. Knockdown of NuMA, LGN or dynactin randomizes spindle orientation in basal cells and shifts the balance from asymmetric towards symmetric divisions [33,34]. These results suggest that the controlled asymmetric localization of LGN (together with NuMA and the dynein complex) determine spindle orientation and thus the switch from proliferative symmetric towards asymmetric basal cell division in the developing skin epidermis (Figures 1D and 2C). This interpretation is validated by findings from *Drosophila* neuroblasts, where lack of the NuMA orthologue Mud results in symmetric neuroblast divisions generating two neuroblasts as opposed to one neuroblast and a differentiating GMC [26]. However, single-cell analysis would be required in mammalian skin to truly demonstrate the causal relationship between spindle orientation and cell fate changes.

Vertebrate neuroepithelial cells also switch from symmetric to asymmetric division modes in order to create differentiating neurons while maintaining a self-renewed neural progenitor [37,66]. How is spindle orientation controlled in the neuroepithelium? Live imaging of chicken neuroepithelial cells revealed that mitotic spindles display a dynamic but stereotypic behavior: during early metaphase, spindles orient themselves parallel to the apical surface. This planar orientation is maintained during late metaphase and anaphase while the spindle is free to revolve randomly around the apical-basal axis [65]. Orienting the mitotic spindle into the plane parallel to the apical surface requires NuMA, LGN and Gαi. Both NuMA and LGN are excluded from the apical and basal cortex but form a cortical lateral belt in the plane parallel to the apical surface. This localization is independent of aPKC but depends on Gαi [40,65]. Randomization of spindle orientation furthermore results in an increase of ectopic progenitors [65]. Thus, an evolutionarily conserved pathway composed of LGN (Pins, GPR1/2), NuMA (Mud, Lin-5) and Gαi seems to play an important role in orienting the mitotic spindle in *Drosophila* neuroblasts, *C. elegans* early zygote, as well as vertebrate neuroepithelial cells and mammalian basal epidermal cells.

This is, however, not the only spindle orientation pathway connecting cell intrinsic polarity with the mitotic spindle. In *Drosophila* neuroblasts, it has been shown earlier that a parallel pathway, consisting of the PDZ protein Discs large (Dlg) and the Gakin orthologue kinesin heavy chain 73 (Khc73), is also involved in spindle orientation [67]. The current model suggests that phosphorylation of Pins through Aurora-A, together with the activation of Pins via Gαi [68], enables recruitment of Dlg to the cortex, where it could anchor microtubules via Khc73 to the apical cortex [69]. Mud, previously shown to bind to the TPR motif of Pins [60–62], could provide a force-generating complex through binding of dynein/dynactin [69]. It would be interesting to see whether these two pathways are also working together in other systems.

CONNECTING EXTRINSIC AND INTRINSIC POLARITY CUES TO CONTROL ORIENTED CELL DIVISION

A central question in oriented cell division concerns the relationship between the PCP and LGN (Pins)/NuMA (Mud)/Gαi spindle orientation pathways. Is there a molecular interaction between these two pathways, and if so, is it universal? An intuitive solution would be to link the PCP pathways (either the core PCP pathway and/or the Ft/Ds system) and the cell intrinsic LGN (Pins)/NuMA (Mud)/Gαi pathway through effector proteins, which could tether the mitotic spindle to the cortex. Alternatively, PCP effector proteins could directly affect the shape of cells, which is instrumental in orienting the mitotic spindle based on Hertwig's 'long axis rule'. Evidence for the former possibility was recently shown in zebrafish epiblast cells and *Drosophila* SOP cells [70]. However, data supporting the latter hypothesis have recently been provided in epithelial cells of the wing imaginal disc (see below) [17].

The SOP lineage in the peripheral *Drosophila* nervous system provides a beautiful example of how asymmetric cell division and oriented cell division cooperate during development (reviewed in [22,71,72]). The approximately 100 SOPs in the fly notum divide multiple times, each producing five cells which give rise to a mechanosensory organ (Figure 3A). The first two divisions in the SOP lineage, the division of the pI cell and one of its daughters, the pIIa cell, occurs strictly along the anterior-posterior axis within the plane

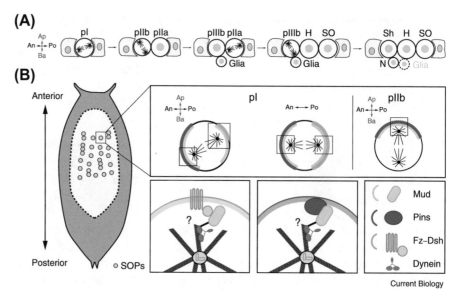

FIGURE 3 **Interplay between PCP and Mud/Pins pathway in the sensory organ precursor lineage in *Drosophila*.**

(A) In *Drosophila*, the multiple divisions of the sensory organ precursor (SOP) lineage gives rise to the hair (H), socket (SO), sheath (Sh), neuron (N) and glia which comprise an external sensory organ. Divisions occur along the anterior (An)-posterior (Po) axis (pI and pIIa) and the apical (Ap)-basal (Ba) axis (pIIb and pIIIb). (B) SOP cells are located on the pupal notum and are organized in a relatively regular pattern. In the pI cell the Pins–Mud complex is localized to the anterior cortex, whereas Fz–Dsh-Mud is bound to the posterior cortex. Note that Fz–Dsh is localized apically on the anterior cortex, co-localizing only partially with Mud. Both molecular complexes exert pulling forces on the mitotic spindle, resulting in a slightly tilted spindle. The Mud–Pins complex is localized apically in the pIIb cell. In the dividing pI cell Fz binds Dsh, which interacts with Mud at the posterior cortex. Mud presumably interacts directly with dynein to establish proper spindle orientation. In both the pI and pIIb cell, Mud interacts with anteriorly localized Pins, orienting the spindle along the apical-basal axis.

of the epithelium [73,74]. The two daughter cells produced by pIIa become the hair and socket of the sensory organ. The pIIb cell (the sibling of pIIa) undergoes two asymmetric divisions along the apical-basal axis, much like the division of the *Drosophila* neuroblast. These divisions result in a basal glia cell and an apical cell, known as pIIIb, followed by the production of an apical sheath cell and a basal neuron [74] (Figure 3A).

Division orientation of the pI cell is mediated by the core PCP pathway. In the pI cell, Fz is localized to the posterior pole, and *fz* mutant pupa lose their anterior-posterior planar polarity and divide with random orientation [75]. Downstream of Fz, Flamingo (Fmi; Celsr1-3 in mouse) is necessary for proper anterior localization of the Numb crescent and positioning of the division axis. Another downstream target of Fz is the Partner of Inscuteable–Discs Large (Pins–Dlg) complex. This complex localizes at the anterior cortex of the pI cell and works together with Fz to localize Bazooka (Baz; Par-3 in vertebrates) to the posterior pole [76]. As in neuroblasts, Pins has been shown to be required for the anterior localization of Mud in SOPs [70]. Interestingly, Mud also localizes to the posterior cortex, where it co-localizes with the key PCP component Dsh. Mud directly interacts with Dsh [70], which is recruited to the membrane via direct interaction with Fz [77]. Proper localization of Mud to both cell poles is necessary for proper anterior-posterior polarity and cell fate specification of the pIIa and pIIb cells [70]. Thus, in SOPs, the core PCP pathway is linked to the cell intrinsic Pins–Dlg–Mud pathway through Dsh, which thus provides a direct molecular connection between extracellular signaling and the mitotic spindle, presumably via dynein (Figure 3). The connection of the Fz–Dsh and Pins–Dlg–Mud pathways is evident in the slight tilt of the spindle of the pI cell. The apical posterior localization of Fz–Dsh pulls the posterior centrosome, and thus the spindle, away from a planar orientation. These data demonstrate how in the pI cell these two pathways work together to orient the spindle along the anterior-posterior and apical-basal axis, respectively (Figure 3B) [70].

Interestingly, the interaction between Dsh and Mud seems to be evolutionarily conserved, as it was shown that NuMA physically interacts with the DEP domain of Dsh in epiblast cells in zebrafish. This interaction is physiologically relevant: knockdown experiments revealed that NuMA affects the orientation of the mitotic spindle in epiblast cells [70].

PCP CONTROL OF CELL SHAPE AND CELL DIVISION ORIENTATION

In addition to the two PCP pathways and the LGN–NuMA–Gαi pathway, physical constraints can affect spindle orientation and oriented cell division. The existence of a connection between cellular shape and division axis has been evident to cell biologists since the 1880s. Cells may take on a wide variety of shapes during development, and in many systems cell shape determines the orientation of the division plane. This correlation is described by Hertwig's 'long axis rule': "*The two poles of the division figure come to lie in the direction of the greatest protoplasmic mass*" [1] (Figure 4A). In the developing *Xenopus* embryo, cells follow the long axis rule until the late blastula and cells isolated from a blastomere will divide along an experimentally induced long axis [3,14]. Similarly, experimental manipulation of cell shape in early mouse embryos influences the division axis [78] (Figure 4A).

In a recent experiment by Minc and colleagues [79], microfabricated wells were used to manipulate sea urchin eggs into several defined shapes, such as stars, ovals, squares and rectangles (Figure 4B). The cells were found to

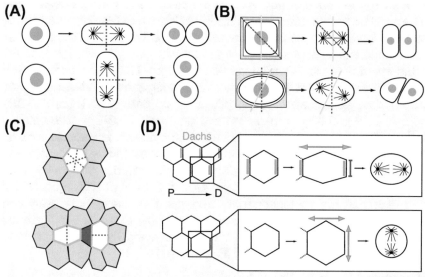

Current Biology

FIGURE 4 Cell shape and division orientation are linked.
(A) The cleavage plane (red dashed line) is oriented perpendicular to the cellular long axis, in accordance with Hertwig's long axis rule. (B) Sea urchin eggs forced into squares or elliptical chambers do not divide according to Hertwig's rule. The position of the actual cleavage plane (yellow line) is shifted from the predicted position (red dashed line). (C) Cells in a hexagonal cell sheet may divide along a number of long axes present in each cell. Addition of a non-hexagonal cell in the cell sheet induces a single long axis in neighboring cells, dictating their division orientation. (D) Planar polarized localization of Dachs to the apical distal membrane in the developing *Drosophila* wing disc constricts the proximal and distal cell membranes, forcing the cell to grow along the proximal-distal (P-D) axis. This P-D elongation dictates the position of the cell's cleavage plane and the direction of tissue growth. In *dachs* mutants cell division orientation is randomized; cell expansion is not confined to the P-D axis.

divide with normal timing, indicating that general physiology was relatively unaffected. Many cell shapes were found to undergo divisions that followed Hertwig's rule, but there were exceptions. The division axis did not follow the long diagonal axis in rectangular cells but instead formed along the largest axis of symmetry. The long axis rule also did not hold for ellipses of small aspect ratio or squares [79] (Figure 4B). When the urchin cells were forced into their new shapes, the nucleus repositioned to the new center of mass and elongated in respect to the future spindle axis. Nuclear centering and elongation are dependent on microtubules but not filamentous actin, and the cell most likely determines the position and orientation of the division axis by 'measuring' the length of microtubules.

In animals, cells are usually embedded in tissues and the shape of a cell is influenced by that of its neighbors. Monolayer cell sheets tend to be composed of three main cell shapes, hexagons, pentagons, and heptagons [80]. Mathematical modeling revealed that manipulating the shape of a cell's neighbor can influence the positioning of its division plane. For example, inserting a non-hexagonal neighbor into a sheet of hexagonal cells will cause the induction of a long axis in a neighboring hexagonal cell. Live imaging studies and mathematical modeling suggested that the presence of a long axis during interphase can influence spindle orientation during mitosis in *Drosophila* wing discs [81] (Figure 4C).

Exceptions to the long axis rule can also be found under normal physiological conditions [8,82]. Nevertheless, cell shape seems to have a strong influence on cell division orientation in certain cell types. Recently, an interesting study connected PCP signaling with the control of cell shape [17]. As mentioned earlier, the *Drosophila* wing epithelium elongates based on oriented cell division orientation along the proximal-distal axis during development (Figure 1B). This elongation is accomplished through the Ft/Ds system and its effector protein Dachs. How is Dachs transducing the positional information provided by Dachsous? Dachs might tether the mitotic spindle to the cortex. Alternatively, Dachs, an atypical myosin, could be controlling cell shape. Analysis of *dachs* mutant cells showed that Dachs exerts a contractile force on apical cell junctions, controlling the cell's shape. Planar polarized Dachs increases tension in the distal cell junction and the proximal junction of that cell's neighbor, forcing the cells to grow along to the proximal-distal axis. This elongation influences the positioning of the division axis, since these cells follow the long axis rule [17]. These data provide a conceptual framework of how positional information provided by PCP signaling is translated into cell shape changes, which influence the orientation of the mitotic spindle, resulting in oriented cell division (Figure 4D). However, mechanistic insight into how cell shape influences spindle orientation is still lacking.

NUMA- AND PCP-INDEPENDENT SPINDLE ORIENTATION PATHWAYS

The evolutionary conserved LGN–NuMA–Gαi and core PCP pathways are utilized repeatedly to orient the mitotic spindle throughout metazoan development. However, recent results in zebrafish revealed that other pathways might also be important for oriented cell division and spindle orientation [11]. As described earlier, in the developing zebrafish neural keel, the mitotic spindle rotates 90 degrees and aligns itself perpendicular to the midline (Figure 1A) [9,11]. Knocking down Vangl2, Wnt11, Wnt5 or Dsh2 does not affect the correct orientation of the mitotic spindle, suggesting that the core PCP pathway is not required for spindle orientation in the neural keel [6,11]. However, removal of the Wnt receptor Fz7 affected the stereotypical division orientation of neural progenitors [6]. Furthermore, scrib mutant zebrafish embryos display a randomization of spindle orientation. The Scrib phenotype is not a consequence of disrupted polarity, as the general organization of the apical cell cortex seemed to be unperturbed and knockdown of Par-6 or aPKC does not compromise spindle orientation [11]. Thus, Scribble-dependent spindle orientation does not act through the core PCP pathway and is distinct form Scribble's role in establishing and maintaining apical-basal polarity [11].

Could Fz7 and Scrib constitute a new cell division orientation pathway working independently of the core PCP pathway? More experiments are needed to resolve this issue. In Drosophila SOP cells [83], Drosophila male germline stem cells [84] and basal cells of the developing epidermis in mice [34], cadherin-based cell–cell contacts play important roles in spindle orientation. Could a similar mechanism be used in the zebrafish neural keel? Zigman and colleagues [11] studied the localization of α-catenin in wild-type fish and found that foci are localized on the cortex and enriched at the presumptive cleavage plane (Figure 1A). Furthermore, knockdown of Scribble resulted in a decrease of cortical α-catenin levels. N-cadherin mutants also display a decrease in cortical α-catenin levels and showed a scribble-like spindle orientation phenotype [11]. Thus, Scribble-dependent assembly of cell–cell adhesion complexes plays an important role in division orientation in the developing neural keel.

DEVELOPMENTAL CONSEQUENCES OF ALTERED CELL DIVISION ORIENTATION

Defective cell division orientation can result in tissue architecture defects, cell fate misspecification and cancer. In the colon and small intestine of mice and humans, stem cells orient their spindles preferentially perpendicularly to the apical surface [85]. In order to see whether oriented cell division is changed in cancer, spindle orientation was measured in the stem cell compartment of *Apc* mutant mice. Mutations in *Apc* are responsible for familial adenomatous polyposis (FAP) and are the most prevalent initiating mutations in colorectal cancers [86]. In contrast to wild-type tissue, there was no spindle orientation bias in *Apc* mutant mice or FAP human intestine; spindle orientation was significantly different between mutant and wild-type tissue [85]. Genetic analysis of one of the two *Drosophila* Apc orthologues, Apc2, suggested that Apc2 could anchor astral microtubules to the cortex [87]. However, other possible mechanisms could account for the *Apc* spindle orientation phenotype. In the mammalian gut, it was shown that *Apc* mutant cells still contain astral microtubules, although they do not fully attach to the cortex [88]. *Apc* mutant stem cells also displayed a widening of the basal region, which altered the shape of these long columnar epithelial cells [85,88]. Thus, changes in cell shape and loss of cortical contact of astral MTs could contribute to a change in spindle orientation. Clearly, more mechanistic data are required to fully understand the role of Apc in spindle orientation and its role in cancer. Furthermore, although spindle misorientation may contribute to tumor formation, not all tumors display spindle orientation defects [86].

Autosomal recessive primary microcephaly is another disease that is, at least in part, associated with defective spindle orientation and cell division orientation. Microcephaly is manifested in the occurrence of small, but structurally normal, brains and mild-to-moderate mental retardation [89,90] (Figure 5). There are at least eight microcephaly loci and five of the affected genes have been cloned [90]. Interestingly, at least three of these genes have proposed roles in spindle orientation and oriented cell division based on molecular genetic analysis of homologues in model systems [90]. CDK5RAP2 (Centrosomin (Cnn) in *Drosophila*), Abnormal spindle-like microcephaly associated protein (ASPM; abnormal spindle (Asp) in *Drosophila*; ASPM-1 in *C. elegans*) and CenpJ (Sas4 in *Drosophila*) are localized to centrosomes and mutations in *cnn*, *Sas4* and *aspm-1* have been associated with spindle orientation defects *in Drosophila*, *C. elegans* and mouse. Loss of Cnn and Sas4 in flies has been directly shown to affect centrioles and centrosomes and also manifests in a lack of astral microtubules [43,63,87,91]. Mutations in these genes uncouple the mitotic spindle from the cortex, resulting in an increase in the stem cell pool, as was shown with both neuroblasts and male germline stem cells [26,87]. Loss of ASPM in *C. elegans* resulted in meiotic spindle orientation defects [92]. In mice, ASPM has been shown to regulate cell division orientation; lack of ASPM resulted in an increase of asymmetric divisions and a loss of neuroepithelial cells abutting the ventricular zone [43]. Thus, a larger proportion of neuroepithelial cell progeny is found in the neuronal layer, associated with a concomitant loss of neuroepithelial cells in the ventricular zone [43]. Microcephaly could thus be caused through spindle orientation defects resulting in a premature shift from symmetric, amplifying divisions towards premature, asymmetric, neurogenic divisions. As a consequence, the neural stem cell pool is reduced. As neurons exit the cell cycle and proliferating neural progenitors are successively lost, microcephalic brains do not contain the same amount of cells as their wild-type counterparts manifested in the small brain phenotype (Figure 5) [90].

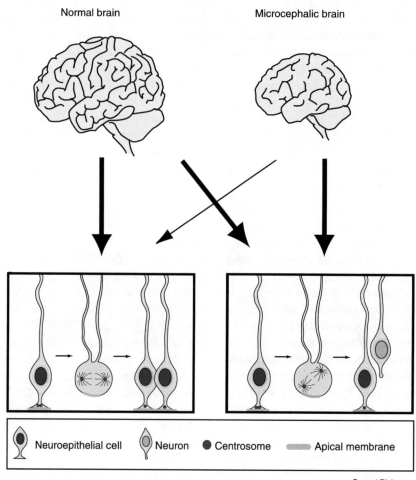

Current Biology

FIGURE 5 Improper cell division orientation contributes to the development of microcephaly. Normal brains contain symmetrically (left panel) and asymmetrically (right panel) dividing progenitors, resulting in the amplification of the progenitor pool but also the production of differentiating siblings. Asymmetric division of these cells produces one neuroepithelial progenitor and one cell which detaches from the apical membrane and becomes a basal progenitor or a neuron. In the microcephalic brain, randomized spindle rotation could result in a premature shift towards asymmetric divisions resulting in a decrease of the neuroepithelial progenitor pool, and limited production of neurons. This leads to a decrease in overall cell number and thus a reduction in brain size.

Misregulated spindle orientation and oriented cell division can also lead to cell differentiation defects in the mammalian epidermis [33]. Knockdown of LGN, NuMA or p150glued resulted in a thinner epidermis with impaired barrier function leading to dehydration and the death of genetically manipulated mice. Careful analysis revealed that spindle orientation defects resulted in an increase in symmetric basal cell divisions, preferentially generating basal progenitors as opposed to wild-type asymmetric divisions, which give rise to one basal progenitor and a suprabasal differentiating cell. In agreement with this observation, fewer suprabasal cells and a reduction in differentiation marker expression were observed [33]. Thus, morphogenesis of the mammalian skin depends on controlled oriented cell division to generate enough differentiating suprabasal cells while maintaining a pool of basal progenitors.

Taken together, these observations suggest that spindle orientation defects can result in altered cell division orientation and cell fate misspecification. Depending on the cellular context, this can lead to tumor initiation and tissue organization defects with impaired function.

CONCLUSION

Oriented cell division in animals has been most intensely studied in systems allowing powerful genetic analyses combined with live imaging. Lately, molecular connections between extracellular signaling pathways and spindle orientation have emerged. Over the next years, it will be interesting to gain more mechanistic insight into how the core PCP signaling pathway and the Ft/Ds system are connected with intrinsic cell polarity but also how this extracellular signaling pathway is translated into cell shape changes. Furthermore, how exactly cell shape controls spindle orientation remains to be seen. As oriented cell division is such a recurring theme during animal development, it will be interesting to see how diverse the mechanism and molecular players are in animals.

REFERENCES

1 Hertwig, O. (1884). Das Problem der Befruchtung und der Isotropie des Eies. Eine Theorie der Vererbung. Jenaische Zeitschrift für Naturwissenschaft *18*, 276–318.

2 Honda, H. (1983). Geometrical models for cells in tissues. Int. Rev. Cytol. *81*, 191–248.

3 Strauss, B., Adams, R.J., and Papalopulu, N. (2006). A default mechanism of spindle orientation based on cell shape is sufficient to generate cell fate diversity in polarised Xenopus blastomeres. Development *133*, 3883–3893.

4 Segalen, M., and Bellaiche, Y. (2009). Cell division orientation and planar cell polarity pathways. Semin. Cell Dev. Biol. *20*, 972–977.

5 Siller, K.H., and Doe, C.Q. (2009). Spindle orientation during asymmetric cell division. Nat. Cell Biol. *11*, 365–374.

6 Quesada-Hernandez, E., Caneparo, L., Schneider, S., Winkler, S., Liebling, M., Fraser, S.E., and Heisenberg, C.P. (2010). Stereotypical cell division orientation controls neural rod midline formation in zebrafish. Curr. Biol. *20*, 1966–1972.

7   Concha, M.L., and Adams, R.J. (1998). Oriented cell divisions and cellular morphogenesis in the zebrafish gastrula and neurula: a time-lapse analysis. Development 125, 983–994.

8   Gong, Y., Mo, C., and Fraser, S.E. (2004). Planar cell polarity signalling controls cell division orientation during zebrafish gastrulation. Nature 430, 689–693.

9   Geldmacher-Voss, B., Reugels, A.M., Pauls, S., and Campos-Ortega, J.A. (2003). A 90-degree rotation of the mitotic spindle changes the orientation of mitoses of zebrafish neuroepithelial cells. Development 130, 3767–3780.

10 Kimmel, C.B., Warga, R.M., and Kane, D.A. (1994). Cell cycles and clonal strings during formation of the zebrafish central nervous system. Development 120, 265–276.

11 Zigman, M., Trinh le, A., Fraser, S.E., and Moens, C.B. (2011). Zebrafish neural tube morphogenesis requires Scribble-dependent oriented cell divisions. Curr. Biol. 21, 79–86.

12 Ciruna, B., Jenny, A., Lee, D., Mlodzik, M., and Schier, A.F. (2006). Planar cell polarity signalling couples cell division and morphogenesis during neurulation. Nature 439, 220–224.

13 Tawk, M., Araya, C., Lyons, D.A., Reugels, A.M., Girdler, G.C., Bayley, P.R., Hyde, D.R., Tada, M., and Clarke, J.D. (2007). A mirror-symmetric cell division that orchestrates neuroepithelial morphogenesis. Nature 446, 797–800.

14 Chalmers, A.D., Strauss, B., and Papalopulu, N. (2003). Oriented cell divisions asymmetrically segregate aPKC and generate cell fate diversity in the early Xenopus embryo. Development 130, 2657–2668.

15 da Silva, S.M., and Vincent, J.P. (2007). Oriented cell divisions in the extending germband of Drosophila. Development 134, 3049–3054.

16 Baena-Lopez, L.A., Baonza, A., and Garcia-Bellido, A. (2005). The orientation of cell divisions determines the shape of Drosophila organs. Curr. Biol. 15, 1640–1644.

17 Mao, Y., Tournier, A.L., Bates, P.A., Gale, J.E., Tapon, N., and Thompson, B.J. (2011). Planar polarization of the atypical myosin Dachs orients cell divisions in Drosophila. Genes Dev. 25, 131–136.

18 Knoblich, J.A. (2010). Asymmetric cell division: recent developments and their implications for tumour biology. Nat. Rev. Mol. Cell Biol. 11, 849–860.

19 Gonczy, P. (2008). Mechanisms of asymmetric cell division: flies and worms pave the way. Nature reviews. Mol. Cell Biol. 9, 355–366.

20 Hyenne, V., Chartier, N.T., and Labbe, J.C. (2010). Understanding the role of asymmetric cell division in cancer using C. elegans. Dev. Dyn. 239, 1378–1387.

21 Munro, E., and Bowerman, B. (2009). Cellular symmetry breaking during Caenorhabditis elegans development. Cold Spring Harb. Perspect. Biol. 1, a003400.

22 Furman, D.P., and Bukharina, T.A. (2011). Drosophila mechanoreceptors as a model for studying asymmetric cell division. Int. J. Dev. Biol. 55, 133–141.

23 Galli, M., and van den Heuvel, S. (2008). Determination of the cleavage plane in early C. elegans embryos. Annu. Rev. Genet. 42, 389–411.

24 Schubert, C.M., Lin, R., de Vries, C.J., Plasterk, R.H., and Priess, J.R. (2000). MEX-5 and MEX-6 function to establish soma/germline asymmetry in early C. elegans embryos. Mol. Cell 5, 671–682.

25 Doe, C.Q. (2008). Neural stem cells: balancing self-renewal with differentiation. Development 135, 1575–1587.

26 Cabernard, C., and Doe, C.Q. (2009). Apical/basal spindle orientation is required for neuroblast homeostasis and neuronal differentiation in Drosophila. Dev. Cell 17, 134–141.

27 Kaltschmidt, J.A., Davidson, C.M., Brown, N.H., and Brand, A.H. (2000). Rotation and asymmetry of the mitotic spindle direct asymmetric cell division in the developing central nervous system. Nat. Cell Biol. 2, 7–12.

28 Rebollo, E., Roldan, M., and Gonzalez, C. (2009). Spindle alignment is achieved without rotation after the first cell cycle in Drosophila embryonic neuroblasts. Development *136*, 3393–3397.

29 Rebollo, E., Sampaio, P., Januschke, J., Llamazares, S., Varmark, H., and Gonzalez, C. (2007). Functionally unequal centrosomes drive spindle orientation in asymmetrically dividing Drosophila neural stem cells. Dev. Cell *12*, 467–474.

30 Rusan, N.M., and Peifer, M. (2007). A role for a novel centrosome cycle in asymmetric cell division. J. Cell Biol. *177*, 13–20.

31 Siegrist, S.E., and Doe, C.Q. (2006). Extrinsic cues orient the cell division axis in Drosophila embryonic neuroblasts. Development *133*, 529–536.

32 Januschke, J., and Gonzalez, C. (2010). The interphase microtubule aster is a determinant of asymmetric division orientation in Drosophila neuroblasts. J. Cell Biol. *188*, 693–706.

33 Williams, S.E., Beronja, S., Pasolli, H.A., and Fuchs, E. (2011). Asymmetric cell divisions promote Notch-dependent epidermal differentiation. Nature *470*, 353–358.

34 Lechler, T., and Fuchs, E. (2005). Asymmetric cell divisions promote stratification and differentiation of mammalian skin. Nature *437*, 275–280.

35 Poulson, N.D., and Lechler, T. (2010). Robust control of mitotic spindle orientation in the developing epidermis. J. Cell Biol. *191*, 915–922.

36 Farkas, L.M., and Huttner, W.B. (2008). The cell biology of neural stem and progenitor cells and its significance for their proliferation versus differentiation during mammalian brain development. Curr. Opin. Cell Biol. *20*, 707–715.

37 Gotz, M., and Huttner, W.B. (2005). The cell biology of neurogenesis. Nat. Rev. Mol. Cell Biol. *6*, 777–788.

38 Shioi, G., Konno, D., Shitamukai, A., and Matsuzaki, F. (2009). Structural basis for self-renewal of neural progenitors in cortical neurogenesis. Cerebral Cortex *19* (*Suppl 1*), i55–61.

39 Konno, D., Shioi, G., Shitamukai, A., Mori, A., Kiyonari, H., Miyata, T., and Matsuzaki, F. (2008). Neuroepithelial progenitors undergo LGN-dependent planar divisions to maintain self-renewability during mammalian neurogenesis. Nat. Cell Biol. *10*, 93–101.

40 Morin, X., Jaouen, F., and Durbec, P. (2007). Control of planar divisions by the G-protein regulator LGN maintains progenitors in the chick neuroepithelium. Nat. Neurosci. *10*, 1440–1448.

41 Silver, D.L., Watkins-Chow, D.E., Schreck, K.C., Pierfelice, T.J., Larson, D.M., Burnetti, A.J., Liaw, H.J., Myung, K., Walsh, C.A., Gaiano, N., et al. (2010). The exon junction complex component Magoh controls brain size by regulating neural stem cell division. Nat. Neurosci. *13*, 551–558.

42 Yingling, J., Youn, Y.H., Darling, D., Toyo-Oka, K., Pramparo, T., Hirotsune, S., and Wynshaw-Boris, A. (2008). Neuroepithelial stem cell proliferation requires LIS1 for precise spindle orientation and symmetric division. Cell *132*, 474–486.

43 Fish, J.L., Kosodo, Y., Enard, W., Paabo, S., and Huttner, W.B. (2006). Aspm specifically maintains symmetric proliferative divisions of neuroepithelial cells. Proc. Natl. Acad. Sci. USA *103*, 10438–10443.

44 Kimmel, C.B., Ballard, W.W., Kimmel, S.R., Ullmann, B., and Schilling, T.F. (1995). Stages of embryonic development of the zebrafish. Dev. Dyn. *203*, 253–310.

45 Witzel, S., Zimyanin, V., Carreira-Barbosa, F., Tada, M., and Heisenberg, C.P. (2006). Wnt11 controls cell contact persistence by local accumulation of Frizzled 7 at the plasma membrane. J. Cell Biol. *175*, 791–802.

46 Goodrich, L.V., and Strutt, D. (2011). Principles of planar polarity in animal development. Development *138*, 1877–1892.

47 Li, W., Kale, A., and Baker, N.E. (2009). Oriented cell division as a response to cell death and cell competition. Curr. Biol. *19*, 1821–1826.

48 Rogulja, D., Rauskolb, C., and Irvine, K.D. (2008). Morphogen control of wing growth through the Fat signaling pathway. Dev. Cell *15*, 309–321.

49 Simon, M.A. (2004). Planar cell polarity in the Drosophila eye is directed by graded Four-jointed and Dachsous expression. Development *131*, 6175–6184.

50 Matakatsu, H., and Blair, S.S. (2004). Interactions between Fat and Dachsous and the regulation of planar cell polarity in the Drosophila wing. Development *131*, 3785–3794.

51 Grill, S.W., Gonczy, P., Stelzer, E.H., and Hyman, A.A. (2001). Polarity controls forces governing asymmetric spindle positioning in the Caenorhabditis elegans embryo. Nature *409*, 630–633.

52 Cheng, N.N., Kirby, C.M., and Kemphues, K.J. (1995). Control of cleavage spindle orientation in Caenorhabditis elegans: the role of the genes par-2 and par-3. Genetics *139*, 549–559.

53 Tsou, M.F., Hayashi, A., DeBella, L.R., McGrath, G., and Rose, L.S. (2002). LET-99 determines spindle position and is asymmetrically enriched in response to PAR polarity cues in C. elegans embryos. Development *129*, 4469–4481.

54 Couwenbergs, C., Labbe, J.C., Goulding, M., Marty, T., Bowerman, B., and Gotta, M. (2007). Heterotrimeric G protein signaling functions with dynein to promote spindle positioning in C. elegans. J. Cell Biol. *179*, 15–22.

55 Nguyen-Ngoc, T., Afshar, K., and Gonczy, P. (2007). Coupling of cortical dynein and G alpha proteins mediates spindle positioning in Caenorhabditis elegans. Nat. Cell Biol. *9*, 1294–1302.

56 Park, D.H., and Rose, L.S. (2008). Dynamic localization of LIN-5 and GPR-1/2 to cortical force generation domains during spindle positioning. Dev. Biol. *315*, 42–54.

57 Schmidt, D.J., Rose, D.J., Saxton, W.M., and Strome, S. (2005). Functional analysis of cytoplasmic dynein heavy chain in Caenorhabditis elegans with fast-acting temperature-sensitive mutations. Mol. Biol. Cell *16*, 1200–1212.

58 Pecreaux, J., Roper, J.C., Kruse, K., Julicher, F., Hyman, A.A., Grill, S.W., and Howard, J. (2006). Spindle oscillations during asymmetric cell division require a threshold number of active cortical force generators. Curr. Biol. *16*, 2111–2122.

59 Grill, S.W., Howard, J., Schaffer, E., Stelzer, E.H., and Hyman, A.A. (2003). The distribution of active force generators controls mitotic spindle position. Science *301*, 518–521.

60 Bowman, S.K., Neumuller, R.A., Novatchkova, M., Du, Q., and Knoblich, J.A. (2006). The Drosophila NuMA Homolog Mud regulates spindle orientation in asymmetric cell division. Dev. Cell *10*, 731–742.

61 Izumi, Y., Ohta, N., Hisata, K., Raabe, T., and Matsuzaki, F. (2006). Drosophila Pins-binding protein Mud regulates spindle-polarity coupling and centrosome organization. Nat. Cell Biol. *8*, 586–593.

62 Siller, K.H., Cabernard, C., and Doe, C.Q. (2006). The NuMA-related Mud protein binds Pins and regulates spindle orientation in Drosophila neuroblasts. Nat. Cell Biol. *8*, 594–600.

63 Siller, K.H., and Doe, C.Q. (2008). Lis1/dynactin regulates metaphase spindle orientation in Drosophila neuroblasts. Dev. Biol. *319*, 1–9.

64 Merdes, A., Ramyar, K., Vechio, J.D., and Cleveland, D.W. (1996). A complex of NuMA and cytoplasmic dynein is essential for mitotic spindle assembly. Cell *87*, 447–458.

65 Peyre, E., Jaouen, F., Saadaoui, M., Haren, L., Merdes, A., Durbec, P., and Morin, X. (2011). A lateral belt of cortical LGN and NuMA guides mitotic spindle movements and planar division in neuroepithelial cells. J. Cell Biol. *193*, 141–154.

66 Huttner, W.B., and Kosodo, Y. (2005). Symmetric versus asymmetric cell division during neurogenesis in the developing vertebrate central nervous system. Curr. Opin. Cell. Biol. *17*, 648–657.

67 Siegrist, S.E., and Doe, C.Q. (2005). Microtubule-induced Pins/Galphai cortical polarity in Drosophila neuroblasts. Cell *123*, 1323–1335.

68 Nipper, R.W., Siller, K.H., Smith, N.R., Doe, C.Q., and Prehoda, K.E. (2007). Galphai generates multiple Pins activation states to link cortical polarity and spindle orientation in Drosophila neuroblasts. Proc. Natl. Acad. Sci. USA *104*, 14306–14311.

69 Johnston, C.A., Hirono, K., Prehoda, K.E., and Doe, C.Q. (2009). Identification of an Aurora-A/PinsLINKER/Dlg spindle orientation pathway using induced cell polarity in S2 cells. Cell *138*, 1150–1163.

70 Segalen, M., Johnston, C.A., Martin, C.A., Dumortier, J.G., Prehoda, K.E., David, N.B., Doe, C.Q., and Bellaiche, Y. (2010). The Fz-Dsh planar cell polarity pathway induces oriented cell division via Mud/NuMA in Drosophila and zebrafish. Dev. Cell *19*, 740–752.

71 Wang, H., and Chia, W. (2005). Drosophila neural progenitor polarity and asymmetric division. Biol. Cell *97*, 63–74.

72 Knoblich, J.A. (2008). Mechanisms of asymmetric stem cell division. Cell *132*, 583–597.

73 Gho, M., and Schweisguth, F. (1998). Frizzled signalling controls orientation of asymmetric sense organ precursor cell divisions in Drosophila. Nature *393*, 178–181.

74 Roegiers, F., Younger-Shepherd, S., Jan, L.Y., and Jan, Y.N. (2001). Two types of asymmetric divisions in the Drosophila sensory organ precursor cell lineage. Nat. Cell Biol. *3*, 58–67.

75 Bellaiche, Y., Gho, M., Kaltschmidt, J.A., Brand, A.H., and Schweisguth, F. (2001). Frizzled regulates localization of cell-fate determinants and mitotic spindle rotation during asymmetric cell division. Nat. Cell Biol. *3*, 50–57.

76 Roegiers, F., Younger-Shepherd, S., Jan, L.Y., and Jan, Y.N. (2001). Bazooka is required for localization of determinants and controlling proliferation in the sensory organ precursor cell lineage in Drosophila. Proc. Natl. Acad. Sci. USA *98*, 14469–14474.

77 Gao, C., and Chen, Y.G. (2010). Dishevelled: The hub of Wnt signaling. Cell Signal *22*, 717–727.

78 Gray, D., Plusa, B., Piotrowska, K., Na, J., Tom, B., Glover, D.M., and Zernicka-Goetz, M. (2004). First cleavage of the mouse embryo responds to change in egg shape at fertilization. Curr. Biol. *14*, 397–405.

79 Minc, N., Burgess, D., and Chang, F. (2011). Influence of cell geometry on division-plane positioning. Cell *144*, 414–426.

80 Gibson, M.C., Patel, A.B., Nagpal, R., and Perrimon, N. (2006). The emergence of geometric order in proliferating metazoan epithelia. Nature *442*, 1038–1041.

81 Gibson, W.T., Veldhuis, J.H., Rubinstein, B., Cartwright, H.N., Perrimon, N., Brodland, G.W., Nagpal, R., and Gibson, M.C. (2011). Control of the mitotic cleavage plane by local epithelial topology. Cell *144*, 427–438.

82 Thery, M., and Bornens, M. (2006). Cell shape and cell division. Curr. Opin. Cell Biol. *18*, 648–657.

83 Le Borgne, R., Bellaiche, Y., and Schweisguth, F. (2002). Drosophila E-cadherin regulates the orientation of asymmetric cell division in the sensory organ lineage. Curr. Biol. *12*, 95–104.

84 Inaba, M., Yuan, H., Salzmann, V., Fuller, M.T., and Yamashita, Y.M. (2010). E-cadherin is required for centrosome and spindle orientation in Drosophila male germline stem cells. PLoS One *5*, e12473.

85 Quyn, A.J., Appleton, P.L., Carey, F.A., Steele, R.J., Barker, N., Clevers, H., Ridgway, R.A., Sansom, O.J., and Nathke, I.S. (2010). Spindle orientation bias in gut epithelial stem cell compartments is lost in precancerous tissue. Cell Stem Cell 6, 175–181.

86 Pease, J.C., and Tirnauer, J.S. (2011). Mitotic spindle misorientation in cancer - out of alignment and into the fire. J. Cell Sci. 124, 1007–1016.

87 Yamashita, Y.M., Jones, D.L., and Fuller, M.T. (2003). Orientation of asymmetric stem cell division by the APC tumor suppressor and centrosome. Science 301, 1547–1550.

88 Fleming, E.S., Temchin, M., Wu, Q., Maggio-Price, L., and Tirnauer, J.S. (2009). Spindle misorientation in tumors from APC(min/+) mice. Mol. Carcinog. 48, 592–598.

89 Mochida, G.H. (2009). Genetics and biology of microcephaly and lissencephaly. Semin. Pediatr. Neurol. 16, 120–126.

90 Thornton, G.K., and Woods, C.G. (2009). Primary microcephaly: do all roads lead to Rome? Trends Genet. 25, 501–510.

91 Basto, R., Lau, J., Vinogradova, T., Gardiol, A., Woods, C.G., Khodjakov, A., and Raff, J.W. (2006). Flies without centrioles. Cell 125, 1375–1386.

92 van der Voet, M., Berends, C.W., Perreault, A., Nguyen-Ngoc, T., Gonczy, P., Vidal, M., Boxem, M., and van den Heuvel, S. (2009). NuMA-related LIN-5, ASPM-1, calmodulin and dynein promote meiotic spindle rotation independently of cortical LIN-5/GPR/Galpha. Nat. Cell Biol. 11, 269–277.

CHAPTER 6

Trends in Cell Biology

Cell Polarity: Mechanochemical Patterning

Nathan W. Goehring[1,2,3,*], Stephan W. Grill[1,2,*]

[1]Max Planck Institute of Molecular Cell Biology and Genetics (MPI-CBG), Pfotenhauerstraße 108, 01307 Dresden, Germany, [2]Max Planck Institute for the Physics of Complex Systems (MPI-PKS), Nöthnitzer Straße 38, 01187 Dresden, Germany, [3]Cancer Research UK London Research Institute, Lincoln's Inn Fields Laboratories, 44 Lincoln's Inn Fields, London WC2A 3LY, UK

*Correspondence: nate.goehring@cancer.org.uk, grill@mpi-cbg.de

Trends in Cell Biology, Vol. 23, No. 2, February 2013 © 2013 Elsevier Inc.
http://dx.doi.org/10.1016/j.tcb.2012.10.009

SUMMARY

Nearly every cell type exhibits some form of polarity, yet the molecular mechanisms vary widely. Here we examine what we term 'chemical systems' where cell polarization arises through biochemical interactions in signaling pathways, 'mechanical systems' where cells polarize due to forces, stresses and transport, and 'mechanochemical systems' where polarization results from interplay between mechanics and chemical signaling. To reveal potentially unifying principles, we discuss mathematical conceptualizations of several prototypical examples. We suggest that the concept of local activation and global inhibition – originally developed to explain spatial patterning in reaction–diffusion systems – provides a framework for understanding many cases of cell polarity. Importantly, we find that the core ingredients in this framework – symmetry breaking, self-amplifying feedback, and long-range inhibition – involve processes that can be chemical, mechanical, or even mechanochemical in nature.

UBIQUITY AND UNIVERSALITY DURING CELL POLARIZATION

Polarity, the ability of a cell to define a geometric axis, is critical to a broad variety of cell functions, including cell migration, directional cell growth, and asymmetric cell division. It is likely to be an ancient property of cells, perhaps

115

having initially evolved to ensure differential inheritance at cell divisions, a role maintained to this day [1]. Polarity is also critical for cells in multicellular environments. In epithelial tissue, for example, apical–basal polarity drives the opposing surfaces of the cell to acquire distinct functions and molecular components [2], whereas planar cell polarity (PCP) aligns cells and cellular structures such as hairs and bristles within the epithelial plane [3]. Consistent with the fundamental importance of polarity in the organization of cells, the ability to polarize is ubiquitous in prokaryotes, plants, fungi, protozoa, and animals [4] and there are even reports of asymmetric cell divisions in *Archaea* [5].

The end point of cell polarity networks is to differentiate molecularly one side of the cell from the other. This molecular differentiation both defines a polarity axis and allows cellular processes to be regulated differentially along this axis. In this way, cells can target specific proteins to a growing tip (e.g., yeast) or to leading edge (e.g., crawling cells). They can set up intracellular signaling gradients or selectively segregate molecules, such as fate determinants, during cell division (e.g., stem cells).

Despite its ubiquity and potentially ancient appearance, the ability to polarize appears to have evolved repeatedly, yielding a diversity of seemingly unrelated molecular mechanisms. Some polarity pathways rely on the ability of interactions between diffusible species to drive pattern formation. We will refer to these as chemical systems because they can be understood primarily within the language of reaction rates and molecular diffusion [6]. We distinguish these from mechanical systems, in which polarization requires consideration of force, stress, and elastic and viscous properties of biological materials such as cytoskeletal elements and membranes. In many cases, polarity may be considered mechanochemical, requiring complex interplay between both chemical and mechanical networks [7,8].

How can we reconcile the universality of cell polarity as a concept in biology with such complexity and diversity of molecular mechanisms? Here we examine a broad range of cell polarity systems to highlight one basic paradigm for cell polarity based on coupling symmetry breaking to local signal amplification and long-range inhibition. This paradigm is a robust, general mechanism for inducing and maintaining cell polarity and can convert potentially weak, transient, or even stochastically arising signals into persistent cellular asymmetry. A central point of this review is to demonstrate that this conceptual framework applies equally well to mechanical, chemical, and even coupled mechanochemical polarity networks, thereby providing a shared framework for understanding a diverse range of polarization processes despite wide variation in their molecular mechanisms. We argue that in the current drive towards a systems level understanding of development, uncovering such key organizing principles of morphogenic processes will increasingly require consideration of both chemical and mechanical processes.

SYMMETRY BREAKING AND PATTERN FORMATION IN CHEMICAL SYSTEMS

The paradigm of pattern formation through local symmetry breaking, signal amplification, and long-range inhibition traces its roots to seminal work by Alan Turing [6]. It was he who first described how chemical patterns can emerge from a uniform starting condition through an instability that arises from stochastic fluctuations combined with interactions between chemical species (morphogens) that diffuse at different rates; hence, pattern formation by reaction–diffusion. Subsequent work by Gierer and Meinhardt codified the idea of competing short- and long-range feedback in the stabilization of patterns, including polarization [9]. Local activation typical takes the form of self-enhancing feedback, which amplifies local, potentially random increases in signaling activity (Figure 1a). On its own, such self-amplifying feedback will tend to convert more and more of a system to an 'on' state. This activation must therefore be coupled to an inhibitory process that is long range. Long-range inhibition is typically provided by a rapidly diffusing inhibitor produced by the activator (Figure 1a). The inhibitor spreads faster than the activator, allowing it to suppress the activator outside the immediate region of activation (Figure 1b). In classic Turing-like systems, the resulting spatial pattern is specified by differences in the diffusivities of the slower activating molecule and the faster inhibitor, leading to polarized (Figure 1b) or repeating patterns (Figure 1c). In fact, a diffusible inhibitor is not strictly required, because its role can be served by a freely diffusing component that is depleted as part of the signal amplification process (Figure 1d,e) [9,10]. The diffusivity of this component effectively specifies the extent of the region over which the component is depleted and thus signal amplification inhibited. As long as the region of inhibition extends beyond that of activation, the region of activation will be constrained.

Although initially stimulated by a need to understand spontaneous pattern formation, the basic principle of local activation and global inhibition underlies a broad range of models to describe chemically polarizing systems (by which we mean polarization in biochemical signaling pathways, in contrast to mechanical; see below) [10]. Indeed, a system that can amplify infinitesimally small fluctuations should also be able to amplify signals emerging from dedicated spatial cues. At the same time, by suppressing amplification everywhere except at the initial location of the cue, the combination of local amplification and global inhibition can transform a weak or noisy chemical

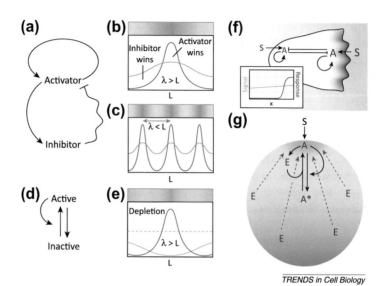

TRENDS in Cell Biology

FIGURE 1 Pattern formation in chemical systems.
(a) In an activator–inhibitor system, a slow-diffusing activator stimulates production of both itself via autocatalysis and a fast-diffusing inhibitor. **(b)** Because the inhibitor spreads faster than the activator, it allows for activation in the central zone, but suppresses activation beyond. The diffusivities of the two molecules together with the reaction kinetics define the scale of periodicity of the pattern. Typically, this length scale is related to the distance over which the inhibitor can effectively exert its activity. If this length scale is larger than the system ($\lambda > L$), inhibition will spread throughout the system, leading to a single activation peak. **(c)** Conversely, if $\lambda < L$, a new region of activation can emerge at a distance λ from the original zone of activation, resulting in a periodic pattern. **(d)** Long-range inhibition can also occur through depletion of a diffusible molecule; for example, if a molecule that stimulates its own conversion from an inactive, rapidly diffusing state to an active, slowly diffusing state. Rapid diffusion ensures that local conversion events result in system-wide reduction in the concentration of inactive molecules, again restricting activation to a single zone **(e)**. **(f)** Activation–inhibition can amplify weak signals. The signal, S, induces autocatalytic accumulation of activator, A. Competition between zones of activation result from A-dependent long-range inhibition. Weak signal asymmetries ensure that the zone of activation at the front wins, resulting in a sharp front-to-back response. **(g)** An activator-depletion mechanism underlies actin-independent polarity in *Saccharomyces cerevisiae*. The bud scar signal, Rsr1 (S), stimulates local recruitment of active membrane-associated Cdc42-GTP(A). A autocatalytically stimulates activation of Cdc42-GDP(A*) via the GTP exchange factor (GEF) complex (E). Long-range inhibition is provided by depletion of E from the cytoplasm.

asymmetry into a steep signaling gradient (Figure 1f). Numerous variations on this theme exist and include mechanisms that are highly reversible, persisting only as long as the signal persists, as well as those that are highly irreversible, locking in a stable response to a transient cue.

Polarization during budding by the yeast *Saccharomyces cerevisiae* is a particularly illustrative case of the principles of local activation and global inhibition in chemical pattern formation (Figure 1g). In *S. cerevisiae*, reproduction involves formation of a bud on the surface of the mother cell that grows to become the new daughter cell. Bud site selection is under control of the small Rho-like GTPase Cdc42, which localizes in highly polarized fashion to a cap near the bud site and directs recruitment of several downstream pathways involved in bud growth [11]. Symmetry breaking can occur either in response to a localized cue or spontaneously if this cue is absent [12]. In most cases, the GTPase Rsr1 acts as a localized cue [13], specifically recruiting Cdc42-GTP near the remnant of the previous budding event, the so-called 'bud scar'. In cells lacking Rsr1, Cdc42 undergoes spontaneous polarization, forming a cap at a random position through a classic Turing-like mechanism [14]. In both cases, the formation of a stable, highly enriched patch of Cdc42-GTP requires that the initial symmetry-breaking event be reinforced through a Cdc42-dependent positive feedback circuit. Positive feedback requires a complex containing the GTP exchange factor (GEF) for Cdc42, Cdc24, the scaffold protein Bem1, and the Cdc42 effector PAK [15]. By binding active Cdc42, PAK recruits this complex to the existing patch, where Cdc24 can catalyze the local activation of additional Cdc42 [16]. Spreading of the Cdc42 cap and the formation of additional caps are inhibited by depletion of free cytoplasmic Bem1 as it is incorporated into the Cdc42 cap [14,17,18]. Because free Bem1 diffuses rapidly, local membrane recruitment of Bem1 rapidly depletes Bem1 from the rest of the cell. In other words, rapid Bem1 diffusion ensures that the effect of local membrane accumulation is global suppression of further membrane recruitment. Thus, the basic paradigm of local amplification and global inhibition within a chemical network serves to reinforce and sustain an initial asymmetry, whether spontaneous or induced, to yield a robustly polarized system.

Self-organizing chemical systems can be both much simpler and much more complex. For example, theory suggests that one can achieve polarity with a single molecular species if there is a positive feedback loop in which an active membrane-associated molecule recruits additional copies of itself from a cytoplasmic pool, provided the system is operating within a stochastic regime and molecule number is limited [19]. Positive feedback effectively stabilizes local fluctuations, allowing them to persist for long periods of time. Limiting molecule numbers are required to ensure that multiple fluctuations compete with one another for unbound molecules, thus yielding a single dominant front. At the other end of the spectrum is *Dictyostelium*, where the core of the polarity pathway is a complicated set of positive and negative feedback loops that involve multiple phosphatidylinositol lipid signaling molecules and a set of phosphatidylinositol lipid-modifying enzymes that interconvert these two species [20,21]. Moreover, because cells can still polarize when this core system is compromised, polarity in *Dictyostelium* is likely to reflect interaction among multiple, partially redundant pathways, each potentially tuned to particular cellular or environmental conditions [22,23].

MECHANICALLY POLARIZING SYSTEMS

For reasons of simplicity, Turing omitted mechanics in his formulation of pattern formation by reaction–diffusion. However, as he noted at the time, force, stress, active movement, and material properties such as viscosity and elasticity cannot be ignored when seeking to understand morphogenetic processes [8]. Indeed, examples of purely mechanical polarizing systems exist. Here we describe two classic cases involving polarized actin networks: the formation of actin comet tails by intracellular pathogens (Figure 2a) and the directional motility of keratocytes (Figure 2b) [24]. In both cases, the mechanical properties of an actin cytoskeletal network appear sufficient for polarization, which can be triggered by stochastic or induced asymmetries in the mechanical network. What are the organizing principles of these systems and how do they relate to the principles espoused by Turing?

Polarized actin tails are used by intracellular pathogens such as *Listeria*, *Rickettsia*, *Shigella*, and vaccinia virus to promote their intracellular transport and spread between cells [25]. In each case, formation of comet tails relies on a pathogen-encoded protein that nucleates branched actin networks on the pathogen's surface. Reconstitution of the process *in vitro* by coating beads with actin nucleators proved that uniform surface nucleation is sufficient to promote motility and provided insight into the mechanisms of symmetry breaking and actin tail formation, summarized in Figure 2a [26,27]. Actin nucleation on the bead surface forms a shell. As additional actin is nucleated at the bead's surface, the outer layers expand isotropically, inducing stress in the actin network. Stochastic variation in the shell results in spontaneous local breakage of actin filaments, the so-called symmetry-breaking event. This initial site of weakening is then amplified: As filaments begin to break, the local actin network weakens, leaving it less able to bear stress and favoring further expansion of the existing rupture. This expansion of the rupture is coupled to a relaxation of tension elsewhere in the actin shell. Importantly, because mechanical stress in a material propagates at roughly the speed of sound, tension relaxation will occur nearly simultaneously with shell rupture, making ruptures of the shell at secondary sites immediately less likely. Thus tension relaxation acts as a long-range inhibitor to confine shell rupture to a single site.

Fish epithelial keratocytes are highly motile cells characterized by a well developed lamellipodium and rapid, persistent, directional migration along surfaces. Like actin tails, the actin network in keratocytes is mechanically self-organizing (Figure 2b). Theoretical models suggest that two components

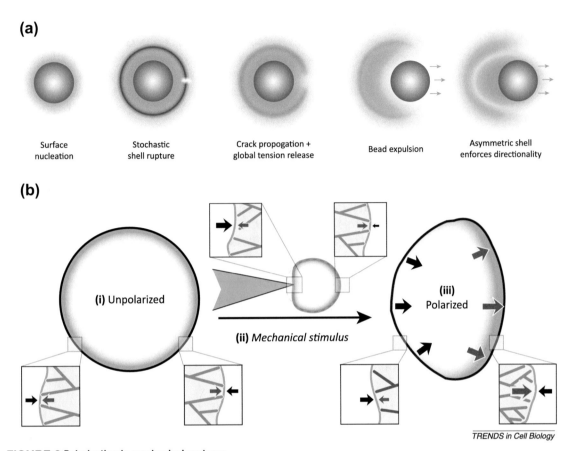

(a)

| Surface nucleation | Stochastic shell rupture | Crack propogation + global tension release | Bead expulsion | Asymmetric shell enforces directionality |

(b)

(i) Unpolarized

(ii) *Mechanical stimulus*

(iii) Polarized

TRENDS in Cell Biology

FIGURE 2 Polarization in mechanical systems.
(a) Actin tail-mediated propulsion. Stochastic instability in the shell of actin nucleated from the surface of nucleator-coated bead results in shell rupture. Shell remnants asymmetrically stabilize the newly synthesized shell at the rear, ensuring that subsequent rupture always occurs at the front. **(b)** Polarization of keratocytes. Protrusive force generated by actin polymerization at the membrane (red arrows) is opposed by membrane tension (black arrows). Stochastic or induced spatial asymmetry in polymerization is reinforced and propagated by the self-amplifying nature of actin polymerization (local positive feedback) and membrane tension, which introduces competition between polymerization and protrusion at different sites in the cell (long-range inhibition).

are critical for polarization. Arp2/3-mediated branched actin assembly at the cell membrane is thought to be autocatalytic and thus locally self-amplifying, leading to a protrusive force on the membrane due to polymerization [24,28]. This protrusive force is resisted by plasma membrane tension. Membrane tension is critical for two things. First, it is required to use some of the protrusive force generated by actin polymerization to move the cell body. By resisting protrusive force at the leading-edge membrane, tension causes the actin network to be displaced rearward as it polymerizes, a phenomenon known as retrograde flow. Physical linkage between the flowing actin network and sites of adhesion to the substrate allow retrograde flow to exert traction forces on the substrate to push the cell body forward [29]. Second, membrane tension is essential for cell polarity itself due to its role as a long-range inhibitor of protrusion [30]. As discussed above, mechanical stress propagates extremely rapidly. Therefore, a local increase in tension due to membrane protrusion at the leading edge will induce a near-simultaneous increase in membrane tension throughout the rest of the cell. This increased tension makes it increasingly difficult for actin polymerization to induce protrusion of the membrane. Thus, membrane tension serves as a long-range mechanical inhibitory signal to limit protrusion to a single leading edge. Ultimately, additional factors are at play beyond these minimal ingredients, including myosin-based contraction at the cell rear, which may play a critical role in initial symmetry breaking [24,31,32]. Notably, though, mechanical stimuli alone are sufficient to trigger polarization [33].

In conclusion, what appear to be largely mechanical polarity systems share core organizing principles with the chemical reaction–diffusion systems described above, including symmetry breaking, local self-amplification, and long-range inhibition, despite not relying on chemical diffusion. Perhaps one should not be surprised that these basic principles apply to mechanical systems, because examples of pattern formation based on these principles exist throughout the physical world (Box 1) [34].

COUPLED MECHANOCHEMICAL SYSTEMS

Although cell polarity can emerge from systems that are largely either chemical or mechanical, in many, if not most, cases, cell polarization depends critically on the interplay between the two. Numerous mechanisms exist for coupling mechanical and chemical processes (examples of which are shown in Boxes 2 and 3). In the simplest cases, asymmetry in one network acts to induce or reinforce asymmetry in the other. In other cases, asymmetry in the two networks is interrelated, with each contributing links in the core feedback circuits that drive polarization. In the following section we consider several examples of coupled mechanochemical systems that highlight the complexity of this interplay during polarization. Only by considering both can the core mechanisms that induce and reinforce asymmetry be understood.

Mechanical Symmetry Breaking

One well-studied example of mechanochemical polarization is the establishment of the anterior–posterior (AP) axis in one-celled *Caenorhabditis elegans* embryos, which depends on both mechanical (actin/myosin) and biochemical (partitioning-defective [PAR] protein) networks.

BOX 1 SAND DUNES FORM VIA LOCAL ACTIVATION AND LONG-RANGE INHIBITION

Because sand transport is highly sensitive to wind speed, local changes in wind velocity will alter the pattern of sand deposition (Figure I). Local accumulation may be triggered by small stochastic variations, a larger object, or even local surface variation, which affect air speed. Once an accumulation forms, it creates a wind shadow – a region of suppressed wind velocity. As sand particles transported by the wind encounter this region, their velocity slows and they settle onto the surface, thereby 'amplifying' the dune. Dune growth in turn increases the wind-shelter effect and enhances sand deposition even further. Long-range inhibition arises because slowed air currents downstream of the dune cannot drive significant sand transport. Thus, the presence of a dune suppresses formation of additional dunes in its immediate vicinity, contributing to their characteristic spacing.

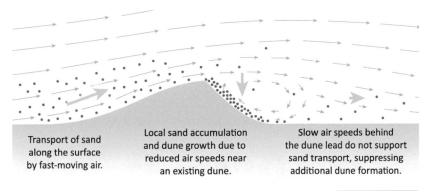

Transport of sand along the surface by fast-moving air.

Local sand accumulation and dune growth due to reduced air speeds near an existing dune.

Slow air speeds behind the dune lead do not support sand transport, suppressing additional dune formation.

TRENDS in Cell Biology

FIGURE I Sand dunes form via local activation and long-range inhibition.

At the core of the polarization machinery are the so-called PAR proteins [35]. In a given cell, there are typically two groups of antagonistic PAR proteins that are segregated to opposite halves of the cell. In *C. elegans*, these include PAR-3, PAR-6, and an atypical protein kinase C (aPKC), which localize to the anterior membrane (anterior PARs), and PAR-1, PAR-2, and lethal giant larva (LGL), which localize to the posterior (posterior PARs) [36]. Recent experimental and theoretical analysis suggests that two ingredients are critical for the ability of PAR proteins to form patterns in the cell: the ability of each group of PAR proteins to displace the other from the membrane, so-called mutual antagonism, and the presence of limiting amounts of PAR protein in the cell (Figure 3a) [37–39]. Mutually antagonistic negative feedback between the anterior and posterior PARs yields a locally self-amplifying feedback loop, because a molecule's inhibition of its own inhibitor effectively constitutes an

BOX 2 CHEMICAL CUES SHAPE CELLULAR MECHANICS

Both the structure of a cell's mechanical elements and their activity can be shaped by local biochemical signals (Figure I). Signaling molecules can polarize the mechanical elements of a cell through local control of the assembly of cytoskeletal polymers. In polarized cells (a), the activity of membrane-associated Rho-GTPases such as Cdc42 and Rac (red) simultaneously stimulate actin nucleators such as formin (green) and Arp2/3 (purple) and suppress destabilizing factors (blue) [50,64]. Similarly, microtubules are nucleated from microtubule organizing centers (MTOCs) such as the centrosome, resulting in polarity of the outgrowing network [65], and in migrating cells leading edge signals stimulate the capture and stabilization of microtubule plus ends by microtubule-associated proteins (MAPs) such as EB1 and APC (purple) [66]. These signals may also inhibit molecules that promote microtubule destabilization (orange), further focusing microtubule networks toward the leading edge. Asymmetries in mechanical networks can also arise through spatial regulation of cytoskeletal motor activity. For example, in the C. elegans one-cell embryo (b), membrane-anchored dynein pulling motors (blue motors) are regulated by the PAR polarity proteins (red/cyan), which are distributed asymmetrically along the anterior–posterior axis. Consequently, the number of active force generators at the posterior exceeds that at the anterior, resulting in an asymmetry of force applied to the spindle [67–69].

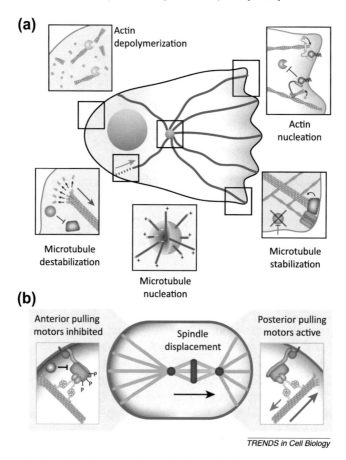

TRENDS in Cell Biology

FIGURE I Chemical cues shape cellular mechanics.

BOX 3 CELL MECHANICS SHAPE SPATIAL SIGNALING PATTERNS

The activity of molecular motors and the assembly dynamics of the cytoskeleton itself can create patterns of inhomogeneous stresses and asymmetrically distributed molecular forces that can shape the local distribution or activity of signaling molecules [70] (Figure I). Motors are typically guided by intrinsically polarized cytoskeletal filaments [24,71] (a). In the presence of even subtle biases in the polarity of the cytoskeleton, motor-driven transport can yield robust asymmetries [48,72]. An alternative to direct, motor-driven transport is advection, the nonspecific transport of material due to bulk flow, similar to the transport of objects in a flowing river (b). Within cells, such bulk-material transport can be driven by flows of cytoplasm [45,73–75]. Bulk flow also occurs at the tissue scale, resulting from changes in cell position, orientation, and shape that are induced by applied stress (c). Such cell flows can have significant effects on the local orientation of polarized cells relative to their environments

[59,76]. Bulk flows can have very molecule-specific effects, because the efficiency of advection depends on the relative effect of material flow versus the diffusivity and turnover of the molecules of interest. Direct coupling between applied mechanical stress and local biochemical activity can also be achieved through stress-sensitive molecules that undergo tension-induced conformational change [77]. For example, at focal adhesions, talin molecules crosslink extracellular matrix (ECM)-associated integrins and actin (d). Force applied to talin-bound actin filaments induce a conformational change in talin, exposing previously cryptic vinculin-binding sites. Newly bound vinculin can then recruit additional actin filaments, strengthening the mechanical linkage between the ECM and the cell cytoskeleton. Such 'stress sensing' is not limited by individual stress-sensitive molecules, but may also arise from more complex mechanisms of feedback within the cell cytoskeleton [78,79].

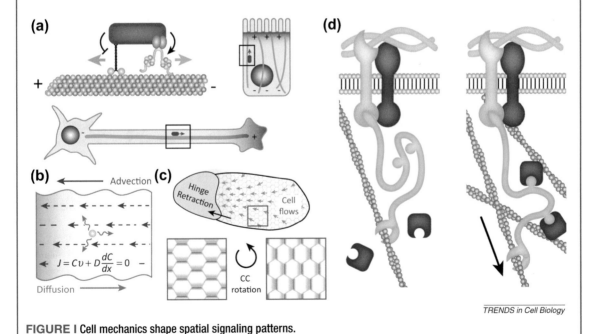

TRENDS in Cell Biology

FIGURE I Cell mechanics shape spatial signaling patterns.

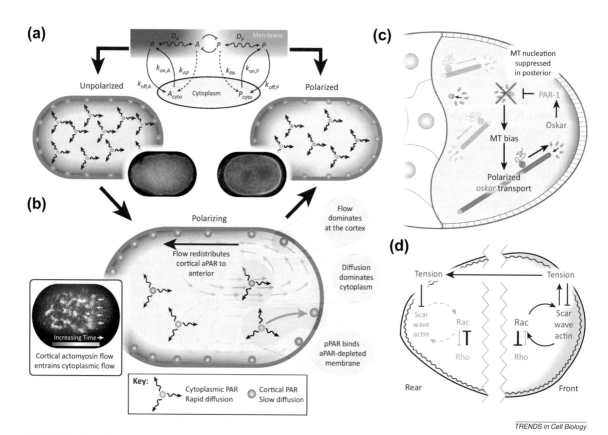

FIGURE 3 Coupling of mechanical and chemical elements in cell polarization.
(a) Polarization of the *Caenorhabditis elegans* embryo. A reaction–diffusion network comprising two antagonistic groups of partitioning-defective (PAR) polarity proteins results in a self-organizing system capable of sustaining either an unpolarized or a polarized state. The polarized pattern relies on local reciprocal negative feedback between the PAR species (local self-enhancing feedback) and depletion of the pool of rapidly diffusing inactive PAR proteins in the cytoplasm as they are converted into their active, membrane-associated forms (long-range inhibition). From [38]. Reprinted with permission from AAAS. **(b)** Pattern formation in the embryo is induced by advection of membrane-associated anterior PAR proteins by actomyosin-dependent cytoplasmic flow. This allows membrane binding of posterior PARs, resulting in an asymmetry that can be amplified by the self-organizing characteristics of the PAR reaction–diffusion system described in (a). **(c)** Polarization of the *Drosophila* oocyte. Posterior localized PAR-1 locally inhibits microtubule (MT) nucleation, leading to an enrichment of MT plus ends at the posterior. Due to this bias, polarized transport of *oskar* mRNA along MTs results in accumulation of *oskar* mRNA and, consequently, Oskar protein at the posterior pole. Oskar, in turn, stimulates additional recruitment of PAR-1, completing formation of a locally self-amplifying mechanochemical feedback circuit. **(d)** Long-range inhibition through tension. Rac activity and leading edge actin polymerization comprise a self-amplifying positive feedback loop resulting in membrane protrusion. Protrusion increases membrane tension (purple), which propagates through the cell to suppress Rac activity. Presumably, the inhibitory effect of tension is global, but is not strong enough to counter the self-amplifying Rac activity that already exists at the leading edge.

auto-activation pathway. This self-amplifying feedback ensures that the membrane will tend to be dominated locally by one or the other PAR complex, but cannot on its own specify whether domains will exist or where the boundaries between domains will occur. Such spatial organization requires long-range inhibition, which in this case is provided by the limiting pools of PAR proteins [38]. As PAR proteins are recruited into a membrane domain, the pool of free cytoplasmic PAR proteins is decreased, thereby reducing the tendency of the domain to grow further. A stably polarized state results when the sizes of the two domains are such that their tendency to grow is balanced.

But how is symmetry broken? In other words, why does the system proceed to a polarized state rather than remaining in its initial unpolarized state in which anterior PARs are enriched throughout the cell membrane and posterior PARs are confined to the cytoplasm, a situation equally supported by the antagonistic process described above. Symmetry breaking in the PAR system relies on the mechanical activity of a thin, contractile cytoskeletal layer under the cell membrane known as the actomyosin cortex (Figure 3b). Recent work points to a contractile asymmetry in this network along the AP axis [40] that is induced by the centrosome, most likely via local modulation of Rho activity [41–44]. This contractile asymmetry results in a long-range flow of cortex from posterior to anterior [40,42], which in turn entrains the motion of the cell cytoplasm, creating a fluid flow towards the anterior along the inner surface of the membrane [45]. As it turns out, the diffusive properties of anterior PAR proteins at the membrane are such that these flows can induce a significant redistribution of PAR proteins within the cell through a process known as advection (Box 3) [38]. Consequently, anterior PARs, which are initially enriched at the membrane, will be preferentially transported towards the anterior, thus depleting anterior PARs from the posterior membrane (Figure 3b). Freely diffusing posterior PARs in the cytoplasm can then take advantage of this local depletion to associate with the posterior membrane. Once asymmetry is established, biochemical reaction–diffusion processes can take over to drive the system to the stably polarized state. Thus, PAR polarity in *C. elegans* can be understood as a largely chemical pattern-forming system based on local self-amplification and long-range inhibition processes that are intrinsic to the PAR network, combined with a mechanical symmetry-breaking event.

Mechanochemical Self-Amplifying Feedback

A key step in the establishment of AP polarity during mid-oogenesis in *Drosophila* is the targeted accumulation of oskar mRNA to the posterior pole, which relies on a self-enhancing feedback loop involving polarized microtubule (MT) arrays, the polarity protein PAR-1, and both mRNA and protein derived from the *oskar* gene [46] (Figure 3c). PAR-1 promotes polarization of the MT cytoskeleton by locally suppressing MT nucleation [47]. This underlying MT polarity combined with kinesin-mediated transport of *oskar* mRNA towards MT plus ends drives accumulation of *oskar* mRNA at the posterior [48]. Local translation of these mRNAs yields Oskar protein, which promotes further recruitment of PAR-1, thereby completing the self-amplifying positive feedback loop [49]. Thus, in *Drosophila*, the central self-amplifying feedback loop involved in AP polarity comprises both mechanical (MT transport) and chemical elements (PAR-1, Oskar activity). In contrast to *C. elegans*, symmetry breaking is not mechanical but chemical: A small group of posterior follicle cells, specified earlier in development, signal into the oocyte to promote the initial recruitment of a small pool of PAR-1 to the posterior pole, thereby initiating the feedback loop [46]. However, similar to *C. elegans*, in the *Drosophila* oocyte depletion of a core component, in this case *oskar* mRNA, is likely to provide for long-range inhibition, because overexpression of *oskar* results in ectopic localization [49].

Long Range Mechanical Inhibition of Signaling Networks

In neutrophils, asymmetry in a biochemical signaling network of Rho-family GTPases, specifically asymmetric Rac activity, is critical for locally promoting the formation of branched actin networks and stimulating leading edge protrusion [50]. This asymmetry is thought to arise through coupled positive and negative feedback between Rho-family GTPases as well as phosphoinositide-based pathways. These chemical feedback circuits amplify asymmetries that arise from extracellular chemical gradients and/or stochastic fluctuations to yield a robustly polarized cell with a clear front and back. For example, Rac is thought to suppress Rho activity at the front, whereas Rho suppresses Rac activity at the rear. Similar to what we have seen for Cdc42 polarity in yeast and for PAR proteins in *C. elegans*, long-range inhibition has been proposed to be due to chemical, diffusion-mediated processes such as a fast-diffusing 'global' inhibitor or depletion of a critical effector such as the pool of inactive GTPase [51,52]. Indeed, such schemes are sufficient to yield polarization in theoretical models [10]. However, in highly elongated but still polarized cells, such diffusive processes are too slow to serve as viable long-range cues [53]. Rather, in these cases long-range inhibition may be transmitted via changes in membrane tension: Asymmetric actin-driven membrane protrusion at the cell front leads to a global increase in membrane tension, which suppresses Rac activity [53]. Consistent with this hypothesis, disruption of tension resulted in cells with multiple protrusions. Thus, similar to keratocytes, neutrophils rely on the propagation of membrane tension from the leading edge to the rest of the cell to mediate long-range inhibition. However, in this case, tension appears to exert its effects not solely through physical inhibition of actin polymerization, but via suppression of Rac signaling, implying that a mechanosensitive pathway must exist to transduce the long-range mechanical cue into a local chemical signal. Interestingly, Rac activity and SCAR/WAVE-mediated actin polymerization appear to be coupled, possibly comprising their own positive feedback loop. By physically resisting SCAR/WAVE-dependent actin protrusion, tension could prevent activation of this Rac activation feedback loop. The fact that Rac remains active at the leading edge under conditions of high tension reinforces how important balance between feedback circuits is for polarization: tension must be able to suppress activation of the Rac activation feedback loop at the cell rear, yet not overcome the already activated feedback loop at the front.

CONCLUDING REMARKS

By extracting core features of prototypical examples of polarizing systems, we have shown here how the general framework of symmetry breaking, self-amplifying feedback, and long-range inhibition, originally postulated by Turing to account for pattern formation in chemical systems, can be used to understand many examples of cell polarity, including those that rely in part or entirely on mechanical processes in the cell.

Examples certainly do exist that fall outside the paradigm that we have focused on in this review. For example, diffusion barriers can prevent the spread of molecules produced within or delivered to a particular region of the cell [54]. However, on close examination, other seemingly unrelated phenomena can be found to share core features. For example, one could imagine a physical phase-transition model for polarity in which membrane constituents segregate into distinct phases on opposite sides of the cell, much like oil and water. Mixtures of saturated and non-saturated lipids in membrane bilayers tend to segregate into liquid-ordered (saturated) and liquid-disordered (non-saturated) phases under appropriate conditions, segregating the membrane into two or more domains [55]. Segregation depends on molecular attraction and repulsion. Similar lipids group together driven by their packing order, simultaneously excluding dissimilar lipids; hence, phase segregation is effectively locally self-reinforcing. Long-range inhibition comes into play here as well, in the form of depletion: in a closed system, the fraction of the membrane occupied by a given lipid species will ultimately be limited by the relative amount of that lipid in the system.

We have also neglected some level of complexity in certain cases to highlight core features. For example, in C. elegans, there is a secondary chemical cue that operates in parallel to the mechanical flow-dependent cue that relies on local modulation of the antagonism between the two groups of PAR proteins to drive symmetry breaking [56]. In Dictyostelium, although actin is dispensable for certain aspects of cell polarity and direction sensing, actin-dependent processes are required to shape the polarity response and drive directed cell motility and thus must be integrated into any comprehensive model of cell polarity [21].

Given the diversity of polarizing systems driven by an equally diverse and complex set of often molecularly unrelated mechanisms, a comprehensive survey here would be impossible. Rather, we hope that we have highlighted how combining a consideration of both cell signaling and cell mechanics together with analysis of mathematical conceptualizations of core features has helped to identify some broadly applicable design principles for polarizing systems.

We should note that, in morphogenesis, many of the concepts discussed here repeat at scales much larger than the cellular scales discussed so far. Chemical signaling networks pattern embryos [57]. Molecules are secreted by cells and can diffuse or be actively transported through tissues [58]. Cells also exert forces on their neighbors and the environment, giving rise to patterns of stresses and forces that can be used to restructure tissue and drive cell flows [59]. Importantly, both signals and forces operate at a range of length and time scales, which is crucial for defining morphogenic processes.

Some of these tissue-scale phenomena fall into the paradigm of symmetry breaking reinforced by the self-amplification and long-range inhibition discussed above. In fact, Alan Turing introduced the basic concepts of chemical pattern formation to explain precisely such types of large-scale developmental patterns and recent work identified at least several cases of such systems where a Turing-like framework appears to apply [60–62]. Given that morphogenesis is an inherently mechanical process, it is also not surprising that mechanics can provide both local and long-range signals during developmental pattern formation [63].

The complexity and variety of developmental forms ensures that the underlying principles that govern morphogenesis will be significantly more diverse than those governing polarity. Yet, just as combined analysis of chemical reaction–diffusion systems and mesoscale mechanics at the intracellular level has provided a richer understanding of cell polarization, the identification of both the biophysical laws by which tissues deform, reshape, and flow, as well as the overlaying chemical processes that govern and are in turn governed by such mechanical processes, will undoubtedly be critical to uncovering general principles that define the mechanochemical basis of morphogenesis.

ACKNOWLEDGMENTS

We thank J. Bois and several anonymous referees for their critical comments on the manuscript. Financial support was provided by the Alexander von Humboldt Foundation (N.W.G.), a Marie Curie Grant (219286) from the European Commission (N.W.G.), the Max Planck Society (N.W.G. and S.W.G.), the ARCHES Minerva Foundation (S.W.G.), the European Molecular Biology Organization Young Investigator Programme (S.W.G.), and the European Research Council (S.W.G.).

REFERENCES

1 Macara, I.G. and Mili, S. (2008) Polarity and differential inheritance–universal attributes of life? *Cell* 135, 801–812

2 St Johnston, D. and Ahringer, J. (2010) Cell polarity in eggs and epithelia: parallels and diversity. *Cell* 141, 757–774

3 McNeill, H. (2010) Planar cell polarity: keeping hairs straight is not so simple. *Cold Spring Harb. Perspect. Biol.* 2, a003376

4 Li, R. and Bowerman, B. (2010) Symmetry breaking in biology. *Cold Spring Harb. Perspect. Biol.* 2, a003475

5 Hamamoto, T. *et al*. (1988) Asymmetric cell division of a triangular halophilic archaebacterium. *FEMS Microbiol. Lett.* 56, 221–224

6 Turing, A.M. (1952) The chemical basis of morphogenesis. *Philos. Trans. R. Soc. Lond. B: Biol. Sci.* 237, 37–72

7 Bois, J.S. *et al*. (2011) Pattern formation in active fluids. *Phys. Rev. Lett.* 106, 028103

8 Howard, J. *et al*. (2011) Turing's next steps: the mechanochemical basis of morphogenesis. *Nat. Rev. Mol. Cell Biol.* 12, 392–398

9 Gierer, A. and Meinhardt, H. (1972) A theory of biological pattern formation. *Kybernetik* 12, 30–39

10 Jilkine, A. and Edelstein-Keshet, L. (2011) A comparison of mathematical models for polarization of single eukaryotic cells in response to guided cues. *PLoS Comput. Biol.* 7, e1001121

11 Johnson, J.M. *et al*. (2011) Symmetry breaking and the establishment of cell polarity in budding yeast. *Curr. Opin. Genet. Dev.* 21, 740–746

12 Wedlich-Soldner, R. *et al*. (2003) Spontaneous cell polarization through actomyosin-based delivery of the Cdc42 GTPase. *Science* 299, 1231–1235

13 Chant, J. and Herskowitz, I. (1991) Genetic control of bud site selection in yeast by a set of gene products that constitute a morphogenetic pathway. *Cell* 65, 1203–1212

14 Goryachev, A.B. and Pokhilko, A.V. (2008) Dynamics of Cdc42 network embodies a Turing-type mechanism of yeast cell polarity. *FEBS Lett.* 582, 1437–1443

15 Irazoqui, J.E. *et al*. (2003) Scaffold-mediated symmetry breaking by Cdc42p. *Nat. Cell Biol.* 5, 1062–1070

16 Kozubowski, L. *et al*. (2008) Symmetry-breaking polarization driven by a Cdc42p GEF-PAK complex. *Curr. Biol.* 18, 1719–1726

17 Howell, A.S. *et al*. (2009) Singularity in polarization: rewiring yeast cells to make two buds. *Cell* 139, 731–743

18 Howell, A.S. *et al*. (2012) Negative feedback enhances robustness in the yeast polarity establishment circuit. *Cell* 149, 322–333

19 Altschuler, S.J. *et al*. (2008) On the spontaneous emergence of cell polarity. *Nature* 454, 886–889

20 Arai, Y. *et al*. (2010) Self-organization of the phosphatidylinositol lipids signaling system for random cell migration. *Proc. Natl. Acad. Sci. U.S.A.* 107, 12399–12404

21 Iglesias, P.A. and Devreotes, P.N. (2012) Biased excitable networks: how cells direct motion in response to gradients. *Curr. Opin. Cell Biol.* 24, 245–253

22 King, J.S. and Insall, R.H. (2009) Chemotaxis: finding the way forward with *Dictyostelium*. *Trends Cell Biol.* 19, 523–530

23 Afonso, P.V. and Parent, C.A. (2011) PI3K and chemotaxis: a priming issue? *Sci. Signal.* 4, pe22

24 Mullins, R.D. (2010) Cytoskeletal mechanisms for breaking cellular symmetry. *Cold Spring Harb. Perspect. Biol.* 2, a003392

25 Haglund, C.M. and Welch, M.D. (2011) Pathogens and polymers: microbe-host interactions illuminate the cytoskeleton. *J. Cell Biol.* 195, 7–17

26 Loisel, T.P. *et al.* (1999) Reconstitution of actin-based motility of *Listeria* and *Shigella* using pure proteins. *Nature* 401, 613–616

27 Dayel, M.J. *et al.* (2009) In silico reconstitution of actin-based symmetry breaking and motility. *PLoS Biol.* 7, e1000201

28 Carlsson, A.E. (2010) Dendritic actin filament nucleation causes traveling waves and patches. *Phys. Rev. Lett.* 104, 228102

29 Gardel, M.L. *et al.* (2010) Mechanical integration of actin and adhesion dynamics in cell migration. *Annu. Rev. Cell Dev. Biol.* 26, 315–333

30 Bershadsky, A.D. and Kozlov, M.M. (2011) Crawling cell locomotion revisited. *Proc. Natl. Acad. Sci. U.S.A.* 108, 20275–20276

31 Mogilner, A. and Keren, K. (2009) The shape of motile cells. *Curr. Biol.* 19, R762–R771

32 Yam, P.T. *et al.* (2007) Actin-myosin network reorganization breaks symmetry at the cell rear to spontaneously initiate polarized cell motility. *J. Cell Biol.* 178, 1207–1221

33 Verkhovsky, A.B. *et al.* (1999) Self-polarization and directional motility of cytoplasm. *Mol. Cell* 9, 11–20

34 Meinhardt, H. (1982) *Models of Biological Pattern Formation*. Academic Press

35 Goldstein, B. and Macara, I.G. (2007) The PAR proteins: fundamental players in animal cell polarization. *Dev. Cell* 13, 609–622

36 Nance, J. and Zallen, J.A. (2011) Elaborating polarity: PAR proteins and the cytoskeleton. *Development* 138, 799–809

37 Goehring, N.W. *et al.* (2011) PAR proteins diffuse freely across the anterior-posterior boundary in polarized *C. elegans* embryos. *J. Cell Biol.* 193, 583–594

38 Goehring, N.W. *et al.* (2011) Polarization of PAR proteins by advective triggering of a pattern-forming system. *Science* 334, 1137–1141

39 Dawes, A.T. and Munro, E.M. (2011) PAR-3 oligomerization may provide an actin-independent mechanism to maintain distinct par protein domains in the early *Caenorhabditis elegans* embryo. *Biophys. J.* 101, 1412–1422

40 Mayer, M. *et al.* (2010) Anisotropies in cortical tension reveal the physical basis of polarizing cortical flows. *Nature* 467, 617–621

41 Jenkins, N. *et al.* (2006) CYK-4/GAP provides a localized cue to initiate anteroposterior polarity upon fertilization. *Science* 313, 1298–1301

42 Munro, E. and Bowerman, B. (2009) Cellular symmetry breaking during *Caenorhabditis elegans* development. *Cold Spring Harb. Perspect. Biol.* 1, a003400

43 Motegi, F. and Sugimoto, A. (2006) Sequential functioning of the ECT-2 RhoGEF, RHO-1 and CDC-42 establishes cell polarity in *Caenorhabditis elegans* embryos. *Nat. Cell Biol.* 8, 978–985

44 Schonegg, S. and Hyman, A.A. (2006) CDC-42 and RHO-1 coordinate acto-myosin contractility and PAR protein localization during polarity establishment in *C. elegans* embryos. *Development* 133, 3507–3516

45 Hird, S.N. and White, J.G. (1993) Cortical and cytoplasmic flow polarity in early embryonic cells of *Caenorhabditis elegans*. *J. Cell Biol.* 121, 1343–1355

46 Roth, S. and Lynch, J.A. (2009) Symmetry breaking during *Drosophila* oogenesis. *Cold Spring Harb. Perspect. Biol.* 1, a001891

47 Parton, R.M. *et al.* (2011) A PAR-1-dependent orientation gradient of dynamic microtubules directs posterior cargo transport in the *Drosophila* oocyte. *J. Cell Biol.* 194, 121–135

48 Zimyanin, V.L. *et al.* (2008) In vivo imaging of oskar mRNA transport reveals the mechanism of posterior localization. *Cell* 134, 843–853

49 Zimyanin, V. *et al.* (2007) An oskar-dependent positive feedback loop maintains the polarity of the *Drosophila* oocyte. *Curr. Biol.* 17, 353–359

50 Ridley, A.J. (2011) Life at the leading edge. *Cell* 145, 1012–1022

51 Parent, C.A. and Devreotes, P.N. (1999) A cell's sense of direction. *Science* 284, 765–770

52 Meinhardt, H. (1999) Orientation of chemotactic cells and growth cones: models and mechanisms. *J. Cell Sci.* 112, 2867–2874

53 Houk, A.R. *et al.* (2012) Membrane tension maintains cell polarity by confining signals to the leading edge during neutrophil migration. *Cell* 148, 175–188

54 Caudron, F. and Barral, Y. (2009) Septins and the lateral compartmentalization of eukaryotic membranes. *Dev. Cell* 16, 493–506

55 Lingwood, D. and Simons, K. (2010) Lipid rafts as a membrane-organizing principle. *Science* 327, 46–50

56 Motegi, F. *et al.* (2011) Microtubules induce self-organization of polarized PAR domains in *Caenorhabditis elegans* zygotes. *Nat. Cell Biol.* 13, 1361–1367

57 Rogers, K.W. and Schier, A.F. (2011) Morphogen gradients: from generation to interpretation. *Annu. Rev. Cell Dev. Biol.* 27, 377–407

58 Yu, S.R. *et al.* (2009) Fgf8 morphogen gradient forms by a source-sink mechanism with freely diffusing molecules. *Nature* 461, 533–536

59 Eaton, S. and Jülicher, F. (2011) Cell flow and tissue polarity patterns. *Curr. Opin. Genet. Dev.* 21, 747–752

60 Kondo, S. and Miura, T. (2010) Reaction-diffusion model as a framework for understanding biological pattern formation. *Science* 329, 1616–1620

61 Nakamura, T. *et al.* (2006) Generation of robust left-right asymmetry in the mouse embryo requires a self-enhancement and lateral-inhibition system. *Dev. Cell* 11, 495–504

62 Müller, P. *et al.* (2012) Differential diffusivity of Nodal and Lefty underlies a reaction-diffusion patterning system. *Science* 336, 721–724

63 Mammoto, T. and Ingber, D.E. (2010) Mechanical control of tissue and organ development. *Development* 137, 1407–1420

64 Li, R. and Gundersen, G.G. (2008) Beyond polymer polarity: how the cytoskeleton builds a polarized cell. *Nat. Rev. Mol. Cell Biol.* 9, 860–873

65 Sugioka, K. and Sawa, H. (2012) Formation and functions of asymmetric microtubule organization in polarized cells. *Curr. Opin. Cell Biol.* 24, 517–525

66 Jiang, K. and Akhmanova, A. (2011) Microtubule tip-interacting proteins: a view from both ends. *Curr. Opin. Cell Biol.* 23, 94–101

67 Grill, S.W. *et al*. (2001) Polarity controls forces governing asymmetric spindle positioning in the *Caenorhabditis elegans* embryo. *Nature* 409, 630–633

68 Colombo, K. *et al*. (2003) Translation of polarity cues into asymmetric spindle positioning in *Caenorhabditis elegans* embryos. *Science* 300, 1957–1961

69 Galli, M. *et al*. (2011) aPKC phosphorylates NuMA-related LIN-5 to position the mitotic spindle during asymmetric division. *Nat. Cell Biol.* 13, 1132–1138

70 Howard, J. (2001) *Mechanics of Motor Proteins and the Cytoskeleton*. Sinauer Associates

71 Mallik, R. and Gross, S.P. (2004) Molecular motors: strategies to get along. *Curr. Biol.* 14, R971–R982

72 Amrute-Nayak, M. and Bullock, S.L. (2012) Single-molecule assays reveal that RNA localization signals regulate dynein-dynactin copy number on individual transcript cargoes. *Nat. Cell Biol.* 14, 416–423

73 Gutzeit, H. and Koppa, R. (1982) Time-lapse film analysis of cytoplasmic streaming during late oogenesis of *Drosophila*. *J. Embryol. Exp. Morphol.* 67, 101–111

74 Wolke, U. *et al*. (2007) Actin-dependent cytoplasmic streaming in *C. elegans* oogenesis. *Development* 134, 2227–2236

75 Verchot-Lubicz, J. and Goldstein, R.E. (2010) Cytoplasmic streaming enables the distribution of molecules and vesicles in large plant cells. *Protoplasma* 240, 99–107

76 Aigouy, B. *et al*. (2010) Cell flow reorients the axis of planar polarity in the wing epithelium of *Drosophila*. *Cell* 142, 773–786

77 Hoffman, B.D. *et al*. (2011) Dynamic molecular processes mediate cellular mechanotransduction. *Nature* 475, 316–323

78 Kee, Y-S. *et al*. (2012) A mechanosensory system governs myosin II accumulation in dividing cells. *Mol. Biol. Cell* 23, 1510–1523

79 Trichet, L. *et al*. (2012) Evidence of a large-scale mechanosensing mechanism for cellular adaptation to substrate stiffness. *Proc. Natl. Acad. Sci. U.S.A.* 109, 6933–6938

CHAPTER 7

Encoding and Decoding Cellular Information through Signaling Dynamics

Jeremy E. Purvis[1], Galit Lahav[1,*]

[1]Department of Systems Biology, Harvard Medical School, Boston, MA 02115, USA
*Correspondence: galit@hms.harvard.edu

Cell, Vol. 152, No. 5, February 28, 2013 © 2013 Elsevier Inc.
http://dx.doi.org/10.1016/j.cell.2013.02.005

SUMMARY

A growing number of studies are revealing that cells can send and receive information by controlling the temporal behavior (dynamics) of their signaling molecules. In this Review, we discuss what is known about the dynamics of various signaling networks and their role in controlling cellular responses. We identify general principles that are emerging in the field, focusing specifically on how the identity and quantity of a stimulus is encoded in temporal patterns, how signaling dynamics influence cellular outcomes, and how specific dynamical patterns are both shaped and interpreted by the structure of molecular networks. We conclude by discussing potential functional roles for transmitting cellular information through the dynamics of signaling molecules and possible applications for the treatment of disease.

INTRODUCTION

A unifying theme in biology is that function is reflected in structure. Consider, for example, the highly specialized structure of a bird's wing. The sparsely arranged bones and feather patterning create a high surface-to-mass ratio that enables flight. Or examine the folded conformation of an enzyme—its three-dimensional structure indicates which substrate molecules it is capable of binding and which reactions it may catalyze. Perhaps the most prevalent example of a biological structure that predicts physiological function is the genome. By knowing the sequence structure of coding DNA, one can infer whether it encodes a protein domain, a binding site, a conserved motif, or a hairpin structure. These examples demonstrate that functional information is encoded in the structural components of a

135

cell. One may argue that all relevant information is embedded in cellular structures, if only we could measure them in sufficient detail. But is this the only way that biological information may be encoded? Are there aspects of biological function that cannot be discovered by simply looking at static structures?

In this Review, we discuss an emerging trend in cell biology that suggests an additional mode for transmitting information in cells—through the *dynamics* of signaling molecules (Behar and Hoffmann, 2010). Here, dynamics is defined as the shape of the curve describing how the concentration, activity, modification state, or localization of a molecule changes over time (Figure 1A). This mode of signaling encodes information in the frequency, amplitude, duration, or other features of the temporal signal (Figure 1B). It is therefore more rich and complex than transmitting information through

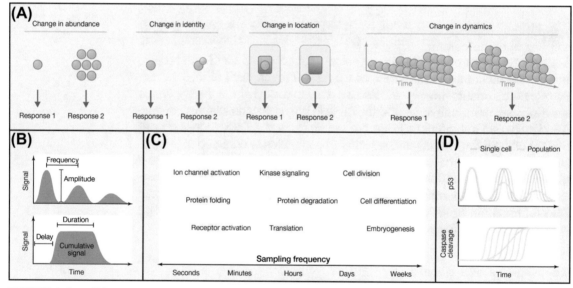

FIGURE 1 Quantifying the Dynamics of Signaling Molecules in Living Systems

(A) Different inputs may be distinguished by differences in static quantities such as the abundance, identity (e.g., posttranslational modifications, binding of a cofactor), or location of signaling molecules. However, not only the absolute number matters (e.g., how much of a specific protein is found in a cell at a specific time) but also the temporal pattern of these variables (the shape of the curve describing changes in concentration, localization, and modifications over time).

(B) Examples of measureable features of a dynamic signal, including amplitude, frequency, duration, delay, and cumulative level.

(C) Cellular processes occur with characteristic timescales ranging from subsecond to several days. Taking measurements at the appropriate timescale is crucial for capturing the true dynamical behavior.

(D) Measurements of cell populations can obscure dynamics of individual cells. For example, pulses of p53 in response to DNA damage have a fixed height and width. Different number of pulses and loss of synchrony among individual cells gives the appearance of damped oscillations in the population. Similarly, the cleavage of caspase substrates during apoptosis appears to occur gradually in a population of cells. Single-cell imaging reveals that cleavage is rapid but with a variable delay from cell to cell.

the state of a signaling molecule at only a single point in time. We present a broad survey of what is known about the dynamics of different systems across biology, focusing on well-studied systems that have been analyzed using multiple quantitative measurement and perturbation approaches. Through these examples, we extract general principles about the role of dynamics in biology and what advantages may be conferred by transmitting information through the dynamics of signaling molecules.

QUANTIFYING DYNAMICS IN LIVING SYSTEMS

Understanding the dynamics of biological responses requires collecting high-quality time series data. An important consideration when measuring the dynamics of a signal is the appropriate timescale of measurement. Some processes, such as ion transport or calcium release, occur in seconds. Others, including changes in protein levels during the cell cycle, occur over minutes or hours. Changes in some observable phenotypes such as cell morphology or expression of cell-surface markers can take days or longer. Thus, a good understanding of the timescale of a particular system is crucial for determining the appropriate sampling frequency to ensure that critical information is not missed (Figure 1C). For example, when the levels of the phosphorylated kinase ATM (ATM-P) were measured at high frequency during the first hour after DNA damage, the conclusion was that ATM is rapidly phosphorylated and reaches a maximal level within 5 min after damage, followed by a slow decrease (Jazayeri et al., 2006). When the levels of ATM-P were measured every hour for 10 hr, it became clear that it shows a series of oscillations after DNA damage, an observation that led to a new model for the control of ATM and the tumor suppressor p53 in response to DNA breaks (Batchelor et al., 2008).

The dynamics of a signal can be measured across a population of cells or in individual cells. The development of fluorescent sensors that allow high-resolution time-lapse imaging in living cells has improved our ability to quantify the dynamics of biological responses in single cells. These include chemical sensors that report activation of a signaling molecule (Welch et al., 2011) as well as sensors that participate directly in the functional response, such as fluorescent fusion proteins (e.g., Albeck et al., 2008; Bakstad et al., 2012). A collective observation from these and additional studies is that individual cells differ widely in their dynamical responses even when challenged with the same stimulus (Cohen et al., 2008; Lee et al., 2009). As a result, the average dynamical behavior of a population often represents a distorted version of individual patterns that can lead to misinterpretations. For example, p53 dynamics in response to DNA damage were originally described as damped oscillations when measured by western blot (Lev Bar-Or et al., 2000). Observation of single cells, however, revealed that these are actually pulses with fixed height and duration (Lahav et al., 2004). Varying number of pulses and loss of synchronization among individual cells over time led to an apparent widening and shortening of successive pulses in the population (Figure 1D). Similarly, the "switch-like" responses of individual cells to certain signals, such as the mitogen-activated protein (MAP) kinase activity in developing oocytes (Ferrell and Machleder, 1998) or the cleavage of caspase substrates during apoptosis (Tyas et al., 2000), give the appearance of a gradual increase in measurements of an averaged population (Figure 1D). These examples underscore the importance of tracking these responses at the single-cell level.

Because tagged reporters represent significant perturbations to the cell, it is important to establish that the introduction of a reporter into a cell line does not alter its dynamical properties. This can be accomplished through control experiments that compare the rates of induction and degradation between the tagged and endogenous proteins using immunoblots or flow cytometry. For example, rapid accumulation of a protein in live cells often appears as distinct subpopulations in flow cytometry because the protein spends relatively little time in the intermediate state (for an example, see Albeck et al. [2008]). However, overinterpretation of the underlying dynamics from flow cytometry should be avoided because simulations show that even graded individual responses can sometimes lead to bimodal populations (Birtwistle et al., 2012). When fluorescent reporters are used to study cell-to-cell variation, the use of clonal stably transfected cell lines is desirable because transiently transfected cells often express varying amounts of the effector, which may alter the dynamics and cause artificial variation between cells (Barken et al., 2005).

IMPACT OF IDENTITY AND STRENGTH OF UPSTREAM STIMULI

One of the first concepts to emerge from studying the temporal behavior of signaling molecules is that different upstream signals can lead to different dynamical patterns of the same molecule. An early example of this behavior was found in the extracellular signal-regulated kinase (ERK) pathway (here, ERK refers to the signaling module comprising both Erk1 and Erk2). It was originally observed that two separate growth factors trigger different cell fates of rat neuronal precursors; nerve growth factor (NGF) leads to differentiation, whereas epidermal growth factor (EGF) leads to cell proliferation. At first glance, one might conclude that a separate signaling pathway is induced in response to each of these stimuli, resulting in different fates.

Closer examination, however, revealed that both stimuli activate ERK but with distinct dynamical patterns (Gotoh et al., 1990; Nguyen et al., 1993; Traverse et al., 1992). Specifically, EGF triggers a transient response, whereas NGF induces sustained ERK activation (Figure 2A). These observations led to the idea that PC-12 differentiation was not strictly ligand specific but was instead governed by the dynamics of ERK activity (Marshall, 1995).

Additional signaling molecules have been shown to encapsulate upstream signals in their dynamics. For example, different inflammatory stimuli induce distinct temporal profiles of the transcription factor NF-κB (Figure 2B). Under resting conditions, NF-κB is continuously shuttled between nuclear and cytosolic compartments. Activation of NF-κB by tumor necrosis factor-α (TNFα) results in prolonged occupation in the nucleus and transcription of its negative regulator IκBα. This negative feedback loop generates oscillations of transcriptionally active NF-κB (Hoffmann et al., 2002; Nelson et al., 2004; Sung et al., 2009; Tay et al., 2010). In contrast, bacterial lipopolysaccharide (LPS) leads to slower accumulation and a single prolonged wave of NF-κB activity (Covert et al., 2005; Lee et al., 2009; Werner et al., 2005).

In various systems, both the identity and strength of the stimulus have been shown to alter the dynamics of the same protein. One example is the yeast transcription factor Msn2, which responds to stress by translocation to the nucleus (Figure 2C). Recent single-cell studies reveal that, in response to glucose limitation or high osmolarity, nuclear Msn2 shows a transient increase with a dose-dependent duration and fixed amplitude (Hao and O'Shea, 2012). In contrast, oxidative stress leads to prolonged nuclear Msn2 accumulation with amplitude that increases with higher concentration of H_2O_2. Closer observation in single cells reveals that, following the initial pulse, glucose limitation and osmotic stress lead to a series of Msn2 bursts. The frequency of these pulses depends on the intensity of the signal in glucose limitation but is not affected by the intensity of the osmotic stress (Hao and O'Shea, 2012).

The tumor suppressor p53 also shows both stimulus- and dose-dependent dynamics (Figure 2D). Double-strand breaks (DSBs) caused by γ-radiation trigger a series of p53 pulses with fixed amplitude and duration. Higher doses of radiation increase the number of pulses without affecting their amplitude or duration (Geva-Zatorsky et al., 2010; Lahav et al., 2004). In contrast, UV triggers a single p53 pulse with a dose-dependent amplitude and duration (Batchelor et al., 2011). Lastly, stimulus strength affects the dynamics of NF-κB activity. Increasing the concentration of TNFα leads to a shortened delay in NF-κB nuclear translocation (Cheong et al., 2006; Tay et al., 2010), and increasing the frequency of TNFα stimulation leads to smaller amplitude oscillations (Ashall et al., 2009). The emerging picture from these examples

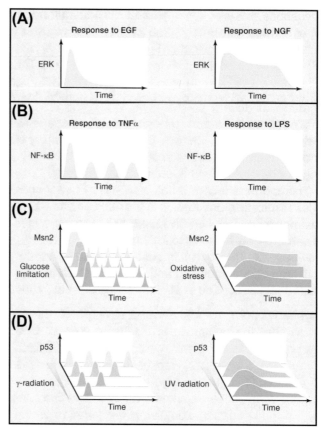

FIGURE 2 The Identity and Strength of Upstream Stimuli Can Be Encoded in the Dynamics of Signaling Molecules

(A) Dynamics of ERK activation in response to growth factors. Stimulation of mammalian cells with EGF or NGF results in transient or sustained ERK activation, respectively. Dynamics represent population responses.

(B) Dynamics of NF-κB in response to TNFα or LPS. Stimulation with TNFα results in oscillatory pattern of repeated nuclear accumulation followed by nuclear export. LPS stimulation causes a sustained level of NF-κB translocation after a short delay. Dynamics represent single-cell responses.

(C) Dynamics of yeast transcription factor Msn2. Yeast respond to glucose limitation with a coordinated burst of Msn2 translocation to the nucleus followed by a series of sporadic bursts of Msn2 activity. Increasing strength of these stresses lengthens the duration of the initial burst and increases the frequency of sporadic bursting. Oxidative stress triggers a sustained nuclear accumulation of Msn2. Increased oxidative stress intensity results in a higher amplitude and shorter delay until the signal peak. Dynamics represent single-cell responses.

(D) Dynamics of p53 in response to DNA damage. γ-radiation causes double-strand DNA breaks and leads to repeated pulses of p53. Increasing damage leads to more pulses. UV radiation triggers a single pulse of p53 that increases in amplitude and duration in proportion to the UV dose. Dynamics represent single-cell responses.

is that the dynamics of a signaling molecule can capture both the identity and quantity of upstream stimuli.

Dynamical patterns can also reflect a combination of two or more stimuli administered simultaneously or sequentially. For example, simultaneous treatment with multiple drugs can have an additive effect on the resulting dynamical pattern of downstream signaling proteins; that is, the individual dynamical patterns are effectively superimposed (Geva-Zatorsky et al., 2010). In other cases, different stimuli interact synergistically or antagonistically to produce a temporal profile in which certain dynamical features are either enhanced or silenced, respectively (Garmaroudi et al., 2010; Werner et al., 2008). This is often the case for sequentially administered stimuli when cells show desensitization to repeated stimulation (Ashall et al., 2009). For example, treatment of human platelets with thrombin produces a characteristic temporal pattern of intracellular calcium release. If preceded by treatment with ADP, however, the thrombin-induced pattern is attenuated (Chatterjee et al., 2010). This implies that dynamics can reflect cellular "memory" to previous stimuli and also suggests crosstalk between pathways.

DYNAMICS ASSOCIATED WITH SPECIFIC DOWNSTREAM RESPONSES

Because the dynamics of various proteins vary with the stimulus, it seems plausible that downstream elements may respond to these different dynamical profiles. In fact, there are a number of examples in which the dynamics of a signaling molecule are associated with, or at least precede specific cellular outcomes. As mentioned previously, the transient activation of ERK in response to EGF allows continued proliferation of neuronal precursors, whereas sustained ERK in response to NGF leads to differentiation of sympathetic-like neurons (Marshall, 1995) (Figure 3A).

The development of highly sensitive calcium dyes in the 1980s (Grynkiewicz et al., 1985) revealed a vast variety of dynamical behaviors of calcium molecules—from oscillations induced by fertilization of mammalian eggs (Malcuit et al., 2006) to noisy spikes observed in the tiny volume of a single human platelet (Heemskerk et al., 2001). Careful study of these behaviors reveals that calcium can activate different responses based solely on its dynamical waveform. A brief spike of calcium induces prolonged activation of NF-κB and JNK that lasts well after the decay in calcium. In contrast, calcium spikes evoke only transient nuclear translocation of NFAT (nuclear factor of activated T cells), whereas prolonged NFAT translocation requires sustained calcium levels (Dolmetsch et al., 1997). These results suggest that NFAT may distinguish between different dynamical patterns of calcium. The possibility of such a mechanism has been revived by a recent study showing

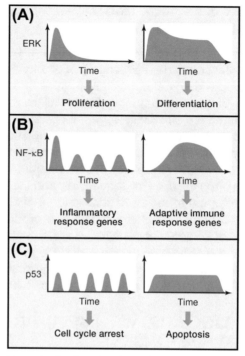

FIGURE 3 Signaling Dynamics Are Associated with Specific Downstream Responses
(A) Transient activation of ERK leads to proliferation of neuronal precursor cells. Sustained ERK levels precede differentiation into neurons.
(B) Transient nuclear accumulation of NF-κB triggers expression of nonspecific inflammatory response genes. Sustained nuclear NF-κB levels lead to expression of additional cytokines and chemokines required for adaptive immune response.
(C) p53 pulses in response to γ-irradiation are associated with cell-cycle arrest. Prolonged p53 signaling, as in response to UV radiation, leads to apoptosis.

that two isoforms of NFAT, NFAT1 and NFAT4, show different nuclear localization dynamics in response to static calcium stimulation (Yissachar et al., 2013). Whether these kinetics are responsible for decoding the calcium signal, however, will require characterizing NFAT1/4 dynamics in response to different calcium dynamics.

The dynamics of NF-κB nuclear localization and DNA binding activity control both the specificity and levels of target gene expression (Hoffmann et al., 2002; Nelson et al., 2004). Studies performed in cell populations show that activation of NF-κB in response to TNFα (which produces NF-κB oscillations) induces expression of multiple inflammatory response genes, whereas sustained NF-κB levels induced by LPS lead to similar expression patterns but also induce additional cytokine secretion as well as genes associated with the adaptive immune response (Figure 3B) (Werner et al., 2005). The dynamics of p53 have also been associated with specific cellular responses. p53 pulses following γ-irradiation are associated with transient cell-cycle arrest and recovery, whereas a single prolonged pulse after UV radiation precedes apoptosis (Figure 3C) (Purvis et al., 2012).

The high-level conclusion that might arise from these studies is that cells are able to "translate" different dynamical patterns of the same signaling molecule into specific outcomes. However, an important caveat to this claim is that, in addition to altering dynamics, different stimuli also affect other pathway components that may be responsible for the observed changes in downstream responses. This concern must ultimately be addressed through direct and careful perturbation of the dynamics using genetic or pharmacological strategies.

TARGETED PERTURBATIONS OF SIGNALING DYNAMICS

The observation that distinct dynamical patterns are correlated with certain cellular responses does not prove that dynamics are the causal agents behind these responses. How can one test whether dynamics are actually driving cellular responses? Similar to the way researchers examine the role of a specific gene by mutating it and testing the resultant behavior, a sound approach to examining the role of dynamics is to artificially perturb the dynamics of the system and test how this affects downstream outcomes. Each method of perturbation offers varying strengths and weaknesses, with the best-characterized systems using multiple approaches.

One of the first examples of controlled perturbation of dynamics was used to study the effect of calcium dynamics on gene expression. Alternating treatments of calcium-carrying ionophores and calcium-sequestering chelating agents has been used to study the effect of different calcium frequencies on gene expression. This "patch-clamp" setup has revealed that different frequencies of calcium activate none, some, or all of the transcription factors NF-κB, NFAT, and Oct/OAP (Dolmetsch et al., 1998). Similarly, the use of photoactivatable inositol 1,4,5-trisphosphate, the intracellular trigger for calcium release, led to the same striking conclusion: specific frequencies of intracellular calcium release could optimize gene expression (Li et al., 1998). Though preceded by earlier indications that the dynamics of second messengers are functional (Darmon et al., 1975), these rational perturbations of intracellular calcium dynamics provided direct evidence that specific dynamical patterns carry functional information and execute specific outcomes.

Perturbations of signaling dynamics can be achieved by inhibiting key components of the circuitry through either small molecules or genetic manipulation. All of these strategies have been employed in turn to study the role of NF-κB dynamics on target gene specificity. Knockout mouse embryonic fibroblasts (MEFs) lacking NF-κB's negative regulator, IκBα, show sustained rather than transient NF-κB activity upon TNFα treatment (Hoffmann et al., 2002). This perturbation reveals that specific genes such as RANTES require sustained activity of nuclear NF-κB. A similar approach identified a component necessary for stimulus-specific NF-κB activity under LPS stimulation. Treatment with LPS in *Tnf*-deficient MEFs reveals that de novo TNFα production is responsible for the sustained phase of NF-κB activity (Werner et al., 2005).

As an example of pharmacological perturbation of NF-κB dynamics, treatment of cells with leptomycin B (LMB) blocks nuclear export, thereby trapping the inactive NF-κB-IκBα complex in the nucleus (Nelson et al., 2004) (Figure 4A). As a result, nuclear localization of the NF-κB protein is sustained, but its transcriptional activity is only transient. In contrast, natural oscillations of NF-κB trigger a monotonic increase in a fluorescent reporter gene. This led to the hypothesis that NF-κB oscillations function to deliver newly activated NF-κB from the cytoplasm into the nucleus (Nelson et al., 2004). In support of this view, a more recent study shows that LMB has no effect on the expression of early genes but leads to inhibition of intermediate and late target genes (Sung et al., 2009).

A combination of theory and perturbation experiments reveals the specific role of ERK dynamics in driving cell fate decisions (Santos et al., 2007). Building on previous observations (Grammer and Blenis, 1997), a pair of small molecules was used to alter ERK dynamics and reverse the effects of EGF and NGF on PC-12 cell fate. Specifically, treatment with phorbol-12-myristate-13-acetate (PMA), which stimulates protein kinase C (PKC) activation and introduces positive feedback from ERK to Raf, results in sustained ERK activation and differentiation in response to EGF (Figure 4B). Conversely, treatment with the PKC inhibitor Gö7874 results in transient ERK activation and increased proliferation following NGF treatment (Figure 4C).

Although genetic and single-treatment perturbations have proved useful in revealing the role of dynamics in these and several other contexts, another desirable way to alter protein dynamics is to deliver precise and timed perturbations to the molecule under study during the response. Our lab recently used such an approach to show that the dynamics of p53 control the selection and timing of gene expression in response to DNA damage (Purvis et al., 2012). We studied cells that naturally show pulses of p53 in response to γ-radiation. These cells typically recover from moderate doses

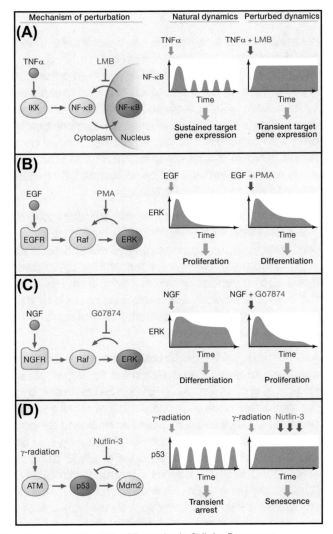

FIGURE 4 Targeted Perturbations Reveal the Role of Dynamics in Cellular Responses

(A) Perturbation of NF-κB translocation dynamics alters gene expression. Stimulation with TNFα triggers IKK-dependent activation of NF-κB and targeting to the nucleus. Subsequent export leads to oscillations of NF-κB nuclear activity. Blocking nuclear export with LMB results in sustained accumulation of nuclear NF-κB. This leads, counterintuitively, to a shift from sustained to transient target gene expression because the negative regulator IκB is also held in the nucleus.

(B and C) Altering ERK dynamics change phenotypic responses. (B) EGF stimulation leads to transient ERK activation and allows proliferation of PC-12 cells. The addition of PMA, an activator of PKC, increases positive feedback from ERK to Raf and sustains the levels of activated ERK in response to EGF. The resulting profile, which resembles ERK dynamics after NGF stimulation, promotes differentiation.

(C) NGF stimulation triggers sustained activation of ERK and leads to cellular differentiation. Inhibition of the positive feedback from ERK to Raf with the PKC inhibitor Gö7874 produces a transient-like ERK response similar to that induced by NGF. This leads to a switch from differentiation to proliferation.

(D) Artificially sustained p53 pulses promote cellular senescence. γ-irradiation induces double-strand DNA breaks and activation of ATM kinase. The resulting pulses of p53 are driven in part by negative feedback from Mdm2 to p53. When the ubiquitin ligase activity of Mdm2 is blocked by the small molecule Nutlin-3, p53 levels accumulate. A sequence of Nutlin-3 doses that sustain p53 dynamics leads to cellular senescence.

of radiation after arresting the cell cycle and repairing their DNA. Using carefully timed doses of the small molecule Nutlin-3, which stabilizes p53 levels, we artificially switched p53 dynamics from pulsed to sustained. This switch in p53 dynamics led to activation of genes associated with irreversible cellular fates such as apoptosis and senescence and pushed cells toward senescence (Figure 4D). As with all pharmacological agents, cross-reactivity of the drug with other components in the cell should be carefully characterized. Although Nutlin-3 is highly selective for p53 (Tovar et al., 2006), use of more promiscuous agents should be compared with genetic perturbations to substantiate any claims about function.

It is worthwhile to note that the artificially sustained dynamics in response to γ-radiation lead to a different cellular outcome (senescence) than would be predicted from the comparable dynamics produced naturally by UV treatment (apoptosis). This shows that similar dynamical patterns can have different consequences when they arise from different stimuli. It also implies that cell fate decisions are determined not only by the dynamics of the signal but by a combination of additional factors such as posttranslational modifications or spatial localization.

Another fine-grained perturbation of dynamics has been used to investigate the effect of yeast Msn2 dynamics on target gene expression. A mutant isoform of protein kinase A that can be controlled by a small-molecule inhibitor modulates nuclear accumulation of Msn2 (Hao and O'Shea, 2012). This setup, which includes a microfluidic device to dynamically administer inhibitor treatment, alters the amplitude, frequency, and duration of Msn2 nuclear localization. A fluorescent reporter of Msn2 transcriptional activity reveals different expression patterns correlated with different dynamical features. Specifically, gene expression exhibits a Hill function-like response to Msn2 amplitude, a linear relationship with the duration of Msn2 nuclear localization, and a nonlinear increase with increasing Msn2 pulse frequency. This example of perturbation has many important advantages: it directly influences the signaling molecule under question (as opposed to altering an upstream ligand), it can be continuously administered and therefore offers control over all parameters of the dynamical waveform, and it allows the ability to record the dynamics in individual cells.

These last two studies employ a similar analysis to determine whether different dynamical signals can be interpreted by cells. The analysis involves calculating the cumulative signal, or area under the curve (Figure 1B), and comparing downstream responses (e.g., gene expression) to the cumulative signal for individual cells. The level of target gene expression in response to Msn2 oscillations is lower than under sustained Msn2 even for similar levels of cumulative Msn2 (Hao and O'Shea,

2012). Similarly, sustained p53 leads to greater expression of senescence genes than pulsed p53 even at the same cumulative p53 signal (Purvis et al., 2012). These findings show a nonlinear relationship between the cumulative level of a transcription factor and the activation of its target genes, suggesting complex machinery for decoding protein dynamics into specific outcomes.

LINKING DYNAMICS WITH NETWORK STRUCTURE

The identification of network motifs in transcription networks and the comprehensive study of their dynamics in various systems have revealed a strong relationship between motif structure, dynamics, and specific function (Alon, 2007; Yosef and Regev, 2011). For example, feedforward loops can generate a pulse of activity or protect against brief fluctuations, depending on the nature of their interactions. Many of the examples discussed here demonstrate that dynamics play a functional role in driving cellular responses, but they do not always explain how dynamics are regulated or interpreted at the molecular level. In this section, we address two questions: (1) what are the molecular mechanisms that give rise to specific dynamical patterns and (2) how can different dynamics of the same molecule be interpreted by downstream components?

Encoding Dynamics

Studying the dynamics of signaling molecules in response to different stimuli can help to reveal the functional feedbacks responsible for shaping the observed dynamics. For example, the differences in ERK dynamics in response to EGF or NGF arise in part because of a negative feedback between ERK and Son of Sevenless (SOS) in the EGF pathway. In addition, NGF, but not EGF, signaling continues after receptor internalization, which contributes to the sustained activation of ERK (Sasagawa et al., 2005). There is also evidence for positive feedback on ERK activation through PKC (Santos et al., 2007). The implication here is that distinct responses to EGF and NGF, which are mediated by the dynamics of ERK, are brought about by differences in the identity and connectivity of various pathway components (Figure 5A).

The difference between TNFα and LPS-induced NF-κB activation dynamics (Figure 2B) is also attributed to specific network structures. The transient activation of NF-κB in response to TNFα is mediated by a negative feedback loop involving NF-κB and one of its target gene products, IκB. Activation of the TNF receptor activates the IκB kinase complex, which phosphorylates IκB and triggers its subsequent degradation through ubiquitination.

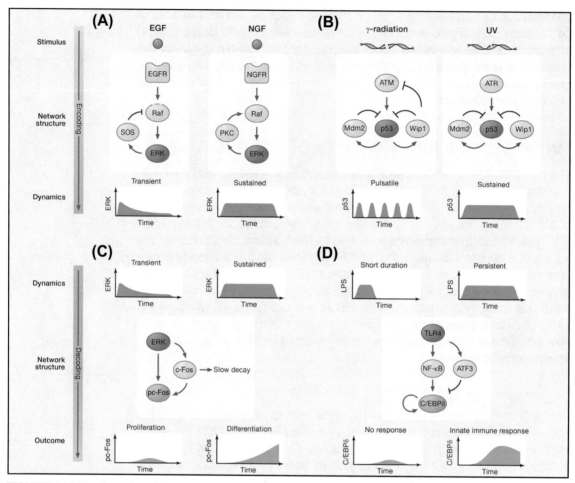

FIGURE 5 Linking Dynamics with Network Structure: Encoding and Decoding Mechanisms

(A and B) Differences in network architecture shape dynamical responses. (A) Transient activation of ERK in response to EGF is facilitated in part by negative feedback through SOS. Sustained ERK activation by NGF relies on positive feedback through PKC, which is not activated downstream of EGFR. (B) γ-radiation causes double-strand DNA breaks and leads to p53 pulses. Negative feedback through the phosphatase Wip1 attenuates the damage signal by dephosphorylating ATM and thereby controls the amplitude and duration of p53 pulses. UV radiation activates ATR kinase. The lack of negative feedback between Wip1 and ATR in the UV pathway is responsible for the difference in p53 dynamics.

(C and D) Network structure selectively interprets dynamics. (C) A network of early responding gene products, such as c-Fos, are induced by activated ERK. Transient ERK activation is not sufficient to productively accumulate c-Fos, whereas sustained ERK activation leads to accumulation of c-Fos. c-Fos is phosphorylated by ERK (pc-Fos) and leads to expression of prodifferentiation genes. Thus, the accumulation of early gene products such as c-Fos serves as a persistence detector for sustained ERK activation. (D) A gene regulatory circuit discriminates transient from persistent TLR4 signals. NF-κB and C/EBPδ form a coherent feedforward loop to stimulate maximum expression of *Il6* transcription. Attenuation of transient LPS signals is mediated by inhibition through ATF3, whereas the dramatic increase in *Il6* under persistent LPS stimulation is due in part to positive feedback through autoregulation of C/EBPδ.

Degradation of IκB allows free NF-κB to bind its target genes, including IκB, resulting in subsequent inhibition of NF-κB. The long-term dynamics of NF-κB in response to persistent TNFα stimulation are controlled by another target gene product, A20. The A20 protein has a longer half-life and acts farther upstream than IκB, which explains why it dampens the long-term phase of NF-κB dynamics (Basak et al., 2012; Werner et al., 2008). In contrast, sustained activation of NF-κB in response to LPS is attributed to positive feedback through an autocrine pathway that involves de novo TNFα production. Activation of the Toll-like receptor 4 (TLR4) by LPS triggers synthesis of TNFα and activation of the TNF receptor. The delay between TRL4- and TNF-dependent activation of NF-κB is proposed to stagger these responses in time and give rise to the stability of LPS-induced NF-κB activation (Covert et al., 2005).

Similarly, specific feedbacks in the DNA damage network are responsible for the differential dynamics of p53 in response to γ-irradiation and UV (Batchelor et al., 2011). In both networks, PI3 kinase-related kinases (ATM or ATR) relay the damage signal to p53, activating two core negative-feedback loops, one between p53 and the E3 ubiquitin ligase Mdm2 and the second between p53 and the phosphatase Wip1. An important difference, however, is that the network responding to γ-radiation includes an additional negative feedback between p53 and ATM mediated by Wip1 (Figure 5B). This feedback is essential for triggering p53 pulses in response to γ-radiation because silencing Wip1 after γ-radiation produces UV-like dynamics (Batchelor et al., 2008). In addition, the response to γ-radiation, but not to UV, is excitable, in which low transient inputs are sufficient for triggering a full p53 pulse. The current model only partially recapitulates the excitability observed experimentally (Batchelor et al., 2011), and additional work is required for identifying the mechanism of excitability in the response to γ-radiation.

Decoding Dynamics

The second question that arises when considering the functional role of dynamics is how cells interpret different dynamical patterns. That is, what molecular mechanisms are necessary to detect time-dependent features and translate these patterns into distinct phenotypic responses? Although many studies have identified functional roles for specific temporal behaviors, only a small fraction of these have determined precisely how different dynamical patterns are distinguished at the molecular level to trigger different downstream responses (Behar et al., 2007). Identifying the mechanisms that decode dynamics remains one of the most challenging goals for the field.

One of the simplest mechanisms proposed for interpreting dynamics is based on the sensitivity of downstream effectors for the molecule displaying dynamics. Under this mechanism, low-affinity effectors require sustained input levels in order to show significant activation, whereas high-affinity effectors can respond to rapidly changing input levels. There is some evidence for this mechanism in the differential activation of JNK, NF-κB, and NFAT in response to transient or sustained calcium. JNK and NF-κB, which respond to strong transient calcium bursts, have a low affinity for calcium and therefore require high concentrations for activity. This property, combined with a slow rate of degradation, allows these downstream factors to stay elevated after a brief stimulation with calcium. NFAT, in contrast, has a high affinity for calcium and a rapid rate of degradation. Thus, low and sustained calcium levels will preferentially activate NFAT over JNK and NF-κB (Dolmetsch et al., 1997). A similar mechanism was proposed to decode dynamics of the yeast stress response factor Msn2. Differences in transcription factor binding properties and the kinetics of promoter transitions govern the response to different dynamical patterns of Msn2 (Hao and O'Shea, 2012). Importantly, these mechanisms do not involve additional factors but rely solely on the strengths of association between the upstream regulator and its effectors.

More complex mechanisms for decoding temporal signals are based on specific network motifs in the responding network that sense time-dependent changes in an upstream regulator. Examples of this type of decoding mechanism have been especially difficult to identify, with two notable exceptions. In the ERK pathway, transient and persistent ERK dynamics are distinguished by a set of "immediate early gene products" that accumulate in response to activated ERK (Murphy et al., 2002, 2004). When ERK activation is transient, gene products such as c-Fos are induced but then undergo rapid degradation. When ERK levels are persistent, however, newly synthesized c-Fos is directly phosphorylated by the still-active kinase, which stabilizes c-Fos in the nucleus. Many of these immediate early gene products are transcription factors that control cell-cycle progression and other cell fate expression programs and possess ERK docking sites (Amit et al., 2007; Murphy et al., 2004). Thus, a feedforward loop comprised of a fast arm (ERK activation) and a slow arm (c-Fos accumulation) serves as a persistence detector for the duration of ERK activation (Figure 5C). More recent work in the ERK system has shown that these two arms of the system act not only at different time scales but also in different compartments of the cell (Nakakuki et al., 2010). Thus, ERK dynamics are decoded by a finely tuned spatiotemporal network controlling cell fate decisions.

The second example of a specific network structure that decodes dynamics is the control of the inflammatory response by TLR4 signaling (Litvak et al.,

2009). Expression of key inflammatory response genes such as interleukin 6 (*Il6*) requires persistent TLR4 activation, whereas transient TLR4 stimulation is effectively filtered out. Using a time series of gene expression profiles in response to the TLR4-stimulating agonist LPS, two waves of transcription have been identified in which a pair of gene products in the first wave, NF-κB and ATF3, control expression of an inflammatory regulator in the second cluster, C/EBPδ. Persistent activation of NF-κB induces expression of C/EBPδ, which regulates itself to provide strong positive feedback. In addition, C/EBPδ synergizes with NF-κB to allow productive expression of *Il6* and other inflammation genes (Figure 5D). This sophisticated decoding mechanism controls not only *Il6* but tens of additional genes associated with host defense against infection (Litvak et al., 2009).

Notably, both the ERK and TLR4 decoding networks involve some type of a feedforward network structure. It is possible that a feedforward loop motif may also decode p53 dynamics to control activation of senescence (Figure 4D). In this scenario, one of p53's target genes might serve as an intermediate factor that is required for activating senescence with p53. If such a factor decays with a timescale close to the timescale of the pulses, it will not accumulate during the pulses but only during a sustained p53 response, resulting in senescence. This mechanism would explain the accelerated expression of senescence genes under sustained p53 signaling (Purvis et al., 2012). Identification of such a factor, however, will require characterizing the kinetics of p53 target gene induction in combination with knockdown studies to identify which transcripts are required for expression of key senescence markers.

Decoding of temporal patterns is not limited to intracellular signals. A recent study proposes a model for how different temporal patterns of blood insulin are decoded by the AKT signaling network in insulin-sensing cells such as those found in the liver (Kubota et al., 2012). The authors first note that intracellular AKT activation follows the same dynamical trends as external insulin levels. By subjecting cells to different dynamical patterns of insulin, they identify downstream effectors that decode different features of the temporal profile of AKT. These results are consistent with a computational model in which different kinetics and connectivity within the signaling network allow each molecule to detect specific parts of the temporal profile. Although perturbation experiments will be necessary to validate this mechanism, the study provides an attractive model in which different dynamical patterns of insulin release are translated into appropriate metabolic responses. For example, ribosomal protein S6 kinase (S6K), which is involved in protein synthesis, responded to the transient insulin response that might appear after a meal. Glucose-6-phosphatase (G6Pase), which is involved in gluconeogenesis, responds to low insulin concentrations that may be present during fasting.

The identification of molecular circuits that decode signaling dynamics remains a major challenge for the field. Decoding mechanisms promise to provide critical answers about the function of temporal signals because they represent the connection between signal patterns and functional responses (Behar and Hoffmann, 2010). Computational approaches have been helpful in understanding the connection between topology, dynamics, and function. For example, Ma et al. (2009) performed a computational search for all possible three-node enzyme network topologies to identify those that could achieve biochemical adaptation, a dynamical response that returns to baseline levels regardless of stimulus strength. A similar approach has been applied to identify networks capable of achieving other emergent behaviors such as interpreting morphogen gradients (Cotterell and Sharpe, 2010). In a similar vein, Modular Response Analysis (Kholodenko et al., 2002), a method for extracting the strength and topology of dynamical subnetworks, is used to identify structural differences between NGF- and EGF-induced MAPK network topology (Santos et al., 2007). Such approaches are a valuable resource because they help narrow down the search for molecular participants that may regulate and interpret signaling dynamics.

FUNCTIONS ACHIEVED THROUGH MODULATION OF SIGNALING DYNAMICS

The examples presented thus far suggest that controlling the temporal behavior of signaling molecules may represent a unique signaling strategy for cells. For example, the conversion of stimulus strength to signal duration, as shown for Msn2 and p53, may be a general feature of cell signaling networks. By converting stimulus dose to signal duration, signaling networks can detect a greater range of stimulus concentrations even beyond the apparent saturation limit (Behar et al., 2008). However, there is indication that the full scope of functionality provided by signaling dynamics remains to be discovered. We now present some recent examples that illustrate the rich functional behaviors enabled by controlling signaling dynamics.

A highlighted example in the study of temporal behaviors is the manner in which transcription factor dynamics may control gene expression. In the canonical model of transcriptional activation, expression of target genes is controlled by the abundance of the transcription factor, usually with a Hill-like or linear dose-response curve (Figure 6A). For transcription factors with multiple gene targets, an increase in transcription factor levels will have different effects on each promoter because, in general, the size and shape of these response curves differs for each promoter. However, by controlling the frequency rather than the absolute level of a transcription factor, cells work within the same range of concentration and thus have a consistent effect

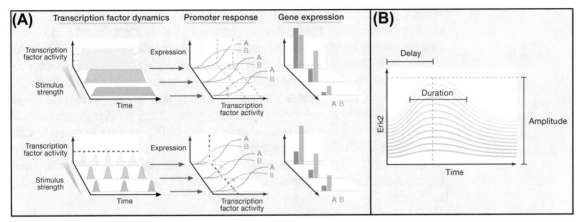

FIGURE 6 Specific Control Mechanisms Achieved through Modulation of Dynamics

(A) (Top) Different amplitudes of a transcription factor lead to different expression of target genes, depending on their promoter response curves. Different amplitudes of the transcription factor are marked by green (lowest) to yellow (highest) dotted lines. Promoter response curves for two hypothetical genes, A and B, are shown as red and blue lines. (Bottom) Frequency modulated transcription factor dynamics maintain relative proportion of target gene expression. Regardless of stimulus strength, transcription factor activity reaches the same level (gray dotted line) and therefore activates target gene promoters at the same location in the response curves. Stimulus strength affects the frequency of the transcription factor activation; higher frequency (yellow) will strike the promoters more often than lower frequency activation (green). This leads to the accumulation of target genes at the same relative proportion (right). See main text and Cai et al. (2008) for further details.

(B) Timing and fold change of ERK2 response is more conserved between individual cells than absolute levels. Individual cells vary considerably in the absolute levels of ERK2 under basal conditions as well as after stimulation with EGF. Certain parameters that describe the timing of the response, however, show less variability. The delay until peak activation, signal duration, and fold change are among the most conserved parameters. See Cohen-Saidon et al. (2009).

on target promoters. This allows coregulated genes to be expressed in the same relative proportion regardless of promoter affinities (Figure 6A). Such behavior was discovered by studying the dynamics of the yeast transcription factor Crz1, which shows bursts of nuclear localization in response to calcium (Cai et al., 2008). The concentration of calcium controls the frequency of Crz1 bursts—an analog-to-digital conversion reminiscent of the yeast response to glucose limitation (Figure 2C). Further examination showed that the frequency of Crz1 activation ensures that target genes are transcribed in the same proportion regardless of promoter affinities for Crz1 (Figure 6A).

The ability to measure not only the temporal features of a signal but also its precise intracellular location has shown that dynamics sometimes operate in specific parts of a cell. A prominent example of spatiotemporal signaling occurs in the Msn2 and NF-κB pathways, in which the transcription factor is shuttled in and out of the nucleus. This use of compartmentalization stands in contrast to the p53 network in which pulsatile dynamics are governed by repeated accumulation and degradation of total protein levels. Additional studies have shown how dynamics play a role in spatially distributed intracellular networks (Kholodenko et al., 2010 and references therein). In rat hippocampal neurons, for example, the long dendritic spines allow accumulation of membrane-generated signals. With appropriately tuned temporal behavior, information about the spatial structure of the spine can be transmitted to distal parts of the cell (Neves et al., 2008). Similarly, gradients in signal concentration have been shown to control tip project in mating yeast (Maeder et al., 2007) and the diffusion of Ras from the plasma membrane (Chandra et al., 2012). In each instance, the interplay between intracellular location and temporal behavior is necessary to carry out specific signaling processes.

Dynamics can also reflect information about the resting state of a cell. Studies on the dynamics of ERK2 translocation in individual human cells reveal a large variation in basal ERK2 nuclear levels (Cohen-Saidon et al., 2009). Upon EGF stimulation, cells show widely varying peak levels of ERK2. However, when the fold change in ERK is quantified relative to the starting levels, the dynamics of the responses are very similar (Figure 6B). This presents an elegant example of how cells can achieve a standardized response in the background of natural noise.

Pulsatile patterns can function as temporal rulers in which each pulse represents a fixed length of time. This phenomenon is demonstrated in the study of spore formation in *Bacillus subtilis* (Levine et al., 2012). The bacterium can defer sporulation for extended time periods by first undergoing multiple rounds of growth and proliferation. How does the bacterium measure this length of time? Time-lapse imaging of the master-regulator Spo0A in individual cells reveals that the deferral time is controlled by a positive feedback loop that allows Spo0A to accumulate to a critical level during multiple cell-cycle generations. This dynamical behavior may increase the bacterium's chance of survival, perhaps by allowing the accumulation of additional nutrients or the proliferation of additional offspring before sporulation occurs.

CONCLUSIONS AND FUTURE DIRECTIONS

We have presented a thematic overview of how cells store information in temporal signaling patterns, focusing on functional outcomes connected to each dynamical pattern. In each of these cases, however, dynamics probably represent only one layer of regulation within a complex signaling response that executes different cellular outcomes. In fact, we have seen that different dynamical patterns arise because of differences in network structure or the kinetics of individual molecular interactions. Thus, changes in the identity and strength of other pathway activities, such as posttranslational modifications, are likely to work with dynamics to induce stimulus-specific responses.

A better understanding of how signaling dynamics are regulated and how they affect cellular responses may provide new insights for manipulating them in a controlled way. In turn, this may enable new pharmacological strategies for altering cell fate. Oscillations of p53, for example, have been shown to occur in mice after total body irradiation (Hamstra et al., 2006). In principle, the same perturbation of p53 pulses used to induce senescence in cell culture (Purvis et al., 2012) could be administered in vivo. This may be useful in situations in which the dynamics of healthy and diseased cells are expected to differ (e.g., Francisco et al., 2008). In this scenario, dynamics represent the phenotype that distinguishes cells and may be targeted by small molecules or other perturbations.

A major theme of this review is the use of perturbations to control dynamical patterns. Such strategies hold promise as an engineering tool for use in synthetic biology. There has been recent work demonstrating light-based perturbations to cellular dynamics (Levskaya et al., 2009; Toettcher et al., 2011), which could provide exceptionally noninvasive and precise control over temporal signaling.

The number of studies that are focused on the dynamics of biological responses is growing and well exceeds the number of studies we could mention in this Review. As fluorescent labeling and time-lapse technology become better and cheaper, it may soon become clear that the vast majority of signals (if not all of them) are transferred through specific dynamical patterns of their components. If so, the study of signaling dynamics promises to provide rich and complex insights about circuit structure and function that could not be otherwise revealed. This is the case for calcium, p53, Msn2, NF-κB, and nearly all other systems mentioned in this review; study of their dynamical properties revealed previously unappreciated regulatory roles. The same applies to the bird's wing—the structure of the wing may give excellent clues about its potential function, but there is no substitute for observing the fluid motion of a wing in flight.

ACKNOWLEDGMENTS

We thank S.J. Rahi, W. Forrester, L. Murphy, R. Dolmetsch, A. Hoffmann, S. Santos, A. Aderam, V. Litvak, M.R. White, and all members of our laboratory for helpful discussions, comments, and reference suggestions. We thank our colleagues and friends at Harvard Medical School and the Department of Systems Biology for creating an inspiring environment that encourages thinking about temporal aspects of cell signaling. This research was supported by the Novartis Institutes for Biomedical Research and by National Institutes of Health grants GM083303 and K99-GM102372 (J.E.P.).

REFERENCES

Albeck, J.G., Burke, J.M., Spencer, S.L., Lauffenburger, D.A., and Sorger, P.K. (2008). Modeling a snap-action, variable-delay switch controlling extrinsic cell death. PLoS Biol. *6*, 2831–2852.

Alon, U. (2007). Network motifs: theory and experimental approaches. Nat. Rev. Genet. *8*, 450–461.

Amit, I., Citri, A., Shay, T., Lu, Y., Katz, M., Zhang, F., Tarcic, G., Siwak, D., Lahad, J., Jacob-Hirsch, J., et al. (2007). A module of negative feedback regulators defines growth factor signaling. Nat. Genet. *39*, 503–512.

Ashall, L., Horton, C.A., Nelson, D.E., Paszek, P., Harper, C.V., Sillitoe, K., Ryan, S., Spiller, D.G., Unitt, J.F., Broomhead, D.S., et al. (2009). Pulsatile stimulation determines timing and specificity of NF-kappaB-dependent transcription. Science *324*, 242–246.

Bakstad, D., Adamson, A., Spiller, D.G., and White, M.R. (2012). Quantitative measurement of single cell dynamics. Curr. Opin. Biotechnol. *23*, 103–109.

Barken, D., Wang, C.J., Kearns, J., Cheong, R., Hoffmann, A., and Levchenko, A. (2005). Comment on "Oscillations in NF-kappaB signaling control the dynamics of gene expression". Science *308*, 52.

Basak, S., Behar, M., and Hoffmann, A. (2012). Lessons from mathematically modeling the NF-kappaB pathway. Immunol. Rev. *246*, 221–238.

Batchelor, E., Mock, C.S., Bhan, I., Loewer, A., and Lahav, G. (2008). Recurrent initiation: a mechanism for triggering p53 pulses in response to DNA damage. Mol. Cell *30*, 277–289.

Batchelor, E., Loewer, A., Mock, C., and Lahav, G. (2011). Stimulus-dependent dynamics of p53 in single cells. Mol. Syst. Biol. *7*, 488.

Behar, M., and Hoffmann, A. (2010). Understanding the temporal codes of intra-cellular signals. Curr. Opin. Genet. Dev. *20*, 684–693.

Behar, M., Dohlman, H.G., and Elston, T.C. (2007). Kinetic insulation as an effective mechanism for achieving pathway specificity in intracellular signaling networks. Proc. Natl. Acad. Sci. USA *104*, 16146–16151.

Behar, M., Hao, N., Dohlman, H.G., and Elston, T.C. (2008). Dose-to-duration encoding and signaling beyond saturation in intracellular signaling networks. PLoS Comput. Biol. *4*, e1000197.

Birtwistle, M.R., Rauch, J., Kiyatkin, A., Aksamitiene, E., Dobrzyński, M., Hoek, J.B., Kolch, W., Ogunnaike, B.A., and Kholodenko, B.N. (2012). Emergence of bimodal cell population responses from the interplay between analog single-cell signaling and protein expression noise. BMC Syst. Biol. *6*, 109.

Cai, L., Dalal, C.K., and Elowitz, M.B. (2008). Frequency-modulated nuclear localization bursts coordinate gene regulation. Nature *455*, 485–490.

Chandra, A., Grecco, H.E., Pisupati, V., Perera, D., Cassidy, L., Skoulidis, F., Ismail, S.A., Hedberg, C., Hanzal-Bayer, M., Venkitaraman, A.R., et al. (2012). The GDI-like solubilizing factor PDEδ sustains the spatial organization and signalling of Ras family proteins. Nat. Cell Biol. *14*, 148–158.

Chatterjee, M.S., Purvis, J.E., Brass, L.F., and Diamond, S.L. (2010). Pairwise agonist scanning predicts cellular signaling responses to combinatorial stimuli. Nat. Biotechnol. *28*, 727–732.

Cheong, R., Bergmann, A., Werner, S.L., Regal, J., Hoffmann, A., and Levchenko, A. (2006). Transient IkappaB kinase activity mediates temporal NF-kappaB dynamics in response to a wide range of tumor necrosis factor-alpha doses. J. Biol. Chem. *281*, 2945–2950.

Cohen, A.A., Geva-Zatorsky, N., Eden, E., Frenkel-Morgenstern, M., Issaeva, I., Sigal, A., Milo, R., Cohen-Saidon, C., Liron, Y., Kam, Z., et al. (2008). Dynamic proteomics of individual cancer cells in response to a drug. Science *322*, 1511–1516.

Cohen-Saidon, C., Cohen, A.A., Sigal, A., Liron, Y., and Alon, U. (2009). Dynamics and variability of ERK2 response to EGF in individual living cells. Mol. Cell *36*, 885–893.

Cotterell, J., and Sharpe, J. (2010). An atlas of gene regulatory networks reveals multiple three-gene mechanisms for interpreting morphogen gradients. Mol. Syst. Biol. *6*, 425.

Covert, M.W., Leung, T.H., Gaston, J.E., and Baltimore, D. (2005). Achieving stability of lipopolysaccharide-induced NF-kappaB activation. Science *309*, 1854–1857.

Darmon, M., Brachet, P., and Da Silva, L.H. (1975). Chemotactic signals induce cell differentiation in Dictyostelium discoideum. Proc. Natl. Acad. Sci. USA *72*, 3163–3166.

Dolmetsch, R.E., Lewis, R.S., Goodnow, C.C., and Healy, J.I. (1997). Differential activation of transcription factors induced by Ca2+ response amplitude and duration. Nature *386*, 855–858.

Dolmetsch, R.E., Xu, K., and Lewis, R.S. (1998). Calcium oscillations increase the efficiency and specificity of gene expression. Nature *392*, 933–936.

Ferrell, J.E., Jr., and Machleder, E.M. (1998). The biochemical basis of an all-or-none cell fate switch in Xenopus oocytes. Science *280*, 895–898.

Francisco, D.C., Peddi, P., Hair, J.M., Flood, B.A., Cecil, A.M., Kalogerinis, P.T., Sigounas, G., and Georgakilas, A.G. (2008). Induction and processing of complex DNA damage in human breast cancer cells MCF-7 and nonmalignant MCF-10A cells. Free Radic. Biol. Med. *44*, 558–569.

Garmaroudi, F.S., Marchant, D., Si, X., Khalili, A., Bashashati, A., Wong, B.W., Tabet, A., Ng, R.T., Murphy, K., Luo, H., et al. (2010). Pairwise network mechanisms in the host signaling response to coxsackievirus B3 infection. Proc. Natl. Acad. Sci. USA *107*, 17053–17058.

Geva-Zatorsky, N., Dekel, E., Cohen, A.A., Danon, T., Cohen, L., and Alon, U. (2010). Protein dynamics in drug combinations: a linear superposition of individual-drug responses. Cell *140*, 643–651.

Gotoh, Y., Nishida, E., Yamashita, T., Hoshi, M., Kawakami, M., and Sakai, H. (1990). Microtubule-associated-protein (MAP) kinase activated by nerve growth factor and epidermal growth factor in PC12 cells. Identity with the mitogen-activated MAP kinase of fibroblastic cells. Eur. J. Biochem. *193*, 661–669.

Grammer, T.C., and Blenis, J. (1997). Evidence for MEK-independent pathways regulating the prolonged activation of the ERK-MAP kinases. Oncogene *14*, 1635–1642.

Grynkiewicz, G., Poenie, M., and Tsien, R.Y. (1985). A new generation of Ca2+ indicators with greatly improved fluorescence properties. J. Biol. Chem. *260*, 3440–3450.

Hamstra, D.A., Bhojani, M.S., Griffin, L.B., Laxman, B., Ross, B.D., and Rehemtulla, A. (2006). Real-time evaluation of p53 oscillatory behavior in vivo using bioluminescent imaging. Cancer Res. *66*, 7482–7489.

Hao, N., and O'Shea, E.K. (2012). Signal-dependent dynamics of transcription factor translocation controls gene expression. Nat. Struct. Mol. Biol. *19*, 31–39.

Heemskerk, J.W., Willems, G.M., Rook, M.B., and Sage, S.O. (2001). Ragged spiking of free calcium in ADP-stimulated human platelets: regulation of puff-like calcium signals in vitro and ex vivo. J. Physiol. *535*, 625–635.

Hoffmann, A., Levchenko, A., Scott, M.L., and Baltimore, D. (2002). The IkappaB-NF-kappaB signaling module: temporal control and selective gene activation. Science *298*, 1241–1245.

Jazayeri, A., Falck, J., Lukas, C., Bartek, J., Smith, G.C., Lukas, J., and Jackson, S.P. (2006). ATM- and cell cycle-dependent regulation of ATR in response to DNA double-strand breaks. Nat. Cell Biol. *8*, 37–45.

Kholodenko, B.N., Kiyatkin, A., Bruggeman, F.J., Sontag, E., Westerhoff, H.V., and Hoek, J.B. (2002). Untangling the wires: a strategy to trace functional interactions in signaling and gene networks. Proc. Natl. Acad. Sci. USA *99*, 12841–12846.

Kholodenko, B.N., Hancock, J.F., and Kolch, W. (2010). Signalling ballet in space and time. Nat. Rev. Mol. Cell Biol. *11*, 414–426.

Kubota, H., Noguchi, R., Toyoshima, Y., Ozaki, Y., Uda, S., Watanabe, K., Ogawa, W., and Kuroda, S. (2012). Temporal coding of insulin action through multiplexing of the AKT pathway. Mol. Cell *46*, 820–832.

Lahav, G., Rosenfeld, N., Sigal, A., Geva-Zatorsky, N., Levine, A.J., Elowitz, M.B., and Alon, U. (2004). Dynamics of the p53-Mdm2 feedback loop in individual cells. Nat. Genet. *36*, 147–150.

Lee, T.K., Denny, E.M., Sanghvi, J.C., Gaston, J.E., Maynard, N.D., Hughey, J.J., and Covert, M.W. (2009). A noisy paracrine signal determines the cellular NF-kappaB response to lipopolysaccharide. Sci. Signal. *2*, ra65.

Lev Bar-Or, R., Maya, R., Segel, L.A., Alon, U., Levine, A.J., and Oren, M. (2000). Generation of oscillations by the p53-Mdm2 feedback loop: a theoretical and experimental study. Proc. Natl. Acad. Sci. USA *97*, 11250–11255.

Levine, J.H., Fontes, M.E., Dworkin, J., and Elowitz, M.B. (2012). Pulsed feedback defers cellular differentiation. PLoS Biol. *10*, e1001252.

Levskaya, A., Weiner, O.D., Lim, W.A., and Voigt, C.A. (2009). Spatiotemporal control of cell signalling using a light-switchable protein interaction. Nature *461*, 997–1001.

Li, W., Llopis, J., Whitney, M., Zlokarnik, G., and Tsien, R.Y. (1998). Cell-permeant caged InsP3 ester shows that Ca2+ spike frequency can optimize gene expression. Nature *392*, 936–941.

Litvak, V., Ramsey, S.A., Rust, A.G., Zak, D.E., Kennedy, K.A., Lampano, A.E., Nykter, M., Shmulevich, I., and Aderem, A. (2009). Function of C/EBPdelta in a regulatory circuit that discriminates between transient and persistent TLR4-induced signals. Nat. Immunol. *10*, 437–443.

Ma, W., Trusina, A., El-Samad, H., Lim, W.A., and Tang, C. (2009). Defining network topologies that can achieve biochemical adaptation. Cell *138*, 760–773.

Maeder, C.I., Hink, M.A., Kinkhabwala, A., Mayr, R., Bastiaens, P.I., and Knop, M. (2007). Spatial regulation of Fus3 MAP kinase activity through a reaction-diffusion mechanism in yeast pheromone signalling. Nat. Cell Biol. *9*, 1319–1326.

Malcuit, C., Kurokawa, M., and Fissore, R.A. (2006). Calcium oscillations and mammalian egg activation. J. Cell. Physiol. *206*, 565–573.

Marshall, C.J. (1995). Specificity of receptor tyrosine kinase signaling: transient versus sustained extracellular signal-regulated kinase activation. Cell *80*, 179–185.

Murphy, L.O., Smith, S., Chen, R.H., Fingar, D.C., and Blenis, J. (2002). Molecular interpretation of ERK signal duration by immediate early gene products. Nat. Cell Biol. *4*, 556–564.

Murphy, L.O., MacKeigan, J.P., and Blenis, J. (2004). A network of immediate early gene products propagates subtle differences in mitogen-activated protein kinase signal amplitude and duration. Mol. Cell. Biol. *24*, 144–153.

Nakakuki, T., Birtwistle, M.R., Saeki, Y., Yumoto, N., Ide, K., Nagashima, T., Brusch, L., Ogunnaike, B.A., Okada-Hatakeyama, M., and Kholodenko, B.N. (2010). Ligand-specific c-Fos expression emerges from the spatiotemporal control of ErbB network dynamics. Cell *141*, 884–896.

Nelson, D.E., Ihekwaba, A.E., Elliott, M., Johnson, J.R., Gibney, C.A., Foreman, B.E., Nelson, G., See, V., Horton, C.A., Spiller, D.G., et al. (2004). Oscillations in NF-kappaB signaling control the dynamics of gene expression. Science *306*, 704–708.

Neves, S.R., Tsokas, P., Sarkar, A., Grace, E.A., Rangamani, P., Taubenfeld, S.M., Alberini, C.M., Schaff, J.C., Blitzer, R.D., Moraru, I.I., and Iyengar, R. (2008). Cell shape and negative links in regulatory motifs together control spatial information flow in signaling networks. Cell *133*, 666–680.

Nguyen, T.T., Scimeca, J.C., Filloux, C., Peraldi, P., Carpentier, J.L., and Van Obberghen, E. (1993). Co-regulation of the mitogen-activated protein kinase, extracellular signal-regulated kinase 1, and the 90-kDa ribosomal S6 kinase in PC12 cells. Distinct effects of the neurotrophic factor, nerve growth factor, and the mitogenic factor, epidermal growth factor. J. Biol. Chem. *268*, 9803–9810.

Purvis, J.E., Karhohs, K.W., Mock, C., Batchelor, E., Loewer, A., and Lahav, G. (2012). p53 dynamics control cell fate. Science *336*, 1440–1444.

Santos, S.D., Verveer, P.J., and Bastiaens, P.I. (2007). Growth factor-induced MAPK network topology shapes Erk response determining PC-12 cell fate. Nat. Cell Biol. *9*, 324–330.

Sasagawa, S., Ozaki, Y., Fujita, K., and Kuroda, S. (2005). Prediction and validation of the distinct dynamics of transient and sustained ERK activation. Nat. Cell Biol. *7*, 365–373.

Sung, M.H., Salvatore, L., De Lorenzi, R., Indrawan, A., Pasparakis, M., Hager, G.L., Bianchi, M.E., and Agresti, A. (2009). Sustained oscillations of NF-kappaB produce distinct genome scanning and gene expression profiles. PLoS ONE *4*, e7163.

Tay, S., Hughey, J.J., Lee, T.K., Lipniacki, T., Quake, S.R., and Covert, M.W. (2010). Single-cell NF-kappaB dynamics reveal digital activation and analogue information processing. Nature *466*, 267–271.

Toettcher, J.E., Gong, D., Lim, W.A., and Weiner, O.D. (2011). Light-based feedback for controlling intracellular signaling dynamics. Nat. Methods *8*, 837–839.

Tovar, C., Rosinski, J., Filipovic, Z., Higgins, B., Kolinsky, K., Hilton, H., Zhao, X., Vu, B.T., Qing, W., Packman, K., et al. (2006). Small-molecule MDM2 antagonists reveal aberrant p53 signaling in cancer: implications for therapy. Proc. Natl. Acad. Sci. USA *103*, 1888–1893.

Traverse, S., Gomez, N., Paterson, H., Marshall, C., and Cohen, P. (1992). Sustained activation of the mitogen-activated protein (MAP) kinase cascade may be required for differentiation of PC12 cells. Comparison of the effects of nerve growth factor and epidermal growth factor. Biochem. J. *288*, 351–355.

Tyas, L., Brophy, V.A., Pope, A., Rivett, A.J., and Tavaré, J.M. (2000). Rapid caspase-3 activation during apoptosis revealed using fluorescence-resonance energy transfer. EMBO Rep. *1*, 266–270.

Welch, C.M., Elliott, H., Danuser, G., and Hahn, K.M. (2011). Imaging the coordination of multiple signalling activities in living cells. Nat. Rev. Mol. Cell Biol. *12*, 749–756.

Werner, S.L., Barken, D., and Hoffmann, A. (2005). Stimulus specificity of gene expression programs determined by temporal control of IKK activity. Science *309*, 1857–1861.

Werner, S.L., Kearns, J.D., Zadorozhnaya, V., Lynch, C., O'Dea, E., Boldin, M.P., Ma, A., Baltimore, D., and Hoffmann, A. (2008). Encoding NF-kappaB temporal control in response to TNF: distinct roles for the negative regulators IkappaBalpha and A20. Genes Dev. *22*, 2093–2101.

Yissachar, N., Sharar Fischler, T., Cohen, A.A., Reich-Zeliger, S., Russ, D., Shifrut, E., Porat, Z., and Friedman, N. (2013). Dynamic response diversity of NFAT isoforms in individual living cells. Mol. Cell *49*, 322–330.

Yosef, N., and Regev, A. (2011). Impulse control: temporal dynamics in gene transcription. Cell *144*, 886–896.

nds in Cell Biology

Directed Cytoskeleton Self-Organization

Timothée Vignaud, Laurent Blanchoin, Manuel Théry*

Laboratoire de Physiologie Cellulaire et Végétale, Institut de Recherche en Technologies et Sciences pour le Vivant, CNRS/UJF/INRA/CEA, 17 Rue des Martyrs, 38054, Grenoble, France
*Correspondence: manuel.thery@cea.fr

Trends in Cell Biology, Vol. 22, No. 12, December 2012 © 2012 Elsevier Inc.
http://dx.doi.org/10.1016/j.tcb.2012.08.012

SUMMARY

The cytoskeleton architecture supports many cellular functions. Cytoskeleton networks form complex intracellular structures that vary during the cell cycle and between different cell types according to their physiological role. These structures do not emerge spontaneously. They result from the interplay between intrinsic self-organization properties and the conditions imposed by spatial boundaries. Along these boundaries, cytoskeleton filaments are anchored, repulsed, aligned, or reoriented. Such local effects can propagate alterations throughout the network and guide cytoskeleton assembly over relatively large distances. The experimental manipulation of spatial boundaries using microfabrication methods has revealed the underlying physical processes directing cytoskeleton self-organization. Here we review, step-by-step, from molecules to tissues, how the rules that govern assembly have been identified. We describe how complementary approaches, all based on controlling geometric conditions, from *in vitro* reconstruction to *in vivo* observation, shed new light on these fundamental organizing principles.

SETTING BOUNDARIES

The reproducible shape and spatial organization of organs imply the existence of deterministic rules directing the assembly of complex biological structures. Organ shape depends on cell architecture, which is supported by cytoskeleton networks. The formation of defined and geometrically controlled intracellular structures relies on the self-organization properties

161

of the cytoskeleton. The contribution of self-organization in cell biology is vast and now well documented [1]. Cytoskeleton self-organization is a process in which the consumption (physicists would say dissipation) of energy brings the cytoskeleton away from its thermodynamic equilibrium (i.e., a disordered mixture of poorly dynamic filaments) toward defined and reproducible steady states. This differs from the process of self-assembly, in which components assemble spontaneously – without an external energy source – to form a structure corresponding to the thermodynamic equilibrium. Depending on the rules regulating the interaction of cytoskeleton components, complex structures may self-organize in a robust manner. The purpose of much of the research described in this review has been to identify and formulate these rules to understand how physical principles direct biological morphogenesis.

Cytoskeleton self-organization is partially regulated by the action of proteins modulating the biochemical rules of filament growth and interactions. The combination of simple biochemical rules can lead to the formation of complex structures [2]. Robust patterns can emerge from oriented displacements of cytoskeleton filaments by molecular motors in the absence of any external guidance [3–5]. However, these autonomous self-organization processes are extremely sensitive to the presence of spatial boundary conditions (SBCs). An SBC is an external geometrical cue, within or at the periphery of the network, that can locally affect the self-organization of the network. For tissues, an SBC can be a frontier with an external fluid or a contact with bone, muscle, or other organ. For a cell, an SBC can be a neighboring cell or extracellular matrix (ECM). For intracellular cytoskeleton networks, an SBC can be a cell adhesion for the actin network, a centrosome for the microtubule (MT) network, or a frontier such as the plasma membrane or an intracellular organelle.

How an SBC can direct an autonomous self-organization process is the subject of this review. We describe recent advances in the understanding of the role of SBCs in the self-organization of actin networks and MT arrays, how these processes are integrated in the internal organization of a cell, and how this in turn affects tissue architecture. In the formation of cytoskeleton networks, an SBC can bias monomer diffusion and thereby the assembly process ([6] and references therein). Here, we focus on the role of geometrical constraints on the growth, orientation, anchorage, and production of mechanical forces during cytoskeleton assembly.

ACTIN NETWORK SELF-ORGANIZATION

Actin is an asymmetric protein that can self-assemble to form polarized actin filaments [7]. This spontaneous process can be accelerated and temporally regulated by the energy liberated from the release of a phosphate group from the nucleotide triphosphate bound to actin [8]. Actin filaments can interact to form actin networks. Actin networks can self-organize into several types of structures in cells: bundles comprising aligned long filaments and mesh-works comprising branched and intermingled short filaments. Bundles and meshworks form such complex intricate networks in cells [9] that it is diffi-cult to identify the principles of their self-organization.

Biochemists have developed alternative methods to analyze self-organization in controlled conditions *in vitro* by mixing, in defined proportions, the individ-ual components (either purified from tissues or from recombinant bacteria or yeasts). The kinetic parameters of actin polymerization measured *in vitro* and how these parameters vary in response to the presence of actin-associated proteins has provided key information about the regulation of actin assembly dynamics [10]. However, the rules guiding the spatial organization of the net-work can be identified only by using controlled geometric boundary conditions.

Symmetry Break

Mechanical constraints in an actin network can induce a symmetry break (i.e., the sudden occurrence of a singular axis in isotropic conditions in which all directions were previously equivalent). This propensity for sym-metry breaking in actin networks was elegantly revealed using a spherical glass bead coated with actin nucleation factors as a simple SBC [11,12]. Actin nucleation is induced from the bead, and the presence of capping proteins, which block filament elongation from their fast growing end, ensures that the actin filaments are short and form a dense branched meshwork. As the actin filaments grow at the bead surface, material accumulates and the stress increases in the network up to a critical value inducing its rupture [13]. The rupture creates an asymmetry in the pres-sure applied on the bead such that the bead is displaced (Figure 1a). Repetition of this sequence of events induces saltatory propulsion of the bead [14,15].

In this experimental system, the SBC can easily be manipulated by chang-ing its dimensional parameters. For example, the larger the bead, the shal-lower the curvature of the bead surface, leading to an increase in the critical value of network thickness before rupture [16,17] and the periodicity of the saltatory propulsion (Figure 1b). An asymmetric SBC can be created using ellipsoidal beads. The difference in surface curvature of the bead biases the location of network rupture, which occurs preferentially in line with or orthogonal to the long axis of the bead [18] (Figure 1c). Higher aspect ratios, obtained by actin nucleation on small glass rods, further increase the spatial bias and branched network growth is restricted to being orthogonal to the long axis of the rod [15]. As the rod length increases, several independent

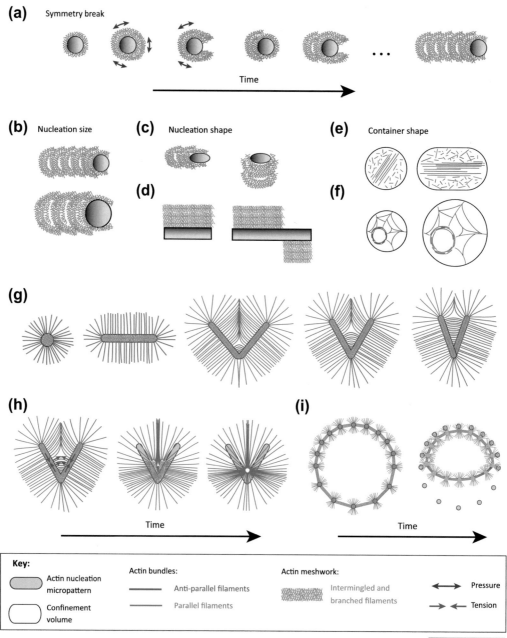

FIGURE 1 Actin network self-organization.

(a) Actin meshwork polymerization around beads leads to symmetry breaking, meshwork rupture, and bead propulsion. **(b)** Bead size regulates the period and size of meshwork rupture. **(c)** Bead asymmetry orients meshwork growth. **(d)** Bar length affects network coherence. **(e)** Long filaments self-align to form bundles, which become oriented along the long axis of the container. **(f)** Inward flow of filaments networks can form, revealing the existence of a critical length for subnetwork interconnections (Figure 1d). Interestingly, symmetry break and asymmetric force production are not restricted to branched meshworks of actin, but can also be induced by the bundling and alignment of individual filaments polymerizing against the bead surface [19].

Filament Alignment

Several self-organization processes can induce actin filament alignment in response to an SBC. Filaments can become aligned by steric interactions. When two long filaments come close to each other, they prevent the insertion of a short filament between them. Long filaments will be further forced to align by the steric interactions of short filaments around them. Steric interactions between long filament bundles will then promote their orientation in line with the long axis of the volume in which they are confined [20] (Figure 1e). Steric interaction of filaments freely moving on a layer of molecular motors can also result in their alignment along each other [4] and along the SBC [21]. Filaments can become aligned by membrane tension. Two filaments pushing orthogonally to a deformable membrane will coalesce and align to reduce the elastic energy of the membrane [22]. Preassembled filaments can become aligned by defining the anchorage positions with regular arrays of beads or micropillars and adding filamin to crosslink filaments [23,24].

Filaments can become aligned in parallel or antiparallel configurations by controlling the orientation of their growth. Surface micropatterning can be used to manipulate precisely the geometrical boundary conditions of filament growth and orientation [25]. Selective adsorption of actin nucleation-promoting factors on micropatterned regions induces localized formation of a branched meshwork. Only non-branched filaments grow out of the micropattern, with their barbed ends reproducibly oriented outward. Steric interactions force growing filaments to align parallel to each other, orthogonal to the nucleation region (Figure 1g). Distant from the nucleation region, two filaments growing toward each other in nearly opposite directions tend to form antiparallel bundles; whereas two filaments growing toward each other but at an oblique angle tend to form parallel bundles (Figure 1g). However, these tendencies can be biased because adjacent filaments sterically affect each other. The reorientation of filaments during bundle formation guides

nucleated at the vesicle periphery leads to the formation of a ring, the size of which is in proportion with the vesicle diameter. **(g)** Filament nucleation and growth of micropatterned branched meshworks. The filament interaction angle modulates the probability of association in either parallel (blue filaments) or antiparallel (red filaments) configurations. **(h)** Myosins induce the specific contraction and disassembly of antiparallel bundles and branched meshworks while leaving parallel bundles unaffected. **(i)** Asymmetric distribution of the ratio between branched and antiparallel networks leads to asymmetric contraction.

adjacent filaments also to align with the bundle (Figure 1g). Bundle formation is thus a combination of local probabilistic events, governed by filament flexibility, and the propagation of the alignment configuration to adjacent filaments by steric interactions [25].

In egg extracts, biochemical conditions are less well defined but closer to intracellular conditions. Encapsulation of egg extracts in membrane vesicles revealed that filaments nucleated at the periphery move inward and align to form a central ring. Interestingly, the ring can form only when nucleation is restricted to the vesicle periphery and not distributed evenly throughout the entire volume. A scaling law appears to regulate the ring size in proportion to the vesicle diameter [26] (Figure 1f).

Network Contraction

Myosins are oriented motors moving toward a defined extremity of actin filaments. Thus, they have specific actions depending on actin network architecture. They walk along parallel filaments, whereas they slide along antiparallel filaments in opposite directions relative to each other and thus contract the network [27] (Figure 1i). Myosins can also induce the contraction of branched meshworks, because these networks also contain antiparallel filaments. However, the rate of contraction is reduced due to the resistance associated with branches and network anchoring to nucleation regions [27]. It has been shown, based on the use of actomyosin bundles connecting beads, that the contraction rate is proportional to bundle length [28]. In more complex structures comprising various types of network, the contraction rate is determined by the local proportion of parallel and antiparallel bundles and branched meshwork [27]. Variations of these proportions in a given architecture will induce anisotropic contraction, although myosins are present throughout the network (Figure 1i). Therefore, an SBC can define the type of network architecture, which in turn can define its pattern of contraction.

MT NETWORK SELF-ORGANIZATION

Similar to the formation of actin filaments from the self-assembly of actin monomers, tubulin forms asymmetric dimers that can self-assemble into MTs. However, the release of tubulin-bound nucleotide triphosphate is required to accelerate the process [29]. Compared with actin filaments, MTs are much more rigid and almost straight in the dimensions of a single cell. MTs can sustain higher compression forces than actin filaments. They can form bundles, but they cannot form branched networks. MTs are not as numerous as actin filaments in the cell cytoskeleton. MT growth is characterized by long growth phases alternated with short periods of rapid shortening. The 'plus-end' of the MT is much more dynamic than the 'minus-end', which can be attached to a MT-organizing center (MTOC). In most animal cells, the MT network forms as an aster in which MTs radiate from the MTOC. As cells divide, the MTOC is duplicated and the network forms a bipolar spindle.

Centering

The most straightforward way to investigate MT aster positioning in response to an SBC has been to purify MTOCs from cells and place them in micro-fabricated chambers of defined dimensions [30]. Hence, the boundaries of the chamber can serve as an external SBC. As MT plus-ends grow and push against the edges of a square chamber, MTs are subjected to compression forces that push the aster toward the geometrical center of the chamber [30] (Figure 2a). When fluctuations cause the MTOC to become off-center in a given direction, MT curvature and pressure increases in that direction and pushes the MTOC back toward the center. Thus, an isotropic array of MTs pushing on peripheral barriers is sufficient to maintain the aster at the center of the volume in which it is confined. However, MTs sliding along the periphery could affect the stability of this centering mechanism by reorienting MTs. In such conditions, both pushing and pulling forces, by minus-end-directed motors attached to the periphery, are necessary to ensure efficient stabilization of the MT aster at the geometrical center of the SBC [31] (Figure 2a).

By contrast, asters with opposite polarities (i.e., with MT plus-ends at the center of the aster) cannot adopt the same steady state. As long as MTs contacting the periphery are short enough to release their elastic energy by straightening, they gently push the aster toward the center. As they get longer, the compression forces in bent MTs increase. The clustering of dynamic plus-ends by kinesins at the aster center is not strong enough to resist these forces and so the aster fragments. The MT network then switches to highly robust vortex-like structures [32] (Figure 2b).

Symmetry Break

When an aster is trapped in a water droplet encapsulated in oil, MTs cannot attach to the periphery. The spherical water–oil interface has minimal tangential resistance and is an effective SBC along which MT can slide easily. In these conditions, symmetry breaks in the aster configuration can occur [33] (Figure 2c). In a relatively large spherical volume, few MTs reach the boundaries and the aster is stabilized close to the geometrical center. As the size of the spherical volume is reduced, MTs tend to be longer than the container radius. To minimize their curvature and relax their elastic energy, MTs slip along the edges and align with the SBC [33,34]. This produces an

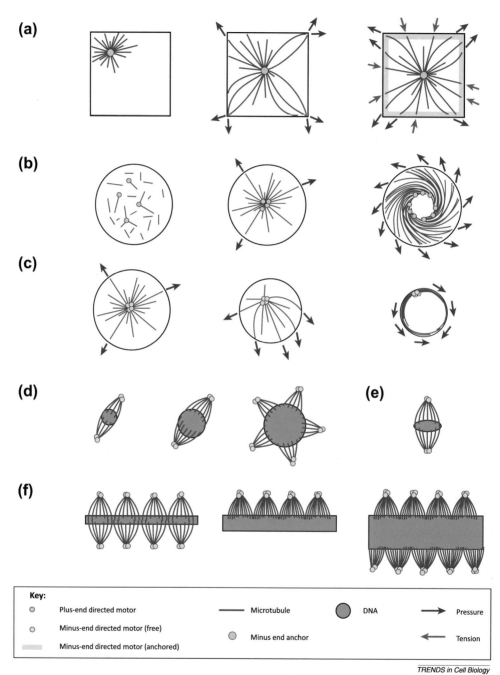

TRENDS in Cell Biology

FIGURE 2 Microtubule (MT) network self-organization.
(a) Aster off-centering with short MTs in a large container (left). Aster centering by MT sliding and pushing on the container corners (middle). Highly efficient aster centering by pushing and pulling forces (right). **(b)** MT length-dependent aster formation and centering. Short

asymmetric redistribution of MTs that pushes the MTOC to the periphery of the droplet [33] (Figure 2c). Interestingly, when the rigidity of the SBC is reduced, clustered MTs push and deform it to the extent that a tubular protrusion can be formed [33,35,36].

Alignment and Spindle Formation

The formation of bipolar mitotic spindles also depends on geometrical boundary conditions defined by DNA and cell shape. Two mechanisms contribute to mitotic spindle assembly around DNA: the focusing of the minus-ends of MTs that are associated with large DNA clusters to form spindle poles, and the antiparallel alignment of the plus-ends of MTs that are anchored at the two MTOCs such that the aster MTs overlap [37].

Multimeric minus-end-directed motors, such as dyneins, induce the formation of spindle poles. DNA provides guidance cues for initial MT alignment and thus biases bipolar spindle formation [38,39]. MTs tend to align parallel to the surface of a DNA-coated bead. The intrinsic molecular machinery supporting spindle pole focusing and mitotic spindle spatial organization is robust and initially appeared insensitive to configuration of the DNA complex [40]. However, extensive manipulations of the amount of DNA and its spatial distribution using microcontact printing revealed the DNA directing role in spindle assembly [41]. The increase in size of DNA aggregates induces spindle lengthening (Figure 2d). Above a critical size, large DNA aggregates can induce the formation of multiple poles [41] (Figure 2d). Moderately asymmetric distribution of DNA is sufficient to orient spindle formation [40,41] (Figure 2e). Long bars coated with DNA result in the formation of multiple repeats of spindles along the length of the bar and thus revealed the existence of an intrinsic spindle width (Figure 2f). This intrinsic spindle width seems to be defined by the balance between motors forcing the focusing MT ends and the elastic reaction force due to MT bending. In a certain range of parameters defined by the ratio between the DNA aggregate width and MT length, the symmetry is broken and all MTs collapse on one side of the DNA, resulting in an asymmetric configuration of spindles with respect to the long axis of the bar [41] (Figure 2f). Below this critical range, antiparallel MTs from opposite poles (with the bar in between) interact to stabilize the

MT 'plus-end' coalescence by motors (left). Aster centering by a few MT 'minus-ends' reaching and pushing on container edges (middle). Aster fragmentation and vortex formation by pushing forces exerted by long MTs on container edges (right). (c) MT length-dependent aster off-centering. Aster centering by few MT plus-ends reaching and pushing on container edges (left). Symmetry break and aster off-centering by a few, sliding MTs pushing on container edges (middle). Cortical alignment of MTs and peripheral localization of MT-organizing center (MTOC) due to numerous MTs sliding and pushing on container edges (right). (d) DNA cluster size regulates spindle size and pole formation. (e) DNA cluster asymmetry regulates spindle orientation. (f) DNA cluster width regulates spindle symmetry.

formation of symmetric bipolar spindles; above this range, the two spindle configurations on opposing sides of the DNA bar are independent and both form independent monopolar spindles.

CELLULAR SELF-ORGANIZATION

In cells, the organizing principles described above appear applicable but more difficult to reveal and investigate. Both actin and MT networks are regulated by hundreds of different types of binding protein. In addition, the assembly of actin filaments and MTs are affected by each other through physical and biochemical interactions. Cytoskeleton network assembly is regulated at the scale of a cell and is no longer solely dependent on local biochemical and geometrical conditions. The implication of biochemical signals forces the system to break its symmetry and define an axis of polarity. Although actin or MT network assembly is more complex in the cellular context than *in vitro*, some self-organizing principles have been identified.

In simple conditions as near to cellular conditions as can be achieved in experiments *in vitro*, similar self-organized structures can be observed. Cytoskeleton networks in cells from lymphatic lines or in other cells or cell fragments on non-adhesive substrates are subjected to no other geometrical constraints than the flexible plasma membrane. In the absence of MT networks, the actin network contracts and breaks symmetry after a local rupture occurs in the network. With the symmetry breaking, an over-contracted region propagates in the network [42–44]. The process of rupture is similar to what happens in branched meshworks around beads [12], except the occurrence of the rupture results from the contractile force generated by myosins rather than by the pushing force associated with actin polymerization. In the absence of actin networks, MTs pushing on a deformable membrane coalesce, align, and break symmetry by forming a long tubular protrusion [42,43], reminiscent of their behavior in vesicles [33,35,36]. However, in cellular conditions, actin and MT networks interact and the SBCs are more complex that a freely fluctuating plasma membrane. The ECM and cell neighbors can represent adhesive SBCs. Hence, the precise control and manipulation of cell adhesions, which are cellular structures that interact with the cell's structural microenvironment, reveal how these SBCs could direct intrinsic cytoskeleton self-organizing properties.

Directed Shape

Cells spreading on a defined regular array of adhesion spots revealed that the size and spacing between spots was a critical regulator of cell shape. Cells need a minimum spot size to assemble focal adhesions and cannot extend over a maximal distance between these spots [45–48]. Cell shape appears to result from the competition between the force from adhesion-induced spreading and a reaction force from the cell's elasticity and other internal contraction forces [49]. However, some cells, such as fibroblasts, have an intrinsic mechanism to regulate the length of their long axis regardless of their width, which seems to implicate tight crosstalk between actin and the MT network [50,51].

Although cell shape elongation, cytoskeletal alignment, and internal cell polarity orientation are usually correlated, cell shape does not determine actin and MT organization. Modifying the actin network by fluid flow while maintaining constant shape reorients the MT network [52]. Similarly, modifying the MT network independently of cell shape reorients the actin network [53]. Rather, there is an intricate coupling between actin and MT networks that affects their respective spatial organizations and the axis of cell polarity.

Directed Actin Network Architecture

The cellular actin network is organized by a balance between the assembly of a contractile network of aligned filaments and the polymerization of a non-contractile branched meshwork. This balance appears to be finely regulated by the degree of cell adhesion [54].

The branched meshwork assembles at the cell periphery. It is preferentially developed along convex rather than concave cell edges [55]; thus, it promotes the formation of larger membrane deformations at a cell apex [56] (Figure 3a), the size of which increases as the angle of the apex is reduced [57].

Contractile bundles of antiparallel filaments are present throughout the cytoplasm. Peripheral bundles and more interior bundles have distinct dynamics and contraction properties. Components of peripheral bundles move toward the bundle center, whereas components of interior bundles remain static with respect to the bundle organization [58] (Figure 3b). This probably reveals key differences in the polarity of filaments and thus specific contraction properties of these two types of bundle. As cell spreading or the cell aspect ratio increases, cell contraction increases [59–61]. The cell aspect ratio increase induces the alignment of contractile bundles, which form structures such as stress fibers or myofibrils (Figure 3c). Aligned stress fibers and the associated anisotropic contraction along the cell's basal surface are coupled to the assembly of similar structures and force distribution along the cell's apical surface [62,63]. Aligned myofibrils tend to organize their banding patterns in register [64].

Asymmetric SBCs, defined in cell culture by micropatterned adhesion sites, can lead to the development of asymmetric actin networks. Bundles accumulate preferentially along concave rather than convex cell edges [65]. As

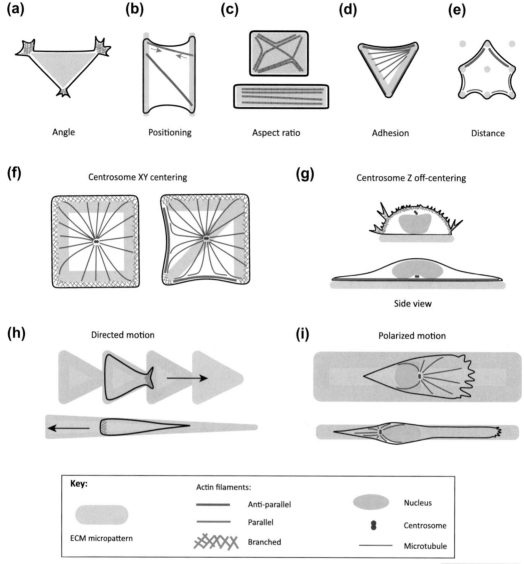

TRENDS in Cell Biology

FIGURE 3 Cellular self-organization.
(a) Branched meshwork polymerization in acute-angled regions of the cell periphery. (b) Inward treadmilling (arrows) in peripheral actin bundles and absence of treadmilling of internal bundles may reveal differences in filament polarities. (c) Alignment of myofibrils in response to cell shape elongation. (d) Formation of conspicuous actin bundles along non-adhesive edges and thin actin bundles along adhesive edges. (e) Longer peripheral bundles are also thicker. (f) Microtubules (MTs) adapt their growth to local actin structures. The centrosome maintains its central position in symmetric (left) and asymmetric environments (right). (g) Centrosome positioned above the nucleus, close to branched actin meshwork, in spatially confined cells (top). Centrosome positioned below the nucleus, close to actin bundles, in spread cells (bottom). (h) Cells move toward confined spaces above a certain threshold (top) and toward open spaces below that threshold (bottom). (i) Spread cells move with the centrosome toward the front (top), whereas confined cells move with the centrosome toward the back (bottom).

the cell spreads over an adhesive region, conspicuous contractile bundles are formed that connect this region to other adhesive regions separated by non-adhesive regions [47,66,61], revealing the development of larger traction forces [67] (Figure 3d). A relatively larger distance between adhesion sites leads to a reduced edge curvature and thicker bundles and so probably reflects a larger force between these sites [66,68] (Figure 3e).

Directed MT Network

The MT network adapts its dynamics to the various configurations of the actin network. MTs bend and grow along actin contractile bundles, but stop growing when they reach a branched actin meshwork [69]. Interestingly, although an asymmetric actin network will lead to asymmetric organization of MTs, the MTOC remains at its central location (Figure 3f). Centrosome positioning appears to depend on generation of forces by dyneins on MTs [70,71], but also on the forces generated by the less characterized connections with the actomyosin network [71,72]. Centrosome central positioning is even more remarkable given that a large part of the cytoplasm is occupied by the nucleus, on which MTs can also push and pull. The robust mechanism by which the centrosome becomes positioned at the geometrical center of the contour that describes the cell shape, where the actin network is asymmetric and the nucleus occupies a large part of the cytoplasmic volume, remains to be elucidated.

The positioning of the nucleus in a cell in culture is off-center and distal from the cell's adhesion to the ECM and the actin branched meshwork, but is proximal to the contractile bundles [69]. Therefore, the internal polarity, as revealed by the nucleus–centrosome vector, is oriented with respect to ECM and actin network asymmetries [55,69]. Biochemical disruption of the actomyosin network, the MT network or the nucleus–cytoskeleton connections can perturb polarity orientation with respect to cell–ECM SBCs [71,73,75].

The centrosome and nucleus are often described as being in the cell spreading plane in culture, but they can be positioned on an axis orthogonal to this. Moreover, their relative positions can be switched on this axis, depending on the degree of confinement imposed by the available spreading area. Indeed, the centrosome is positioned toward the apical curved surface, above the nucleus, in confined cells and below the nucleus in cells that have spread extensively [76] (Figure 3g).

Directed Migration

The asymmetric cytoskeleton organization in response to an asymmetric SBC can affect the direction of a motile cell. The relationship between external asymmetry and oriented motility is not straightforward, because it seems to depend on cell type and the degree of asymmetry. A linear track of repeated micropatterns in the shape of isosceles triangles can lead cells to move toward the acute angle apices of these triangles [77,78]. By contrast, an elongated isosceles triangle with a very acute angle (several degrees only) can lead cells away from the acute angle apex [79] (Figure 3h).

Without an external bias such as those created by asymmetric SBC, the direction of motility is defined by intrinsic cell polarization mechanisms and can be observed in motile cells on adhesive micropatterns in the shape of bars. However, bar width affects actin network organization. Variations in actin network assembly in response to bar width are cell type-specific because keratocytes need large transversal spreading to move relatively fast [80], whereas fibroblast speed is greater when the bar width is narrower [81]. Interestingly, the orientation of internal cell polarity, revealed by the position of the centrosome with respect to the nucleus, also depends on the width of the adhesive micropattern. A cell migrates on a wide bar with the centrosome nearer the leading edge, whereas on a narrow bar, the centrosome is nearer the trailing edge [82] (Figure 3i). How this centrosome positioning is related to the different types of actin organization remains to be investigated.

The speed of migrating cells and their persistence in moving in a given direction are both affected in cells whose internal polarity orientation process is defective [75]. The systematic connection between the actin network machinery powering cell migration and the degree of stability of spatial organization of the internal cell polarity was further supported by the observation in around a hundred different cell types of a correlation between cell speed and the persistence of the cell in maintaining their direction of migration [83].

Directed Cell Division

The adaptation of MT network architecture in relation to the actin network architecture and to cell shape is manifest during cell division. The tensions in astral MTs, radiating from the spindle pole toward the cell cortex, exert a torque on the spindle and direct its orientation. The tension in these astral MTs is regulated by the presence of cortical cues associated with the actin network, which orient the cell division axis with respect to cell adhesion cues and the architecture of the actin network [84–87]. Tension can also be exerted throughout the cytoplasm and therefore be proportional to astral MT length; differences in astral MT length can differ with respect to cell shape elongation and these variations can direct the orientation of the division axis accordingly [88,89].

TISSUE SELF-ORGANIZATION

At the level of tissue organization, the complexity of the system increases with greater numbers of components. Nevertheless, precise manipulation of the geometries of SBCs has proven useful in identifying consistent self-organization rules.

Directed Cell Positioning

Self-organization of cells in a given space depends on the balance of mechanical forces between the cells and the surrounding matrix. Two cells in contact constitute a minimal multicellular structure where cells can form cell–matrix adhesions (CMAs) and cell–cell adhesions (CCAs). When confined on a homogeneous micropattern (i.e., when the cell basal surface is in contact with a continuous layer of ECM), endothelial cells forming CCAs move regularly around each other in the plane of the culture dish, whereas fibroblasts, which cannot form CCAs, do not [90]. Therefore, the formation of CCAs appears to modulate the capacity of the two cells to reach a mechanical balance. The two adhesive systems – CCA and CMA – within a cell can mutually affect their respective localizations [91]. Two cells of a given epithelial cell type confined on micropatterned ECM within a defined area can move or adopt a stationary position in response to subtle changes in ECM geometry [92] (Figure 4a). Indeed, the production of tensional forces on the CCA depends on the spatial organization of the ECM. Intercellular force is higher when the CCA is close to the ECM. This directs the CCA away from the ECM and stabilizes the cell position in this configuration, which corresponds to global minimization of the overall contractile energy [92] (Figure 4a). Conversely, the formation of a CCA prevents the formation of proximal CMAs [74,93]. The mutual exclusions of the two adhesion systems lead to their spatial segregation [91] and directs cell positioning.

Directed Collective Motion

The collective motion of a large multicellular group depends on the production of intercellular forces, the spatial distribution of which directs the migration of cells with respect to their neighbors. The sudden removal of SBCs, allowing a previously confined group to migrate, revealed that intercellular forces propagate from the migrating front to the group's rear [94] (Figure 4b). Intercellular forces did not appear to pull cells forward but rather to orient the traction force field they develop on the ECM to migrate.

An intriguing recent work revealed that global coherence can emerge in the spatial organization and collective motion of large cell groups [95]. Cells plated as multicellular groups on micropatterned discs do not display any coherent global motion nor specific cell orientation. However, on a torus-shaped micropattern, there is a clear asymmetry in cell orientations such that the long cell axes tilt at similar angles with respect to the torus center (Figure 4c). This appears to be reflected in the direction cells adopt when motile on the torus.

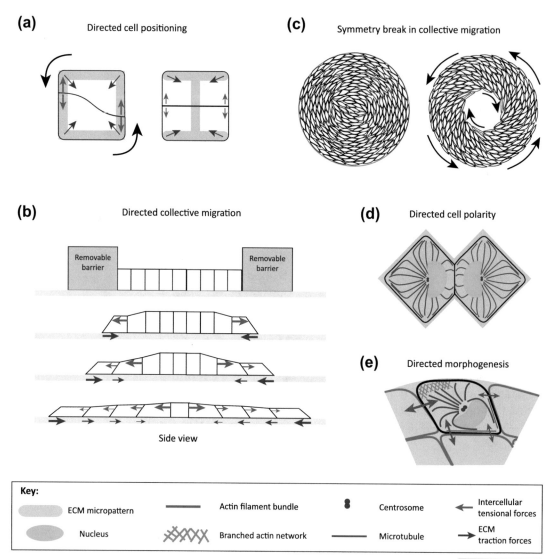

FIGURE 4 Tissue self-organization.
(a) Two cells move regularly around each other (black arrows) when extracellular matrix (ECM) is present all along the periphery (left), whereas they stop moving when the extremities of their ECM contact plane reach a region without ECM (right). The presence and absence of ECM regulate intra- and intercellular forces in opposite ways. **(b)** Intercellular forces propagate from the front to the center of a migrating cell group. **(c)** A large multicellular group on a disk of ECM displays no geometrical bias (left), whereas on a torus, symmetry is broken and cells bias their orientation and move (black arrows) in a coherent fashion (right). **(d)** Two adherent cells orient their internal polarity away from their contact plane. **(e)** Speculation on coherent tissue polarity establishment. Symmetry break first occurs in the actin network, followed by microtubule (MT) rerouting and internal polarity reorientation. The asymmetric distribution of internal forces associated with these changes is counterbalanced by asymmetric intercellular forces, which further affect polarity in adjacent cells and propagate asymmetric orientation cues.

The angular direction of cell motility at the peripheral edge of the torus (with positive curvature) tends to be opposite to that at the interior edge of the torus (with negative curvature) (Figure 4c). Therefore, it appears that the symmetry break imposed by the torus arises from this directional property of cell motility at the edges of the torus that is propagated throughout the entire group of cells. Surprisingly, the angular bias of endothelial cell orientation is clockwise, whereas with myoblasts it is counterclockwise. Thus, variations in intracellular parameters presumably can be manifested as specific asymmetries for different cell types. However, no explanation has yet been proposed for this geometrically simple organization resulting from a probably quite complex mechanism. One area where a mechanism may be identified is in the regulation of cell polarity and its relationship to oriented cell motility.

Directed Cell Polarity

The relationship between the locations of CCAs and CMAs affects nucleus–centrosome axis orientation. The centrosome, with respect to the nucleus, tends to adopt a more distal position from CCAs and a more proximal position to CMAs [73,74,96] (Figure 4d). Thus, the asymmetric locations of both CCAs and CMAs are sufficient to bias the nucleus–centrosome axis [73,74]. CCAs seem to regulate centrosome positioning [73,96], whereas CMAs seem to regulate nucleus off-centering [69,73]. Both actin filaments [96] and MTs [73] have been shown to be involved in the regulation of centrosome positioning away from CCAs. Therefore, the mechanisms by which the cytoskeleton affects centrosome and nucleus positioning remain unclear. In addition, the orientation of cell polarity not only depends on the position of CCAs, but also on the orientation of intercellular force fields [97].

Given that the self-organization of actin filament and MT networks is highly sensitive to SBCs and to the distribution of mechanical constraints, and that both types of network have intrinsic capacities to break symmetry, perhaps biased collective directional motility [95] results from symmetry break in the intracellular actin networks and the consequent asymmetric orientation of internal cell organization [98] (Figure 4e). How these polarized signals propagate to adjacent cells and result in collective oriented motility remains to be elucidated. Particularly, the role of internal mechanics and intercellular force transmission could be the key elements supporting intracellular integration of spatial signals and the establishment of coherent cell polarities in dynamic multicellular structures.

CONCLUDING REMARKS

SBCs play a major role in directing intrinsic cytoskeleton self-organization properties, from the architecture of macromolecular structures to the distribution of cells in tissues. Investigations at each scale – on isolated cytoskeleton components, more complex cell extracts, or entire cells – provide complementary information. All contribute to the establishment of a working framework, which should ultimately allow us to formulate the exact rules of cytoskeleton self-organization during morphogenesis. However, our understanding of the self-organization of minimal molecular systems *in vitro* is not sufficient to account for genuine cellular architectures and dynamics. Additional efforts need to be initiated to connect *in vitro* and *in vivo* self-organized cytoskeleton networks and fully to benefit from the former in understanding the latter.

There is currently a gap between the few self-organized structures that have been characterized *in vitro* and the myriad different structures observed in cells. Efforts should be made to reconstitute all of these structures *in vitro*. This will become possible by: (i) using more complex protein mixtures *in vitro* to recapitulate their effects on cytoskeleton networks observed in cells; (ii) identifying ways to engineer controlled SBCs mimicking actual biological membrane; and (iii) modulating biochemical signaling.

The regulation of network disassembly is as important as the regulation of assembly in network dynamics. There is a critical need to further understand how this network disassembly is modulated by SBCs. Progress in this direction should allow the reconstitution of dynamic steady states in which manipulation of SBCs and network assembly–disassembly could lead to conditions in which the network persistently self-renews, with its overall structure remaining unaffected. Technological developments are also required to modulate SBCs in real time [47], especially for analyzing dynamic systems and cytoskeleton adaptation to external changes.

However, the considerable efforts made to understand the regulation of the self-organization properties of actin filament or MT networks will not be sufficient to understand their self-organization in a cellular context, because the two networks are not independent of each other. Instead, the two networks are physically and biochemically coupled. It is necessary to design new, controlled *in vitro* biochemical assays in which the two networks can interact and regulate each other. Such assays should offer the possibility to manipulate the geometry of network interactions as well as the spatial distribution of crosslinking proteins and regulating enzymes such as Rho-GTPases. Physical SBCs need to be completed by biochemical SBCs comprising surface-grafted, but also soluble and diffusible, cues.

Notably, understanding of the basic laws governing cytoskeleton assembly can not only provide insights into cell and tissue morphogenesis, but may also have technological applications in the development of microdevices requiring complex and dynamic architectures. A structure whose precise architecture is regulated by deterministic assembly rules, that can grow and self-repair because it self-renews, has advantages over a fixed structure that would have to be repaired or replaced by a prefabricated static component. This new sort of manufacturing would be a useful way to prepare novel biomaterials and should find promising applications in microelectronics and robotics.

ACKNOWLEDGMENTS

We apologize to authors whose work on cytoskeleton self-organization was instructive and influential but not cited here because the purpose was to focus on the specific role of SBCs. We thank all members of the Physics of the Cytoskeleton and Morphogenesis Laboratory for their experimental work and discussions. This work was supported by grants from the Human Frontier Science Programs (RGP0004/2011 to L.B. and RGY0088/2012 to M.T.) and Institut National du Cancer (PLBIO 2011-141 to M.T.).

REFERENCES

1 Karsenti, E. (2008) Self-organization in cell biology: a brief history. *Nat. Rev. Mol. Cell Biol.* 9, 255–262

2 Huber, F. and Käs, J. (2011) Self-regulative organization of the cytoskeleton. *Cytoskeleton (Hoboken)* 68, 259–265

3 Surrey, T. *et al.* (2001) Physical properties determining self-organization of motors and microtubules. *Science* 292, 1167–1171

4 Schaller, V. *et al.* (2010) Polar patterns of driven filaments. *Nature* 467, 73–77

5 Sumino, Y. *et al.* (2012) Large-scale vortex lattice emerging from collectively moving microtubules. *Nature* 483, 448–452

6 Cortès, S. *et al.* (2006) Microtubule self-organisation by reaction-diffusion processes in miniature cell-sized containers and phospholipid vesicles. *Biophys. Chem.* 120, 168–177

7 Pollard, T.D. and Cooper, J.A. (1986) Actin and actin-binding proteins. A critical evaluation of mechanisms and functions. *Annu. Rev. Biochem.* 55, 987–1035

8 De La Cruz, E.M. *et al.* (2000) Polymerization and structure of nucleotide-free actin filaments. *J. Mol. Biol.* 295, 517–526

9 Xu, K. *et al.* (2012) Dual-objective STORM reveals three-dimensional filament organization in the actin cytoskeleton. *Nat. Methods* 9, 185–188

10 Pollard, T.D. (2007) Regulation of actin filament assembly by Arp2/3 complex and formins. *Annu. Rev. Biophys. Biomol. Struct.* 36, 451–477

11 Yarar, D. *et al.* (1999) The Wiskott-Aldrich syndrome protein directs actin-based motility by stimulating actin nucleation with the Arp2/3 complex. *Curr. Biol.* 9, 555–558

12 Oudenaarden, A.V. and Theriot, J.A. (1999) Cooperative symmetry-breaking by actin polymerization in a model for cell motility. *Nat. Cell Biol.* 1, 493–499

13 van der Gucht, J. *et al.* (2005) Stress release drives symmetry breaking for actin-based movement. *Proc. Natl. Acad. Sci. U.S.A.* 102, 7847–7852

14 Bernheim-Groswasser, A. *et al.* (2005) Mechanism of actin-based motility: a dynamic state diagram. *Biophys. J.* 89, 1411–1419

15 Achard, V. *et al.* (2010) A "primer"-based mechanism underlies branched actin filament network formation and motility. *Curr. Biol.* 20, 423–428

16 Noireaux, V. *et al.* (2000) Growing an actin gel on spherical surfaces. *Biophys. J.* 78, 1643–1654

17 Bernheim-Groswasser, A. *et al.* (2002) The dynamics of actin-based motility depend on surface parameters. *Nature* 417, 308–311

18 Lacayo, C.I. *et al.* (2012) Choosing orientation: influence of cargo geometry and ActA polarization on actin comet tails. *Mol. Biol. Cell* 23, 614–629

19 Michelot, A. *et al.* (2007) Actin-filament stochastic dynamics mediated by ADF/cofilin. *Curr. Biol.* 17, 825–833

20 Soares e Silva, M. *et al.* (2011) Self-organized patterns of actin filaments in cell-sized confinement. *Soft Matter* 7, 10631

21 Månsson, A. *et al.* (2012) Self-organization of motor-propelled cytoskeletal filaments at topographically defined borders. *J. Biomed. Biotechnol.* 2012, 647265

22 Liu, A.P. *et al.* (2008) Membrane-induced bundling of actin filaments. *Nat. Phys.* 4, 789–793

23 Roos, W.H. *et al.* (2003) Freely suspended actin cortex models on arrays of microfabricated pillars. *Chemphyschem* 4, 872–877

24 Uhrig, K. *et al.* (2009) Optical force sensor array in a microfluidic device based on holographic optical tweezers. *Lab Chip* 9, 661–668

25 Reymann, A-C. *et al.* (2010) Nucleation geometry governs ordered actin networks structures. *Nat. Mater.* 9, 827–832

26 Pinot, M. *et al.* (2012) Confinement induces actin flow in a meiotic cytoplasm. *Proc. Natl. Acad. Sci. U.S.A.* 109, 11705–11710

27 Reymann, A-C. *et al.* (2012) Actin network architecture can determine myosin motor activity. *Science* 336, 1310–1314

28 Thoresen, T. *et al.* (2011) Reconstitution of contractile actomyosin bundles. *Biophys. J.* 100, 2698–2705

29 Shelanski, M.L. (1973) Chemistry of the filaments and tubules of brain. *J. Histochem. Cytochem.* 21, 529–539

30 Holy, T.E. *et al.* (1997) Assembly and positioning of microtubule asters in microfabricated chambers. *Proc. Natl. Acad. Sci. U.S.A.* 94, 6228–6231

31 Laan, L. *et al.* (2012) Cortical dynein controls microtubule dynamics to generate pulling forces that position microtubule asters. *Cell* 148, 502–514

32 Nedelec, F. *et al.* (1997) Self-organization of microtubules and motors. *Nature* 389, 305–308

33 Pinot, M. *et al.* (2009) Effects of confinement on the self-organization of microtubules and motors. *Curr. Biol.* 19, 954–960

34 Cosentino Lagomarsino, M. *et al.* (2007) Microtubule organization in three-dimensional confined geometries: evaluating the role of elasticity through a combined in vitro and modeling approach. *Biophys. J.* 92, 1046–1057

35 Emsellem, V. *et al.* (1998) Vesicle deformation by microtubules: a phase diagram. *Phys. Rev. E* 58, 4807–4810

36 Fygenson, D. *et al.* (1997) Mechanics of microtubule-based membrane extension. *Phys. Rev. Lett.* 79, 4497–4500

37 Karsenti, E. and Vernos, I. (2001) The mitotic spindle: a self-made machine. *Science* 294, 543–547

38 Heald, R. *et al.* (1996) Self-organization of microtubules into bipolar spindles around artificial chromosomes in *Xenopus* egg extracts. *Nature* 382, 420–425

39 Halpin, D. *et al.* (2011) Mitotic spindle assembly around RCC1-coated beads in *Xenopus* egg extracts. *PLoS Biol.* 9, e1001225

40 Gaetz, J. *et al*. (2006) Examining how the spatial organization of chromatin signals influences metaphase spindle assembly. *Nat. Cell Biol.* 8, 924–932

41 Dinarina, A. *et al*. (2009) Chromatin shapes the mitotic spindle. *Cell* 138, 502–513

42 Bornens, M. *et al*. (1989) The cortical microfilament system of lymphoblasts displays a periodic oscillatory activity in the absence of microtubules: implications for cell polarity. *J. Cell Biol.* 109, 1071–1083

43 Bailly, E. *et al*. (1991) The cortical actomyosin system of cytochalasin D-treated lymphoblasts. *Exp. Cell Res.* 196, 287–293

44 Paluch, E. *et al*. (2005) Cortical actomyosin breakage triggers shape oscillations in cells and cell fragments. *Biophys. J.* 89, 724–733

45 Cavalcanti-Adam, E.A. *et al*. (2007) Cell spreading and focal adhesion dynamics are regulated by spacing of integrin ligands. *Biophys. J.* 92, 2964–2974

46 Schvartzman, M. *et al*. (2011) Nanolithographic control of the spatial organization of cellular adhesion receptors at the single-molecule level. *Nano Lett.* 11, 1306–1312

47 Vignaud, T. *et al*. (2012) Reprogramming cell shape with laser nano-patterning. *J. Cell Sci.* 125, 2134–2140

48 Lehnert, D. *et al*. (2004) Cell behaviour on micropatterned substrata: limits of extracellular matrix geometry for spreading and adhesion. *J. Cell Sci.* 117, 41–52

49 Vianay, B. *et al*. (2010) Single cells spreading on a protein lattice adopt an energy minimizing shape. *Phys. Rev. Lett.* 105, 3–6

50 Levina, E.M. *et al*. (2001) Cytoskeletal control of fibroblast length: experiments with linear strips of substrate. *J. Cell Sci.* 114, 4335–4341

51 Picone, R. *et al*. (2010) A polarised population of dynamic microtubules mediates homeostatic length control in animal cells. *PLoS Biol.* 8, e1000542

52 Vartanian, K.B. *et al*. (2008) Endothelial cell cytoskeletal alignment independent of fluid shear stress on micropatterned surfaces. *Biochem. Biophys. Res. Commun.* 371, 787–792

53 Terenna, C.R. *et al*. (2008) Physical mechanisms redirecting cell polarity and cell shape in fission yeast. *Curr. Biol.* 18, 1748–1753

54 Bergert, M. *et al*. (2012) Cell mechanics control rapid transitions between blebs and lamellipodia during migration. *Proc. Natl. Acad. Sci. U.S.A.* 666, 1–7

55 James, J. *et al*. (2008) Subcellular curvature at the perimeter of micropatterned cells influences lamellipodial distribution and cell polarity. *Cell Motil. Cytoskeleton* 65, 841–852

56 Parker, K.K. *et al*. (2002) Directional control of lamellipodia extension by constraining cell shape and orienting cell tractional forces. *FASEB J.* 16, 1195–1204

57 Brock, A. *et al*. (2003) Geometric determinants of directional cell motility revealed using microcontact printing. *Langmuir* 19, 1611–1617

58 Rossier, O.M. *et al*. (2010) Force generated by actomyosin contraction builds bridges between adhesive contacts. *EMBO J.* 29, 1055–1068

59 Tan, J.L. *et al*. (2003) Cells lying on a bed of microneedles: an approach to isolate mechanical force. *Proc. Natl. Acad. Sci. U.S.A.* 100, 1484–1489

60 Rape, A.D. *et al*. (2011) The regulation of traction force in relation to cell shape and focal adhesions. *Biomaterials* 32, 2043–2051

61 Kilian, K. *et al*. (2010) Geometric cues for directing the differentiation of mesenchymal stem cells. *Proc. Natl. Acad. Sci. U.S.A.* 107, 4872–4877

62 Hu, S. *et al*. (2004) Mechanical anisotropy of adherent cells probed by a 3D magnetic twisting device. *Am. J. Physiol. Cell Physiol.* 287 C1884–C1191

63 Khatau, S.B. *et al*. (2009) A perinuclear actin cap regulates nuclear shape. *Proc. Natl. Acad. Sci. U.S.A.* 106, 19017–19022

64 Bray, M-A. *et al*. (2008) Sarcomere alignment is regulated by myocyte shape. *Cell Motil. Cytoskeleton* 651, 641–651

65 Xu, J. *et al*. (2011) Effects of micropatterned curvature on the motility and mechanical properties of airway smooth muscle cells. *Biochem. Biophys. Res. Commun.* 415, 591–596

66 Théry, M. *et al*. (2006) Cell distribution of stress fibres in response to the geometry of the adhesive environment. *Cell Motil. Cytoskeleton* 63, 341–355

67 Tseng, Q. *et al*. (2011) A new micropatterning method of soft substrates reveals that different tumorigenic signals can promote or reduce cell contraction levels. *Lab Chip* 11, 2231–2240

68 Bischofs, I.B. *et al*. (2008) Filamentous network mechanics and active contractility determine cell and tissue shape. *Biophys. J.* 95, 3488–3496

69 Théry, M. *et al*. (2006) Anisotropy of cell adhesive microenvironment governs cell internal organization and orientation of polarity. *Proc. Natl. Acad. Sci. U.S.A.* 103, 19771–19776

70 Wu, J. *et al*. (2011) Effects of dynein on microtubule mechanics and centrosome positioning. *Mol. Biol. Cell* 22, 4834–4841

71 Hale, C.M. *et al*. (2011) SMRT analysis of MTOC and nuclear positioning reveals the role of EB1 and LIC1 in single-cell polarization. *J. Cell Sci.* 124, 4267–4285

72 Zhu, J. *et al*. (2010) finding the cell center by a balance of dynein and myosin pulling and microtubule pushing: a computational study. *Mol. Biol. Cell* 21, 4418–4427

73 Dupin, I. *et al*. (2009) Classical cadherins control nucleus and centrosome position and cell polarity. *J. Cell Biol.* 185, 779–786

74 Camand, E. *et al*. (2012) N-cadherin expression level modulates integrin-mediated polarity and strongly impacts on the speed and directionality of glial cell migration. *J. Cell Sci.* 125, 844–857

75 Lombardi, M.L. *et al*. (2011) The interaction between nesprins and sun proteins at the nuclear envelope is critical for force transmission between the nucleus and cytoskeleton. *J. Biol. Chem.* 286, 26743–26753

76 Pitaval, A. *et al*. (2010) Cell shape and contractility regulate ciliogenesis in cell cycle-arrested cells. *J. Cell Biol.* 191, 303–312

77 Mahmud, G. *et al*. (2009) Directing cell motions on micropatterned ratchets. *Nat. Phys.* 5, 606–612

78 Kushiro, K. *et al*. (2012) Modular design of micropattern geometry achieves combinatorial enhancements in cell motility. *Langmuir* 28, 4357–4362

79 Yoon, S-H. *et al*. (2011) A biological breadboard platform for cell adhesion and detachment studies. *Lab Chip* 11, 3555–3562

80 Csucs, G. *et al*. (2007) Locomotion of fish epidermal keratocytes on spatially selective adhesion patterns. *Cell Motil. Cytoskeleton* 64, 856–867

81 Doyle, A.D. *et al*. (2009) One-dimensional topography underlies three-dimensional fibrillar cell migration. *J. Cell Biol.* 184, 481–490

82 Pouthas, F. *et al*. (2008) In migrating cells, the Golgi complex and the position of the centrosome depend on geometrical constraints of the substratum. *J. Cell Sci.* 121, 2406–2414

83 Maiuri, P. *et al*. (2012) The world first cell race. *Curr. Biol.* 22, R673–R675

84 Théry, M. *et al*. (2007) Experimental and theoretical study of mitotic spindle orientation. *Nature* 447, 493–496

85 Fink, J. *et al*. (2011) External forces control mitotic spindle positioning. *Nat. Cell Biol.* 13, 771–778

86 Samora, C.P. *et al*. (2011) MAP4 and CLASP1 operate as a safety mechanism to maintain a stable spindle position in mitosis. *Nat. Cell Biol.* 13, 1040–1050

87 Kiyomitsu, T. and Cheeseman, I.M. (2012) Chromosome- and spindle-pole-derived signals generate an intrinsic code for spindle position and orientation. *Nat. Cell Biol.* 14, 311–317

88 Minc, N. *et al*. (2011) Influence of cell geometry on division-plane positioning. *Cell* 144, 414–426

89 Minc, N. and Piel, M. (2012) Predicting division plane position and orientation. *Trends Cell Biol.* 22, 193–200

90 Huang, S. *et al*. (2005) Symmetry-breaking in mammalian cell cohort migration during tissue pattern formation: role of random-walk persistence. *Cell Motil. Cytoskeleton* 61, 201–213

91 Burute, M. and Thery, M. (2012) Spatial segregation of cell–cell and cell–matrix adhesions. *Curr. Opin. Cell Biol.* 24, 628–636

92 Tseng, Q. *et al*. (2012) Spatial organization of the extracellular matrix regulates cell-cell junction positioning. *Proc. Natl. Acad. Sci. U.S.A.* 109, 1506–1511

93 McCain, M.L. *et al*. (2012) Cooperative coupling of cell-matrix and cell-cell adhesions in cardiac muscle. *Proc. Natl. Acad. Sci. U.S.A.* 109, 9881–9886

94 Serra-Picamal, X. *et al*. (2012) Mechanical waves during tissue expansion. *Nat. Phys.* 8, 628–634

95 Wan, L.Q. *et al*. (2011) Micropatterned mammalian cells exhibit phenotype-specific left-right asymmetry. *Proc. Natl. Acad. Sci. U.S.A.* 108, 12295–12300

96 Desai, R.A. *et al*. (2009) Cell polarity triggered by cell-cell adhesion via E-cadherin. *J. Cell Sci.* 122, 905–911

97 Reffay, M. *et al*. (2011) Orientation and polarity in collectively migrating cell structures: statics and dynamics. *Biophys. J.* 100, 2566–2575

98 Mullins, R.D. (2010) Cytoskeletal mechanisms for breaking cellular symmetry. *Cold Spring Harb. Perspect. Biol.* 2, a003392

velopmental Cell

Mitochondria in Apoptosis: Bcl-2 Family Members and Mitochondrial Dynamics

Jean-Claude Martinou[1,2,*], Richard J. Youle[2,*]

[1]Department of Cell Biology, University of Geneva, Faculty of Sciences, 30 quai Ernest-Ansermet, 1211 Geneva 4, Switzerland, [2]Surgical Neurology Branch, NINDS, National Institute of Health, Bethesda, MD 20892, USA

*Correspondence: jean-claude.martinou@unige.ch (J.-C.M.), youler@ninds.nih.gov (R.J.Y.)

Developmental Cell, Vol. 21, No. 1, July 19, 2011 © 2011 Elsevier Inc.
http://dx.doi.org/10.1016/j.devcel.2011.06.017

SUMMARY

Mitochondria participate in apoptosis through a range of mechanisms that vary between vertebrates and invertebrates. In vertebrates, they release intermembrane space proteins, such as cytochrome *c*, to promote caspase activation in the cytosol. This process is the result of the loss of integrity of the outer mitochondrial membrane caused by proapoptotic members of the Bcl-2 family. This event is always accompanied by a fissioning of the organelle. Fission of mitochondria has also been reported to participate in apoptosis in *Drosophila* and *Caenorhabditis elegans*. However, in these organisms, mitochondrial membrane permeabilization does not occur and the mechanism by which mitochondrial dynamics participates in cell death remains elusive.

INTRODUCTION

Mitochondria are essential organelles because they supply the cell with metabolic energy in the form of ATP generated by oxidative phosphorylation. In addition they perform a number of other key metabolic reactions. Their shape, from spherical to elongated, is continually remodeled by fusion and fission events that link all the organelles within a cell into a continuum over time. Although why mitochondria are so dynamic is not known, the process is essential for mitochondrial maintenance in yeast (Hoppins et al., 2007) and mammals (Chen et al., 2003). Recently, mitochondrial fission has been linked to the cellular death program of apoptosis. Mitochondria are involved

185

CellPress

in the so-called intrinsic pathway of apoptosis where they release soluble proteins, including cytochrome *c*, from the intermembrane space to initiate caspase activation in the cytosol (Kroemer et al., 2007; Vaux, 2011). The release of these proteins is a consequence of the integrity of the mitochondrial outer membrane (OMM) being compromised, a process called mitochondrial outer membrane permeabilization (MOMP). So far, this process has only been validated in vertebrates; it does not seem to be required in *C. elegans* and is debated in *Drosophila*. As reviewed here, in these invertebrates, mitochondria may be involved in apoptosis through distinct mechanisms. In vertebrates, MOMP is under the control of the proapoptotic Bcl-2 family members. Here, we will focus on the mechanisms by which Bcl-2 family members elicit MOMP during apoptosis. We will also review the link that connects some Bcl-2 family proteins with fission and fusion of the organelle in apoptosis and in healthy cells.

BCL-2 FAMILY MEMBERS

Bcl-2 family proteins are subdivided into three groups on the basis of their pro- or antiapoptotic action and the Bcl-2 Homology (BH) domains they possess (Figure 1A) (Schinzel et al., 2004; Youle and Strasser, 2008). Antiapoptotic Bcl-2-like proteins (e.g., Bcl-2, Bcl-x_L, Bcl-w, Mcl-1, and A1/Bfl-1) and proapoptotic Bax-like proteins (e.g., Bax, Bak, and Bok/Mtd) display four BH domains (Kvansakul et al., 2008). The proapoptotic BH3-only proteins (e.g., Bid, Bim/Bod, Bad, Bmf, Bik/Nbk, Blk, Noxa, Puma/Bbc3, and Hrk/DP5), on the other hand, possess only a short motif called the BH3 domain as their name indicates. BH3 proteins integrate and transmit death signals that emanate from defective cellular processes to other Bcl-2 family members. Through their BH3 domain, these proteins either interact with antiapoptotic proteins to inhibit their function and/or to interact directly with multidomain proteins such as Bax or Bak to stimulate their activity. The former are often referred to as "sensitizers," whereas the later are classified as "activators" (reviewed in Giam et al., 2008). The multidomain proapoptotic proteins, Bax and Bak, and perhaps Bok in some tissues, are responsible for MOMP and are the master effectors of apoptosis as cells lacking Bax and Bak fail to undergo MOMP and apoptosis in response to many death stimuli (Wei et al., 2001).

MECHANISMS OF ACTIVATION OF PROAPOPTOTIC BCL-2 FAMILY MEMBERS

Activation of the proapoptotic Bcl-2 family protein Bax results from a highly regulated multistep process involving its translocation from the cytosol to the OMM where it inserts and oligomerizes. In contrast to Bax, Bak is

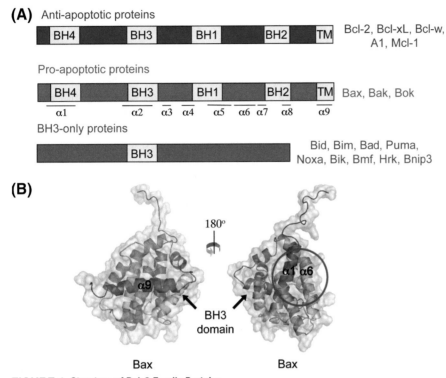

FIGURE 1 Structure of Bcl-2 Family Proteins
(A) The Bcl-2 family is divided into three groups based on their Bcl-2 Homology domains (BH). Pro- and antiapoptotic proteins contain 4 BH domains, while BH3-only proteins, as their name indicates, contain only the BH3 domain.
(B) Structure of Bax showing the α helix 9 (transmembrane domain) embedded in the hydrophobic groove (left structure). A 180° rotation (structure on the right) shows the α helices 1 and 6. The red circle represents the binding site for the Bim BH3 domain.

constitutively inserted in the OMM by a C-terminal transmembrane domain. Its insertion can be facilitated by the voltage-dependent anion channel iso-form 2 with which it was found to interact (Cheng et al., 2003; Lazarou et al., 2010; Roy et al., 2009). In this section, we will describe what maintains Bax in the cytosol in healthy cells and how BH3-only proteins, together with the lipid bilayer, cooperate to promote oligomerization of Bax and Bak in the OMM during apoptosis.

Bax Travels Back and Forth from the Cytosol to the OMM in Healthy Cells

In healthy cells, Bax is mainly cytosolic with a minor fraction loosely attached to mitochondria. In contrast to what is observed in cultured cells, the amount of membrane-bound Bax is negligible in tissues such as liver or brain. This suggests that the mitochondrial sub-population of Bax at the mitochondria could be a measure of the stress experienced by the cells, which increases with in vitro culture. Using fluorescence loss in photobleaching, Edlich et al. (2011) observed that Bax constantly travels back and forth from the cytoplasm to mitochondria (Figure 2). Once attached to the OMM, it can be retrotranslocated to the cytosol by Bcl-x_L by a mechanism that is not yet fully understood. Bax retrotranslocation by Bcl-x_L and possibly by other prosurvival Bcl-2 proteins, would ensure that Bax does not chronically accumulate at the OMM to reach a critical level that could promote its autoactivation. It is still unclear what allows Bax to attach to the OMM specifically in healthy cells. The most likely explanation is that Bax undergoes discrete and reversible conformational changes that would expose not only its BH3 domain but also its N- and C-terminal domains. Consistently, it was previously observed that during detachment of cells from the extracellular matrix, the N terminus of Bax undergoes a conformational change that can be reversed upon reattachment of cells to substrate (Gilmore et al., 2000). Understanding what triggers and controls changes in Bax conformation in healthy cells requires further investigations. A reversible conformational change in Bax could be induced by a simple contact of the protein with the lipid bilayer of the OMM, as suggested by experiments performed with liposomes (Yethon et al., 2003). Other triggers, including BH3-only proteins, prostaglandins (Lalier et al., 2011), p53 (Chipuk et al., 2004), pH variations (Khaled et al.,

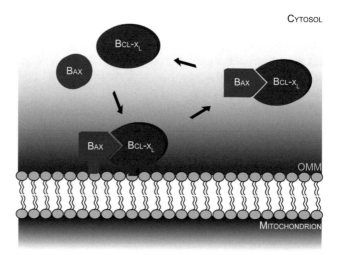

FIGURE 2 Bax Moves Back and Forth from the Cytosol to the Mitochondrial Outer Membrane
In healthy cells, Bax (blue) and Bclx-L (red) are translocated from the cytosol to the outer mitochondrial membrane (OMM) where they attach loosely. The triangle in Bax represents its BH3 domain, which needs to be exposed in order for Bcl-x_L-mediated retrotranslocation to occur. OMM: outer mitochondrial membrane.

1999), or posttranslational modifications (Kutuk and Letai, 2008), all could also be responsible for sending Bax to the OMM. But what would prevent a complete conformational change in Bax in healthy cells? The structure of the OMM may counteract Bax activation. Its cholesterol content has been shown to prevent Bax insertion or oligomerization in the OMM (Christenson et al., 2008; Lucken-Ardjomande et al., 2008). Moreover, previous data have shown that stress-induced mitochondrial hyperfusion, which leads to highly elongated organelles, delays Bax activation upon exposure of cells to apoptotic stimuli (Tondera et al., 2009). Thus, the lipid content and the shape of the OMM may be refractory to complete Bax activation in healthy cells.

The Role of BH3-Only Proteins and of the Lipid Bilayer

The protein tBid has been originally described as a direct activator of Bax, able to induce a conformational change of the Bax N terminus, as well as modulate Bax insertion and oligomerization in the OMM (Desagher et al., 1999; Eskes et al., 2000). Many groups have confirmed and extended these observations to other BH3-only proteins, including Bim and Puma (Gallenne et al., 2009; Kim et al., 2009; Kuwana et al., 2005; Mérino et al., 2009; Ren et al., 2010). Importantly, these studies have also provided details about the mechanisms by which BH3-only proteins trigger conformational changes in Bax, leading to its insertion and oligomerization in the OMM. Most BH3-only proteins, except Bid, appear to be intrinsically unstructured proteins. However, their BH3 domain is known to form an α helix upon binding to the hydrophobic groove that is on the surface of antiapoptotic proteins. In Bax, this hydrophobic groove is occupied by α helix 9 (the transmembrane domain) (Figure 1B). Until recently, it was unclear where BH3 activators impact Bax to induce the structural changes required for its insertion and assembly in the OMM. To answer this question, Walenski and colleagues used peptides whose α helicity was stabilized by a chemical modification termed "hydrocarbon stapling." They derived staple peptides from the BH3 domain of Bid or Bim and reported the first evidence of a direct interaction between the BH3 domain of Bim and Bid with Bax (Gavathiotis et al., 2008; Walensky et al., 2006). Moreover, consistent with previous data (Cartron et al., 2004), their studies localized the binding site of the BH3 domain of Bim to the junction of α helices 1 and 6 on the N-terminal side of Bax, i.e., on the opposite side of the hydrophobic groove (Gavathiotis et al., 2008) (Figure 1B). Thus, BH3-only proteins appear to interact with different sites in Bax and in prosurvival Bcl-2 proteins in which the hydrophobic groove is the receiving domain. According to a model derived from recent NMR studies, upon interaction with Bim, Bax structural metamorphosis begins with the displacement of its α helices 1 and 2 from a closed to an open position, which leads to the release of the α helix 9 from the hydrophobic groove and exposure of the BH3 domain (Gavathiotis et al., 2010). Whether the results obtained with peptides in solution in these studies faithfully reflect what occurs in the cell, at the OMM, will require further investigation (discussed in Czabotar et al., 2009). Moreover, it will be important to clarify where BH3-only proteins impact Bak, as a previous report defined the hydrophobic groove as the receiving domain of the tBid BH3 domain (Moldoveanu et al., 2006).

In contrast to the stapled Bim peptide, tBid is unable to interact with full-length Bax or Bcl-x_L in solution (Lovell et al., 2008) begging the question of where in the cell do interactions between BH3-only proteins and Bax or Bak occur. Following its cleavage by caspase 8, the N terminus of Bid (N-Bid) remains attached to the C-terminal portion of the protein (C-Bid), possibly masking the BH3 domain in C-Bid. However, in the presence of a lipid bilayer with a phospholipid composition mimicking that of the OMM, it was shown that only C-Bid inserts while N-Bid remains in solution (Lovell et al., 2008). Thus, upon interaction with the membrane, tBid is fully operational to recruit Bax to the OMM as shown by fluorescence resonance energy transfer (Lovell et al., 2008). These results emphasize the importance of the lipid environment for activation of these Bcl-2 proteins and prompted Andrews and colleagues to propose the "membrane-embedded together" model. According to this model, tBid first inserts in the membrane, then recruits Bax that in turn inserts and oligomerizes. According to these authors, Bcl-x_L and possibly other antiapoptotic proteins such as Bcl-w that are in part cytosolic, are also recruited and activated at the membrane by both the activator and sensitizer BH3-only proteins (Shamas-Din et al., 2011). This part of the model confers on BH3-only proteins an ambiguous role since, in theory, they could recruit both pro- or antiapoptotic proteins. If this is the case, and if the role of antiapoptotic Bcl-2 proteins in the OMM is to prevent Bax or Bak oligomerization, then BH3-only proteins should be classified as antiapoptotic proteins, which is not consistent with their known function. Therefore, it is unlikely that the BH3-only proteins recruit antiapoptotic Bcl-2 family members into the OMM to inhibit Bax or Bak function. Rather, BH3-only proteins may recruit antiapoptotic proteins and remain tightly bound to them to inhibit their function. Soluble and membrane bound antiapoptotic Bcl-2 proteins could therefore be inhibited either in the cytosol or at the membrane level by soluble or membrane-associated BH3-only proteins, respectively. Moreover, antiapoptotic Bcl-2 proteins, whether constitutively present in membranes (such as Bcl-2), or cytosolic (such as Bcl-x_L and Bcl-w), have the possibility to directly interact with the BH3-only domain of Bax or Bak during the process of their oligomerization. This implies that the affinity of antiapoptotic proteins for the BH3 domain of Bax or Bak must be very high to compete with other Bax or Bak molecules and prevent their homo-oligomerization.

Is the Lipid Composition of the Membrane Crucial for tBid-Induced Bax Oligomerization?

Using liposomes containing phosphatidylcholine and phosphatidylglycerol, it was previously found that tBid can recruit Bax to the membrane and drive formation of small pores that could allow passage of small molecules (such as carboxyfluorescein) but were impermeable to cytochrome c (Roucou et al., 2002). Analysis of the quaternary structure of Bax revealed that the protein was monomeric under this condition. At the same time, Kuwana et al. (2002) and later others (Lucken-Ardjomande et al., 2008; Terrones et al., 2004), reported that cardiolipin is required for the formation of tBid-induced Bax oligomers and of giant pores. Thus, the presence of cardiolipin in the membrane appears to promote tBid-induced Bax oligomerization and formation of cytochrome c-permeable pores. Cardiolipin is a negatively charged phospholipid, exclusively found in mitochondria, mainly in the IMM where it is important for the activity of several complexes of the respiratory chain and of some transporters (Hasler, 2011; Schlame et al., 2000; Schug and Gottlieb, 2009). Subfractionation of the mitochondrial membranes, together with the accessibility of cardiolipin to enzymatic degradation by Phospholipase 2 added to isolated mitochondria, indicate that small amounts of this phospholipid are also present in the OMM, probably enriched at contact sites between the OMM and the IMM (Ardail et al., 1990; Daum, 1985; Epand et al., 2007; Gebert et al., 2009; Hovius et al., 1990). tBid was found to bind cardiolipin through central α helices, which would explain its recruitment to mitochondria, principally at contact sites between the IMM and the OMM (Kim et al., 2004; Lutter et al., 2000). In addition to promoting membrane insertion of tBid, cardiolipin could be involved in the Bax oligomerization process itself as suggested by the experiment described above (Roucou et al., 2002). Besides cardiolipin, Gross and colleagues recently identified MTCH2/MIMP, a protein similar to mitochondrial carriers of the adenine nucleotide translocator family, as a receptor for tBid (Zaltsman et al., 2010). In the absence of this protein, tBid is recruited less efficiently to the OMM and apoptosis is delayed when cells are exposed to tBid-dependent death stimuli. Thus, MTCH2/MIMP seems to accelerate the recruitment of tBid to the OMM and may cooperate with cardiolipin to provide the OMM with an optimal concentration of tBid. Further experiments, in particular, using proteoliposomes prepared with recombinant MTCH2/MIMP, are required to determine how tBid binds this carrier protein. Moreover, whether cardiolipin is needed for Bax activation and MOMP in the cell requires further investigation as several reports have shown that this phospholipid can be dispensable for Bax activation under some conditions (Gonzalvez et al., 2008; Iverson et al., 2003; Polcic et al., 2005). Importantly, in most studies using liposomes, the process of Bax oligomerization induced by tBid does not appear to be optimal as only a small fraction of Bax oligomerizes, even when high concentrations of tBid are used and despite the presence of cardiolipin in membranes (Kuwana et al., 2002; Schafer et al., 2009). This suggests that, in vitro, either important components that facilitate Bax insertion and/or oligomerization are missing or that the lipid bilayer structure in these experimental setups is not optimal to allow efficient Bax oligomerization.

The Role of Mitochondrial Fission

Mitochondria are dynamic organelles that fuse and divide continuously. The core components of the fusion and fission machineries have been identified as the dynamin-related proteins Drp1, mitofusins (Mfn) 1 and 2, and Opa1 (Box 1) (reviewed in Westermann, 2010). Fission of mitochondria is a constant in apoptosis (Figure 3) (Martinou and Youle, 2006). The mechanism underlying this event appears to involve Drp1, the most downstream component of the fission machinery (Frank et al., 2001). The precise mechanism by which Drp1 is recruited to the mitochondria during apoptosis remains vague. Under normal conditions, Drp1 moves back and forth from the cytosol to the OMM. Following Bax activation, Drp1 stably associates with the OMM due to Bax/Bak-dependent SUMO modification (Wasiak et al., 2007). However, the mechanistic link between activation of Bax or Bak and Drp1 SUMOylation is still missing. Importantly, it should be mentioned that inhibition of Drp1, delays, but does not completely prevent, mitochondrial fission (Ishihara et al., 2009). This suggests that a Drp1-independent mechanism also seems to participate in the fission of the organelle during apoptosis. A link between mitochondrial fission and Bax activation was suggested by the detection of active Bax in discrete foci at mitochondrial fission sites (Karbowski et al., 2002). Bak, which initially surrounds the organelle, also coalesces into foci at mitochondrial fission sites during apoptosis. In most

---

BOX 1 DYNAMIN-RELATED PROTEINS INVOLVED IN THE FUSION AND FISSION MITOCHONDRIAL MACHINERIES

(1) Drp1 is a large GTPase of the dynamin family that is required for mitochondrial fission. The protein is cytosolic and translocates to the mitochondria where it binds to receptors such as Mff and Fis1. It assembles into rings around mitochondria and constricts and breaks down their membrane by a mechanism that is not completely understood.

(2) Mitofusins: two mitofusins (Mfn), Mfn1, and Mfn2, are required in the fusion of the outer mitochondrial membrane. These proteins are integrated in the OMM, facing the cytosol, and mediate the docking of mitochondria to one another during fusion by engaging in trans-homo-oligomeric and hetero-oligomeric complexes.

(3) Opa1 (Optic atrophy 1): As its name implies, this protein was found to be mutated in the disease dominant optic atrophy, which leads to degeneration of the optic nerve. Opa1 is required for fusion of the inner mitochondrial membrane. Different isoforms of this protein that form a complex integrated in the IMM, facing the intermembrane space of mitochondria, have been identified.

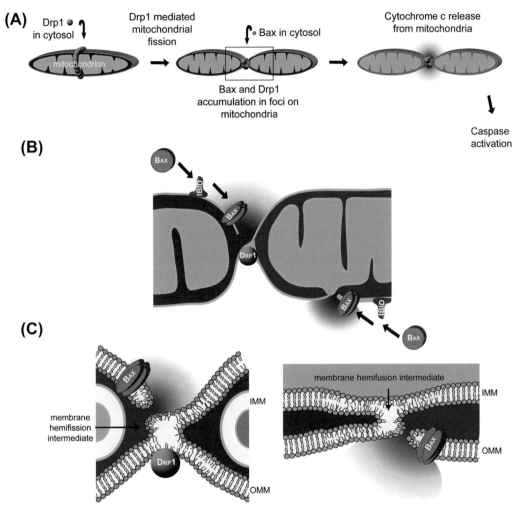

FIGURE 3 Model for Bax Activation and Mitochondrial Outer Membrane Permeabilization: The Role of Membrane Hemifission and Hemifusion Intermediates

(A) During apoptosis, Drp1 and Bax are recruited to mitochondrial foci. As a result MOMP occurs concomitantly with mitochondrial fission. The release of cytochrome *c* promotes caspase activation in the cytosol.

(B) A magnification of what occurs at the level of mitochondrial fission sites. According to the "embedded together model," tBid first inserts in the outer mitochondrial membrane (OMM) to expose its BH3 domain. Upon interaction with the Bid BH3 domain, Bax is, in turn, recruited to the OMM where it inserts and oligomerizes. This process does not occur randomly at the OMM surface. Bax mainly oligomerizes at mitochondrial fission sites and at contact sites between the OMM and the IMM.

(C) (Left) A detailed structure of the hemifission intermediate that forms before fission of the organelle. The membrane forms a nonbilayer structure that represents a privileged site for the formation of a membrane hole. Upon its oligomerization in the OMM, Bax forms a lipidic pore or triggers the formation of a hole at the edge of the hemifission intermediate. (Right) A detailed structure of the membrane at OMM and inner mitochondrial membrane (IMM) contact sites. The model postulates that Bax forms a pore in the membrane at the edges of membrane hemifusion intermediates that may form between the OMM and the IMM.

cases, (although not always), these foci appear to be located in proximity to Drp1 and Mfn2 and form independently of Drp1 activity (Frank et al., 2001; Ishihara et al., 2009; Montessuit et al., 2010). By inhibiting Drp1 and delaying mitochondrial fission, Karbowski et al. (2002) could observe these Bax spots at the surface of elongated mitochondria. Interestingly, some of them were located at constriction sites that the authors described as "aberrant mitochondrial scission attempts." This observation further supports the existence of a Drp1-independent mechanism responsible for constricting, and possibly fissioning, the organelle during apoptosis. Such constriction sites have also been observed in healthy Drp1-deficient cells, not only on mitochondria (Ishihara et al., 2009), but also on peroxisomes, which rely on Drp1 for division (Koch et al., 2004). These sites may provide the appropriate membrane curvature, or lipid composition, for the assembly of Drp1 rings and the formation of Bax foci during apoptosis.

It was recently found that the formation of membrane hemifusion intermediates is a process able to promote tBid-induced Bax oligomerization (Montessuit et al., 2010). This membrane remodeling can be promoted by Drp1 in a cell free system and may recapitulate the membrane remodeling that occurs at mitochondrial fission sites in cells undergoing apoptosis (Figure 3). When membranes divide or fuse, contacting leaflets fuse to form a stalk. The stalk then expands to form a metastable nonbilayer membrane intermediate also called hemifission or hemifusion intermediate for membrane fission and fusion respectively (Chernomordik and Kozlov, 2008; Kozlovsky and Kozlov, 2003). For fusion, this transient state would evolve toward the formation of a fusion pore, while for membrane fission it would decay spontaneously into two separate membranes, thereby completing the fission process. The mechanism by which these membrane intermediates promote Bax oligomerization is still unclear. Hemifusion or hemifission intermediates are expected to occur during fusion or fission of the organelle, respectively. These structures may also possibly form at contact sites between the OMM and the IMM, which are enriched in cardiolipin and phosphatidylethanolamine, two phospholipids that have a propensity to form nonbilayer structures (Figure 3). The limited number of such membrane structures at the surface of mitochondria may explain why Bax or Bax oligomers are not randomly distributed in the OMM in apoptotic cells.

HOW DOES BAX OR BAK PERMEABILIZE THE OUTER MITOCHONDRIAL MEMBRANE?

The mechanism by which Bax, or Bak, permeabilizes membranes to allow efflux of large proteins is still unclear, although several models having been proposed and challenged. These models predict that Bax, alone or combined with other proteins, forms channels large enough to allow passage of cytochrome c and other proteins. Alternatively, Bax or Bak could modulate the opening of existing channels such as the so-called permeability transition pore, to induce MOMP. However, opening of the permeability transition pore has been questioned by a number of genetic studies that have excluded the requirement of some of its major components (reviewed in Tait and Green, 2010; Westphal et al., 2011). It is therefore unlikely that the permeability transition pore plays a role in Bax and Bak-induced MOMP.

Proteinaceous or Lipidic Pores

The model of Bax or Bak pore formation finds its roots in the determination of the 3D structure of Bcl-x_L (Muchmore et al., 1996). This protein exhibits a structural similarity to bacterial pore-forming toxins, certain colicins, and the translocation domain of diphtheria toxin. Other members of the Bcl-2 family were later determined to have a similar structure. Consistent with this structural resemblance, Bax and Bcl-2 were found to form pores in planar lipid bilayers (Antonsson et al., 1997; Schlesinger et al., 1997). The ability of Bax and Bak to induce MOMP seems to reside at least in part in the nature of their central α helices 5 and 6. Thus, replacing the α helix 5 of Bcl-x_L by the equivalent in Bax is sufficient to turn Bcl-x_L into a "killer" protein (George et al., 2007). In addition, differences in the ability of pro- and antisurvival proteins to oligomerize could explain their capacity to induce MOMP. Biochemical data support a model in which Bak and Bax oligomers are formed by two interfaces involving the BH3:groove and α helix 6: α helix 6 interfaces, while α helices 5, 6, and 9 are embedded in the membrane (Dewson et al., 2008; Zhang et al., 2010). Although a model for Bax or Bak oligomerization is beginning to emerge, the atomic details of membrane-embedded Bax and Bak oligomer are still missing. In particular, the number of units present in the Bax or Bak oligomers remains elusive and estimates vary considerably depending on the approaches used. Assessing the minimal size of the Bax or Bak oligomer that is responsible for MOMP is critical, but difficult. Indeed, once oligomerization has been initiated, self-oligomerization proceeds which may lead to the recruitment of hundreds of Bax molecules, which then protrude out of the OMM (Nechushtan et al., 2001b). Whether all these molecules are required for pore-formation in the membrane is questionable (Düssmann et al., 2010). Such heterogeneity of oligomers may hamper resolution of the 3D structure of these proteins within a membrane. Moreover, the nature of the pore formed by Bax—whether proteinaceous or lipidic (Box 2)—is still under debate. Although some reports favor formation of a proteinaceous channel (Dejean et al., 2005; Epand et al., 2002), other results are consistent with formation of a lipidic

BOX 2 PROTEINACEOUS VERSUS LIPIDIC PORES

Many toxins are known to spontaneously induce trans-membrane pores in lipid bilayers under certain conditions. Depending on their structure, they can form either protein-aceous or lipidic pores (Yang et al., 2001).

Proteinaceous pores (also called barrel-stave pores): tox-ins form a barrel-like pore that spans the membrane. The pore lumen is lined by peptides that are perpendicularly inserted in the membrane.

Lipidic pores (also called toroidal pores): insertion of the peptides in the membrane triggers lipid monolayer bending such that the outer and inner leaflets of the membrane are continuous. The pore is lined by both the peptides and the lipid head groups.

pore (Basañez et al., 1999; García-Sáez et al., 2007; García-Sáez et al., 2006). In favor of the latter model, α helix 5 of Bax, which is sufficient to perforate membranes, was found to form lipidic pores in a synthetic membrane by X-ray diffraction (Qian et al., 2008). However, one must be cautious with overall structural predictions of Bax based on data obtained with only minimal domains of the protein. Nevertheless, there are other reasons to favor the lipidic pore model for Bax and Bak. Recently, analysis of Bax-permeabilized liposomes by cryoelectron microscopy revealed large openings of the membrane, from 25 to 100 nm, compatible with the release of megadalton dextrans. Importantly, the edges of these pores were devoid of protein (Bleicken et al., 2010; Schafer et al., 2009). Whether pores of this type are formed in the OMM of mitochondria under-going MOMP remains now to be shown.

MECHANISTIC LINK BETWEEN MITOCHONDRIAL DYNAMICS AND MOMP: THE ROLE OF MEMBRANE HEMIFUSION/HEMIFISSION INTERMEDIATES

The notion that MOMP and mitochondrial fission are mechanistically linked remains controversial. In some reports, inhibition of mitochondrial fission or stimulation of fusion, significantly delayed apoptosis (Cassidy-Stone et al., 2008; Frank et al., 2001; Germain et al., 2005; Merrill et al., 2011). Puzzlingly, cells from the neural crest failed to die in Drp1-knockout mouse embryos (Wakabayashi et al., 2009), whereas mouse embryonic fibro-blast cell lines generated from these Drp1-deficient mice displayed little (Wakabayashi et al., 2009) or partial (Ishihara et al., 2009) resistance to apoptosis. In other studies, inhibiting mitochondrial fission in HeLa cells had no or only a minor impact on the kinetics of MOMP and cell death (Estaquier and Arnoult, 2007; Ishihara et al., 2009; Parone et al., 2006; Sheridan et al., 2008). Moreover, some studies showed that it was pos-sible to dissociate mitochondrial fission from MOMP. James et al. (2003)

reported that Bcl-x_L could inhibit MOMP and apoptosis due to overexpression of the putative Drp1 receptor hFis1, without preventing Fis-1-induced fission. On the other hand, Sheridan et al. (2008) observed that Bcl-x_L, as well as other members of the apoptosis-inhibitory subset of the Bcl-2 family, antagonized Bax and/or Bak-induced cytochrome c release but failed to block mitochondrial fragmentation associated with Bax/Bak activation. Several studies also reported that inhibition of mitochondrial fission did not prevent the efflux of Smac/DIABLO, a protein present in the intermembrane space of mitochondria, although it could delay cytochrome c release (Estaquier and Arnoult, 2007; Parone et al., 2006). These data favor a model in which mitochondrial fission inhibition could retard cytochrome c release by preventing the remodeling of mitochondria cristae, a process that is thought to be required for efficient cytochrome c mobilization (Cipolat et al., 2006; Yamaguchi et al., 2008). Re-evaluation of these conflicting data, in particular, in light of the new results by Montessuit et al. (2010) on the role of mitochondrial membrane hemifusion/hemifission intermediates in Bax oligomerization, provides at least two explanations for why inhibiting Drp1 can only delay cytochrome c release and cell death. First, this is because the inhibition of known components of the mitochondrial fission machinery cannot completely prevent mitochondrial fission as previously mentioned. Second, this is because it appears that rather than the fission of mitochondria per se, it is the formation of membrane hemifission intermediates that may play a central role in Bax oligomerization and MOMP. According to theoretical studies, a place where the cost in energy to form a membrane hole would be minimal is at the edge of a membrane hemifusion structure (Katsov et al., 2004). The observation that Bax is confined at discrete foci at the surface of mitochondria, some of which coincide with mitochondrial fission sites, together with the new role of membrane hemifusion or hemifission intermediates in the oligomerization of Bax during apoptosis, raise the possibility that Bax may opportunistically target those sites that are optimally designed for it to oligomerize and to perforate the membrane (Figure 3).

THE ROLE OF MITOCHONDRIA IN *DROSOPHILA* CELL APOPTOSIS

Steps of apoptosis upstream of caspases are regulated differently in flies and mammals (Salvesen and Abrams, 2004). Indeed, the role of Bcl-2 family members in *Drosophila* in apoptosis is relatively minor (Sevrioukov et al., 2007). In contrast, the Hid/Grim/Reaper gene products that derepress the IAP proteins that block caspases through the ubiquitin proteosome pathway are potent inducers of fly cell apoptosis (Bader and Steller, 2009). Recent work in flies links mitochondrial fission and fusion processes to Hid, Grim, and Reaper, as well as to the Bcl-2 family members Buffy and DEBCL and to apoptosis. As derepressors of cytosolic IAP, there had been no explanation why Hid (Haining et al., 1999), Grim (Clavería et al., 2002), and Reaper (Olson et al., 2003) localize to mitochondria until recent work has linked Reaper to the mitochondrial fusion machinery. Although caution should be taken in interpreting these results due to the expression in heterologous cells of fly proteins that lack known mammalian homologs, Reaper expressed in HeLa cells localizes to the OMM and inhibits mitochondrial fusion inducing mitochondrial fragmentation. Reminiscent of mammalian proapoptotic Bcl-2 family members (Nechushtan et al., 2001a), Reaper forms concentrated foci on mitochondria (Thomenius et al., 2011). Overexpression in HeLa cells of Mfn2, a dynamin family member that mediates mitochondrial fusion, changes Reaper localization from foci to more evenly coat mitochondria suggesting they interact, consistent with data that a peptide of Reaper can coimmunoprecipitate with *Xenopus* Mfn2 (Thomenius et al., 2011). Mitochondrial localization, mitochondrial fragmentation activity and Mfn2 interaction required the GH3 domain of Reaper indicating that Reaper binds Mfn2 on mitochondria through the GH3 domain that is needed to induce apoptosis (Thomenius et al., 2011). Alternatively or additionally, mitochondrial Hid may recruit Reaper to mitochondria (Sandu et al., 2010). Interestingly, mitochondrial Reaper and Hid can recruit IAPs to mitochondria where they may induce IAP autoubiquitination and proteosomal degradation. Mitochondria fragment during apoptosis in *Drosophila* (Goyal et al., 2007; Abdelwahid et al., 2007) possibly through Reaper interaction with mitofusin and inhibition of mitochondria fusion (Thomenius et al., 2011). Preventing this mitochondrial fragmentation by inhibition of Drp1 prevents death of fly cells in vitro and in vivo (Abdelwahid et al., 2007; Goyal et al., 2007).

APOPTOSIS IN THE WORM

CED-9, the sole multidomain Bcl-2 family member in *C. elegans,* is localized to the mitochondria (Chen et al., 2000). Although it primarily inhibits apoptosis through the binding and sequestration of the caspase activator, CED-4, CED-9 can promote apoptosis in weak loss of function CED-3 caspase mutants (Hengartner and Horvitz, 1994). As in mammalian and *Drosophila* cells, mitochondria fragment in the worm during apoptosis and this fragmentation is dependent on CED-9 and the BH3-only protein, EGL-1 (Jagasia et al., 2005). Although there is no evidence of cytochrome c release or MOMP in the worm, inhibition of mitochondrial fragmentation by expression of a dominant negative inhibitor of Drp1 prevents 20% of the normal developmental cell death in *C. elegans* (Jagasia et al., 2005).

BCL-2 FAMILY PROTEINS AND MITOCHONDRIAL DYNAMICS

The link between mitochondrial fission and apoptosis has been further strengthened by several studies that show Bcl-2 family members alter mitochondrial dynamics even in healthy cells (Delivani et al., 2006; Karbowski et al., 2006; Rolland et al., 2009; Tan et al., 2008). Overexpression of Bcl-2 family proteins CED-9 and Bcl-x$_L$ can induce mitochondrial fusion in mammalian cells (Delivani et al., 2006) and in *C. elegans* (Rolland et al., 2009; Yamaguchi et al., 2008). Bcl-x$_L$ can also accelerate mitochondrial fission in mammalian neurons thereby accelerating mitochondrial dynamics (Berman et al., 2009). The proapoptotic Bax and Bak, although having the opposite effect on cell viability relative to Bcl-x$_L$, are required for the normal rate of mitochondrial fusion in cells (Karbowski et al., 2006). Recent work also shows that in a cell free system of mitochondrial fusion, recombinant Bax can stimulate the process specifically through Mfn2 and not Mfn1 (Hoppins et al., 2011). Tethering Bax with disulfides to hinder conformational change abolishes the mitochondrial fusion activity indicating that at least subtle conformational changes are involved. Promotion of mitochondrial fusion may be a consequence of direct interaction between Bcl-2 family members and Mfn1/2 (Brooks et al., 2007; Delivani et al., 2006; Rolland et al., 2009; Cleland et al., 2011) although how mitofusins may be activated remains unclear. Recent work links Bcl-2 family members to mitochondrial dynamics also in *Drosophila*, during fly oogenesis. Loss of the fly Bcl-2 family members, Buffy and/or DEBCL, disrupts nurse cell death, which occurs after the nurse cells transfer some of their cytosol and organelles into the developing oocyte. Buffy and/or DEBCL loss also increases mitochondrial network formation and mitochondrial clumping in the surviving cells, suggesting that in certain tissues in the fly, Bcl-2 family members affect mitochondrial dynamics upstream of apoptosis (Tanner et al., 2011). Thus, links between Bcl-2 family members and mitochondrial dynamics have been identified in mammals, insects, and roundworms. How the effects of the Bcl-2 family members on mitochondrial dynamics in healthy cells may relate to mitochondrial fragmentation during apoptosis (subsequent to Bcl-2 family member conformational change and insertion into mitochondrial membranes) is an important question that remains to be resolved.

CONCLUDING REMARKS

Over the past decade, our understanding of the mechanism of action of Bcl-2 family members has expanded significantly. The 3D structure of many members of the family, in solution, has been solved and the mechanism by which these proteins interact is better understood, even though some controversies remain as to how BH3-only proteins regulate multidomain Bcl-2 family members. Importantly, the lipid composition and structural organization of the OMM has emerged as one of the central players in the regulation of Bcl-2 family member activity. However, the precise 3D structure that many Bcl-2 proteins adopt in this membrane is still unknown. Without this essential information, it is impossible to understand precisely how MOMP occurs. The challenge for the coming years will be to solve the structure of membrane-embedded key Bcl-2 family members and to understand how the lipid composition and structure of the OMM controls activation of Bcl-2 family members and MOMP.

Recent work from several laboratories has also unraveled new functions for many pro- and antiapoptotic members of the Bcl-2 family, besides their role in apoptosis. Although our review has been focused on the involvement of Bcl-2 family members in mitochondrial dynamics, it is clear that these proteins play additional roles in other cellular processes. These functions, and new ones that may be discovered in the future, will certainly help decipher the complex role Bcl-2 family members play in life and death of the cell.

ACKNOWLEDGMENTS

We thank Drs. Megan Cleland and Soojay Banerjee for help with artwork. J.-C.M is supported by the Swiss National Foundation (subsidy 31003A-124968/1), Oncosuisse, and the Geneva Department of Education. R.J.Y. is supported by the Intramural Program of the NINDS, NIH.

REFERENCES

Abdelwahid, E., Yokokura, T., Krieser, R.J., Balasundaram, S., Fowle, W.H., and White, K. (2007). Mitochondrial disruption in Drosophila apoptosis. Dev. Cell 12, 793–806.

Antonsson, B., Conti, F., Ciavatta, A., Montessuit, S., Lewis, S., Martinou, I., Bernasconi, L., Bernard, A., Mermod, J.J., Mazzei, G., et al. (1997). Inhibition of Bax channel-forming activity by Bcl-2. Science 277, 370–372.

Ardail, D., Privat, J.P., Egret-Charlier, M., Levrat, C., Lerme, F., and Louisot, P. (1990). Mitochondrial contact sites. Lipid composition and dynamics. J. Biol. Chem. 265, 18797–18802.

Bader, M., and Steller, H. (2009). Regulation of cell death by the ubiquitin-proteasome system. Curr. Opin. Cell Biol. 21, 878–884.

Basañez, G., Nechushtan, A., Drozhinin, O., Chanturiya, A., Choe, E., Tutt, S., Wood, K.A., Hsu, Y., Zimmerberg, J., and Youle, R.J. (1999). Bax, but not Bcl-xL, decreases the lifetime of planar phospholipid bilayer membranes at subnanomolar concentrations. Proc. Natl. Acad. Sci. USA 96, 5492–5497.

Berman, S.B., Chen, Y.B., Qi, B., McCaffery, J.M., Rucker, E.B., 3rd, Goebbels, S., Nave, K.A., Arnold, B.A., Jonas, E.A., Pineda, F.J., and Hardwick, J.M. (2009). Bcl-x L increases mitochondrial fission, fusion, and biomass in neurons. J. Cell Biol. 184, 707–719.

Bleicken, S., Classen, M., Padmavathi, P.V., Ishikawa, T., Zeth, K., Steinhoff, H.J., and Bordignon, E. (2010). Molecular details of Bax activation, oligomerization, and membrane insertion. J. Biol. Chem. 285, 6636–6647.

Brooks, C., Wei, Q., Feng, L., Dong, G., Tao, Y., Mei, L., Xie, Z.J., and Dong, Z. (2007). Bak regulates mitochondrial morphology and pathology during apoptosis by interacting with mitofusins. Proc. Natl. Acad. Sci. USA *104*, 11649–11654.

Cartron, P.F., Gallenne, T., Bougras, G., Gautier, F., Manero, F., Vusio, P., Meflah, K., Vallette, F.M., and Juin, P. (2004). The first alpha helix of Bax plays a necessary role in its ligand-induced activation by the BH3-only proteins Bid and PUMA. Mol. Cell *16*, 807–818.

Cassidy-Stone, A., Chipuk, J.E., Ingerman, E., Song, C., Yoo, C., Kuwana, T., Kurth, M.J., Shaw, J.T., Hinshaw, J.E., Green, D.R., and Nunnari, J. (2008). Chemical inhibition of the mitochondrial division dynamin reveals its role in Bax/Bak-dependent mitochondrial outer membrane permeabilization. Dev. Cell *14*, 193–204.

Chen, F., Hersh, B.M., Conradt, B., Zhou, Z., Riemer, D., Gruenbaum, Y., and Horvitz, H.R. (2000). Translocation of C. elegans CED-4 to nuclear membranes during programmed cell death. Science *287*, 1485–1489.

Chen, H., Detmer, S.A., Ewald, A.J., Griffin, E.E., Fraser, S.E., and Chan, D.C. (2003). Mitofusins Mfn1 and Mfn2 coordinately regulate mitochondrial fusion and are essential for embryonic development. J. Cell Biol. *160*, 189–200.

Cheng, E.H., Sheiko, T.V., Fisher, J.K., Craigen, W.J., and Korsmeyer, S.J. (2003). VDAC2 inhibits BAK activation and mitochondrial apoptosis. Science *301*, 513–517.

Chernomordik, L.V., and Kozlov, M.M. (2008). Mechanics of membrane fusion. Nat. Struct. Mol. Biol. *15*, 675–683.

Chipuk, J.E., Kuwana, T., Bouchier-Hayes, L., Droin, N.M., Newmeyer, D.D., Schuler, M., and Green, D.R. (2004). Direct activation of Bax by p53 mediates mitochondrial membrane permeabilization and apoptosis. Science *303*, 1010–1014.

Christenson, E., Merlin, S., Saito, M., and Schlesinger, P. (2008). Cholesterol effects on BAX pore activation. J. Mol. Biol. *381*, 1168–1183.

Cipolat, S., Rudka, T., Hartmann, D., Costa, V., Serneels, L., Craessaerts, K., Metzger, K., Frezza, C., Annaert, W., D'Adamio, L., et al. (2006). Mitochondrial rhomboid PARL regulates cytochrome c release during apoptosis via OPA1-dependent cristae remodeling. Cell *126*, 163–175.

Clavería, C., Caminero, E., Martínez-A, C., Campuzano, S., and Torres, M. (2002). GH3, a novel proapoptotic domain in Drosophila Grim, promotes a mitochondrial death pathway. EMBO J. *21*, 3327–3336.

Cleland, M.M., Norris, K.L., Karbowski, M., Wang, C., Suen, D.F., Jiao, S., George, N.M., Luo, X., Li, Z., and Youle, R.J. (2011). Bcl-2 family interaction with the mitochondrial morphogenesis machinery. Cell Death Differ. *18*, 235–247.

Czabotar, P.E., Colman, P.M., and Huang, D.C. (2009). Bax activation by Bim? Cell Death Differ. *16*, 1187–1191.

Daum, G. (1985). Lipids of mitochondria. Biochim. Biophys. Acta *822*, 1–42.

Dejean, L.M., Martinez-Caballero, S., Guo, L., Hughes, C., Teijido, O., Ducret, T., Ichas, F., Korsmeyer, S.J., Antonsson, B., Jonas, E.A., and Kinnally, K.W. (2005). Oligomeric Bax is a component of the putative cytochrome c release channel MAC, mitochondrial apoptosis-induced channel. Mol. Biol. Cell *16*, 2424–2432.

Delivani, P., Adrain, C., Taylor, R.C., Duriez, P.J., and Martin, S.J. (2006). Role for CED-9 and Egl-1 as regulators of mitochondrial fission and fusion dynamics. Mol. Cell *21*, 761–773.

Desagher, S., Osen-Sand, A., Nichols, A., Eskes, R., Montessuit, S., Lauper, S., Maundrell, K., Antonsson, B., and Martinou, J.-C. (1999). Bid-induced conformational change of Bax is responsible for mitochondrial cytochrome c release during apoptosis. J. Cell Biol. *144*, 891–901.

Dewson, G., Kratina, T., Sim, H.W., Puthalakath, H., Adams, J.M., Colman, P.M., and Kluck, R.M. (2008). To trigger apoptosis, Bak exposes its BH3 domain and homodimerizes via BH3:groove interactions. Mol. Cell 30, 369–380.

Düssmann, H., Rehm, M., Concannon, C.G., Anguissola, S., Würstle, M., Kacmar, S., Völler, P., Huber, H.J., and Prehn, J.H. (2010). Single-cell quantification of Bax activation and mathematical modelling suggest pore formation on minimal mitochondrial Bax accumulation. Cell Death Differ. 17, 278–290.

Edlich, F., Banerjee, S., Suzuki, M., Cleland, M.M., Arnoult, D., Wang, C., Neutzner, A., Tjandra, N., and Youle, R.J. (2011). Bcl-x(L) retrotranslocates Bax from the mitochondria into the cytosol. Cell 145, 104–116.

Epand, R.F., Martinou, J.C., Montessuit, S., Epand, R.M., and Yip, C.M. (2002). Direct evidence for membrane pore formation by the apoptotic protein Bax. Biochem. Biophys. Res. Commun. 298, 744–749.

Epand, R.F., Schlattner, U., Wallimann, T., Lacombe, M.L., and Epand, R.M. (2007). Novel lipid transfer property of two mitochondrial proteins that bridge the inner and outer membranes. Biophys. J. 92, 126–137.

Eskes, R., Desagher, S., Antonsson, B., and Martinou, J.C. (2000). Bid induces the oligomerization and insertion of Bax into the outer mitochondrial membrane. Mol. Cell. Biol. 20, 929–935.

Estaquier, J., and Arnoult, D. (2007). Inhibiting Drp1-mediated mitochondrial fission selectively prevents the release of cytochrome c during apoptosis. Cell Death Differ. 14, 1086–1094.

Frank, S., Gaume, B., Bergmann-Leitner, E.S., Leitner, W.W., Robert, E.G., Catez, F., Smith, C.L., and Youle, R.J. (2001). The role of dynamin-related protein 1, a mediator of mitochondrial fission, in apoptosis. Dev. Cell 1, 515–525.

Gallenne, T., Gautier, F., Oliver, L., Hervouet, E., Noël, B., Hickman, J.A., Geneste, O., Cartron, P.F., Vallette, F.M., Manon, S., and Juin, P. (2009). Bax activation by the BH3-only protein Puma promotes cell dependence on antiapoptotic Bcl-2 family members. J. Cell Biol. 185, 279–290.

García-Sáez, A.J., Coraiola, M., Serra, M.D., Mingarro, I., Müller, P., and Salgado, J. (2006). Peptides corresponding to helices 5 and 6 of Bax can independently form large lipid pores. FEBS J. 273, 971–981.

García-Sáez, A.J., Chiantia, S., Salgado, J., and Schwille, P. (2007). Pore formation by a Bax-derived peptide: effect on the line tension of the membrane probed by AFM. Biophys. J. 93, 103–112.

Gavathiotis, E., Suzuki, M., Davis, M.L., Pitter, K., Bird, G.H., Katz, S.G., Tu, H.C., Kim, H., Cheng, E.H., Tjandra, N., and Walensky, L.D. (2008). BAX activation is initiated at a novel interaction site. Nature 455, 1076–1081.

Gavathiotis, E., Reyna, D.E., Davis, M.L., Bird, G.H., and Walensky, L.D. (2010). BH3-triggered structural reorganization drives the activation of proapoptotic BAX. Mol. Cell 40, 481–492.

Gebert, N., Joshi, A.S., Kutik, S., Becker, T., McKenzie, M., Guan, X.L., Mooga, V.P., Stroud, D.A., Kulkarni, G., Wenk, M.R., et al. (2009). Mitochondrial cardiolipin involved in outer-membrane protein biogenesis: implications for Barth syndrome. Curr. Biol. 19, 2133–2139.

George, N.M., Evans, J.J., and Luo, X. (2007). A three-helix homo-oligomerization domain containing BH3 and BH1 is responsible for the apoptotic activity of Bax. Genes Dev. 21, 1937–1948.

Germain, M., Mathai, J.P., McBride, H.M., and Shore, G.C. (2005). Endoplasmic reticulum BIK initiates DRP1-regulated remodelling of mitochondrial cristae during apoptosis. EMBO J. 24, 1546–1556.

Giam, M., Huang, D.C., and Bouillet, P. (2008). BH3-only proteins and their roles in programmed cell death. Oncogene 27 (Suppl 1), S128–S136.

Gilmore, A.P., Metcalfe, A.D., Romer, L.H., and Streuli, C.H. (2000). Integrin-mediated survival signals regulate the apoptotic function of Bax through its conformation and subcellular localization. J. Cell Biol. 149, 431–446.

Gonzalvez, F., Schug, Z.T., Houtkooper, R.H., MacKenzie, E.D., Brooks, D.G., Wanders, R.J., Petit, P.X., Vaz, F.M., and Gottlieb, E. (2008). Cardiolipin provides an essential activating platform for caspase-8 on mitochondria. J. Cell Biol. 183, 681–696.

Goyal, G., Fell, B., Sarin, A., Youle, R.J., and Sriram, V. (2007). Role of mitochondrial remodeling in programmed cell death in Drosophila melanogaster. Dev. Cell 12, 807–816.

Haining, W.N., Carboy-Newcomb, C., Wei, C.L., and Steller, H. (1999). The proapoptotic function of Drosophila Hid is conserved in mammalian cells. Proc. Natl. Acad. Sci. USA 96, 4936–4941.

Hasler, S.L.A. (2011). Cholesterol, cardiolipin, and mitochondria permeabilisation. Anticancer Agents Med. Chem., in press. Published online May 9, 2011.

Hengartner, M.O., and Horvitz, H.R. (1994). Activation of C. elegans cell death protein CED-9 by an amino-acid substitution in a domain conserved in Bcl-2. Nature 369, 318–320.

Hoppins, S., Edlich, F., Cleland, M.M., Banerjee, S., McCaffery, J.M., Youle, R.J., and Nunnari, J. (2011). The soluble form of Bax regulates mitochondrial fusion via MFN2 homotypic complexes. Mol. Cell 41, 150–160.

Hoppins, S., Lackner, L., and Nunnari, J. (2007). The machines that divide and fuse mitochondria. Annu. Rev. Biochem. 76, 751–780.

Hovius, R., Lambrechts, H., Nicolay, K., and de Kruijff, B. (1990). Improved methods to isolate and subfractionate rat liver mitochondria. Lipid composition of the inner and outer membrane. Biochim. Biophys. Acta 1021, 217–226.

Ishihara, N., Nomura, M., Jofuku, A., Kato, H., Suzuki, S.O., Masuda, K., Otera, H., Nakanishi, Y., Nonaka, I., Goto, Y., et al. (2009). Mitochondrial fission factor Drp1 is essential for embryonic development and synapse formation in mice. Nat. Cell Biol. 11, 958–966.

Iverson, S.L., Enoksson, M., Gogvadze, V., Ott, M., and Orrenius, S. (2003). Cardiolipin is not required for Bax-mediated cytochrome c release from yeast mitochondria. J. Biol. Chem. 279, 1100–1107.

Jagasia, R., Grote, P., Westermann, B., and Conradt, B. (2005). DRP-1-mediated mitochondrial fragmentation during EGL-1-induced cell death in C. elegans. Nature 433, 754–760.

James, D.I., Parone, P.A., Mattenberger, Y., and Martinou, J.C. (2003). hFis1, a novel component of the mammalian mitochondrial fission machinery. J. Biol. Chem. 278, 36373–36379.

Karbowski, M., Lee, Y.J., Gaume, B., Jeong, S.Y., Frank, S., Nechushtan, A., Santel, A., Fuller, M., Smith, C.L., and Youle, R.J. (2002). Spatial and temporal association of Bax with mitochondrial fission sites, Drp1, and Mfn2 during apoptosis. J. Cell Biol. 159, 931–938.

Karbowski, M., Norris, K.L., Cleland, M.M., Jeong, S.Y., and Youle, R.J. (2006). Role of Bax and Bak in mitochondrial morphogenesis. Nature 443, 658–662.

Katsov, K., Müller, M., and Schick, M. (2004). Field theoretic study of bilayer membrane fusion. I. Hemifusion mechanism. Biophys. J. 87, 3277–3290.

Khaled, A.R., Kim, K., Hofmeister, R., Muegge, K., and Durum, S.K. (1999). Withdrawal of IL-7 induces Bax translocation from cytosol to mitochondria through a rise in intracellular pH. Proc. Natl. Acad. Sci. USA 96, 14476–14481.

Kim, H., Tu, H.C., Ren, D., Takeuchi, O., Jeffers, J.R., Zambetti, G.P., Hsieh, J.J., and Cheng, E.H. (2009). Stepwise activation of BAX and BAK by tBID, BIM, and PUMA initiates mitochondrial apoptosis. Mol. Cell 36, 487–499.

Kim, T.H., Zhao, Y., Ding, W.X., Shin, J.N., He, X., Seo, Y.W., Chen, J., Rabinowich, H., Amoscato, A.A., and Yin, X.M. (2004). Bid-cardiolipin interaction at mitochondrial contact site contributes to mitochondrial cristae reorganization and cytochrome C release. Mol. Biol. Cell 15, 3061–3072.

Koch, A., Schneider, G., Lüers, G.H., and Schrader, M. (2004). Peroxisome elongation and constriction but not fission can occur independently of dynamin-like protein 1. J. Cell Sci. 117, 3995–4006.

Kozlovsky, Y., and Kozlov, M.M. (2003). Membrane fission: model for intermediate structures. Biophys. J. 85, 85–96.

Kroemer, G., Galluzzi, L., and Brenner, C. (2007). Mitochondrial membrane permeabilization in cell death. Physiol. Rev. 87, 99–163.

Kutuk, O., and Letai, A. (2008). Regulation of Bcl-2 family proteins by posttranslational modifications. Curr. Mol. Med. 8, 102–118.

Kuwana, T., Mackey, M.R., Perkins, G., Ellisman, M.H., Latterich, M., Schneiter, R., Green, D.R., and Newmeyer, D.D. (2002). Bid, Bax, and lipids cooperate to form supramolecular openings in the outer mitochondrial membrane. Cell 111, 331–342.

Kuwana, T., Bouchier-Hayes, L., Chipuk, J.E., Bonzon, C., Sullivan, B.A., Green, D.R., and Newmeyer, D.D. (2005). BH3 domains of BH3-only proteins differentially regulate Bax-mediated mitochondrial membrane permeabilization both directly and indirectly. Mol. Cell 17, 525–535.

Kvansakul, M., Yang, H., Fairlie, W.D., Czabotar, P.E., Fischer, S.F., Perugini, M.A., Huang, D.C., and Colman, P.M. (2008). Vaccinia virus anti-apoptotic F1L is a novel Bcl-2-like domain-swapped dimer that binds a highly selective subset of BH3-containing death ligands. Cell Death Differ. 15, 1564–1571.

Lalier, L., Cartron, P.F., Olivier, C., Logé, C., Bougras, G., Robert, J.M., Oliver, L., and Vallette, F.M. (2011). Prostaglandins antagonistically control Bax activation during apoptosis. Cell Death Differ. 18, 528–537.

Lazarou, M., Stojanovski, D., Frazier, A.E., Kotevski, A., Dewson, G., Craigen, W.J., Kluck, R.M., Vaux, D.L., and Ryan, M.T. (2010). Inhibition of Bak activation by VDAC2 is dependent on the Bak transmembrane anchor. J. Biol. Chem. 285, 36876–36883.

Lovell, J.F., Billen, L.P., Bindner, S., Shamas-Din, A., Fradin, C., Leber, B., and Andrews, D.W. (2008). Membrane binding by tBid initiates an ordered series of events culminating in membrane permeabilization by Bax. Cell 135, 1074–1084.

Lucken-Ardjomande, S., Montessuit, S., and Martinou, J.C. (2008). Bax activation and stress-induced apoptosis delayed by the accumulation of cholesterol in mitochondrial membranes. Cell Death Differ. 15, 484–493.

Lutter, M., Fang, M., Luo, X., Nishijima, M., Xie, X., and Wang, X. (2000). Cardiolipin provides specificity for targeting of tBid to mitochondria. Nat. Cell Biol. 2, 754–761.

Martinou, J.C., and Youle, R.J. (2006). Which came first, the cytochrome c release or the mitochondrial fission? Cell Death Differ. 13, 1291–1295.

Mérino, D., Giam, M., Hughes, P.D., Siggs, O.M., Heger, K., O'Reilly, L.A., Adams, J.M., Strasser, A., Lee, E.F., Fairlie, W.D., and Bouillet, P. (2009). The role of BH3-only protein Bim extends beyond inhibiting Bcl-2-like prosurvival proteins. J. Cell Biol. 186, 355–362.

Merrill, R.A., Dagda, R.K., Dickey, A.S., Cribbs, J.T., Green, S.H., Usachev, Y.M., and Strack, S. (2011). Mechanism of Neuroprotective Mitochondrial Remodeling by PKA/AKAP1. PLoS Biol. 9, e1000612.

Moldoveanu, T., Liu, Q., Tocilj, A., Watson, M., Shore, G., and Gehring, K. (2006). The X-ray structure of a BAK homodimer reveals an inhibitory zinc binding site. Mol. Cell 24, 677–688.

Montessuit, S., Somasekharan, S.P., Terrones, O., Lucken-Ardjomande, S., Herzig, S., Schwarzenbacher, R., Manstein, D.J., Bossy-Wetzel, E., Basañez, G., Meda, P., and Martinou, J.C. (2010). Membrane remodeling induced by the dynamin-related protein Drp1 stimulates Bax oligomerization. Cell *142*, 889–901.

Muchmore, S.W., Sattler, M., Liang, H., Meadows, R.P., Harlan, J.E., Yoon, H.S., Nettesheim, D., Chang, B.S., Thompson, C.B., Wong, S.L., et al. (1996). X-ray and NMR structure of human Bcl-xL, an inhibitor of programmed cell death. Nature *381*, 335–341.

Nechushtan, A., Smith, C.L., Lamensdorf, I., Yoon, S.-H., and Youle, R.J. (2001a). Bax and Bak coalesce into novel mitochondria-associated clusters during apoptosis. J. Cell Biol. *153*, 1265–1276.

Nechushtan, A., Smith, C.L., Lamensdorf, I., Yoon, S.H., and Youle, R.J. (2001b). Bax and Bak coalesce into novel mitochondria-associated clusters during apoptosis. J. Cell Biol. *153*, 1265–1276.

Olson, M.R., Holley, C.L., Gan, E.C., Colón-Ramos, D.A., Kaplan, B., and Kornbluth, S. (2003). A GH3-like domain in reaper is required for mitochondrial localization and induction of IAP degradation. J. Biol. Chem. *278*, 44758–44768.

Parone, P.A., James, D.I., Da Cruz, S., Mattenberger, Y., Donzé, O., Barja, F., and Martinou, J.C. (2006). Inhibiting the mitochondrial fission machinery does not prevent Bax/Bak-dependent apoptosis. Mol. Cell. Biol. *26*, 7397–7408.

Polcic, P., Su, X., Fowlkes, J., Blachly-Dyson, E., Dowhan, W., and Forte, M. (2005). Cardiolipin and phosphatidylglycerol are not required for the in vivo action of Bcl-2 family proteins. Cell Death Differ. *12*, 310–312.

Qian, S., Wang, W., Yang, L., and Huang, H.W. (2008). Structure of transmembrane pore induced by Bax-derived peptide: evidence for lipidic pores. Proc. Natl. Acad. Sci. USA *105*, 17379–17383.

Ren, D., Tu, H.C., Kim, H., Wang, G.X., Bean, G.R., Takeuchi, O., Jeffers, J.R., Zambetti, G.P., Hsieh, J.J., and Cheng, E.H. (2010). BID, BIM, and PUMA are essential for activation of the BAX- and BAK-dependent cell death program. Science *330*, 1390–1393.

Rolland, S.G., Lu, Y., David, C.N., and Conradt, B. (2009). The BCL-2-like protein CED-9 of C. elegans promotes FZO-1/Mfn1,2- and EAT-3/Opa1-dependent mitochondrial fusion. J. Cell Biol. *186*, 525–540.

Roucou, X., Rostovtseva, T., Montessuit, S., Martinou, J.C., and Antonsson, B. (2002). Bid induces cytochrome c-impermeable Bax channels in liposomes. Biochem. J. *363*, 547–552.

Roy, S.S., Ehrlich, A.M., Craigen, W.J., and Hajnóczky, G. (2009). VDAC2 is required for truncated BID-induced mitochondrial apoptosis by recruiting BAK to the mitochondria. EMBO Rep. *10*, 1341–1347.

Salvesen, G.S., and Abrams, J.M. (2004). Caspase activation - stepping on the gas or releasing the brakes? Lessons from humans and flies. Oncogene *23*, 2774–2784.

Sandu, C., Ryoo, H.D., and Steller, H. (2010). Drosophila IAP antagonists form multimeric complexes to promote cell death. J. Cell Biol. *190*, 1039–1052.

Schafer, B., Quispe, J., Choudhary, V., Chipuk, J.E., Ajero, T.G., Du, H., Schneiter, R., and Kuwana, T. (2009). Mitochondrial outer membrane proteins assist Bid in Bax-mediated lipidic pore formation. Mol. Biol. Cell. *20*, 2276–2285.

Schinzel, A., Kaufmann, T., and Borner, C. (2004). Bcl-2 family members: integrators of survival and death signals in physiology and pathology [corrected]. Biochim. Biophys. Acta *1644*, 95–105.

Schlame, M., Rua, D., and Greenberg, M.L. (2000). The biosynthesis and functional role of cardiolipin. Prog. Lipid Res. *39*, 257–288.

Schlesinger, P.H., Gross, A., Yin, X.M., Yamamoto, K., Saito, M., Waksman, G., and Korsmeyer, S.J. (1997). Comparison of the ion channel characteristics of proapoptotic BAX and antiapoptotic BCL-2. Proc. Natl. Acad. Sci. USA 94, 11357–11362.

Schug, Z.T., and Gottlieb, E. (2009). Cardiolipin acts as a mitochondrial signalling platform to launch apoptosis. Biochim. Biophys. Acta 1788, 2022–2031.

Sevrioukov, E.A., Burr, J., Huang, E.W., Assi, H.H., Monserrate, J.P., Purves, D.C., Wu, J.N., Song, E.J., and Brachmann, C.B. (2007). Drosophila Bcl-2 proteins participate in stress-induced apoptosis, but are not required for normal development. Genesis 45, 184–193.

Shamas-Din, A., Brahmbhatt, H., Leber, B., and Andrews, D.W. (2011). BH3-only proteins: Orchestrators of apoptosis. Biochim Biophys Acta. 1813, 508–520.

Sheridan, C., Delivani, P., Cullen, S.P., and Martin, S.J. (2008). Bax- or Bak-induced mitochondrial fission can be uncoupled from cytochrome C release. Mol. Cell 31, 570–585.

Tait, S.W., and Green, D.R. (2010). Mitochondria and cell death: outer membrane permeabilization and beyond. Nat. Rev. Mol. Cell Biol. 11, 621–632.

Tan, F.J., Husain, M., Manlandro, C.M., Koppenol, M., Fire, A.Z., and Hill, R.B. (2008). CED-9 and mitochondrial homeostasis in C. elegans muscle. J. Cell Sci. 121, 3373–3382.

Tanner, E.A., Blute, T.A., Brachmann, C.B., and McCall, K. (2011). Bcl-2 proteins and autophagy regulate mitochondrial dynamics during programmed cell death in the Drosophila ovary. Development 138, 327–338.

Terrones, O., Antonsson, B., Yamaguchi, H., Wang, H.G., Liu, J., Lee, R.M., Herrmann, A., and Basañez, G. (2004). Lipidic pore formation by the concerted action of proapoptotic BAX and tBID. J. Biol. Chem. 279, 30081–30091.

Thomenius, M., Freel, C.D., Horn, S., Krieser, R., Abdelwahid, E., Cannon, R., Balasundaram, S., White, K., and Kornbluth, S. (2011). Mitochondrial fusion is regulated by Reaper to modulate Drosophila programmed cell death. Cell Death Differ., in press. Published online April 8, 2011. 10.1038/cdd.2011.26.

Tondera, D., Grandemange, S., Jourdain, A., Karbowski, M., Mattenberger, Y., Herzig, S., Da Cruz, S., Clerc, P., Raschke, I., Merkwirth, C., et al. (2009). SLP-2 is required for stress-induced mitochondrial hyperfusion. EMBO J. 28, 1589–1600.

Vaux, D.L. (2011). Apoptogenic factors released from mitochondria. Biochim. Biophys. Acta 1813, 546–550.

Wakabayashi, J., Zhang, Z., Wakabayashi, N., Tamura, Y., Fukaya, M., Kensler, T.W., Iijima, M., and Sesaki, H. (2009). The dynamin-related GTPase Drp1 is required for embryonic and brain development in mice. J. Cell Biol. 186, 805–816.

Walensky, L.D., Pitter, K., Morash, J., Oh, K.J., Barbuto, S., Fisher, J., Smith, E., Verdine, G.L., and Korsmeyer, S.J. (2006). A stapled BID BH3 helix directly binds and activates BAX. Mol. Cell 24, 199–210.

Wasiak, S., Zunino, R., and McBride, H.M. (2007). Bax/Bak promote sumoylation of DRP1 and its stable association with mitochondria during apoptotic cell death. J. Cell Biol. 177, 439–450.

Wei, M.C., Zong, W.X., Cheng, E.H., Lindsten, T., Panoutsakopoulou, V., Ross, A.J., Roth, K.A., MacGregor, G.R., Thompson, C.B., and Korsmeyer, S.J. (2001). Proapoptotic BAX and BAK: a requisite gateway to mitochondrial dysfunction and death. Science 292, 727–730.

Westermann, B. (2010). Mitochondrial fusion and fission in cell life and death. Nat. Rev. Mol. Cell Biol. 11, 872–884.

Westphal, D., Dewson, G., Czabotar, P.E., and Kluck, R.M. (2011). Molecular biology of Bax and Bak activation and action. Biochim. Biophys. Acta 1813, 521–531.

Yamaguchi, R., Lartigue, L., Perkins, G., Scott, R.T., Dixit, A., Kushnareva, Y., Kuwana, T., Ellisman, M.H., and Newmeyer, D.D. (2008). Opa1-mediated cristae opening is Bax/Bak and BH3 dependent, required for apoptosis, and independent of Bak oligomerization. Mol. Cell *31*, 557–569.

Yang, L., Harroun, T.A., Weiss, T.M., Ding, L., and Huang, H.W. (2001). Barrel-stave model or toroidal model? A case study on melittin pores. Biophys. J. *81*, 1475–1485.

Yethon, J.A., Epand, R.F., Leber, B., Epand, R.M., and Andrews, D.W. (2003). Interaction with a membrane surface triggers a reversible conformational change in Bax normally associated with induction of apoptosis. J. Biol. Chem. *278*, 48935–48941.

Youle, R.J., and Strasser, A. (2008). The BCL-2 protein family: opposing activities that mediate cell death. Nat. Rev. Mol. Cell Biol. *9*, 47–59.

Zaltsman, Y., Shachnai, L., Yivgi-Ohana, N., Schwarz, M., Maryanovich, M., Houtkooper, R.H., Vaz, F.M., De Leonardis, F., Fiermonte, G., Palmieri, F., et al. (2010). MTCH2/MIMP is a major facilitator of tBID recruitment to mitochondria. Nat. Cell Biol. *12*, 553–562.

Zhang, Z., Zhu, W., Lapolla, S.M., Miao, Y., Shao, Y., Falcone, M., Boreham, D., McFarlane, N., Ding, J., Johnson, A.E., et al. (2010). Bax forms an oligomer via separate, yet interdependent, surfaces. J. Biol. Chem. *285*, 17614–17627.

rrent Biology

Golgi Membrane Dynamics and Lipid Metabolism

Vytas A. Bankaitis[1,2,*], Rafael Garcia-Mata[1], Carl J. Mousley[1,2]

[1]Department of Cell and Developmental Biology, University of North Carolina School of Medicine, Chapel Hill, NC 27599-7090, USA, [2]Lineberger Comprehensive Cancer Center, University of North Carolina School of Medicine, Chapel Hill, NC 27599-7090, USA

*Correspondence: vytas@med.unc.edu

Current Biology, Vol. 22, No. 10, R414–R424, May 22, 2012 © 2012 Elsevier Inc.
http://dx.doi.org/10.1016/j.cub.2012.03.004

SUMMARY

The striking morphology of the Golgi complex has fascinated cell biologists since its discovery over 100 years ago. Yet, despite intense efforts to understand how membrane flow relates to Golgi form and function, this organelle continues to baffle cell biologists and biochemists alike. Fundamental questions regarding Golgi function, while hotly debated, remain unresolved. Historically, Golgi function has been described from a protein-centric point of view, but we now appreciate that conceptual frameworks for how lipid metabolism is integrated with Golgi biogenesis and function are essential for a mechanistic understanding of this fascinating organelle. It is from a lipid-centric perspective that we discuss the larger question of Golgi dynamics and membrane trafficking. We review the growing body of evidence for how lipid metabolism is integrally written into the engineering of the Golgi system and highlight questions for future study.

INTRODUCTION

The Golgi apparatus is a central station for the sorting and transport of proteins and lipids that transit the secretory pathway. This organelle also serves as a biochemical factory where anterograde cargo is subject to serial post-translational modifications before being sorted at the *trans*-Golgi network (TGN) for delivery to the appropriate destinations. As such, the Golgi system plays a central role in eukaryotic cell biology. At steady state, Golgi membranes are typically organized in a stack of flattened cisternae with

209

dilated rims [1]. Such an organization has been argued to reflect the logic for ordering the biochemical activities of the system. That is, one simply generates stable compartments in the context of the cisternal arrangement. This steady-state morphology is deceptive, however. The Golgi complex is a dynamic organelle subject to enormous membrane flux in its capacity as an intermediate station between the endoplasmic reticulum (ER) and the distal compartments of the secretory pathway. These fluxes are bidirectional because the Golgi system directs retrograde trafficking pathways for purposes of retrieval and recycling of Golgi and ER components and receives cargo from the plasma membrane and endosomes [1].

The structural plasticity of the Golgi system is evident at multiple levels. In mammalian cells, this organelle disassembles in mitosis and subsequently reassembles into a functional unit upon completion of cell division [2]. Golgi structural plasticity is also evident when the system is subjected to a variety of perturbations [3]. Disruption of Golgi morphology interferes with the modification, sorting and delivery of proteins and with wider cellular processes, such as ciliogenesis, cell polarity, cell migration, stress responses and apoptosis [4–6]. Thus, the forces that shape Golgi morphology exert unexpectedly broad effects on cell physiology. Perhaps reflective of these wider cellular functions, individual Golgi stacks are often laterally interconnected to form a reticular ribbon positioned in the perinuclear region of the cell in proximity to the vertebrate centrosome.

Yet, the Golgi system is resilient. It exhibits remarkable capacities for self-organization that allow it to recover from catastrophic structural derangements. As an example, induction of the collapse of the Golgi system into the ER by treatment with the drug brefeldin A is followed by reformation of a functional organelle upon drug removal [7]. Thus, the steady-state form of the Golgi system portrays an illusion of compartmental stability. The very existence of the organelle is balanced on a knife's edge of competing forces that create it and consume it. It is the remarkable dynamics of the Golgi system that have, over the past decade, fueled a re-evaluation of the fundamental nature of this organelle, and maturation models now supplant stable compartment models as favored mechanisms for Golgi biogenesis and function [8].

Initial studies of Golgi membrane trafficking and dynamics were exclusively protein-centric [9,10]. It is now appreciated that lipid metabolism is integrally written into the fabric of the transport carrier cycle and of Golgi function. Since the first demonstrations to this effect in permeabilized cell systems and in yeast [11–14], we now understand that the interface of lipid metabolism with membrane trafficking is complex. This interface is a major factor in controlling Golgi morphology and dynamics. It also involves a large cast of interesting proteins and enzymes, including lipid transfer proteins

[11,12,15–18], lipid kinases and phosphatases [19–22], phospholipase D (PLD) and phospholipase A_2 (PLA$_2$) [23–25], phospholipid acyl-transferases [26,27], and amino-phospholipid flippases that harness their ATPase activities for topological control of lipid distribution between bilayer leaflets [28–30].

Lipid metabolism interfaces with membrane trafficking in several general ways. First, it helps to create platforms for protein recruitment to, and activation at, appropriate sites on membrane surfaces. The reduced dimensionality achieved by recruiting soluble factors to a surface has a powerful concentrating effect that promotes effective biochemistry in systems governed by modest affinities. In these capacities, lipid metabolism has a signaling role. Second, it facilitates the structural deformations of membranes that accompany vesicle budding, fusion and tubulation. Third, it effects a lateral segregation of molecules, and this partitioning contributes to Golgi function. For example, regulation by lateral segregation is the underlying principle of a rapid partitioning model proposed to account for cargo export kinetics from the Golgi complex [31]. The model is based on a continuous two-phase system; one that can readily be generated by lipid segregation into fluid and relatively less fluid domains. While the continuous two-phase partitioning model is overly simplistic, and some of its basic tenets are at odds with known properties of the Golgi complex [32], the concept illustrates how self-organizing principles linked to lipid metabolism/composition might give rise to complex Golgi functions. Studies suggesting that the transmembrane domains of resident proteins are matched to the physical properties of the membranes in which these reside also support partitioning concepts [33]. In this review, we describe the impact of lipid metabolism on Golgi dynamics and organize the discussion from the perspective of classes of lipids and how these molecules modulate Golgi functions.

Ptdins-4-PHOSPHATE AND TGN FUNCTION

Involvements of phosphatidylinositol (PtdIns), and its phosphorylated derivatives (the phosphoinositides), were the first established examples of lipids having active roles in regulating membrane trafficking [11–14,34,35]. PtdIns-4-phosphate (PtdIns-4-P) is an important phosphoinositide in the operation of the Golgi system [36]. That biologically sufficient production of PtdIns-4-P is integrated with phosphatidylcholine (PtdCho) metabolism provides a striking demonstration of the cross-talk between lipid metabolism and Golgi secretory function [11,12,37,38] (the issue of cross-talk is discussed below in the context of lipid transfer protein function).

Mammalian Golgi membranes harbor two types of PtdIns 4-kinases — PI4KIIIβ and PI4KIIα. Their respective yeast cognates are Pik1 and Lsb6, and Pik1 localizes to yeast Golgi membranes [39]. The PI4KIIIβ enzymes are the best understood and function as heterodimers with a myristoylated Ca^{2+}-binding non-catalytic subunit [40,41]. These PtdIns 4-kinases also engage in direct interactions with the vesicle biogenic machinery; mammalian PI4KIIIβ homes to Golgi membranes by binding to the GTP-bound form of the small GTPase Arf1 [42,43], whereas the yeast ortholog Pik1 targets to Golgi membranes by binding to a guanine nucleotide exchange factor (GEF) for Arf GTPases [44].

PtdIns 4-kinase catalytic activity is clearly important for Golgi function. Acute inactivation of yeast Pik1 kinase activity [21,22], or induction of the degradation of PtdIns-4-P to PtdIns in mammalian Golgi [45], results in trafficking defects. Inactivation of PtdIns-binding proteins, such as Sec14, which potentiate PtdIns 4-kinase activities by presenting PtdIns to the enzyme for efficient modification, also compromises Golgi membrane trafficking [11,38]. The lipid kinase activity is not the sole essential property of PtdIns 4-kinase with respect to Golgi function, however; *Drosophila* PI4KIIIβ binds to the small GTPase Rab11 in the TGN and executes a scaffolding function independent of its catalytic activity [46].

How does PtdIns-4-P potentiate Golgi secretory functions? First, PtdIns-4-P contributes to the recruitment of peripheral membrane proteins important for transport carrier biogenesis (Figure 1). These include Golgi adaptors for clathrin binding, such as AP-1 [47,48], and Arf1-GTP effectors, such as GGA proteins [49,50], Rabs and Rab-GEFs [51,52], and the Arf-GEF GBF1 [53]. Oxysterol binding-related proteins (OSBPs) interface with PtdIns-4-P signaling [16–18], and other lipid binding/transfer proteins that further remodel Golgi membrane lipid composition are also PtdIns-4-P effectors (see below).

Second, PtdIns-4-P modulates protein activities by direct binding mechanisms. One example is the yeast amino-phospholipid flippase Drs2, a type-IV integral membrane ATPase, which translocates phosphatidylserine (PtdSer) and phosphatidylethanolamine (PtdEtn) from the lumenal to the cytosolic leaflets of TGN/endosomal membranes [28–30]. Drs2 flippase activity is stimulated by binding both to PtdIns-4-P and to an Arf1-GEF [54]. Another example of the regulation of the activity of a protein component of the trafficking machinery by PtdIns-4-P binding is described by the coincidence-detection mechanism for the function of yeast Sec2, a GEF for the Rab GTPase Sec4. PtdIns-4-P binding quenches the ability of Sec2 to nucleate assembly of the exocyst complex, which is required for the interaction of secretory vesicles with the plasma membrane [54,55]. By discouraging premature Sec2-mediated recruitment of the exocyst to TGN membranes, PtdIns-4-P helps to preserve the compartmental distinction between transport intermediates and the Golgi system.

Recent studies of the yeast PtdIns-binding protein Sec14 and PtdIns-4-P-binding protein Vps74 indicate that retrograde membrane flow from

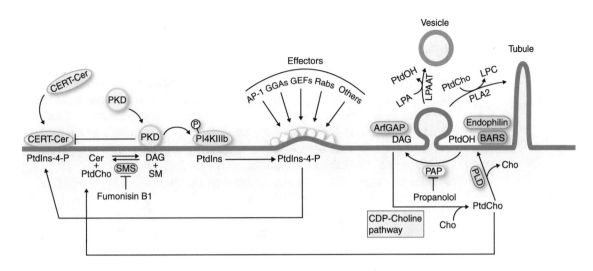

FIGURE 1 Lipid metabolism and formation of transport carriers.

This diagram highlights existing views of the interface between lipid metabolism and components of the protein machinery that drives formation of vesicles or tubular transport carriers in mammals. No obvious ceramide transfer protein (CERT) or protein kinase D (PKD) activities are present in yeast, and yeast do not have obvious counterparts for BARS, endophilin or lysoPtdOH acyltransferases (LPAATs). The dissonance between this diagram and *in vivo* readouts for PLD, BARS and CERT function in mammals is discussed in the text. (Cer, ceramide; LPC, lyso-PtdCho; PAP, PtdOH phosphatase; SM, sphingomyelin; SMS, sphingomyelin synthase.)

endosomes to the TGN and retention of glycosyltransferases within the Golgi system, respectively, is PtdIns-4-P dependent [56,57]. PtdIns-4-P binding by Vps74 coordinates the interactions of this protein with the cytosolic tails of glycosyltransferases and with the coatomer complex, which coats vesicles [57]. Vps74 is an ortholog of the mammalian protein GOLPH3, which collaborates with the non-conventional myosin MYO18A to control Golgi morphology [57]. In addition, GOLPH3 interacts with the retromer complex, which potentiates retrograde membrane trafficking from endosomes [58]. These data suggest that GOLPH3 also functions in cargo sorting and retrieval in mammals.

THE YIN AND YANG OF LIPID TRANSFER PROTEINS AND TGN FUNCTIONS

Extensive involvement of lipids in regulating Golgi function demands close coordination of lipid metabolism with PtdIns-4-P signaling. Lipid transfer proteins are the coupling devices through which this coordination is executed, and PtdIns transfer proteins (PITPs) provide outstanding examples. The major yeast PITP, Sec14, coordinates PtdIns-4-P function in the TGN with the activity of a pathway that generates PtdCho from diacylglycerol (DAG) [11–13,59,60]. The remarkable structural design for how Sec14 differentially binds PtdIns and PtdCho is central to how Sec14 is posited to use heterotypic phospholipid exchange to effect a 'PtdCho-primed' presentation of PtdIns to PtdIns 4-kinases [37,38]. Such an elaborate presentation mechanism is essential for phosphoinositide homeostasis *in vivo* because PtdIns 4-kinases are biologically inadequate interfacial enzymes when asked to modify membrane-incorporated PtdIns, i.e. the presumed natural mode of presentation. By this view, Sec14-stimulated PtdIns-4-P synthesis is primed in response to PtdCho metabolic cues [37,38]. Indeed, analysis of the effects of mammalian disease-associated mutations in Sec14-like proteins suggests that such presentation functions are general properties of these proteins [37,61]. A physical interaction between Sec14 and PtdIns 4-kinases has not been reported, and such an interaction may not be necessary for presentation of PtdIns to these kinases [15,16]. Details of the PtdIns-presentation mechanism remain to be elucidated, and other evidence suggests that some mammalian multidomain Sec14-like proteins involved in vesicle trafficking bind both lipids and proteins [62].

The pro-secretory activities of yeast Sec14 are opposed by Kes1 (also known as Osh4), a member of an unrelated class of lipid transfer proteins (the oxysterol binding related proteins — ORPs) [15,16]. Kes1 binds to PtdIns-4-P and this activity is essential for Kes1's biological function as a trafficking 'brake' [15,16]. The Kes1–Sec14 antagonism plays itself out in the context of PtdIns-4-P signaling [16,18,63], but how this occurs is not clear. Some evidence suggests that ORPs stimulate phosphoinositide phosphatases that degrade PtdIns-4-P (such as Sac1, see below [18]). Other data indicate that Kes1 competes with pro-secretory factors for PtdIns-4-P binding [64]. Kes1 has two PtdIns-4-P-binding sites — one on the protein surface [16], and the other involving the hydrophobic cavity and overlapping with the sterol-binding site [65]. PtdIns-4-P binding is essential for Kes1 localization to TGN/endosomal membranes. Missense substitutions in either of the two PtdIns-4-P binding sites render Kes1 incompetent for targeting to TGN/endosomes [16,64].

With regard to sterol binding, it is now clear that sterol-binding defects enhance Kes1 biological activity as a TGN/endosomal trafficking brake [64,66]. These findings are in direct contradiction to a prominent claim that sterol binding is required for Kes1 function *in vivo* [67]. The dual PtdIns-4-P and sterol-binding activities of Kes1 cooperate in a rheostat mechanism where the interplay between sterol- and PtdIns-4-P-binding controls the amplitude of the Kes1-imposed PtdIns-4-P clamp on TGN/endosomal trafficking [64] (Figure 2). The discovery that the Kes1 sterol-binding site overlaps with a PtdIns-4-P-binding site neatly accounts for how sterol tunes

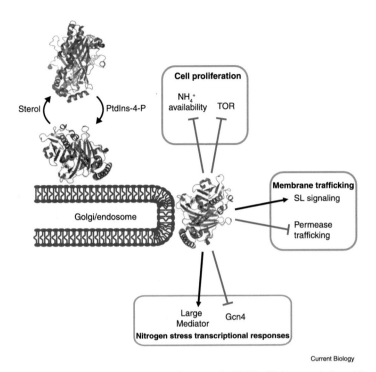

Sterol

PtdIns-4-P

Cell proliferation

NH$_4$$^+$ availability   TOR

Golgi/endosome

Membrane trafficking
SL signaling

Permease trafficking

Large Mediator   Gcn4

Nitrogen stress transcriptional responses

Current Biology

FIGURE 2 Kes1 integrates PtdIns-4-P signaling, sterols, TGN trafficking control, and larger cellular physiological responses.

Kes1 is recruited to Golgi membranes by virtue of its ability to bind PtdIns-4-P where it clamps the availability of this phosphoinositide and functions as a trafficking 'brake' [64]. Sterol binding at the TGN releases Kes1 from the membrane, thereby releasing the trafficking brake. Broader consequences of this negative regulation of membrane trafficking include the control of cell proliferation by TGN/endosomal sphingolipid metabolism, TOR signaling, and execution of nitrogen stress transcriptional responses.

Kes1-mediated inhibition of PtdIns-4-P signaling. When coupled with the demonstration that Kes1 and other ORPs are collectively dispensable for non-vesicular sterol transfer in yeast [68], the data indicate that Kes1 is not a sterol transfer protein *in vivo* as has been previously argued [69,70].

The Kes1 rheostat has broader physiological roles as it sets the gain of endosomal sphingolipid signaling, modulates activation of the serine/threonine kinase TOR by amino acids, regulates the nuclear activity of the major transcriptional activator of the general amino acid control pathway (Gcn4 in yeast and Atf4 in mammals), and administrates a coherent exit from proliferative programs to quiescent states (Figure 2). The transcriptional arm of this novel endosomal–nuclear axis involves the cyclin-dependent kinase module of the Mediator transcriptional coactivator complex [64]. Whether mammalian PITPs and ORPs play similarly opposing functions is a question for future inquiry. The idea that a PITP–ORP 'tug-of war' fine-tunes cell growth regulation and metabolic control as a function of TGN/endosomal trafficking flux has interesting implications for cell entry into post-mitotic fates and for tissue biogenesis [64].

METAZOAN LIPID TRANSFER PROTEINS AND THE GOLGI SYSTEM

The steroidogenic acute regulatory protein-related lipid transfer (StART)-like mammalian PITPs are structurally unrelated to Sec14-like PITPs, and functional depletion of specific isoforms, such as PITPβ, has been reported to compromise retrograde Golgi-to-ER transport mediated by the COP1 coat complex [71]. This defect purportedly comes without compromising anterograde ER-to-Golgi trafficking — a curious result given that retrograde transport is essential for recycling of the v-SNARE vesicle fusion protein required for anterograde ER-to-Golgi transport. With regard to vertebrate PITPβ, zebrafish with strongly reduced PITPβ levels develop normally, but exhibit defects in outer segment biogenesis and/or maintenance in double cone photoreceptor cells [72]. These data argue against house-keeping roles for PITPβ in retrograde Golgi-to-ER trafficking, although such functions might be important in specialized contexts that require high-capacity membrane flux.

Other StART-like PITPs are reported to play roles similar to yeast Sec14 in coordinating Golgi PtdCho and DAG metabolism with PtdIns-4-P signaling [17]. Nir2 is a multidomain PITP reported to collaborate with two other StART-domain lipid transfer proteins, an OSBP and a ceramide transfer protein (CERT), in forming a membrane contact site. The membrane contact site is hypothesized to bridge TGN and ER membranes — a concept that couples TGN activities with those of the ER [73]. For the purposes of discussion, the hypothetical membrane contact site is illustrated in Figure 3. Both OSBP and CERT are PtdIns-4-P-binding proteins and both are substrates of protein kinase D (PKD) [74]. In this model, CERT supplies the TGN with ceramide via a mechanism in which PtdIns-4-P binding mediates CERT interaction with TGN membranes, and CERT phosphorylation by PKD releases CERT from the TGN [75] (Figure 3). Current models propose that CERT fuels DAG production by sphingomyelin synthase in the Golgi at the expense of ceramide. DAG recruits PKD which then activates PI4KIIIβ [73]. Nir2 is posited to co-assemble into the CERT–OSBP membrane contact site for the purpose of transferring PtdIns from the ER to the Golgi to sustain PI4KIIIβ activity [73]. In sum, this specific membrane contact site model assigns essential roles for CERT, Nir2 and OSBPs in promoting membrane trafficking from the Golgi complex [76].

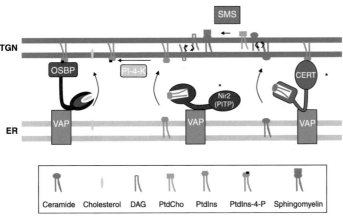

Ceramide Cholesterol DAG PtdCho PtdIns PtdIns-4-P Sphingomyelin

FIGURE 3 **Lipid transfer proteins and a hypothetical membrane contact site.**
CERT, OSBP and Nir2 are proposed to co-assemble into a TGN–ER membrane contact site where these
proteins catalyze ER-to-TGN trafficking of ceramide, sterol and PtdIns-4-P, respectively. CERT and OSBP
interact with the TGN by virtue of their PtdIns-4-P-binding activities. Nir2 is proposed to supply the TGN
with PtdIns from the ER for sustained PtdIns-4-P production. The membrane contact site is envisioned to
be held together by integral membrane proteins of the ER (VAPs) that bind the FFAT motifs of CERT, OSBP
and Nir2, and the lipid transfer activities are proposed to be essential for membrane trafficking from the
Golgi complex. The concepts highlighted by asterisks are not supported by available *in vivo* data (see text).

CERT and Nir2 gene ablation data fail to support the basic hypotheses,
however. *Drosophila* lacking CERT develop normally and reach adulthood.
Although these *cert*[0/0] flies age prematurely due to oxidative plasma membrane damage, they are otherwise remarkably unaffected [77]. By contrast,
cert[0/0] mice expire at embryonic day 11.5 from a failure in cardiac organogenesis. *cert*[0/0] embryonic fibroblasts are viable, although these too are
prone to accelerated senescence [78]. Thus, the evidence consistently highlights a role for CERT in oxidative stress management. While a requirement
for CERT in the trafficking of unusual cargos required for organogenesis
remains a formal possibility, genetic data argue against an involvement of
CERT in core Golgi trafficking functions. It is also yet to be demonstrated
that CERT bridges ER and Golgi membranes *in vivo*, a critical tenet of the
membrane contact site model (Figure 3).

Are Nir2-like PITPs obligatorily required for the function of metazoan Golgi,
as proposed [73]? Available evidence does not support this hypothesis
either. *Drosophila* mutants ablated for their single Nir2 ortholog (RdgB)
survive through adulthood, but do suffer a rapid light-accelerated retinal
degeneration caused by an inability of photoreceptors to terminate the
photoresponse. This degeneration is cured by expression of the isolated
RdgB PITP domain, which presumably does not efficiently assemble into a

membrane contact site, and is associated with a restored photoresponse, even under saturating light conditions [79]. This is an impressive outcome given the enormous phosphoinositide flux demanded by this signaling cascade. Mammals employ Nir2 differently from flies, however, as evidenced by the demonstration that *nir2^{0/0}* mice suffer pre-implantation lethality [80]. The terminal *nir2^{0/0}* phenotype in mice is uncharacterized, so it is unknown whether the lethality stems from Golgi trafficking defects, or not.

There is also no direct evidence to indicate that Nir2 functions in a PtdIns-supply capacity, even though this is the common interpretation. It is difficult to determine in vertebrate systems whether StART-like PITPs operate in cells as PtdIns-presenting scaffolds or as PtdIns carriers. These two modes of action differ in fundamental respects and have recently been discussed from alternative points of view [73,81]. However, vertebrate StART-like PITPs resemble their yeast and plant Sec14-like PITP counterparts in having the intrinsic capacity to potentiate PtdIns 4-kinase activities under conditions in which PtdIns-supply requirements are moot [72]. Moreover, the discovery that PtdIns synthase mobilizes from the ER to sites adjacent to plasma membrane and other organelles to replenish phosphoinositide signaling pools in those membranes challenges the fundamental assumption for why cells would even require *bona fide* PITPs [82].

SAC1 PHOSPHOINOSITIDE PHOSPHATASES AND THE GOLGI SYSTEM

Maintenance of phosphoinositide homeostasis requires a balanced control over the biosynthetic and the degradative aspects of their metabolism. Phosphoinositide turnover is achieved by lipid phosphatases, such as synaptojanins, the oculocerebrorenal Lowe's Syndrome OCRL protein, PTEN, and myotubularins. These enzymes are particularly interesting because of their association with a variety of diseases [83–86]. The mixed-specificity phosphoinositide phosphatase Sac1 localizes to ER and Golgi membranes and is unique amongst the inositol lipid phosphatases in that it is an integral membrane protein [19,87]. This enzyme is incapable of utilizing phosphoinositides with vicinal phosphate groups as substrates [20]. Sac1 executes wider physiological roles on the basis of its modulation of PtdIns-4-P signaling in both yeast and mammals. Under conditions of extreme nutrient or growth factor insufficiency, the Sac1 PtdIns-4-P phosphatase redistributes from the ER to the Golgi complex [88]. Presumably, regulated trafficking of the phosphatase discourages cell proliferation by clamping the activity of the distal secretory pathway (via degradation of Golgi PtdIns-4-P) under suboptimal growth conditions.

Sac1 dysfunction also alleviates the normally essential requirement for PITP activity for Golgi secretory function in yeast [19,87], reflecting the fact that Sac1 constitutes the major PtdIns-4-P degradation activity in yeast [20,60]. Paradoxically, Sac1 specifically consumes the PtdIns-4-P produced by the plasma-membrane-localized PtdIns 4-kinase Stt4 and not the Golgi-localized PtdIns 4-kinase Pik in yeast [89,90]. It is unknown whether this specificity of Sac1 for particular pools of PtdIns-4-P translates to mammals. While Sac1 insufficiencies do not evoke large derangements in bulk PtdIns-4-P mass in mammalian cells, they do result in pre-implantation lethality, morphological derangements of the Golgi, defects in mitotic spindle organization, mis-sorting of Golgi glycosyltransferases, and aberrant protein glycosylation [91,92].

DIACYLGLYCEROL AND THE TRANSPORT CARRIER CYCLE

In addition to PtdIns-4-P, to which we assign primarily a signaling role, a number of other lipids also have key involvements in Golgi functions. Some of these lipids likely play both signaling and structural roles; DAG, a neutral lipid with unusual physical properties, is one of these. The extreme inverted cone shapes assumed by DAG (due to its small headgroup to acyl chain axial area ratio) facilitate adoption of the non-bilayer configurations that lipid molecules assume in strongly deformed membrane regions. Such deformations accompany both vesicle budding and scission [93,94]. Accordingly, DAG regulates vesicle budding at multiple steps in the exocytic pathway. These include transport from the yeast and mammalian TGN [17,59,95–97], and formation of mammalian COP1-bound vesicles for retrograde trafficking from early Golgi cisternae to the ER [65,66].

In some cases, DAG directly regulates the activity of protein components of the trafficking machinery. For example, DAG potentiates Arf GTPase-activating protein (ArfGAP) function in both yeast and mammals [95,97], and DAG exhibits at least two points of action in mammalian COP1-dependent vesicle biogenesis. One is at an early step in formation of buds/tubules when the membrane is first deformed, and another at the scission step where the nascent vesicle is released from its donor membrane [98,99] (Figure 1). DAG involvement in scission requires ArfGAP1 activity, suggesting that DAG potentiates scission both by activating ArfGAP1 and by facilitating the formation of non-bilayer membrane structures that characterize fission intermediates [98]. DAG-activated PKD also promotes vesicle scission in the TGN [100,101]. Although there is as yet no evidence for obligate DAG involvements in early stages of the yeast secretory pathway, roles for DAG in the yeast TGN have been documented [98].

Compartment-specific requirements imply that DAG primarily functions in a signaling capacity. In that regard, DAG recruits PKD isoforms to mammalian TGN membranes [96]. PKD activation serves as the nexus of a larger signaling hub which connects DAG metabolism to downstream lipid metabolic events required for optimal membrane trafficking from the TGN [100–102] (Figure 1). This larger hub includes the recruitment of PI4KIIIβ, which generates PtdIns-4-P in the TGN with the pro-trafficking sequelae detailed above. DAG recruits PKC, Ras guanine nucleotide release proteins [103–106], and PKCη (which phosphorylates PKD and activates the enzyme) to TGN membranes as well [107].

PHOSPHATIDIC ACID METABOLISM AND GOLGI MEMBRANE TRAFFICKING

Pools of DAG generated from phosphatidic acid (PtdOH) by the action of PtdOH phosphatases are required for membrane trafficking through the yeast and mammalian Golgi [98,99]. That PtdOH itself executes pro-secretory functions was suggested by demonstrations that PLD, which hydrolyzes PtdCho to PtdOH and choline, is activated by PtdIns-4,5-P_2 and Arf-GTP [23,25]. Mammals express two PLD isoforms — PLD1 and PLD2 — and numerous studies claim obligatory PLD1 and/or PLD2 involvements in producing PtdOH pools that are essential for membrane trafficking through the mammalian Golgi system [108,109]. Biochemical studies suggest PtdOH acts in concert with its binding proteins endophilin and BARS in the scission of COP1 vesicles, which form in an Arf-GTP-dependent manner [110,111] (Figure 1). *In vitro* studies suggest that endophilin and BARS resolve fission intermediates by physically deforming membranes and that a PtdOH pool generated by PLD2 is required for execution of these functions [112]. The *in vivo* relevance of these findings is suggested by experiments that show that PLD2 depletion affects *cis*-Golgi maintenance and retrieval of KDEL receptors from early Golgi cisternae back to the ER [113]. DAG kinases, which produce PtdOH from DAG, cannot function as PLD2 surrogates in this system [113], suggesting direct roles for PLD-generated PtdOH pools in Golgi function. Whether PtdOH serves as a DAG precursor in these contexts is unresolved.

PTDOH REMODELING ENZYMES AND GOLGI DYNAMICS

Phospholipids are subject to two-stage remodeling reactions that convert one molecular species of a particular phospholipid to another. The first reaction involves removal of the *sn*-2 fatty acid from the glycerol backbone of a phospholipid by a PLA_2 to form a lyso-phospholipid with a single acyl chain. The lyso-phospholipid is a substrate for acyltransferases that incorporate another fatty acid at *sn*-2 to regenerate the original phospholipid, albeit a different molecular species. PtdOH-remodeling enzymes have been suggested to contribute to Golgi dynamics and trafficking in the light of reports showing that endophilin and BARS are lyso-PtdOH acyltransferases (LPAATs) [110,111]. More detailed analyses showed that these proteins have no such activity, however [114]. The evidence indicates a PLA_2–LPAAT cycle regulates tubulation events that potentiate membrane trafficking and cargo sorting in mammalian Golgi [24,26,27,115]. PLA_2-induced Golgi tubulations are enhanced by secretory cargo, and these tubules consolidate what would otherwise be individual Golgi stacks into a Golgi ribbon [24,26,27,115]. Connecting tubules are suggested to represent the portals through which anterograde cargo passes as it transits from one Golgi cisterna to the next.

COP1 or BARS initiates the formation of both tubules and vesicles from mammalian Golgi membranes *in vitro* [116]. Growing tubules are stabilized by cytosolic PLA2 ($cPLA_2$-α) activity on the one hand, and resolved into vesicles by LPAAT-γ on the other. In these assays, tubules score as non-concentrative cargo carriers, while vesicles score as concentrative carriers, suggesting that anterograde trafficking is a passive process while retrograde trafficking is an active one [116]. Because the formation of both tubules and vesicles is dependent on COP1, this idea offers a resolution to the debate of whether COP1-coated membranes define anterograde or retrograde carriers by conceptualizing how COP1 might participate in both pathways [32]. How general $cPLA_2$–LPAAT mechanisms are as a core Golgi trafficking strategy is unclear given that yeast and worms do not have obvious LPAATs. However, yeast have a naturally vesiculated Golgi — a feature that might obviate an LPAAT requirement. Also, some organisms might employ monoacylglycerol-acyltransferases, rather than LPAATs, for vesicle scission.

GENETIC MODELS FOR PLD FUNCTION

Given the pharmacological and biochemical evidence for PLD-generated PtdOH pools in driving multiple aspects of membrane trafficking, it is surprising that mice nullizygous for either *PLD1* or *PLD2* are developmentally normal [117,118]. The PLD1 null model does reveal a *PLD1* requirement in both starvation-induced expansion of autophagosomes and clearance of protein aggregates in brain tissue by macroautophagy [98]. The enzyme relocalizes from endosomes to the outer membrane of autophagosomes in the face of nutrient stress via a mechanism that requires PtdIns 3-kinase activity [98]. *PLD2*-nullizygous mice, while also overtly normal, present enhanced resistance to the neurotoxic effects of amyloid β-peptide [98]. Perhaps most surprising is that PLD1 and PLD2 activities fail to cross-compensate to any significant degree because $pld1^{0/0}$ $pld2^{0/0}$ double mutant mice do not appear to exhibit enhanced phenotypes relative to the respective single mutants (G. DiPaolo, personal communication).

Does DAG-kinase-mediated conversion of DAG to PtdOH compensate for PLD in nullizygous mice and cells? While an unresolved question, compensation by DAG kinases would necessarily operate in the absence of the numerous physical interactions reported between PLD isoforms and membrane trafficking components and proteins involved in lipid signaling [98]. Such functional compensation would also be inconsistent with the conclusions of in vitro experiments that contend that PtdOH pools produced by DAG kinases cannot substitute for those generated by PLD2, at least not for COP1 vesicle budding [113]. Why the dissonance? In vitro reconstitutions, their power notwithstanding, are inefficient systems. Consequently, these might exhibit non-physiological dependencies on particular lipid metabolic pathways for basic operation (i.e. those that preserve a relative robustness in cell-free preparations), even when the in vitro system faithfully reconstitutes a specific lipid requirement.

PLD is also non-essential for core trafficking functions in fungi because the single PtdCho-specific yeast PLD is dispensable in vegetative cells [119]; however, PLD catalytic activity is required for membrane trafficking in mutants lacking certain lipid transfer proteins [120]. A physiological role for PLD is demonstrated by the developmental reorientation of membrane trafficking from the TGN to the nuclear envelope during sporulation. PLD produces a PtdOH pool that recruits and activates a sporulation-specific t-SNARE of the Sec9/SNAP-25 family (Spo20). This t-SNARE re-directs post-Golgi trafficking to the forming nuclear envelopes at the expense of the plasma membrane, and PLD defects prevent post-Golgi vesicle fusion with nascent nuclear envelopes [121]. Thus, PLD-generated PtdOH generated by PLD is required for a developmentally-regulated vesicle fusion process, but not vesicle formation or vesicle scission.

GENETIC MODELS FOR BARS FUNCTION

Are in vivo models consistent with in vitro data regarding a role for endophilin and BARS as PtdOH effectors in membrane trafficking? In the case of endophilin there is good agreement as this protein is indeed essential for fission and uncoating of clathrin-coated vesicles in neurons [122–124]. BARS is reported to be essential for fragmentation of the Golgi ribbon in cultured cells, and ribbon scission is required for cells to negotiate the G2/M boundary. Interestingly, only cells with Golgi ribbons exhibit BARS-regulated Golgi mitotic checkpoints [125]. These findings emphasize the link between Golgi structure, lipid metabolism, and cell-cycle control.

BARS has a curious history, however, as it was first described as a member of the CtBP protein family of transcriptional co-repressors and is a spliceo-form of CtBP1 [126]. BARS/CtBP null ($ctbp1^{0/0}$) mice exhibit various develop-mental phenotypes associated with defects in body size, vascularization and body patterning. These phenotypes primarily reflect the transcriptional func-tions of BARS/CtBP. The $ctbp1^{0/0}$ phenotypes, and the viability of $ctbp1^{0/0}$ embryonic fibroblasts, are not consistent with essential roles for BARS in Golgi housekeeping functions or obligate requirements for BARS in progres-sion through the G2/M Golgi checkpoint [126]. It has been noted that the Golgi system of $ctbp1^{0/0}$ embryonic fibroblasts differs from that of wild-type fibroblasts in that it is not organized as an intact ribbon, and this morpho-logical derangement is argued to relieve $ctbp1^{0/0}$ fibroblasts of a requirement for BARS in cell-cycle progression [125]. This argument begs the question of what activities are responsible for fragmentation of the Golgi ribbon in cells that lack BARS. Also, what activity (if any) compensates for BARS in $ctbp1^{0/0}$ cells? Non-neuronal endophilins are candidates, and *in vitro* data support this notion [127]. But, given the dissonance between *in vitro* and *in vivo* read-outs, this hypothesis must be tested in a suitable *in vivo* context.

AMINO-PHOSPHOLIPIDS AND MEMBRANE TRAFFICKING

Functional involvements of glycerophospholipids in Golgi secretory function are not limited to PtdIns, phosphoinositides, PtdCho, and PtdOH. Roles for ami-no-phospholipids, such as PtdEtn and PtdSer, in membrane trafficking is amply demonstrated by the important roles that amino-phospholipid flippases, such as Drs2, play in controlling membrane trafficking through the yeast TGN/endo-somal system [28–30]. These P4-type ATPases translocate PtdSer and PtdEtn from cytosolic to lumenal membrane leaflets, and these activities interface with PtdIns-4-P signaling and the Arf pathway because yeast Drs2 flippase activ-ity is stimulated by binding both to PtdIns-4-P and to an Arf1-GEF [54]. Yeast P4-type ATPases are also indirectly subject to regulation by sphingolipids via the Fpk protein kinases that phosphorylate (and activate) the flippases [128].

The complexity of the amino-phospholipid flippase involvement in yeast membrane trafficking is emphasized by the overlapping functional redun-dancies of multiple Drs2-like flippases [28–30]. A long-standing idea is that amino-phospholipid flippases promote positive membrane curvature (and therefore vesicle budding) by driving local leaflet asymmetries, both in terms of phospholipid composition and phospholipid distribution between the cytosolic and lumenal TGN/endosomal leaflets [28–30]. While the evi-dence identifies an interface of Drs2 (and Drs2-like flippases) with Arf and clathrin-dependent functions in yeast [36], it remains to be determined how flippase activities potentiate membrane trafficking. The functional significance of removal of PtdEtn or PtdSer from the cytosolic leaflet, and/or enrichment of the lumenal leaflet with PtdEtn or PtdSer, also needs to be explored.

Trafficking functions for PtdSer on the cytosolic leaflets of endosomal membranes are also recognized. This amino-phospholipid is required for retrograde membrane trafficking from mammalian recycling endosomes [129]. The primary, and perhaps exclusive, PtdSer effector in this system is evectin-2. This protein harbors a pleckstrin homology domain that displays an exquisite specificity for PtdSer and does not bind phosphoinositides. PtdSer binding is required for evectin-2 localization to recycling endosomes and for protein function in cells [129], but how evectin-2 regulates trafficking remains to be elucidated.

STEROLS AND THE GOLGI COMPLEX

Membrane sterol content increases progressively through the compartments of the secretory pathway, and this gradient facilitates membrane protein sorting [130]. Sterols organize plasma membrane microdomains that modulate endocytosis and receptor activation and regulate membrane trafficking from the TGN. Biosynthetic trafficking of a subset of yeast plasma membrane proteins is disrupted by defects in sterol biosynthesis [131–133]. A common property of the affected cargos is their incorporation into ergosterol-containing detergent-resistant membranes [134]. Interestingly, compromising late steps in sterol biosynthesis results in mis-sorting of these cargos; although bulk sterol levels are unchanged under these conditions, the chemical profile of the accumulated sterols is altered [135]. These accumulated sterols, while chemically distinct from ergosterol, support the formation of detergent-resistant membrane microdomains. Yet, integral membrane proteins destined for the plasma membrane are mis-sorted, indicating that subtle differences in sterol structure influence trafficking fidelity.

Budding of anterograde vesicles from the TGN is proposed to be driven by the phase separation of sterol and sphingolipids into microdomains where the immiscibility of two liquid phases in lipid bilayers promotes the membrane bending necessary for vesicle budding [136]. Indeed, sterols and sphingolipids are enriched in TGN-derived vesicles relative to the bulk composition of the donor organelle [134,137]. The data suggest that a single lipid-driven sorting process drives biogenesis of TGN-derived vesicles bound for the plasma membrane [134]. This mechanism diverges from that which governs COP1 vesicle budding from bulk Golgi membranes *in vitro*. Those vesicles have a reduced sphingomyelin and cholesterol content relative to the bulk Golgi membranes from which they were formed [138]. Yet, sphingomyelin has an important role as a cofactor in COP1 vesicle formation: Brügger, Wieland and colleagues [139] have reported the remarkable discovery that a single molecule of a specific molecular species of sphingomyelin binds to the transmembrane domain of a COP1 coat subunit (p24), modulates the oligomeric state of p24, and thereby regulates COP1 coat biogenesis.

GLYCOLIPID TRANSFER PROTEINS

Glycolipid transfer proteins bind both sphingoid- and glycerol-based glyco-lipids and mobilize these lipids between membrane bilayers *in vitro* [140]. The glucosylceramide (GlcCer) transfer protein FAPP2 is recruited to Golgi membranes in a PtdIns-4-P-dependent manner, and is required for the production of complex glycosphingolipids for which GlcCer is a precursor. Although FAPP2 is suggested to deliver GlcCer to distal Golgi compart-ments as a lipid carrier [141], others report that FAPP2 promotes retrograde transport of GlcCer from Golgi to the ER [142]. The rationale for the retro-grade pathway is that newly synthesized GlcCer, which resides in the cyto-solic leaflet of Golgi membranes, is mobilized to the ER for the purpose of being flipped into the lumenal ER leaflet (Figure 3). Vesicular trafficking from the ER to the Golgi subsequently introduces the lumenally disposed GlcCer to Golgi-localized glycosyltransferases for maturation into complex glyco-sphingolipids [142]. A GlcCer-independent role for FAPP2 in TGN trafficking has also been suggested by the finding that FAPP2 forms a curved dimer that tubulates membranes in a PtdIns-4-P-dependent manner [143]. These studies describe a mechanism for how FAPP2 potentiates cargo transport from the TGN to apical surfaces of polarized epithelial cells [144].

CONCLUDING THOUGHTS

Much progress has been made in understanding the mechanisms that con-trol Golgi dynamics and architecture since the discovery of this organelle more than 100 years ago. Lipids, lipid-binding proteins, and lipid metabolism are major contributors to plasticity of the Golgi system. However, we have only a rudimentary understanding of the cross-talk between different arms of the Golgi lipid metabolome. 'Systems' approaches to model the land-scape of Golgi lipid metabolism will be necessary for a detailed description of crosstalk mechanisms. These approaches also hold the ultimate promise of unifying lipid biochemical principles with Golgi function.

It still remains unclear why the Golgi adopts its characteristic morphology, given that secretory activity can be insensitive to dramatic structural derange-ments of this organelle. The answer must lie in unappreciated levels of phys-iological regulation associated with the organization of the Golgi, or with the maturation process itself. Insights to this effect are offered by tunable PITP–ORP rheostats, as these suggest mechanisms for integrating TGN/endosomal maturation (and lipid signaling) with the control of cell proliferation and nuclear responses to stress [64]. These circuits reveal an unappreciated physiological plasticity of Golgi/endosomal maturation programs and identify involvements of such rheostats in modulating Golgi plasticity. Such circuits seem ideally suited for chaperoning cell entry into post-mitotic states or in maintaining post-mitotic cell physiology. Perhaps maturation mechanisms for membrane trafficking evolved, in part, because these afford superior instruments for fine-tuning cell-growth regulation and metabolic control than do stable compartment mechanisms. In this regard, the fidelity of mitotic spindle formation and function is also influenced by Golgi organization, and evidence is building that lipid metabolism has a hand in this circuit as well [92,98,141,142]. We anticipate that studies of Golgi lipid metabolism in the developmental context of multicellular organisms will prove a major contributor to the future of Golgi research. The fruits of those studies will undoubtedly yield more surprises from an organelle that has already produced its share.

ACKNOWLEDGEMENTS

This work was supported by NIH grant GM44530 to V.A.B. We are grateful to Gilbert Di Paolo for granting us permission to cite unpublished data. We also thank Lora L. Yanagisawa (Univ. Alabama-Birmingham) and three referees for careful review of the work and for their critical comments. Their input greatly improved the manuscript. The authors declare no financial conflict.

REFERENCES

1 Farquhar, M.G., and Palade, G.E. (1981). The Golgi apparatus (complex)-(1954–1981)-from artifact to center stage. J. Cell Biol. *91*, 77s–103s.

2 Tang, D., Mar, K., Warren, G., and Wang, Y. (2008). Molecular mechanism of mitotic Golgi disassembly and reassembly revealed by a defined reconstitution assay. J. Biol. Chem. *283*, 6085–6094.

3 Lippincott-Schwartz, J., Roberts, T.H., and Hirschberg, K. (2000). Secretory protein trafficking and organelle dynamics in living cells. Annu. Rev. Cell Dev. Biol. *16*, 557–589.

4 Bisel, B., Wang, Y., Wei, J.H., Xiang, Y., Tang, D., Miron-Mendoza, M., Yoshimura, S., Nakamura, N., and Seemann, J. (2008). ERK regulates Golgi and centrosome orientation towards the leading edge through GRASP65. J. Cell Biol. *182*, 837–843.

5 Follit, J.A., San Agustin, J.T., Xu, F., Jonassen, J.A., Samtani, R., Lo, C.W., and Pazour, G.J. (2008). The Golgin GMAP210/TRIP11 anchors IFT20 to the Golgi complex. PLoS Genet. *4*, e1000315.

6 Yadav, S., Puri, S., and Linstedt, A.D. (2009). A primary role for Golgi positioning in directed secretion, cell polarity, and wound healing. Mol. Biol. Cell *20*, 1728–1736.

7 Altan-Bonnet, N., Sougrat, R., and Lippincott-Schwartz, J. (2004). Molecular basis for Golgi maintenance and biogenesis. Curr. Opin. Cell Biol. *16*, 364–372.

8 Glick, B.S., and Nakano, A. (2009). Membrane traffic within the Golgi apparatus. Annu. Rev. Cell Dev. Biol. *25*, 113–132.

9 Rothman, J.E. (1996). The protein machinery of vesicle budding and fusion. Protein Sci. *5*, 185–194.

10 Schekman, R., and Orci, L. (1996). Coat proteins and vesicle budding. Science *271*, 1526–1533.

11 Bankaitis, V.A., Aitken, J.R., Cleves, A.E., and Dowhan, W. (1990). An essential role for a phospholipid transfer protein in yeast Golgi function. Nature *347*, 561–562.

12 Cleves, A., McGee, T., and Bankaitis, V. (1991). Phospholipid transfer proteins: a biological debut. Trends Cell Biol. *1*, 30–34.

13 Cleves, A.E., McGee, T.P., Whitters, E.A., Champion, K.M., Aitken, J.R., Dowhan, W., Goebl, M., and Bankaitis, V.A. (1991). Mutations in the CDP-choline pathway for phospholipid biosynthesis bypass the requirement for an essential phospholipid transfer protein. Cell *64*, 789–800.

14 Eberhard, D.A., Cooper, C.L., Low, M.G., and Holz, R.W. (1990). Evidence that the inositol phospholipids are necessary for exocytosis. Loss of inositol phospholipids and inhibition of secretion in permeabilized cells caused by a bacterial phospholipase C and removal of ATP. Biochem. J. *268*, 15–25.

15 Fang, M., Kearns, B.G., Gedvilaite, A., Kagiwada, S., Kearns, M., Fung, M.K., and Bankaitis, V.A. (1996). Kes1p shares homology with human oxysterol binding protein and participates in a novel regulatory pathway for yeast Golgi-derived transport vesicle biogenesis. EMBO J. *15*, 6447–6459.

16 Li, X., Rivas, M.P., Fang, M., Marchena, J., Mehrotra, B., Chaudhary, A., Feng, L., Prestwich, G.D., and Bankaitis, V.A. (2002). Analysis of oxysterol binding protein homologue Kes1p function in regulation of Sec14p-dependent protein transport from the yeast Golgi complex. J. Cell Biol. *157*, 63–77.

17 Litvak, V., Dahan, N., Ramachandran, S., Sabanay, H., and Lev, S. (2005). Maintenance of the diacylglycerol level in the Golgi apparatus by the Nir2 protein is critical for Golgi secretory function. Nat. Cell Biol. *7*, 225–234.

18 Stefan, C.J., Manford, A.G., Baird, D., Yamada-Hanff, J., Mao, Y., and Emr, S.D. (2011). Osh proteins regulate phosphoinositide metabolism at ER-plasma membrane contact sites. Cell *144*, 389–401.

19 Cleves, A.E., Novick, P.J., and Bankaitis, V.A. (1989). Mutations in the SAC1 gene suppress defects in yeast Golgi and yeast actin function. J. Cell Biol. *109*, 2939–2950.

20 Guo, S., Stolz, L.E., Lemrow, S.M., and York, J.D. (1999). SAC1-like domains of yeast SAC1, INP52, and INP53 and of human synaptojanin encode polyphosphoinositide phosphatases. J. Biol. Chem. *274*, 12990–12995.

21 Hama, H., Schnieders, E.A., Thorner, J., Takemoto, J.Y., and DeWald, D.B. (1999). Direct involvement of phosphatidylinositol 4-phosphate in secretion in the yeast Saccharomyces cerevisiae. J. Biol. Chem. *274*, 34294–34300.

22 Walch-Solimena, C., and Novick, P. (1999). The yeast phosphatidylinositol-4-kinase pik1 regulates secretion at the Golgi. Nat. Cell Biol. *1*, 523–525.

23 Brown, H.A., Gutowski, S., Moomaw, C.R., Slaughter, C., and Sternweis, P.C. (1993). ADP-ribosylation factor, a small GTP-dependent regulatory protein, stimulates phospholipase D activity. Cell *75*, 1137–1144.

24 de Figueiredo, P., Drecktrah, D., Polizotto, R.S., Cole, N.B., Lippincott-Schwartz, J., and Brown, W.J. (2000). Phospholipase A2 antagonists inhibit constitutive retrograde membrane traffic to the endoplasmic reticulum. Traffic *1*, 504–511.

25 Ktistakis, N.T., Brown, H.A., Sternweis, P.C., and Roth, M.G. (1995). Phospholipase D is present on Golgi-enriched membranes and its activation by ADP ribosylation factor is sensitive to brefeldin A. Proc. Natl. Acad. Sci. USA *92*, 4952–4956.

26 Drecktrah, D., Chambers, K., Racoosin, E.L., Cluett, E.B., Gucwa, A., Jackson, B., and Brown, W.J. (2003). Inhibition of a Golgi complex lysophospholipid acyltransferase induces membrane tubule formation and retrograde trafficking. Mol. Biol. Cell *14*, 3459–3469.

27 Schmidt, J.A., and Brown, W.J. (2009). Lysophosphatidic acid acyltransferase 3 regulates Golgi complex structure and function. J. Cell Biol. *186*, 211–218.

28 Natarajan, P., Wang, J., Hua, Z., and Graham, T.R. (2004). Drs2p-coupled aminophospholipid translocase activity in yeast Golgi membranes and relationship to in vivo function. Proc. Natl. Acad. Sci. USA *101*, 10614–10619.

29 Natarajan, P., Liu, K., Patil, D.V., Sciorra, V.A., Jackson, C.L., and Graham, T.R. (2009). Regulation of a Golgi flippase by phosphoinositides and an ArfGEF. Nat. Cell Biol. *11*, 1421–1426.

30 Muthusamy, B.P., Natarajan, P., Zhou, X., and Graham, T.R. (2009). Linking phospholipid flippases to vesicle-mediated protein transport. Biochim. Biophys. Acta *1791*, 612–619.

31 Patterson, G.H., Hirschberg, K., Polishchuk, R.S., Gerlich, D., Phair, R.D., and Lippincott-Schwartz, J. (2008). Transport through the Golgi apparatus by rapid partitioning within a two-phase membrane system. Cell *133*, 1055–1067.

32 Emr, S., Glick, B.S., Linstedt, A.D., Lippincott-Schwartz, J., Luini, A., Malhotra, V., Marsh, B.J., Nakano, A., Pfeffer, S.R., Rabouille, C., *et al.* (2009). Journeys through the Golgi–taking stock in a new era. J. Cell Biol. *187*, 449–453.

33 Sharpe, H.J., Stevens, T.J., and Munro, S. (2010). A comprehensive comparison of transmembrane domains reveals organelle-specific properties. Cell *142*, 158–169.

34 Hay, J.C., and Martin, T.F. (1993). Phosphatidylinositol transfer protein required for ATP-dependent priming of Ca(2+)-activated secretion. Nature *366*, 572–575.

35 Schu, P.V., Takegawa, K., Fry, M.J., Stack, J.H., Waterfield, M.D., and Emr, S.D. (1993). Phosphatidylinositol 3-kinase encoded by yeast VPS34 gene essential for protein sorting. Science *260*, 88–91.

36 Graham, T.R., and Burd, C.G. (2011). Coordination of Golgi functions by phosphatidylinositol 4-kinases. Trends Cell Biol. *21*, 113–121.

37 Bankaitis, V.A., Mousley, C.J., and Schaaf, G. (2010). The Sec14 superfamily and mechanisms for crosstalk between lipid metabolism and lipid signaling. Trends Biochem. Sci. *35*, 150–160.

38 Schaaf, G., Ortlund, E.A., Tyeryar, K.R., Mousley, C.J., Ile, K.E., Garrett, T.A., Ren, J., Woolls, M.J., Raetz, C.R., Redinbo, M.R., *et al.* (2008). Functional anatomy of phospholipid binding and regulation of phosphoinositide homeostasis by proteins of the sec14 superfamily. Mol. Cell *29*, 191–206.

39 Strahl, T., and Thorner, J. (2007). Synthesis and function of membrane phosphoinositides in budding yeast, Saccharomyces cerevisiae. Biochim. Biophys. Acta *1771*, 353–404.

40 Hendricks, K.B., Wang, B.Q., Schnieders, E.A., and Thorner, J. (1999). Yeast homologue of neuronal frequenin is a regulator of phosphatidylinositol-4-kinase. Nat. Cell Biol. *1*, 234–241.

41 Zhao, X., Varnai, P., Tuymetova, G., Balla, A., Toth, Z.E., Oker-Blom, C., Roder, J., Jeromin, A., and Balla, T. (2001). Interaction of neuronal calcium sensor-1 (NCS-1) with phosphatidylinositol 4-kinase beta stimulates lipid kinase activity and affects membrane trafficking in COS-7 cells. J. Biol. Chem. *276*, 40183–40189.

42 Godi, A., Pertile, P., Meyers, R., Marra, P., Di Tullio, G., Iurisci, C., Luini, A., Corda, D., and De Matteis, M.A. (1999). ARF mediates recruitment of PtdIns-4-kinase-beta and stimulates synthesis of PtdIns(4,5)P2 on the Golgi complex. Nat. Cell Biol. *1*, 280–287.

43 Haynes, L.P., Thomas, G.M., and Burgoyne, R.D. (2005). Interaction of neuronal calcium sensor-1 and ADP-ribosylation factor 1 allows bidirectional control of phosphatidylinositol 4-kinase beta and trans-Golgi network-plasma membrane traffic. J. Biol. Chem. *280*, 6047–6054.

44 Gloor, Y., Schone, M., Habermann, B., Ercan, E., Beck, M., Weselek, G., Muller-Reichert, T., and Walch-Solimena, C. (2010). Interaction between Sec7p and Pik1p: the first clue for the regulation of a coincidence detection signal. Eur. J. Cell Biol. *89*, 575–583.

45 Szentpetery, Z., Varnai, P., and Balla, T. (2010). Acute manipulation of Golgi phosphoinositides to assess their importance in cellular trafficking and signaling. Proc. Natl. Acad. Sci. USA *107*, 8225–8230.

46 Polevoy, G., Wei, H.C., Wong, R., Szentpetery, Z., Kim, Y.J., Goldbach, P., Steinbach, S.K., Balla, T., and Brill, J.A. (2009). Dual roles for the Drosophila PI 4-kinase four wheel drive in localizing Rab11 during cytokinesis. J. Cell Biol. *187*, 847–858.

47 Carlton, J.G., and Cullen, P.J. (2005). Coincidence detection in phosphoinositide signaling. Trends Cell Biol. *15*, 540–547.

48 Wang, Y.J., Wang, J., Sun, H.Q., Martinez, M., Sun, Y.X., Macia, E., Kirchhausen, T., Albanesi, J.P., Roth, M.G., and Yin, H.L. (2003). Phosphatidylinositol 4 phosphate regulates targeting of clathrin adaptor AP-1 complexes to the Golgi. Cell *114*, 299–310.

49 Demmel, L., Gravert, M., Ercan, E., Habermann, B., Muller-Reichert, T., Kukhtina, V., Haucke, V., Baust, T., Sohrmann, M., Kalaidzidis, Y., *et al.* (2008). The clathrin adaptor Gga2p is a phosphatidylinositol 4-phosphate effector at the Golgi exit. Mol. Biol. Cell *19*, 1991–2002.

50 Wang, J., Sun, H.Q., Macia, E., Kirchhausen, T., Watson, H., Bonifacino, J.S., and Yin, H.L. (2007). PI4P promotes the recruitment of the GGA adaptor proteins to the trans-Golgi network and regulates their recognition of the ubiquitin sorting signal. Mol. Biol. Cell *18*, 2646–2655.

51 de Graaf, P., Zwart, W.T., van Dijken, R.A., Deneka, M., Schulz, T.K., Geijsen, N., Coffer, P.J., Gadella, B.M., Verkleij, A.J., van der Sluijs, P., *et al.* (2004). Phosphatidylinositol 4-kinasebeta is critical for functional association of rab11 with the Golgi complex. Mol. Biol. Cell *15*, 2038–2047.

52 Mizuno-Yamasaki, E., Medkova, M., Coleman, J., and Novick, P. (2010). Phosphatidylinositol 4-phosphate controls both membrane recruitment and a regulatory switch of the Rab GEF Sec2p. Dev. Cell *18*, 828–840.

53 Dumaresq-Doiron, K., Savard, M.F., Akam, S., Costantino, S., and Lefrancois, S. (2010). The phosphatidylinositol 4-kinase PI4KIIIalpha is required for the recruitment of GBF1 to Golgi membranes. J. Cell Sci. *123*, 2273–2280.

54 Chantalat, S., Park, S.K., Hua, Z., Liu, K., Gobin, R., Peyroche, A., Rambourg, A., Graham, T.R., and Jackson, C.L. (2004). The Arf activator Gea2p and the P-type ATPase Drs2p interact at the Golgi in Saccharomyces cerevisiae. J. Cell Sci. *117*, 711–722.

55 Munson, M., and Novick, P. (2006). The exocyst defrocked, a framework of rods revealed. Nat. Struct. Mol. Biol. *13*, 577–581.

56 Mousley, C.J., Tyeryar, K., Ile, K.E., Schaaf, G., Brost, R.L., Boone, C., Guan, X., Wenk, M.R., and Bankaitis, V.A. (2008). Trans-Golgi network and endosome dynamics connect ceramide homeostasis with regulation of the unfolded protein response and TOR signaling in yeast. Mol. Biol. Cell *19*, 4785–4803.

57 Wood, C.S., Schmitz, K.R., Bessman, N.J., Setty, T.G., Ferguson, K.M., and Burd, C.G. (2009). PtdIns4P recognition by Vps74/GOLPH3 links PtdIns 4-kinase signaling to retrograde Golgi trafficking. J. Cell Biol. *187*, 967–975.

58 Scott, K.L., Kabbarah, O., Liang, M.C., Ivanova, E., Anagnostou, V., Wu, J., Dhakal, S., Wu, M., Chen, S., Feinberg, T., *et al.* (2009). GOLPH3 modulates mTOR signalling and rapamycin sensitivity in cancer. Nature *459*, 1085–1090.

59 Kearns, B.G., McGee, T.P., Mayinger, P., Gedvilaite, A., Phillips, S.E., Kagiwada, S., and Bankaitis, V.A. (1997). Essential role for diacylglycerol in protein transport from the yeast Golgi complex. Nature *387*, 101–105.

60 Rivas, M.P., Kearns, B.G., Xie, Z., Guo, S., Sekar, M.C., Hosaka, K., Kagiwada, S., York, J.D., and Bankaitis, V.A. (1999). Pleiotropic alterations in lipid metabolism in yeast sac1 mutants: relationship to "bypass Sec14p" and inositol auxotrophy. Mol. Biol. Cell *10*, 2235–2250.

61 Nile, A.H., Bankaitis, V.A., and Grabon, A. (2010). Mammalian diseases of phosphatidylinositol transfer proteins and their homologs. Clin. Lipidol. *5*, 867–897.

62 Huynh, H., Bottini, N., Williams, S., Cherepanov, V., Musumeci, L., Saito, K., Bruckner, S., Vachon, E., Wang, X., Kruger, J., et al. (2004). Control of vesicle fusion by a tyrosine phosphatase. Nat. Cell Biol. 6, 831–839.

63 Fairn, G.D., Curwin, A.J., Stefan, C.J., and McMaster, C.R. (2007). The oxysterol binding protein Kes1p regulates Golgi apparatus phosphatidylinositol-4-phosphate function. Proc. Natl. Acad. Sci. USA 104, 15352–15357.

64 Mousley, C., Yuan, P., Gaur, N.A., Trettin, K.D., Nile, A.H., Deminoff, S., Dewar, B.J., Wolpert, M., MacDonald, J.M., Herman, P.K., et al. (2012). A sterol binding protein integrates endosomal lipid metabolism with TOR signaling and nitrogen sensing. Cell 148, 702–715.

65 de Saint-Jean, M., Delfosse, V., Douguet, D., Chicanne, G., Payrastre, B., Bourguet, W., Antonny, B., and Drin, G. (2011). Osh4p exchanges sterols for phosphatidylinositol 4-phosphate between lipid bilayers. J. Cell Biol. 195, 965–978.

66 Alfaro, G., Johansen, J., Dighe, S.A., Duamel, G., Kozminski, K.G., and Beh, C.T. (2011). The sterol-binding protein Kes1/Osh4p is a regulator of polarized exocytosis. Traffic 12, 1521–1536.

67 Im, Y.J., Raychaudhuri, S., Prinz, W.A., and Hurley, J.H. (2005). Structural mechanism for sterol sensing and transport by OSBP-related proteins. Nature 437, 154–158.

68 Georgiev, A.G., Sullivan, D.P., Kersting, M.C., Dittman, J.S., Beh, C.T., and Menon, A.K. (2011). Osh proteins regulate membrane sterol organization but are not required for sterol movement between the ER and PM. Traffic 12, 1341–1355.

69 Raychaudhuri, S., Im, Y.J., Hurley, J.H., and Prinz, W.A. (2006). Nonvesicular sterol movement from plasma membrane to ER requires oxysterol-binding protein-related proteins and phosphoinositides. J. Cell Biol. 173, 107–119.

70 Schulz, T.A., and Prinz, W.A. (2007). Sterol transport in yeast and the oxysterol binding protein homologue (OSH) family. Biochim. Biophys. Acta 1771, 769–780.

71 Carvou, N., Holic, R., Li, M., Futter, C., Skippen, A., and Cockcroft, S. (2010). Phosphatidylinositol- and phosphatidylcholine-transfer activity of PITPbeta is essential for COPI-mediated retrograde transport from the Golgi to the endoplasmic reticulum. J. Cell Sci. 123, 1262–1273.

72 Ile, K.E., Kassen, S., Cao, C., Vihtehlic, T., Shah, S.D., Mousley, C.J., Alb, J.G., Jr., Huijbregts, R.P., Stearns, G.W., Brockerhoff, S.E., et al. (2010). Zebrafish class 1 phosphatidylinositol transfer proteins: PITPbeta and double cone cell outer segment integrity in retina. Traffic 11, 1151–1167.

73 Peretti, D., Dahan, N., Shimoni, E., Hirschberg, K., and Lev, S. (2008). Coordinated lipid transfer between the endoplasmic reticulum and the Golgi complex requires the VAP proteins and is essential for Golgi-mediated transport. Mol. Biol. Cell 19, 3871–3884.

74 Fugmann, T., Hausser, A., Schoffler, P., Schmid, S., Pfizenmaier, K., and Olayioye, M.A. (2007). Regulation of secretory transport by protein kinase D-mediated phosphorylation of the ceramide transfer protein. J. Cell Biol. 178, 15–22.

75 Hanada, K., Kumagai, K., Tomishige, N., and Yamaji, T. (2009). CERT-mediated trafficking of ceramide. Biochim. Biophys. Acta 1791, 684–691.

76 Prinz, W.A. (2010). Lipid trafficking sans vesicles: where, why, how? Cell 143, 870–874.

77 Rao, R.P., Yuan, C., Allegood, J.C., Rawat, S.S., Edwards, M.B., Wang, X., Merrill, A.H., Jr., Acharya, U., and Acharya, J.K. (2007). Ceramide transfer protein function is essential for normal oxidative stress response and lifespan. Proc. Natl. Acad. Sci. USA 104, 11364–11369.

78 Wang, X., Rao, R.P., Kosakowska-Cholody, T., Masood, M.A., Southon, E., Zhang, H., Berthet, C., Nagashim, K., Veenstra, T.K., Tessarollo, L., et al. (2009). Mitochondrial degeneration and not apoptosis is the primary cause of embryonic lethality in ceramide transfer protein mutant mice. J. Cell Biol. 184, 143–158.

79 Milligan, S.C., Alb, J.G., Jr., Elagina, R.B., Bankaitis, V.A., and Hyde, D.R. (1997). The phospha-tidylinositol transfer protein domain of Drosophila retinal degeneration B protein is essential for photoreceptor cell survival and recovery from light stimulation. J. Cell Biol. *139*, 351–363.

80 Lu, C., Peng, Y.W., Shang, J., Pawlyk, B.S., Yu, F., and Li, T. (2001). The mammalian retinal degeneration B2 gene is not required for photoreceptor function and survival. Neurosci-ence *107*, 35–41.

81 Cockcroft, S., and Carvou, N. (2007). Biochemical and biological functions of class I phos-phatidylinositol transfer proteins. Biochim. Biophys. Acta *1771*, 677–691.

82 Kim, Y.J., Guzman-Hernandez, M.L., and Balla, T. (2011). A highly dynamic ER-derived phosphatidylinositol-synthesizing organelle supplies phosphoinositides to cellular mem-branes. Dev. Cell *21*, 813–824.

83 Blero, D., Payrastre, B., Schurmans, S., and Erneux, C. (2007). Phosphoinositide phos-phatases in a network of signalling reactions. Pflugers Arch. *455*, 31–44.

84 Clague, M.J., and Lorenzo, O. (2005). The myotubularin family of lipid phosphatases. Traffic *6*, 1063–1069.

85 Di Paolo, G., and De Camilli, P. (2006). Phosphoinositides in cell regulation and membrane dynamics. Nature *443*, 651–657.

86 Liu, Y., and Bankaitis, V.A. (2010). Phosphoinositide phosphatases in cell biology and dis-ease. Prog. Lipid Res. *49*, 201–217.

87 Whitters, E.A., Cleves, A.E., McGee, T.P., Skinner, H.B., and Bankaitis, V.A. (1993). SAC1p is an integral membrane protein that influences the cellular requirement for phospholipid transfer protein function and inositol in yeast. J. Cell Biol. *122*, 79–94.

88 Blagoveshchenskaya, A., Cheong, F.Y., Rohde, H.M., Glover, G., Knodler, A., Nicolson, T., Boehmelt, G., and Mayinger, P. (2008). Integration of Golgi trafficking and growth factor signaling by the lipid phosphatase SAC1. J. Cell Biol. *180*, 803–812.

89 Foti, M., Audhya, A., and Emr, S.D. (2001). Sac1 lipid phosphatase and Stt4 phosphati-dylinositol 4-kinase regulate a pool of phosphatidylinositol 4-phosphate that functions in the control of the actin cytoskeleton and vacuole morphology. Mol. Biol. Cell *12*, 2396–2411.

90 Nemoto, Y., Kearns, B.G., Wenk, M.R., Chen, H., Mori, K., Alb, J.G., Jr., De Camilli, P., and Bankaitis, V.A. (2000). Functional characterization of a mammalian Sac1 and mutants exhibiting substrate-specific defects in phosphoinositide phosphatase activity. J. Biol. Chem. *275*, 34293–34305.

91 Cheong, F.Y., Sharma, V., Blagoveshchenskaya, A., Oorschot, V.M., Brankatschk, B., Klumperman, J., Freeze, H.H., and Mayinger, P. (2010). Spatial regulation of Golgi phosphatidylinositol-4-phosphate is required for enzyme localization and glycosylation fidelity. Traffic *11*, 1180–1190.

92 Liu, Y., Boukhelifa, M., Tribble, E., Morin-Kensicki, E., Uetrecht, A., Bear, J.E., and Bankaitis, V.A. (2008). The Sac1 phosphoinositide phosphatase regulates Golgi mem-brane morphology and mitotic spindle organization in mammals. Mol. Biol. Cell *19*, 3080–3096.

93 Burger, K.N. (2000). Greasing membrane fusion and fission machineries. Traffic *1*, 605–613.

94 Chernomordik, L., Kozlov, M.M., and Zimmerberg, J. (1995). Lipids in biological mem-brane fusion. J. Membr. Biol. *146*, 1–14.

95 Antonny, B., Huber, I., Paris, S., Chabre, M., and Cassel, D. (1997). Activation of ADP-ribosylation factor 1 GTPase-activating protein by phosphatidylcholine-derived diacylglycerols. J. Biol. Chem. *272*, 30848–30851.

96 Baron, C.L., and Malhotra, V. (2002). Role of diacylglycerol in PKD recruitment to the TGN and protein transport to the plasma membrane. Science *295*, 325–328.

97 Yanagisawa, L.L., Marchena, J., Xie, Z., Li, X., Poon, P.P., Singer, R.A., Johnston, G.C., Randazzo, P.A., and Bankaitis, V.A. (2002). Activity of specific lipid-regulated ADP ribosylation factor-GTPase-activating proteins is required for Sec14p-dependent Golgi secretory function in yeast. Mol. Biol. Cell *13*, 2193–2206.

98 Asp, L., Kartberg, F., Fernandez-Rodriguez, J., Smedh, M., Elsner, M., Laporte, F., Barcena, M., Jansen, K.A., Valentijn, J.A., Koster, A.J., *et al.* (2009). Early stages of Golgi vesicle and tubule formation require diacylglycerol. Mol. Biol. Cell *20*, 780–790.

99 Fernandez-Ulibarri, I., Vilella, M., Lazaro-Dieguez, F., Sarri, E., Martinez, S.E., Jimenez, N., Claro, E., Merida, I., Burger, K.N., and Egea, G. (2007). Diacylglycerol is required for the formation of COPI vesicles in the Golgi-to-ER transport pathway. Mol. Biol. Cell *18*, 3250–3263.

100 Bard, F., and Malhotra, V. (2006). The formation of TGN-to-plasma-membrane transport carriers. Annu. Rev. Cell Dev. Biol. *22*, 439–455.

101 Liljedahl, M., Maeda, Y., Colanzi, A., Ayala, I., Van Lint, J., and Malhotra, V. (2001). Protein kinase D regulates the fission of cell surface destined transport carriers from the trans-Golgi network. Cell *104*, 409–420.

102 Bossard, C., Bresson, D., Polishchuk, R.S., and Malhotra, V. (2007). Dimeric PKD regulates membrane fission to form transport carriers at the TGN. J. Cell Biol. *179*, 1123–1131.

103 Caloca, M.J., Zugaza, J.L., and Bustelo, X.R. (2003). Exchange factors of the RasGRP family mediate Ras activation in the Golgi. J. Biol. Chem. *278*, 33465–33473.

104 Lehel, C., Olah, Z., Jakab, G., Szallasi, Z., Petrovics, G., Harta, G., Blumberg, P.M., and Anderson, W.B. (1995). Protein kinase C epsilon subcellular localization domains and proteolytic degradation sites. A model for protein kinase C conformational changes. J. Biol. Chem. *270*, 19651–19658.

105 Maissel, A., Marom, M., Shtutman, M., Shahaf, G., and Livneh, E. (2006). PKCeta is localized in the Golgi, ER and nuclear envelope and translocates to the nuclear envelope upon PMA activation and serum-starvation: C1b domain and the pseudosubstrate containing fragment target PKCeta to the Golgi and the nuclear envelope. Cell Signal *18*, 1127–1139.

106 Wang, Q.J., Bhattacharyya, D., Garfield, S., Nacro, K., Marquez, V.E., and Blumberg, P.M. (1999). Differential localization of protein kinase C delta by phorbol esters and related compounds using a fusion protein with green fluorescent protein. J. Biol. Chem. *274*, 37233–37239.

107 Diaz Anel, A.M., and Malhotra, V. (2005). PKCeta is required for beta1gamma2/beta3gamma2- and PKD-mediated transport to the cell surface and the organization of the Golgi apparatus. J. Cell Biol. *169*, 83–91.

108 Bi, K., Roth, M.G., and Ktistakis, N.T. (1997). Phosphatidic acid formation by phospholipase D is required for transport from the endoplasmic reticulum to the Golgi complex. Curr. Biol. *7*, 301–307.

109 Roth, M.G. (2008). Molecular mechanisms of PLD function in membrane traffic. Traffic *9*, 1233–1239.

110 Schmidt, A., Wolde, M., Thiele, C., Fest, W., Kratzin, H., Podtelejnikov, A.V., Witke, W., Huttner, W.B., and Soling, H.D. (1999). Endophilin I mediates synaptic vesicle formation by transfer of arachidonate to lysophosphatidic acid. Nature *401*, 133–141.

111 Weigert, R., Silletta, M.G., Spano, S., Turacchio, G., Cericola, C., Colanzi, A., Senatore, S., Mancini, R., Polishchuk, E.V., Salmona, M., *et al.* (1999). CtBP/BARS induces fission of Golgi membranes by acylating lysophosphatidic acid. Nature *402*, 429–433.

112 Ferguson, S.M., Raimondi, A., Paradise, S., Shen, H., Mesaki, K., Ferguson, A., Destaing, O., Ko, G., Takasaki, J., Cremona, O., *et al.* (2009). Coordinated actions of actin and BAR proteins upstream of dynamin at endocytic clathrin-coated pits. Dev. Cell *17*, 811–822.

113 Yang, J.S., Gad, H., Lee, S.Y., Mironov, A., Zhang, L., Beznoussenko, G.V., Valente, C., Turacchio, G., Bonsra, A.N., Du, G., et al. (2008). A role for phosphatidic acid in COPI vesicle fission yields insights into Golgi maintenance. Nat. Cell Biol. 10, 1146–1153.

114 Gallop, J.L., Butler, P.J., and McMahon, H.T. (2005). Endophilin and CtBP/BARS are not acyl transferases in endocytosis or Golgi fission. Nature 438, 675–678.

115 de Figueiredo, P., Drecktrah, D., Katzenellenbogen, J.A., Strang, M., and Brown, W.J. (1998). Evidence that phospholipase A2 activity is required for Golgi complex and trans Golgi network membrane tubulation. Proc. Natl. Acad. Sci. USA 95, 8642–8647.

116 Yang, J.S., Valente, C., Polishchuk, R.S., Turacchio, G., Layre, E., Moody, D.B., Leslie, C.C., Gelb, M.H., Brown, W.J., Corda, D., et al. (2011). COPI acts in both vesicular and tubular transport. Nat. Cell Biol. 13, 996–1003.

117 Dall'Armi, C., Hurtado-Lorenzo, A., Tian, H., Morel, E., Nezu, A., Chan, R.B., Yu, W.H., Robinson, K.S., Yeku, O., Small, S.A., et al. (2010). The phospholipase D1 pathway modulates macroautophagy. Nat. Commun. 1, 142.

118 Oliveira, T.G., Chan, R.B., Tian, H., Laredo, M., Shui, G., Staniszewski, A., Zhang, H., Wang, L., Kim, T.W., Duff, K.E., et al. (2010). Phospholipase d2 ablation ameliorates Alzheimer's disease-linked synaptic dysfunction and cognitive deficits. J. Neurosci. 30, 16419–16428.

119 Rose, K., Rudge, S.A., Frohman, M.A., Morris, A.J., and Engebrecht, J. (1995). Phospholipase D signaling is essential for meiosis. Proc. Natl. Acad. Sci. USA 92, 12151–12155.

120 Xie, Z., Fang, M., Rivas, M.P., Faulkner, A.J., Sternweis, P.C., Engebrecht, J.A., and Bankaitis, V.A. (1998). Phospholipase D activity is required for suppression of yeast phosphatidylinositol transfer protein defects. Proc. Natl. Acad. Sci. USA 95, 12346–12351.

121 Neiman, A.M., Katz, L., and Brennwald, P.J. (2000). Identification of domains required for developmentally regulated SNARE function in Saccharomyces cerevisiae. Genetics 155, 1643–1655.

122 Gad, H., Ringstad, N., Low, P., Kjaerulff, O., Gustafsson, J., Wenk, M., Di Paolo, G., Nemoto, Y., Crun, J., Ellisman, M.H., et al. (2000). Fission and uncoating of synaptic clathrin-coated vesicles are perturbed by disruption of interactions with the SH3 domain of endophilin. Neuron 27, 301–312.

123 Verstreken, P., Koh, T.W., Schulze, K.L., Zhai, R.G., Hiesinger, P.R., Zhou, Y., Mehta, S.Q., Cao, Y., Roos, J., and Bellen, H.J. (2003). Synaptojanin is recruited by endophilin to promote synaptic vesicle uncoating. Neuron 40, 733–748.

124 Bai, J., Hu, Z., Dittman, J.S., Pym, E.C., and Kaplan, J.M. (2010). Endophilin functions as a membrane-bending molecule and is delivered to endocytic zones by exocytosis. Cell 143, 430–441.

125 Colanzi, A., Hidalgo Carcedo, C., Persico, A., Cericola, C., Turacchio, G., Bonazzi, M., Luini, A., and Corda, D. (2007). The Golgi mitotic checkpoint is controlled by BARS-dependent fission of the Golgi ribbon into separate stacks in G2. EMBO J. 26, 2465–2476.

126 Hildebrand, J.D., and Soriano, P. (2002). Overlapping and unique roles for C-terminal binding protein 1 (CtBP1) and CtBP2 during mouse development. Mol. Cell Biol. 22, 5296–5307.

127 Yang, J.S., Zhang, L., Lee, S.Y., Gad, H., Luini, A., and Hsu, V.W. (2006). Key components of the fission machinery are interchangeable. Nat. Cell Biol. 8, 1376–1382.

128 Roelants, F.M., Baltz, A.G., Trott, A.E., Fereres, S., and Thorner, J. (2010). A protein kinase network regulates the function of aminophospholipid flippases. Proc. Natl. Acad. Sci. USA 107, 34–39.

129 Uchida, Y., Hasegawa, J., Chinnapen, D., Inoue, T., Okazaki, S., Kato, R., Wakatsuki, S., Misaki, R., Koike, M., Uchiyama, Y., et al. (2011). Intracellular phosphatidylserine is essential for retrograde membrane traffic through endosomes. Proc. Natl. Acad. Sci. USA 108, 15846–15851.

130 Kaiser, H.J., Orlowski, A., Rog, T., Nyholm, T.K., Chai, W., Feizi, T., Lingwood, D., Vattulainen, I., and Simons, K. (2011). Lateral sorting in model membranes by cholesterol-mediated hydrophobic matching. Proc. Natl. Acad. Sci. USA *108*, 16628–16633.

131 Umebayashi, K., and Nakano, A. (2003). Ergosterol is required for targeting of tryptophan permease to the yeast plasma membrane. J. Cell Biol. *161*, 1117–1131.

132 Bagnat, M., Chang, A., and Simons, K. (2001). Plasma membrane proton ATPase Pma1p requires raft association for surface delivery in yeast. Mol. Biol. Cell *12*, 4129–4138.

133 Bagnat, M., and Simons, K. (2002). Cell surface polarization during yeast mating. Proc. Natl. Acad. Sci. USA *99*, 14183–14188.

134 Surma, M.A., Klose, C., Klemm, R.W., Ejsing, C.S., and Simons, K. (2011). Generic sorting of raft lipids into secretory vesicles in yeast. Traffic *12*, 1139–1147.

135 Proszynski, T.J., Klemm, R.W., Gravert, M., Hsu, P.P., Gloor, Y., Wagner, J., Kozak, K., Grabner, H., Walzer, K., Bagnat, M., *et al*. (2005). A genome-wide visual screen reveals a role for sphingolipids and ergosterol in cell surface delivery in yeast. Proc. Natl. Acad. Sci. USA *102*, 17981–17986.

136 Lipowsky, R. (1993). Domain-induced budding of fluid membranes. Biophys. J. *64*, 1133–1138.

137 Klemm, R.W., Ejsing, C.S., Surma, M.A., Kaiser, H.J., Gerl, M.J., Sampaio, J.L., de Robillard, Q., Ferguson, C., Proszynski, T.J., Shevchenko, A., *et al*. (2009). Segregation of sphingolipids and sterols during formation of secretory vesicles at the trans-Golgi network. J. Cell Biol. *185*, 601–612.

138 Brugger, B., Sandhoff, R., Wegehingel, S., Gorgas, K., Malsam, J., Helms, J.B., Lehmann, W.D., Nickel, W., and Wieland, F.T. (2000). Evidence for segregation of sphingomyelin and cholesterol during formation of COPI-coated vesicles. J. Cell Biol. *151*, 507–518.

139 Contreras, F.X., Ernst, A.M., Haberkant, P., Bjorkholm, P., Lindahl, E., Gonen, B., Tischer, C., Elofsson, A., von Heijne, G., Thiele, C., *et al*. (2012). Molecular recognition of a single sphingolipid species by a protein's transmembrane domain. Nature *481*, 525–529.

140 Brown, R.E., and Mattjus, P. (2007). Glycolipid transfer proteins. Biochim. Biophys. Acta *1771*, 746–760.

141 D'Angelo, G., Polishchuk, E., Di Tullio, G., Santoro, M., Di Campli, A., Godi, A., West, G., Bielawski, J., Chuang, C.C., van der Spoel, A.C., *et al*. (2007). Glycosphingolipid synthesis requires FAPP2 transfer of glucosylceramide. Nature *449*, 62–67.

142 Halter, D., Neumann, S., van Dijk, S.M., Wolthoorn, J., de Maziere, A.M., Vieira, O.V., Mattjus, P., Klumperman, J., van Meer, G., and Sprong, H. (2007). Pre- and post-Golgi translocation of glucosylceramide in glycosphingolipid synthesis. J. Cell Biol. *179*, 101–115.

143 Cao, X., Coskun, U., Rossle, M., Buschhorn, S.B., Grzybek, M., Dafforn, T.R., Lenoir, M., Overduin, M., and Simons, K. (2009). Golgi protein FAPP2 tubulates membranes. Proc. Natl. Acad. Sci. USA *106*, 21121–21125.

144 Vieira, O.V., Verkade, P., Manninen, A., and Simons, K. (2005). FAPP2 is involved in the transport of apical cargo in polarized MDCK cells. J. Cell Biol. *170*, 521–526.

CHAPTER 11

Weaving the Web of ER Tubules

Junjie Hu[1,*], William A. Prinz[2], Tom A. Rapoport[3,*]

[1]Department of Genetics and Cell Biology, College of Life Sciences, and Tianjin Key
Laboratory of Protein Sciences, Nankai University, Tianjin 300071, China,
[2]Laboratory of Cell Biochemistry and Biology, National Institute of Diabetes and Digestive
and Kidney Diseases, National Institutes of Health, Bethesda, MD 20892, USA,
[3]Howard Hughes Medical Institute, Department of Cell Biology, Harvard Medical
School, Boston, MA 02115, USA
*Correspondence: huj@nankai.edu.cn (J.H.), tom_rapoport@hms.harvard.edu (T.A.R.)

Cell, Vol. 147, No. 6, December 9, 2011 © 2011 Elsevier Inc.
http://dx.doi.org/10.1016/j.cell.2011.11.022

SUMMARY

How is the characteristic shape of an organelle generated? Recent work
has provided insight into how the tubular network of the endoplasmic reticulum (ER) is formed. The tubules themselves are shaped by the reticulons
and DP1/Yop1p, whereas their fusion into a network is brought about by
membrane-bound GTPases that include the atlastins, Sey1p, and RHD3.

INTRODUCTION

Membrane-bound organelles in eukaryotic cells have characteristic shapes.
Some are relatively spherical, whereas others consist of flat sheets or a
tubular network. How organelles are shaped is a fundamental problem in
cell biology, but surprisingly, until about a decade ago this question had
not been addressed (Dreier and Rapoport, 2000; Hermann et al., 1998;
Chen et al., 2003). The endoplasmic reticulum (ER) is particularly attractive for studying the morphogenesis of an organelle, as it contains distinct,
but contiguous domains. The ER consists of the nuclear envelope and the
peripheral ER, which in turn is composed of a polygonal network of tubules
and interdispersed sheets (Baumann and Walz, 2001; Shibata et al., 2006).
A recent study of the peripheral ER in yeast revealed that it is a complex
mix of fenestrated sheets of various sizes connected by tubules (West et al.,
2011). The peripheral ER is very dynamic; the tubules continuously form
and disappear, and the sheets rearrange (Du et al., 2004; Griffing, 2010). In

235

mammals, the tubular ER network extends throughout the entire cell, whereas in yeast and plant cells, it is located close to the plasma membrane and is referred to as "cortical ER" (Griffing, 2010; Prinz et al., 2000). In all eukaryotic cells, the ER plays an important role in key processes, such as the biosynthesis of secretory and membrane proteins, lipid synthesis, and Ca^{2+} storage.

The relative amounts of ER sheets and tubules vary greatly among different cell types (Shibata et al., 2006). For example, sheets are prominent in "professional" secretory cells, such as pancreatic cells or B cells, and tubules predominate in adrenal cells. It is likely that peripheral ER sheets correspond to rough ER domains, which are regions of the ER membrane to which polysomes are bound. These ER domains are thus specialized in synthesizing secretory and membrane proteins. The tubular ER contains some bound ribosomes, but it is uncertain whether large polysomes can be accommodated on their highly curved surfaces. Instead, tubular structures may be the preferred site for lipid synthesis, Ca^{2+} signaling, vesicle budding, and contact to other organelles, such as mitochondria and lipid droplets.

Recent results have provided important insight into how the different domains of the ER are generated and maintained, with most of the progress related to the mechanisms by which the tubular ER network is generated. Specifically, we now have some understanding of how the tubules themselves are formed and how they fuse to generate a three-dimensional network. Here we discuss these recent advances and point out unresolved issues.

THE CYTOSKELETON

Many ER tubules are formed on the basis of their connection with the cytoskeleton. In mammalian cells, ER tubules are pulled out of membrane reservoirs by being attached to the tips of growing microtubules or by associating with molecular motors that move along microtubules (Du et al., 2004). This second mechanism, called sliding, is more prominent and significantly faster, and it occurs primarily along acetylated microtubules (Friedman et al., 2010). In yeast and plant cells, the ER is also dynamic, but the movements occur along actin filaments (Griffing, 2010; Prinz et al., 2000). Ultimately, however, the alignment of ER tubules with the cytoskeleton is not perfect. In addition, the ER network does not disassemble upon depolymerization of the actin filaments in yeast cells, and it collapses only with some delay upon depolymerization of microtubules in mammalian cells (Prinz et al., 2000; Terasaki et al., 1986). Finally, it is possible that ER tubules can be generated from vesicles, as observed in vitro (Dreier and Rapoport, 2000). Thus, the cytoskeleton is probably not required for maintaining the shape of ER tubules. Instead, as we will discuss below, curvature-stabilizing proteins play a critical role.

How ER membranes associate with the cytoskeleton or molecular motors remains elusive. The integral ER protein STIM1 has been proposed to interact with the tips of growing microtubules (Grigoriev et al., 2008). It harbors a motif in its cytoplasmic tail with which it interacts with EB1, a microtubule plus-end tracking protein. However, because STIM1 only moves to the plasma membrane upon Ca^{2+} depletion from the ER, it remains unclear whether it is involved in the constitutive dynamics of the ER. The mechanism by which ER membranes attach to molecular motors is entirely unknown. Some proteins implicated in maintaining ER morphology (REEP1, Climp-63, p180, kinectin; Shibata et al., 2010) have microtubule-binding domains, but their significance remains to be clarified.

CURVATURE-STABILIZING PROTEINS

ER tubules are characterized by high membrane curvature in cross-section, as their diameter is relatively small (~30 nm in yeast and ~60 nm in mammals). Thus, there must be mechanisms that stabilize this high-energy state. To identify proteins involved in stabilizing membrane curvature and ER tubule formation, an in vitro assay was used, in which the formation of an ER network could be recapitulated, starting with small vesicles derived from *Xenopus leavis* eggs (Dreier and Rapoport, 2000). Network formation is inhibited by reagents that modify cysteines in membrane proteins. The target of these reagents is reticulon 4a (Rtn4a), confirmed by the inhibitory effect of antibodies on network formation (Voeltz et al., 2006). Rtn4a belongs to the reticulon family, members of which are probably found in every eukaryotic cell. The conserved region is a domain of ~200 amino acids, which contains two relatively long hydrophobic sequences (30–35 amino acids) that each sit in the membrane as a hairpin, with little exposure to the lumenal side of the membrane. The reticulons interact with members of another protein family, called DP1 or REEP5 in mammals and Yop1p in *Saccharomyces cerevisiae* (Voeltz et al., 2006). The DP1/Yop1p family is again ubiquitous. Although not sequence related to the reticulons, these proteins also contain a conserved domain with two hydrophobic hairpins in the membrane.

The reticulons and DP1/Yop1p localize preferentially to the tubular ER and to the highly curved edges of sheets (Shibata et al., 2010; Voeltz et al., 2006). Deletion of these proteins in yeast or their depletion in mammalian cells converts most ER tubules into sheets, indicating that they are required for tubule formation. One of these proteins, if present in adequate amounts, is sufficient to maintain normal ER morphology in yeast (Voeltz et al., 2006). The deletion of the reticulons and Yop1p causes only a moderate growth defect in *S. cerevisiae*, indicating that yeast cells can manage with only a small amount of tubules. However, similar experiments in *Caenorhabditis elegans* result in a marked reduction of embryonic viability (Audhya et al., 2007), suggesting that the tubular ER plays a more important role in higher organisms. The overexpression of the reticulons or DP1/Yop1p leads to long tubules, which become resistant to the collapse that normally follows the depolymerization of microtubules, indicating that these proteins can stabilize ER tubules (Shibata et al., 2008). The reconstitution of purified members of the reticulon and DP1/Yop1p families into proteoliposomes transforms the vesicles into tubules (Hu et al., 2008). Thus, these proteins are both necessary and sufficient for tubule formation.

The reticulons and DP1/Yop1p might stabilize the high curvature of ER tubules seen in cross-section by utilizing two cooperating mechanisms, hydrophobic insertion (wedging) and scaffolding (Shibata et al., 2009) (Figure 1). The hydrophobic insertion mechanism is based on the two hydrophobic hairpins in these proteins, which are proposed to form a wedge-like structure that displaces the lipids preferentially in the outer leaflet of the lipid bilayer, thus causing local curvature. The scaffolding mechanism assumes that the reticulons and DP1/Yop1 form arc-like oligomers that mold the lipid bilayer into tubules (Figure 1). Arc-like scaffolds, rather than rings or spirals that wrap around tubules, are consistent with the observation that the reticulons also localize to sheet edges (Shibata et al., 2010). Because arcs are not entirely encircling

FIGURE 1 The Reticulons and DP1/Yop1p Shape ER Tubules
The cut-away view shows the double hairpin structure of the proteins proposed to form a wedge. Several of these molecules would form an arc-like structure around the tubule (indicated by yellow strips).

a tubule, they would not block the long-distance diffusion of other membrane proteins, explaining the high mobility of these proteins observed in fluorescence recovery after photobleaching (FRAP) experiments. Calculations show that arc-like oligomers would have to occupy only ~10% of the total membrane surface to generate near perfect tubules, in agreement with estimates on the abundance of the curvature-stabilizing proteins in *S. cerevisiae* (Hu et al., 2008). Oligomers of the reticulons and DP1/Yop1p have indeed been identified by crosslinking, gradient centrifugation, and FRAP experiments, and mutants with oligomerization defects do not exclusively localize to tubules and are unable to generate tubules in vitro (Shibata et al., 2008). The membrane-embedded hairpin structures appear to mediate the intermolecular interactions.

ATLASTIN GTPase-MEDIATED HOMOTYPIC ER FUSION

Once ER tubules are formed, they need to be connected into a network, which requires the fusion of identical membranes, a process called homotypic fusion. Much progress has been made in understanding this reaction, which turns out to be quite different from heterotypic fusion mediated by SNAREs or viral proteins.

In metazoans, the homotypic fusion of ER membranes appears to be mediated by the atlastins (ATLs), a class of GTPases that belong to the dynamin family (Hu et al., 2009; Orso et al., 2009). Some species, such as *Drosophila melanogaster*, *C. elegans*, and *Danio rerio*, contain only one ATL gene, but many organisms express several isoforms (Rismanchi et al., 2008). The ATLs are anchored in the membrane by a hairpin of two closely spaced transmembrane segments, exposing both the N-terminal GTPase domain and the C-terminal tail to the cytosol. As in the case of the reticulons and DP1, the hairpin localizes the ATLs specifically to the tubular ER, perhaps by sensing the high curvature of tubules. All isoforms of ATLs interact with different isoforms of the reticulons and DP1 inside the membrane (Hu et al., 2009). A role for the ATLs in ER fusion is suggested by the observation that the depletion of ATLs leads to long, nonbranched ER tubules in tissue culture cells (Hu et al., 2009) and ER fragmentation in *Drosophila* (Orso et al., 2009), which may be caused by insufficient fusion between the tubules. Nonbranched ER tubules are also observed when dominant-negative ATL mutants, which have reduced GTPase activity or lack either the cytosolic or transmembrane regions, are expressed (Hu et al., 2009). In addition, antibodies to ATL inhibit ER network formation in *Xenopus* egg extracts (Hu et al., 2009). Finally, and most convincingly, proteoliposomes containing purified *Drosophila* ATL undergo GTP-dependent fusion in vitro (Orso et al., 2009).

Mutations in human ATL1, the isoform that is abundant in neuronal tissues, cause hereditary spastic paraplegia (HSP) (Zhao et al., 2001), a disease characterized by progressive spasticity and weakening of the lower limbs due to the axonal degeneration of the longest corticospinal motor neurons (Salinas et al., 2008). The depletion of ATL1 in primary neuron cultures leads to axon shortening (Zhu et al., 2006), mimicking the pathological condition, and the deletion or depletion of ATL in *Drosophila* and zebrafish causes neuronal defects (Fassier et al., 2010; Lee et al., 2009). Interestingly, mutations in two other ER proteins also frequently cause HSP (in spastin, a microtubule-severing ATPase, and in REEP1, a member of the DP1/Yop1p protein family). Mutations in the genes encoding ATL, spastin, and REEP1 collectively account for over 50% of all HSP cases (Salinas et al., 2008), suggesting that ER morphology defects may be a major cause of the disease. Defects in the function of these proteins might be particularly manifest in neurons with long axons, which rely to the greatest extent on an extended ER network. Although other HSP mutations affect genes that are not obviously related to ER morphology, they could still affect the ER indirectly, for example, by altering the cytoskeleton.

MECHANISM OF ATL-MEDIATED FUSION

Insight into ATL-mediated fusion was obtained by determining crystal structures of the N-terminal part of human ATL1 (Bian et al., 2011; Byrnes and Sondermann, 2011), which includes the GTPase domain, a linker region, and a three-helix bundle. One crystal structure was obtained in the presence of GDP and high concentrations of inorganic phosphate (Figure 2, state 1). The structure shows a dimer in which the GTPase domains with their bound nucleotides face each other. The helical bundles of the two ATL molecules point in opposite directions, suggesting that in the full-length proteins, the molecules would sit in apposing membranes. This structure thus likely corresponds to a pre-fusion state, in which the two membranes are tethered but not yet fused. A second crystal structure has been obtained in the presence of GDP (Figure 2, state 5). Although the GTPase domains are arranged as in the pre-fusion state, the helical bundles of the two ATL molecules have undergone a major conformational change. They are now parallel to one another, and the C termini come so close that they have to sit in the same membrane. This conformation thus likely corresponds to a post-fusion state. Interestingly, the two ATL molecules cross each other: the helix bundle of one molecule interacts with the GTPase domain of the other. The transition from the pre- to the post-fusion conformation was confirmed by trypsin protection and crosslinking experiments and occurs not only with the cytosolic domain of human ATL but also with the full-length *Drosophila* protein reconstituted into proteoliposomes (Bian et al., 2011).

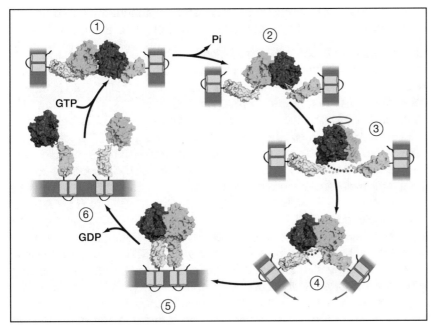

FIGURE 2 The Fusion Reaction Mediated by the Atlastins
The mechanism of membrane tethering and fusion is based on two crystal structures (state 1 [PDB 3QOF] and state 5 [3QNU]) and biochemical experiments. The atlastin (ATL) molecule in one membrane is colored with its GTPase domain in green and the helix bundle in yellow. The ATL molecule in the apposing membrane is colored with its GTPase domain in purple and the helix bundle in cyan. Upon GTP binding, the two ATL molecules dimerize and tether the membranes. GTP hydrolysis then causes conformational changes that pull the membranes together. Finally, GDP is released to reset the fusion machinery.

Based on the structures and biochemical experiments, it was proposed that the first step in the fusion reaction is the GTP-dependent dimerization of ATL molecules sitting in apposing membranes (Figure 2, transition to state 1). Nucleotide-dependent dimerization has been verified experimentally and is indeed a common feature of dynamin-like GTPases. Following GTP hydrolysis and phosphate release, the helical bundle of each ATL molecule would be dislodged from its own GTPase domain (state 2). The linker region would then allow the GTPase domains to rotate freely (state 3). When rotated by 180°, each GTPase domain could capture the helical bundle of the partner molecule (state 4), which would pull the apposing membranes together so that they can fuse (state 5). In principle, it is possible that the GTPase domains are stationary and the helical bundles move instead, but it is difficult to see how this would occur when the bundles are anchored in membranes. The final step would be release of GDP, which would dissociate the ATL dimers (state 6) and set the stage for the next round of fusion. The proposed model is consistent with the location of mutations that cause HSP, as many of them would be expected to disrupt dimer formation or disturb the conformational change deduced from the crystal structures (Bian et al., 2011).

A comparison of the pre- and post-fusion conformations of ATL indicates that the gain in energy is much smaller than observed in fusion reactions caused by SNAREs or viral proteins. This suggests that domains not included in the crystal structures may contribute to ATL-mediated fusion. In fact, deletion of the C-terminal tail following the two transmembrane segments drastically reduces fusion in vitro and causes HSP in humans (Bian et al., 2011). The tail contains a conserved, predicted amphipathic helix that may destabilize the lipid bilayer and thus facilitate fusion. Indeed, point mutants in the hydrophobic face of the amphipathic helix abolish fusion, and synthetic peptides, but not peptides carrying the mutations, stimulate fusion (T. Liu, X. Bian, S. Sun, X. Hu, T.A.R., and J.H., unpublished data). It is also likely that the two transmembrane segments play a role, perhaps by mediating oligomerization. This is supported by the observation of higher oligomers with full-length ATL (Zhu et al., 2006; Rismanchi et al., 2008), by the fact that the expression of the membrane-embedded region plus the C-terminal tail acts as a dominant-negative mutant (Hu et al., 2009), and by the finding that the transmembrane segments cannot be replaced by unrelated hydrophobic sequences (unpublished data). Perhaps, these oligomers allow multiple ATL molecules in each membrane to simultaneously undergo the proposed conformational change. Finally, it should be noted that the experiments so far have not entirely excluded the possibility that the ATLs only mediate hemifusion, rather than the fusion of both leaflets of the lipid bilayer, and that other factors contribute to the fusion reaction in vivo.

ATL-mediated homotypic fusion differs from heterotypic viral and SNARE-mediated fusion, in that nucleotide triphosphate hydrolysis directly drives the reaction. In SNARE-mediated fusion, nucleotide hydrolysis by the NEM-sensitive fusion factor (NSF) is used to reset the fusion machinery by disassembling SNARE complexes, and viral fusion does not require nucleotide hydrolysis at all. On the other hand, the mechanism proposed for ATLs may also apply to the mitofusins (Fzo1p in S. cerevisiae and fuzzy onion in Drosophila). These are proteins that mediate the homotypic fusion of the outer membrane of mitochondria (Hermann et al., 1998; Chen et al., 2003). Like the ATLs, the mitofusins are GTPases of the dynamin family that contain two closely spaced transmembrane segments and a cytoplasmic tail (which is longer than in the ATLs). Thus, it is possible that membrane-bound GTPases of the dynamin family may be generally involved in the homotypic fusion of organelles.

HOMOTYPIC ER FUSION IN YEAST AND PLANTS

Yeast and plant cells do not possess ATLs, but a similar GTPase, called Sey1p in S. cerevisiae and Root Hair Defective 3 (RHD3) in Arabidopsis thaliana, may have an analogous function. All eukaryotic organisms appear to have either ATL or Sey1p homologs, and no species has both. Sey1p was originally identified as a genetic interaction partner of Yop1p (hence the name *synthetic enhancement of YOP1*) (Brands and Ho, 2002), immediately suggesting that it plays a role in formation of the tubular ER. Sey1p has the same topology as ATL, that is, an N-terminal GTPase domain, a predicted helical bundle (which is significantly longer than in ATLs), two closely spaced transmembrane segments, and a C-terminal tail that includes an amphipathic helix. The GTPase domain contains the signature motifs that are typical for this subfamily of dynamin-like proteins. Like the ATLs, Sey1p interacts with the curvature-stabilizing proteins and localizes to the tubular ER (Hu et al., 2009). Mutations in the plant homolog RHD3 cause ER morphology defects (Zheng et al., 2004), similar to those seen after depletion of ATLs. Surprisingly, in yeast the deletion of *SEY1* alone does not abolish the tubular ER or result in ER fragmentation (K. Anwar, R. Klemm, A. Condon, M. Zhang, G. Ghirlando, J.H., T.A.R., and W.A.P., unpublished data); it requires the absence of both Sey1p and one of the curvature-stabilizing proteins Rtn1p or Yop1p to disrupt the tubular ER (Hu et al., 2009). Normal ER morphology can be re-established by expression of wild-type Sey1p, but not by Sey1p defective in GTP binding. Human ATL can partially replace Sey1p. A direct role for Sey1p in ER fusion is suggested by experiments in which the fusion of ER membranes is assessed after mating of haploid yeast cells and by the observation that proteoliposomes containing purified Sey1p undergo GTP-dependent fusion in vitro (unpublished data). All these results suggest that Sey1p functions like ATL to mediate homotypic ER fusion. However, the lack of an ER morphology defect in a *sey1* deletion mutant indicates that there must be an alternative fusion mechanism in yeast. This process might involve ER SNAREs, which had previously been implicated in homotypic ER fusion (Patel et al., 1998). Recent results indicate that cells lacking Sey1p and the ER SNARE Ufe1p have severe ER morphology defects and that there is a strong genetic interaction between *SEY1* and *UFE1* or *USE1*, which encodes another ER SNARE (unpublished data). It is unclear why mammalian and plant cells do not have a back-up fusion mechanism, but perhaps the alternate process occurs only in the cortical ER, which is much more abundant in yeast.

PERSPECTIVE

Despite significant progress over recent years, much remains to be learned. One immediate goal is a molecular understanding of how ATL and Sey1p catalyze homotypic ER fusion. The molecular mechanism of membrane fusion remains a general unresolved issue, but the GTP dependence of ATL-dependent fusion offers the opportunity to analyze intermediate stages of the reaction, something that is not easily done for viral or SNARE-mediated fusion. The role of ATL and Sey1p in vivo also needs further investigation. For example, there may be a link between these GTPases and lipid droplets. Many questions also concern the reticulons and DP1/Yop1p. Are the models for how they shape ER tubules correct? What is the significance of their interaction with ATL and Sey1p? Why are there so many isoforms of these proteins, which often have long segments unrelated to their proposed role in curvature stabilization? Are they regulated during the cell cycle?

Concerning the more general question of how other ER domains are generated, the surface has only been scratched. Recent work has provided first insight into how peripheral ER sheets might be generated (Shibata et al., 2010), but the exact function of the potential sheet-promoting proteins (Climp-63, p180, kinectin) remains to be elucidated, and the mechanism by which sheets are stacked on top of each other is totally unknown. Exciting research topics also include whether and how ER fission occurs, how the nuclear envelope is shaped, and how ER membranes interact with the cytoskeleton and molecular motors. One might hope that answering these questions will also have an impact on the understanding of the morphology of other organelles.

ACKNOWLEDGMENTS

J.H. is supported by the National Basic Research Program of China (973 Program, Grant 2010CB833702). W.A.P. is supported by the Intramural Research Program of the National Institute of Diabetes and Digestive and Kidney Diseases. T.A.R. is a Howard Hughes Medical Institute Investigator.

REFERENCES

Audhya, A., Desai, A., and Oegema, K. (2007). A role for Rab5 in structuring the endoplasmic reticulum. J. Cell Biol. 178, 43–56.

Baumann, O., and Walz, B. (2001). Endoplasmic reticulum of animal cells and its organization into structural and functional domains. Int. Rev. Cytol. 205, 149–214.

Bian, X., Klemm, R.W., Liu, T.Y., Zhang, M., Sun, S., Sui, X., Liu, X., Rapoport, T.A., and Hu, J. (2011). Structures of the atlastin GTPase provide insight into homotypic fusion of endoplasmic reticulum membranes. Proc. Natl. Acad. Sci. USA 108, 3976–3981.

Brands, A., and Ho, T.H. (2002). Function of a plant stress-induced gene, HVA22. Synthetic enhancement screen with its yeast homolog reveals its role in vesicular traffic. Plant Physiol. 130, 1121–1131.

Byrnes, L.J., and Sondermann, H. (2011). Structural basis for the nucleotide-dependent dimerization of the large G protein atlastin-1/SPG3A. Proc. Natl. Acad. Sci. USA 108, 2216–2221.

Chen, H., Detmer, S.A., Ewald, A.J., Griffin, E.E., Fraser, S.E., and Chan, D.C. (2003). Mitofusins Mfn1 and Mfn2 coordinately regulate mitochondrial fusion and are essential for embryonic development. J. Cell Biol. 160, 189–200.

Dreier, L., and Rapoport, T.A. (2000). In vitro formation of the endoplasmic reticulum occurs independently of microtubules by a controlled fusion reaction. J. Cell Biol. 148, 883–898.

Du, Y., Ferro-Novick, S., and Novick, P. (2004). Dynamics and inheritance of the endoplasmic reticulum. J. Cell Sci. *117*, 2871–2878.

Fassier, C., Hutt, J.A., Scholpp, S., Lumsden, A., Giros, B., Nothias, F., Schneider-Maunoury, S., Houart, C., and Hazan, J. (2010). Zebrafish atlastin controls motility and spinal motor axon architecture via inhibition of the BMP pathway. Nat. Neurosci. *13*, 1380–1387.

Friedman, J.R., Webster, B.M., Mastronarde, D.N., Verhey, K.J., and Voeltz, G.K. (2010). ER sliding dynamics and ER-mitochondrial contacts occur on acetylated microtubules. J. Cell Biol. *190*, 363–375.

Griffing, L.R. (2010). Networking in the endoplasmic reticulum. Biochem. Soc. Trans. *38*, 747–753.

Grigoriev, I., Gouveia, S.M., van der Vaart, B., Demmers, J., Smyth, J.T., Honnappa, S., Splinter, D., Steinmetz, M.O., Putney, J.W., Jr., Hoogenraad, C.C., and Akhmanova, A. (2008). STIM1 is a MT-plus-end-tracking protein involved in remodeling of the ER. Curr. Biol. *18*, 177–182.

Hermann, G.J., Thatcher, J.W., Mills, J.P., Hales, K.G., Fuller, M.T., Nunnari, J., and Shaw, J.M. (1998). Mitochondrial fusion in yeast requires the transmembrane GTPase Fzo1p. J. Cell Biol. *143*, 359–373.

Hu, J., Shibata, Y., Voss, C., Shemesh, T., Li, Z., Coughlin, M., Kozlov, M.M., Rapoport, T.A., and Prinz, W.A. (2008). Membrane proteins of the endoplasmic reticulum induce high-curvature tubules. Science *319*, 1247–1250.

Hu, J., Shibata, Y., Zhu, P.P., Voss, C., Rismanchi, N., Prinz, W.A., Rapoport, T.A., and Blackstone, C. (2009). A class of dynamin-like GTPases involved in the generation of the tubular ER network. Cell *138*, 549–561.

Lee, M., Paik, S.K., Lee, M.J., Kim, Y.J., Kim, S., Nahm, M., Oh, S.J., Kim, H.M., Yim, J., Lee, C.J., et al. (2009). Drosophila Atlastin regulates the stability of muscle microtubules and is required for synapse development. Dev. Biol. *330*, 250–262.

Orso, G., Pendin, D., Liu, S., Tosetto, J., Moss, T.J., Faust, J.E., Micaroni, M., Egorova, A., Martinuzzi, A., McNew, J.A., and Daga, A. (2009). Homotypic fusion of ER membranes requires the dynamin-like GTPase atlastin. Nature *460*, 978–983.

Patel, S.K., Indig, F.E., Olivieri, N., Levine, N.D., and Latterich, M. (1998). Organelle membrane fusion: a novel function for the syntaxin homolog Ufe1p in ER membrane fusion. Cell *92*, 611–620.

Prinz, W.A., Grzyb, L., Veenhuis, M., Kahana, J.A., Silver, P.A., and Rapoport, T.A. (2000). Mutants affecting the structure of the cortical endoplasmic reticulum in Saccharomyces cerevisiae. J. Cell Biol. *150*, 461–474.

Rismanchi, N., Soderblom, C., Stadler, J., Zhu, P.P., and Blackstone, C. (2008). Atlastin GTPases are required for Golgi apparatus and ER morphogenesis. Hum. Mol. Genet. *17*, 1591–1604.

Salinas, S., Proukakis, C., Crosby, A., and Warner, T.T. (2008). Hereditary spastic paraplegia: clinical features and pathogenetic mechanisms. Lancet Neurol. *7*, 1127–1138.

Shibata, Y., Voeltz, G.K., and Rapoport, T.A. (2006). Rough sheets and smooth tubules. Cell *126*, 435–439.

Shibata, Y., Voss, C., Rist, J.M., Hu, J., Rapoport, T.A., Prinz, W.A., and Voeltz, G.K. (2008). The reticulon and DP1/Yop1p proteins form immobile oligomers in the tubular endoplasmic reticulum. J. Biol. Chem. *283*, 18892–18904.

Shibata, Y., Hu, J., Kozlov, M.M., and Rapoport, T.A. (2009). Mechanisms shaping the membranes of cellular organelles. Annu. Rev. Cell Dev. Biol. *25*, 329–354.

Shibata, Y., Shemesh, T., Prinz, W.A., Palazzo, A.F., Kozlov, M.M., and Rapoport, T.A. (2010). Mechanisms determining the morphology of the peripheral ER. Cell *143*, 774–788.

Terasaki, M., Chen, L.B., and Fujiwara, K. (1986). Microtubules and the endoplasmic reticulum are highly interdependent structures. J. Cell Biol. *103*, 1557–1568.

Voeltz, G.K., Prinz, W.A., Shibata, Y., Rist, J.M., and Rapoport, T.A. (2006). A class of membrane proteins shaping the tubular endoplasmic reticulum. Cell *124*, 573–586.

West, M., Zurek, N., Hoenger, A., and Voeltz, G.K. (2011). A 3D analysis of yeast ER structure reveals how ER domains are organized by membrane curvature. J. Cell Biol. *193*, 333–346.

Zhao, X., Alvarado, D., Rainier, S., Lemons, R., Hedera, P., Weber, C.H., Tukel, T., Apak, M., Heiman-Patterson, T., Ming, L., et al. (2001). Mutations in a newly identified GTPase gene cause autosomal dominant hereditary spastic paraplegia. Nat. Genet. *29*, 326–331.

Zheng, H., Kunst, L., Hawes, C., and Moore, I. (2004). A GFP-based assay reveals a role for RHD3 in transport between the endoplasmic reticulum and Golgi apparatus. Plant J. *37*, 398–414.

Zhu, P.P., Soderblom, C., Tao-Cheng, J.H., Stadler, J., and Blackstone, C. (2006). SPG3A protein atlastin-1 is enriched in growth cones and promotes axon elongation during neuronal development. Hum. Mol. Genet. *15*, 1343–1353.

velopmental Cell

The ESCRT Pathway

William M. Henne[1,2], Nicholas J. Buchkovich[1,2], Scott D. Emr[1,2,*]

[1]Weill Institute for Cell and Molecular Biology, Cornell University, Weill Hall, Ithaca, NY 14853, USA, [2]Department of Molecular Biology and Genetics, Cornell University, Weill Hall, Ithaca, NY 14853, USA

*Correspondence: sde26@cornell.edu

Developmental Cell, Vol. 21, No. 1, July 19, 2011 © 2011 Elsevier Inc.
http://dx.doi.org/10.1016/j.devcel.2011.05.015

SUMMARY

Multivesicular bodies (MVBs) deliver cargo destined for degradation to the vacuole or lysosome. The ESCRT (endosomal sorting complex required for transport) pathway is a key mediator of MVB biogenesis, but it also plays critical roles in retroviral budding and cytokinetic abscission. Despite these diverse roles, the ESCRT pathway can be simply seen as a cargo-recognition and membrane-sculpting machine viewable from three distinct perspectives: (1) the ESCRT proteins themselves, (2) the cargo they sort, and (3) the membrane they deform. Here, we review ESCRT function from these perspectives and discuss how ESCRTs may drive vesicle budding.

INTRODUCTION

The formation of multivesicular bodies (MVBs) is a key stage in the delivery of cargo destined for degradation in the yeast vacuole or mammalian lysosome. MVBs were discovered in the 1950s by Keith Porter and George Palade when they observed "two large vesicles with smaller vesicles inside" in electron micrographs (Palade, 1955; Sotelo and Porter, 1959). In the late 1970s, Stanley Cohen and colleagues observed the sorting of epidermal growth factor receptor (EGFR) into the intraluminal vesicles (ILVs) of endosomes, similar to the structures previously reported by Porter and Palade (Haigler et al., 1979). Despite these early observations, the machinery responsible for MVB biogenesis was identified only recently.

247

In 2001, the endosomal sorting complex required for transport-I (ESCRT-I) complex was characterized and shown to engage ubiquitinated cargo at the endosome and mediate its sorting into MVBs (Katzmann et al., 2001). A year later, two back-to-back papers in *Developmental Cell* identified the ESCRT-II and ESCRT-III complexes as critical players in the delivery of ubiquitinated cargo to the yeast vacuole (Babst et al., 2002a, 2002b). These studies were preceded by the identification of the AAA ATPase Vps4 (part of a fifth ESCRT complex) in 1997, which is necessary for delivery of cargo to the vacuole (Babst et al., 1997). Together, these papers set the framework establishing the ESCRT proteins as cargo sequestering and sorting machinery that can deform the endosomal-limiting membrane inward to generate MVBs.

These studies were also significant in that they organized and assigned individual functions to the earlier identified "class E" vacuolar protein sorting (*vps*) genes that, when knocked out in yeast, fail to deliver cargo to the vacuole and exhibit aberrant endosome morphology (Raymond et al., 1992). They also initiated work demonstrating that the ESCRTs comprise a pathway of five distinct complexes (ESCRTs -0, -I, -II, and -III, and Vps4) that recognize and sort ubiquitinated cargo through an exquisite division of labor.

Today, however, the ESCRT field includes more than studies of endosomal trafficking. Recent work has elucidated roles for the ESCRTs in numerous biological processes. Beyond MVB formation, ESCRT proteins are well established to function in eukaryotic cell abscission and viral budding (Figure 1), as well as exosome secretion, and autophagy (Carlton et al., 2008; Filimonenko et al., 2007; Lee et al., 2007; Rusten et al., 2007). Although not the focus of this review, ESCRT dysfunction is associated with numerous diseases, including cancer, neurodegeneration, Huntington's, and Parkinson's diseases (for review, see Saksena and Emr, 2009).This review cannot adequately cover all the roles ESCRTs play in the cell. Rather, the purpose here is to emphasize the key findings that define ESCRT function. These discoveries highlight "unifying principles" of the ESCRTs, which are shared in some capacity between processes. At the most basic level, these principles define the ESCRT pathway as a cargo-recognition and membrane-deformation machine that can be viewed from three perspectives: (1) the ESCRTs themselves, (2) the cargo they sort, and (3) the lipid membrane they deform (Figure 2). We have structured this review to focus on each of these three perspectives. At the same time, there is an apparent divergence in ESCRT function regarding these different processes, as well as differences in ESCRT complexes between yeast and metazoans. Divergence is most apparent in the differential requirements for some ESCRT proteins over others in certain biological contexts. Careful analysis of these differences provides insight into the mechanisms of ESCRT function. Finally, we will discuss the leading

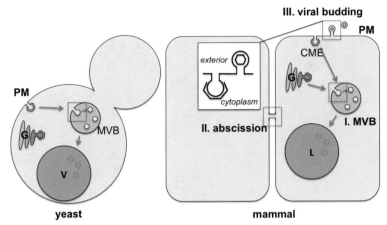

FIGURE 1 ESCRT-Mediated Processes in Yeast and Mammals
The left view shows the ESCRT pathway in yeast delivering cargo from the plasma membrane (PM) or *trans*-Golgi (G) to the vacuole (V) via MVBs. The right view illustrates mammalian cells utilizing ESCRTs in at least three distinct ways. MVB formation (I) delivers cargo to the lysosome (L). ESCRTs are also used in the abscission event of cell cytokinesis (II). Some viruses also use ESCRT proteins to bud from the plasma membrane (III). In the inset, this budding is topologically inverted from clathrin-mediated endocytosis (CME), which forms a vesicle inside the cell.

models explaining how ESCRT-III, the least understood of the ESCRT complexes, achieves vesicle formation. Thus, this review is meant to both reflect on ESCRT discoveries as well as discuss key mysteries remaining in the field. For more in-depth discussion, the reader should refer to several reviews that focus on ESCRT structure, ESCRT roles in diseases, the ubiquitin cycle, and vesicle budding (Hurley et al., 2010; Raiborg and Stenmark, 2009; Saksena and Emr, 2009; Williams and Urbé, 2007).

THE ESCRT COMPLEXES

ESCRTs were first defined as a ubiquitin-dependent protein-sorting pathway in yeast. Earlier studies in the 1980s and 1990s in the Emr and Stevens laboratories had identified numerous genes in *Saccharomyces cerevisiae* yeast that, when lost, displayed *vps* defects (Banta et al., 1988; Robinson et al., 1988; Rothman et al., 1989; Rothman and Stevens, 1986). These *vps* genes were initially categorized into three "classes"— A, B, and C—according to the morphology exhibited by the vacuole in yeast deficient with each gene (Banta et al., 1988). A fourth major class, denoted "class E," was also identified because of the distinctive endosome-derived compartment its members exhibited (Raymond et al., 1992). Through characterization of the trafficking of the vacuolar hydrolase carboxypeptidase S (CPS), it was demonstrated

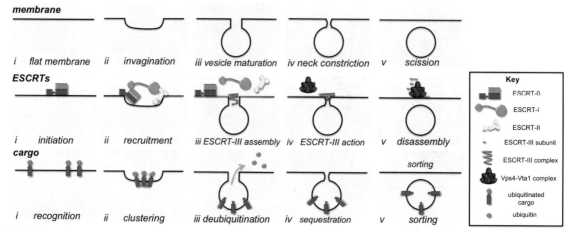

FIGURE 2 Perspectives of ESCRT-Mediated Vesicle Budding

The top view is from the membrane perspective, showing that there are five distinct stages of ILV budding. A flat membrane (i) is invaginated (ii) and matures into a vesicle that is still attached to the limiting membrane by a "neck" region (iii). The neck may undergo constriction (iv) and a scission event (v) to complete vesicle budding. The middle view is from the ESCRT protein perspective, in which ESCRT-0 initiates the pathway by engaging ubiquitinated cargo (i), until it is sequestered and sorted by ESCRT-III (iv and v). ESCRT-I and ESCRT-II complexes bind cargo and each other to create an ESCRT-cargo-enriched zone (ii). ESCRT-II nucleates ESCRT-III assembly (iii), which drives vesicle budding (iv) and is disassembled by the Vps4-Vta1 complex (v). The bottom view is from the cargo perspective, in which ubiquitinated cargo is first recognized by the ESCRT-0 ubiquitin-binding regions (i) and clustered by the ESCRT-0, -I, and -II complexes (ii). ESCRT-III assembly recruits deubiquitination machinery (iii) and packages cargo into the maturing vesicle (iv), which is finally sorted upon vesicle budding (v).

that transmembrane proteins can be sorted into the lumen of the vacuole via the MVB pathway (Odorizzi et al., 1998; Reggiori and Pelham, 2001). Two additional studies using a chimeric protein composed of CPS fused with the histidine biosynthesis enzyme His3 were used to select mutants defective in MVB sorting (taking advantage of the His minus phenotype exhibited by the His3-CPS fusion) (Katzmann et al., 2004; Odorizzi et al., 2003). By screening the various classes of *vps* mutants, Odorizzi et al. (1998) were the first to demonstrate that class E mutants were defective for CPS sorting to the lumen. Intriguingly, whereas His3-CPS sorting was attenuated with class E mutants, the trafficking of the soluble carboxypeptidase Y (CPY) was not. This suggested that the transport of integral membrane proteins followed a distinct route through the endosome (Odorizzi et al., 1998). Extensive biochemical and cell biological studies soon indicated that the class E proteins form distinct complexes that directly mediate this trafficking, and they were named the ESCRT complexes -I, -II, and -III (Asao et al., 1997; Babst et al., 1997, 2002a, 2002b; Katzmann et al., 2001; Shih et al., 2002) (Table 1). These complexes are all required for MVB formation and are recruited from the cytoplasm to late endosomes in a sequential manner.

Table 1 Perspectives of the ESCRT Pathway

Complexes					
Yeast	*Human*	**Structure**	**Membrane Binding**	**Cargo Recognition**	**Intercomplex Interactions**
ESCRT-0					
Vps27	Hrs		with PtdIns3P via FYVE domain on Vps27	via UIM and VHS domains on both Vps27 and Hse1	to Vps23 (ESCRT-I) via a PTAP-like motif on Vps27 (ESCRT-0)
Hse1	STAM1/2				
ESCRT-I					
Vps23	Tsg101		weak electrostatics on N terminus of Vps37	via UEV domain on Vps23 and a novel UBD on Mvb12	to Vps27 (ESCRT-0) through UEV motif of Vps23 (ESCRT-I); to Vps36 (ESCRT-II) via the C terminus of Vps28 (ESCRT-I)
Vps28	hVps28				
Vps37	Vps37A,B,C				
Mvb12	hMvb12A,B				
ESCRT-II					
Vps36	EAP45		with PtdIns3P GLUE domain of Vps36	via GLUE domain on Vps36	to Vps28 (ESCRT-I) via the GLUE domain of Vps36 (ESCRT-II); to Vps20 (ESCRT-III) through C terminus of Vps25 (ESCRT-II)
Vps22	EAP30				
Vps25	EAP20				
ESCRT-III					
Vps20	CHMP6		myristoylation of Vps20; electrostatics on helix-1 of CHMP3	interacts with DUBs to deubiquitinate cargo	to Vps25 (ESCRT-II) through helix-1 of Vps20 (ESCRT-III); to Vps4 through C-terminal MIM domains present on all four ESCRT-III subunits
Snf7	CHMP4A,B,C				
Vps24	CHMP3				
Vps2	CHMP2A,B				
Vps4 Complex					
Vps4	SKD1				to MIM domains of ESCRT-III subunits via MIT domain
Vps60	CHMP5				
Vta1	LIP5				

The yeast and human ESCRT proteins are analyzed from the perspectives of their membrane interactions, their interactions with ubiquitinated cargo, and their interactions with one other. Please refer to text for references. PDB files: ESCRT-0 [3F1I], ESCRT-I [2P22], ESCRT-II [3CUQ], ESCRT-III [3FRT]; Vps4-Vta1 complex derived from Yu et al., 2008.

ESCRT-0

ESCRT-0 consists of two subunits, Hrs (*h*epatocyte growth factor-*r*egulated tyrosine kinase *s*ubstrate) and STAM1/2 (*s*ignal *t*ransducing *a*daptor *m*olecule1/2) (Vps27 and Hse1 in yeast) (Table 1). These subunits interact in a 1:1 ratio via coiled-coil GAT (*G*GAs and *T*om) domains (Asao et al., 1997; Prag et al., 2007). Although both subunits display structural similarities, one difference between them is a FYVE (*F*ab-1, *Y*GL023, *V*ps27, and *E*EA1) zinc finger domain on Hrs (Mao et al., 2000). The ability of the Hrs FYVE domain to bind phosphatidylinositol 3-phosphate, PtdIns3P, provides both membrane recruitment and endosomal specificity for the ESCRT-0 complex (Raiborg et al., 2001). Both subunits also bind ubiquitin, providing an additional targeting module that promotes their binding to cargo-enriched endosomes. The ability to bind both PtdIns(3)P and ubiquitin makes ESCRT-0 a coincidence detection module for initiating the ESCRT pathway at endosomes. This differs from ESCRT-I, which interacts only weakly with membrane through electrostatic interactions at the N terminus of Vps23 (Kostelansky et al., 2007). In fact, ESCRT-I recruitment to membrane is dependent upon ESCRT-0 in both metazoans and yeast (Bache et al., 2003; Katzmann et al., 2003; Lu et al., 2003). Thus, the recruitment of ESCRT-I by ESCRT-0 is essential for initiating MVB-dependent cargo sorting.

ESCRT-I

The first ESCRT complex to be described was ESCRT-I, identified originally in yeast as a heteromeric complex consisting of Vps23 (*tumor susceptibility gene 101*, or Tsg101 in mammals), Vps28, and Vps37 (Katzmann et al., 2001) (Table 1). Mvb12 (*multivesicular body 12*) was later identified as an additional subunit of ESCRT-I (Chu et al., 2006; Curtiss et al., 2007). Mammalian ESCRT-I similarly consists of Tsg101, Vps28, Vps37, and hMvb12, although additional isoforms of both Vps37 and hMvb12 exist (Bache et al., 2004; Bishop and Woodman, 2001; Eastman et al., 2005; Morita et al., 2007; Stuchell et al., 2004). This is a theme repeated among the ESCRT complexes, as multiple isoforms of several mammalian ESCRT subunits exist. To our knowledge, the significance of multiple subunit isoforms is not yet known and may reflect tissue-specific variances among the ESCRT complexes. The crystal structure of the core yeast ESCRT-I complex reveals an elongated heterotetramer of ~20 nm in length, with three subunits intertwined into a long coiled-coil stalk with a globular head group (Kostelansky et al., 2007). ESCRT-I interacts with both ESCRT-0 and ESCRT-II, but at opposite ends of the complex. The interaction with ESCRT-0 occurs through the ubiquitin E2 variant domain (UEV), which corresponds to the N terminus of Vps23 and projects from the ESCRT-I stalk. The UEV binds to the PTAP-like motifs of the ESCRT-0 subunit Vps27/Hrs (Katzmann et al., 2003; Kostelansky et al., 2006), an interaction that is mimicked by the Gag protein of human immunodeficiency virus-1 (HIV-1) (Pornillos et al., 2003).

ESCRT-II

In addition to ESCRT-0, ESCRT-I also interacts with the ESCRT-II complex. ESCRT-II is a Y-shaped heterotetramer consisting of one subunit each of Vps22 (*ELL-associated protein of 30 kDa*, or EAP30) and Vps36 (EAP45), and two subunits of Vps25 (EAP20) (Babst et al., 2002b; Langelier et al.,

2006) (Table 1). Vps22 and Vps36 form the base of the Y and are each bound by one copy of Vps25, which forms the arms of the Y (Hierro et al., 2004; Teo et al., 2004). In yeast, ESCRT-II interacts with nanomolar affinity to ESCRT-I through an interaction between the GLUE (GRAM-*like* *u*biquitin-binding in *E*AP45) domain of Vps36 and the C terminus of Vps28 (Gill et al., 2007; Teo et al., 2006). Two NZF (*N*pl4-type *z*inc *f*inger) domains are inserted into the yeast GLUE domain; the first NZF binds to ESCRT-I, and the second NZF binds ubiquitin (Teo et al., 2006). The human GLUE domain has no NZF insertions but can still bind ubiquitin. Like ESCRT-0, ESCRT-II can also bind with high affinity to PtdIns(3)P via its GLUE domain (Slagsvold et al., 2005). Together with the ESCRT-0 FYVE domain, the GLUE and FYVE domains provide endosomal localization specificity by binding PtdIns(3)P, which is generated at endosomes through the class III PI3 kinase Vps34. Because the mammalian GLUE domain does not contain the NZF inserts required for the interaction with ESCRT-I, to our knowledge, the mammalian link between the ESCRT-I and ESCRT-II complexes remains uncharacterized. In contrast, the link between ESCRT-II and ESCRT-III has been defined—Vps25 binds to Vps20 with high affinity—revealing a critical role for ESCRT-II in initiating ESCRT-III complex formation.

ESCRT-III

Unlike the other ESCRT complexes, ESCRT-III does not form a stable, cytoplasmic complex, and attempts to crystallize an intact ESCRT-III complex have thus far failed. ESCRT-III consists of four core subunits: Vps20, Snf7, Vps24, and Vps2 (Babst et al., 2002a) (Table 1). In mammals these are denoted as *ch*arged *m*ultivesicular body *p*roteins (CHMPs) (CHMP6, CHMP4, CHMP3, CHMP2, and their isoforms respectively). The crystal structure of human Vps24, CHMP3, reveals an ~7 nm structure with five discernable helices; the core consisting of a hairpin formed by the first two helices (Bajorek et al., 2009b; Muzioł et al., 2006). A second structure of the "accessory" ESCRT-III protein Ist1 (*i*ncreased *s*alt *t*olerance-1) was recently solved and displays the same helical fold as CHMP3, suggesting that the architecture of the ESCRT-III subunits is generally conserved between subunits (Bajorek et al., 2009b).

ESCRT-III monomers do not localize to the endosome but exist in an auto-inhibited "closed" state in the cytoplasm. To our knowledge, the exact mechanism for autoinhibition has yet to be determined; however, it is clear that it involves intramolecular interactions between the positively charged ESCRT-III N terminus and negatively charged C terminus (Shim et al., 2007; Zamborlini et al., 2006). Activation of ESCRT-III occurs when the ESCRT-II subunit Vps25 binds to Vps20, initiating ESCRT-III recruitment to the endosome and complex formation (Teo et al., 2004). Vps20 then recruits Snf7, which homo-oligomerizes. Snf7 polymers are thought to be capped by the recruitment of Vps24, which when mixed at a stoichiometric ratio of 10:1 with Snf7 in vitro can attenuate Snf7 homo-oligomerization (Saksena et al., 2009; Teis et al., 2008). Snf7 also recruits the ESCRT-III adaptor protein Bro1/Alix (*B*CK1-like *r*esistance to *o*smotic shock protein-1/*a*poptosis-*l*inked gene-2 *i*nteracting protein *X*), which stabilizes Snf7 filaments and recruits the deubiquitinating enzyme Doa4 (*d*egradation *o*f *a*lpha-4), necessary for cargo deubiquitination (Luhtala and Odorizzi, 2004; Odorizzi et al., 2003). Once at the endosome, Vps24 recruits Vps2, completing ESCRT-III complex assembly. Although the order of ESCRT-III subunit recruitment has been elucidated, to our knowledge, the exact stoichiometry of the ESCRT-III polymer remains uncharacterized and is a key point of emphasis for ongoing ESCRT studies.

In addition to the "core" ESCRT-III subunits, there are several "accessory" ESCRT-III subunits that govern ESCRT-III-Vps4 interactions. The most studied are Ist1, Did2 (*D*oa4-*i*ndependent *d*egradation-2), Vta1 (*V*ps *t*wenty-*a*ssociated-1), and Vps60. These ESCRT factors appear to modulate Vps4 function in three ways: by mediating Vps4 interactions with itself, with ESCRT-III subunits, and by promoting ATP hydrolysis. Vta1 promotes the ATPase activity of Vps4 and has been shown to form a supercomplex with oligomeric Vps4 at a 2:1 Vps4:Vta1 ratio (Azmi et al., 2006; Yu et al., 2008). The core of this supercomplex appears stable with a more flexible "cap" at one end. Vta1 also interacts directly with ESCRT-III subunits, Vps60 in particular, via two MIT domains (*m*icrotubule-*i*nteracting and *t*rafficking) at its own N terminus that bind to MIMs (*MIT-i*nteracting *m*otifs) on the ESCRT-III subunits. Thus, Vta1 may function as an adaptor bridging Vps4 and ESCRT-III (Azmi et al., 2008; Shiflett et al., 2004). Bro1/Alix is also an important ESCRT-III accessory subunit and binds to the extreme C terminus of Snf7. Loss of Bro1/Alix affects ESCRT cargo sorting in yeast (Odorizzi et al., 2003). It is thought to both stabilize Snf7 oligomers and recruit the deubiquitinating enzyme Doa4.

Vps4-Vta1 Complex

Once assembled, the ESCRT-III complex requires energy to disassociate from the membrane. This energy is provided by the class I AAA (*A*TPase *a*ssociated with various cellular *a*ctivities) ATPase Vps4 (Babst et al., 1998). Vps4 is a multimeric mechanoenzyme that binds ESCRT-III subunits via a N-terminal MIT domain that recognizes C-terminal MIMs present in the ESCRT-III subunits (Babst et al., 1997; Scott et al., 2005b) (Table 1). It exists in vitro as a monomer or dimer in the nucleotide-free or ADP-bound state but multimerizes into a stable dodecamer of two hexameric rings and a supercomplex with Vta1 when fully assembled (Babst et al., 1997;

Scott et al., 2005a; Yu et al., 2008). ATP hydrolysis is promoted by multi-merization and association with ESCRT-III subunits and is necessary for removal of ESCRT-III subunits from the membrane because expression of a hydrolysis mutant Vps4 E233Q leads to the accumulation of hyper-oligomeric ESCRT-III subunits on the endosome and a class E phenotype (Babst et al., 1998).

AAA ATPases are involved in a wide array of cellular processes, including membrane trafficking and fusion, DNA replication, proteolysis, and cyto-skeletal reorganization (Barends et al., 2010; Neuwald et al., 1999; Striebel et al., 2009). However, there are common themes shared between AAA ATPases that provide insight into Vps4 function. All share one or two AAA domains of ~250 amino acids that bind ATP in most members. Most mul-timerize and form hexameric rings. Several AAA ATPases alter substrate conformation, including the heat shock-related protein ClpX (caseino-lytic peptidase X) that has been shown to mediate substrate unfolding (Weber-Ban et al., 1999). Protein unfolding is thought to be achieved by capturing and "threading" proteins through a narrow pore in the core of the AAA oligomer (Reid et al., 2001; Weber-Ban et al., 1999). It has been suggested that the Vps4-Vta1 complex mediates ESCRT-III disassembly in this manner. Here, Vps4-Vta1 binds ESCRT-III subunits and threads them through a central pore in an ATP-dependent process. Although only a model, mutation of residues within the Vps4-Vta1 oligomer pore leads to defects in HIV-1 budding (Gonciarz et al., 2008). If this is truly how ESCRT-III is disassembled, then, to our knowledge, how ESCRT-III sub-units are refolded after dissociation remains to be elucidated, although ClpX substrate MuA is thought to spontaneously refold following ClpX activity (Burton et al., 2001). Notably, the Vps20 and Snf7 MIMs (MIM2 type) appear to bind a different area of the Vps4 MIT than the Vps2 and Vps24 MIMs (MIM1 type), and all bind with low affinity, suggesting that multiple ESCRT-III subunits must interact with Vps4 for stable recruit-ment (Shestakova et al., 2010).

Vps4 interacts with the accessory ESCRT-III subunits to further modulate its activity. As discussed above, Vta1 forms a heteromeric complex with Vps4. The presence of MIT domains on Vta1 and on each Vps4 subunits means that the Vps4-Vta1 supercomplex contains up to 24 MIT domains, allowing it to interact with multiple ESCRT-III subunits simultaneously. Did2 and Ist1 form a complex in yeast and together interact with both Vps4 and ESCRT-III (Rue et al., 2008). This complex may function as an endosomal anchor for the Vps4 oligomer, because Vps4 lacking its MIT domain still localizes to endosomes but is redistributed to the cytoplasm in yeast lacking Did2 (Shestakova et al., 2010).

CARGO: ESCRT-MEDIATED CARGO RECOGNITION AND SORTING

Cargo Engagement

A key characteristic required by any cargo-sorting machinery is the ability to recognize and engage cargo. This review now switches perspectives from the ESCRT proteins to the ubiquitinated cargo that they bind. Accordingly, ESCRT complexes have several distinct ubiquitin-binding motifs (Bilodeau et al., 2002; Hirano et al., 2006; Mizuno et al., 2003; Pornillos et al., 2002; Raiborg et al., 2002; Shields et al., 2009; Shih et al., 2002; Slagsvold et al., 2005). Both subunits of the earliest ESCRT complex, ESCRT-0, bind ubiquitin in numerous ways. This highlights ESCRT-0 as a cargo-recognition module. Hrs binds ubiquitin via a double-sided ubiquitin-interacting motif (UIM) and a VHS (Vps27 Hrs STAM) domain (Hirano et al., 2006) (Ren and Hurley, 2010). The yeast homolog of Hrs, Vps27, also contains a VHS domain that binds ubiquitin (Table 1). Interestingly, Vps27 contains two UIMs in tandem instead of a double-sided UIM (Bilodeau et al., 2002). STAM1/2 and its yeast homolog Hse1 contain both a UIM and a VHS domain for binding ubiquitin (Bilodeau et al., 2002; Fisher et al., 2003; Mizuno et al., 2003). Although the ESCRT-0 heterodimer forms GAT domains that have been shown to bind ubiquitin in other proteins, to our knowledge, direct binding of ubiquitin to the ESCRT-0 GAT domain has not been observed (Prag et al., 2007; Ren et al., 2009). However, ESCRT-0 clearly contains several ubiquitin-binding domains (approximately five in yeast), although it is unclear if this is primarily to engage several cargoes simultaneously, to bind tightly to a cargo through avidity interactions, or even bind with high affinity to polyubiquitinated cargo (Ren and Hurley, 2010).

In addition to its ubiquitin-binding motifs, ESCRT-0 may engage cargo through its interactions with the clathrin vesicle machinery at the endosome. Vps27/Hrs binds clathrin heavy chain via its clathrin box motif (Raiborg et al., 2001). STAM contains a canonical clathrin-binding motif, but it is unclear if it binds clathrin in vivo (McCullough et al., 2006). The interaction between ESCRT-0 components and clathrin results in microdomains of flat clathrin lattices, ESCRT-0, and ubiquitinated cargo that appear important for cargo sorting (Raiborg et al., 2006; Sachse et al., 2002).

ESCRT-I and ESCRT-II also contain ubiquitin-binding domains (UBDs). Tsg101 and hMvb12 and their corresponding yeast homologs both bind ubiquitin. In addition, Tsg101 and Vps23 contain a catalytically inactive UEV (Pornillos et al., 2002). Recently, a novel UBD was identified on yeast Mvb12 and human Mvb12A (Shields et al., 2009; Tsunematsu et al., 2010) (Table 1). Thus far, UBDs have been identified on only one subunit of ESCRT-II, Vps36 (EAP45). In yeast Vps36, the GLUE domain contains a NZF motif insertion that binds ubiquitin (Alam et al., 2004). The GLUE domain of the mammalian homolog EAP45 also binds ubiquitin, although the NZF motifs are not present (Slagsvold et al., 2005) (Table 1). Notably, to our knowledge, no UBDs have been identified on the ESCRT-III subunits. This is consistent with a role for ESCRT-III primarily in vesicle budding and scission, and not in cargo recognition, and highlights the division of labor between the ESCRT complexes. Whereas the "early" ESCRT-0, -I, and -II complexes display direct ubiquitin binding and high phosphoinositide specificity, ESCRT-III does not require these characteristics to achieve vesicle formation at sites already enriched with cargo.

Deubiquitination

Before delivery to the vacuole/lysosome, cargo must be deubiquitinated. This event appears to occur just prior to packaging of the cargo into the ILV. In mammals AMSH (associated *molecular* with *SH3* domain of STAM), a ubiquitin isopeptidase, interacts with both ESCRT-0 and ESCRT-III (McCullough et al., 2006) and is responsible for the deubiquitinating step (Kyuuma et al., 2007). A second mammalian ubiquitin isopeptidase Y (UBPY) also interacts with ESCRT-0 and is critical in maintaining cellular ubiquitin levels (Row et al., 2006). Unlike AMSH, which only recognizes Lys-63-linked polyubiquitin chains, UBPY deubiquitinates both Lys-63 and Lys-48 polyubiquitin chains (Row et al., 2006). In yeast, Doa4 deubiquitinates cargo, and its loss depletes ubiquitin pools and alters ubiquitin homeostasis because of its numerous non-ESCRT functions (Amerik et al., 2000; Swaminathan et al., 1999). Thus, ESCRTs have an important role in recruiting the deubiquitinating machinery required to remove ubiquitin from cargo and maintain cellular ubiquitin pools.

In addition to binding ubiquitin, ESCRT proteins can themselves be ubiquitinated. The ESCRT-0 subunit Vps27/Hrs is monoubiquitinated (Polo et al., 2002), although this ubiquitination is not essential for Vps27 function, and to our knowledge, its physiological significance is unknown (Stringer and Piper, 2011). However, Hrs ubiquitination can be inhibitory because the intramolecular interaction of the Hrs UIM and ubiquitin prevents the binding of ubiquitinated cargo (Hoeller et al., 2006). Cargo fused to a single ubiquitin can successfully be sorted by an ESCRT pathway with a deubiquitinating enzyme fused to ESCRT subunits, reinforcing the idea that ESCRT subunit ubiquitination may not be essential for MVB sorting (Stringer and Piper, 2011).

In addition to deubiquitinating cargo, AMSH is responsible for removing the ubiquitin from Hrs (Sierra et al., 2010). Similarly, the ESCRT-I subunit Tsg101 is ubiquitinated by the E3-ubiquitin protein ligase Mahogunin, the disruption of which affects endosome to lysosomal trafficking (Kim et al., 2007). This observation is intriguing because ubiquitination of Tsg101 by another E3 ligase, Tal, is inhibitory and disrupts both MVB formation and viral budding (Amit et al., 2004).

In summary, from the perspective of the cargo, the ESCRT pathway is subdivided into early stages where cargo is recognized and potentially clustered (ESCRT-0, -I, and -II) until it is corralled by the downstream ESCRT-III complex, which lacks ubiquitin binding, but is able to couple ESCRT-III assembly with cargo deubiquitination and vesicle budding.

ESCRT-MEDIATED MEMBRANE DEFORMATION AND SCISSION

Retroviral Budding

This review now moves to focus on the ESCRT pathway from the perspective of the lipid membrane it can deform. ESCRT-mediated MVB biogenesis requires membrane deformation and scission to generate an ILV. When ESCRTs were identified as essential factors in the budding of the HIV-1, it became clear that ESCRT proteins were potentially direct mediators of the membrane remodeling necessary for vesicle or viral particle budding (Garrus et al., 2001). Since then, the ESCRTs have been implicated in the replication of many viruses. Recent reviews have examined in detail the roles of ESCRT proteins in viral budding (Carlton, 2010; Chen and Lamb, 2008). For the purpose of this review, we will focus on the extensively studied role of ESCRTs in HIV-1 budding from the plasma membrane as a model viral system.

Perhaps the most striking difference between MVB biogenesis and viral budding is the differential requirement of ESCRT components in these processes. Unlike MVB formation, HIV-1 budding does not require ESCRT-0, ESCRT-II, or the ESCRT-III subunit hVps20 (CHMP6) (Langelier et al., 2006). In fact, overexpression of a C-terminal fragment of Hrs inhibits HIV particle release, likely due to its sequestration of Tsg101 (Bouamr et al., 2007). Tsg101 mediates HIV-1 budding by binding to the L (late) domain of the viral Gag protein (Garrus et al., 2001; VerPlank et al., 2001). Similar to HIV-1, other viruses contain L-domain proteins that interact directly with the ESCRT machinery (Carpp et al., 2011; Martin-Serrano et al., 2001; Urata et al., 2007; Wirblich et al., 2006). They function in combination with other ESCRT proteins to recruit ESCRT-III and Vps4 to sites of viral budding. Interestingly, the ubiquitination of Gag can also enhance its interaction with the ESCRTs by providing additional binding sites for Tsg101 (Garrus et al., 2001). Accordingly, a role for Nedd4 and the ubiquitin ligase machinery has been identified in the release of enveloped viruses (Strack et al., 2000).

Because ESCRT-II is not required for HIV budding, a question arises as to how ESCRT-III is recruited to sites of viral release. Bro1/Alix is an attractive candidate because it can bind ESCRT-I and ESCRT-III, and localizes to viral bud sites (Odorizzi et al., 2003). Indeed, live cell imaging of EIAV viral budding from *HeLa* cell plasma membranes indicates that GFP-Alix is recruited to viral bud sites very early along with the viral Gag protein (Jouvenet et al., 2011). Future studies will be necessary to dissect how Alix bridges ESCRT-I and ESCRT-III and functions in viral budding.

Cytokinesis

The discovery that ESCRT proteins are necessary for cell abscission, the final stage of cytokinesis in which the cell midbody is constricted and severed, clearly establishes the ESCRTs as a multipurpose machine that can execute numerous topologically equivalent membrane-deforming processes. Cytokinesis requires membrane scission, and the creation of membrane curvature that is topologically consistent with the curvatures needed in MVB sorting and viral budding. Thus, the ESCRTs provide an intuitive mechanism as executioners of the abscission step. However, this link was not always obvious. In addition to the ESCRTs, numerous membrane-trafficking proteins are required for cell division, including dynamin, SNAREs, exocyst components, and Rab proteins (Gromley et al., 2005). Furthermore, both endocytosis and exocytosis are required for cytokinesis, likely because new membrane must be delivered to the midbody during cleavage furrow progression (Danilchik et al., 1998; Gerald et al., 2001).

What then is the role of membrane-trafficking proteins in cell division? Until recently, the leading model was that endosomes were recruited and clustered at the midbody as cytokinesis progressed, then underwent a massive round of homotypic fusion as well as heterotypic fusion with the plasma membrane (Danilchik et al., 1998; Gromley et al., 2005). This would have the net effect of adding membrane to the midbody as it was contracted by the actomyosin ring. The narrow neck remaining after this ingression would finally undergo a fission event, thus completing abscission.

Tsg101 and Bro1/Alix were the first ESCRT proteins found to localize to late-stage midbodies (Carlton and Martin-Serrano, 2007; Spitzer et al., 2006). However, because endosomes localize to the midbody, it was assumed that ESCRTs function in their capacity as MVB-sorting machines during cytokinesis. However, two findings argue against this. One is that Tsg101 and Bro1/Alix are recruited to the midbody by Cep55, a multimeric cell division protein essential for a late stage in cell division (Carlton et al., 2008; Carlton and Martin-Serrano, 2007). Importantly, Cep55 depletion prevents Tsg101 and Bro1/Alix recruitment to the midbody, while not affecting viral budding, demonstrating that these two processes are distinct

(Carlton and Martin-Serrano, 2007). The second is that Crenarchaeota of the genus *Sulfolobus* uses ESCRT-III and Vps4 orthologs directly in cytokinesis (Lindås et al., 2008), suggesting an ancient role for these proteins in cell division. Furthermore, a Vps25-like protein CdvA can interact with and recruit ESCRT-III-like subunits to membranes in *Sulfolobus* (Samson et al., 2011).

Thus, an emerging model is that ESCRT-I and Bro1/Alix are recruited to the midbody by directly binding Cep55 and in turn recruit ESCRT-III to commit the final scission event in cytokinesis (Carlton and Martin-Serrano, 2007). Although direct evidence for ESCRT-III involvement in the actual scission event of abscission is lacking, depletion of Vps4 or expression of dominant negative hVps24 (CHMP3) arrests late cytokinesis (Dukes et al., 2008). Recent studies have also demonstrated that CHMP4B and Tsg101 localize to two distinct rings at opposing ends of the cell midbody (Elia et al., 2011; Guizetti et al., 2011). Using electron tomography, Guizetti et al. (2011) recently visualized protein filaments that spiral around the constriction zone of a dividing cell. ESCRTs were required for the formation of these filaments, and CHMP4b colocalized with them (Guizetti et al., 2011). It is hypothesized that constriction or other conformational changes within these rings drive the final stages of abscission. This ring formation is a distinct late event after cleavage furrow progression because the midbody will form and progress to late stages before being arrested in ESCRT-I or -III depleted cells. Ist1 also localizes to the midbody and is essential for abscission (Agromayor et al., 2009; Bajorek et al., 2009a). How ESCRT-mediated scission is coordinated with actomyosin constriction and potential SNARE-mediated fusion at the midbody is still not understood.

Recent insight into how microtubule disassembly at the midbody occurs demonstrates a second important role for the ESCRTs in cytokinesis. The AAA-ATPase spastin is recruited to the midbody by ESCRT-III subunit CHMP1B and subsequently severs microtubules as abscission progresses (Yang et al., 2008). This suggests that ESCRT-III can recruit at least two AAA ATPases to sites of scission.

Themes of Vesicle Budding

Thus, at present all well-characterized ESCRT-mediated processes involve membrane deformation (i.e., budding) and scission. Furthermore, this membrane deformation is topologically consistent between processes in that it produces membrane curvature that pushes away from the cell cytoplasm (Figure 2). This is in contrast to "classical" budding events like COP-I, COP-II, and clathrin vesicle formation, which invoke vesicle budding into the cytoplasm (Doherty and McMahon, 2009; Lee et al., 2005; Rothman and Wieland, 1996; Shimoni and Schekman, 2002). "Classical" vesicle formation involves a cytoplasmic protein coat that encapsulates the vesicle, adaptor proteins that help to sequester cargo and promote vesicle maturation, and a scission mechanism that liberates the vesicle from the source membrane. During dynamin-mediated endocytosis, this scission event is promoted by the formation of a narrow "neck" of extreme curvature on which helical dynamin oligomers encircle and potentially constrict to achieve scission, although the exact nature of this scission event is still contested (Hinshaw and Schmid, 1995). Thus, from the membrane's perspective, "classical" vesicle budding can be divided into five stages: a flat membrane (1) is invaginated (2), forming a vesicle and neck region (3) that undergoes constriction (4) and scission (5) (Figure 2). These stages appear to be tightly regulated as evidenced by the extreme uniformity of vesicle size for a particular vesicle type.

In ESCRT-mediated vesicle formation, ESCRT complexes cannot coat the vesicle exterior because they are segregated in the cytoplasm away from the luminal vesicle by the limiting membrane of the endosome. It is possible that ESCRT-III could coat the vesicle interior during vesicle maturation, but if so the ESCRTs are efficiently removed prior to scission because they are not consumed in the reaction (Babst, 2005; Wollert and Hurley, 2010). Furthermore, because the vesicle neck and scission mechanism is topologically inverted, the ESCRTs cannot encircle any neck that forms (Figure 2). Despite this, the membrane must undergo analogous stages of progression to achieve ILV formation. This is supported by the observation that ESCRT-mediated vesicle budding is highly regulated because endosomal ILVs retain a consistent diameter (Teis et al., 2010). Thus, "classical" and ESCRT-mediated vesicle budding events are conceptually similar yet mechanistically distinct.

Analyzing ESCRT-Mediated Vesicle Budding from a "Classical" Perspective

Because ILV formation is topologically distinct from "classical" vesicle formation, how then is ESCRT-mediated vesicle formation achieved? Recent studies have defined a key role for the ESCRT-III complex in particular as a membrane deformation and scission machine (Hanson et al., 2008; Saksena et al., 2009; Wollert and Hurley, 2010; Wollert et al., 2009). Although these studies partially reconstituted ESCRT-III scission activity in vitro, the mechanism by which ESCRT-III achieves this remains elusive. Furthermore, it is unclear if there is a division of labor between the ESCRT complexes to mediate the stages of membrane deformation and scission in vivo. Using giant unilamellar vesicles (GUVs), studies by Wollert et al. (2009) demonstrated that ESCRT-III subunits alone were sufficient to bud and release ILVs in vitro. However, ESCRTs -I and -II could deform the membrane, although scission was not observed (Wollert and Hurley, 2010). The observation that ESCRTs -I and -II can generate membrane curvature in vitro is intriguing, but the mechanism by which they do this is also unclear. The membrane-binding portions of these complexes are small. Vps37 contains a small positively charged N terminus that binds acidic liposomes, but this N terminus is not required for ESCRT-I function in vivo (Kostelansky et al., 2007). ESCRT-II binds PtdIns(3)P with high affinity through the Vps36 GLUE domain, and can bind negatively charged lipids electrostatically via a small helix on the N terminus of Vps22, but it is unclear how these domains can generate membrane curvature (Im and Hurley, 2008).

Despite their different topologies, "classical" vesicle budding mechanisms may provide additional insight into ESCRT-mediated vesicle budding. Recent models suggest that dynamin mediates scission by forming an oligomeric scaffold around the vesicle neck, as well as potentially penetrating the lipid bilayer. This has the net effect of promoting fission at the vesicle neck (Ramachandran et al., 2009; Roux et al., 2006; Stowell et al., 1999). In yeast, the membrane-sculpting potential of the heterodimeric Bin/Amphiphysin/ Rvs (BAR) domain-containing complex Rvs161/Rvs167 promotes scission by generating a highly curved vesicle neck that is stressed by the pulling forces of actin polymerization (Kaksonen et al., 2005). BAR superfamily proteins can promote vesicle budding by binding electrostatically to negatively charged membranes and bending them to match the intrinsically bent shape of their dimers (Gallop and McMahon, 2005; Henne et al., 2007; Peter et al., 2004). N-BAR domain-containing proteins can also generate high membrane curvature via their N-terminal amphipathic helices that penetrate the lipid bilayer (Peter et al., 2004). In COP vesicle budding, Sar1 and possibly Arf1 promote vesicle formation by membrane insertion of protein residues (Lee et al., 2005; Long et al., 2010; Lundmark et al., 2008). Thus, in other protein-mediated budding systems, vesicle formation is promoted by at least three distinct mechanisms: (1) individual protein scaffolding via electrostatic protein-membrane interactions, (2) oligomeric scaffolding, and (3) membrane insertion of hydrophobic residues. Can these mechanisms be applied to ESCRT-mediated ILV formation?

Snf7 Oligomer-Mediated Vesicle Budding

ESCRT-III subunits share at least two features with proteins that promote "classical" vesicle budding: they bind to lipid membrane electrostatically via their positively charged N termini, and some subunits can oligomerize (Hanson et al., 2008; Muzioł et al., 2006; Teis et al., 2008). Like dynamin, Snf7 forms oligomeric assemblies (Ghazi-Tabatabai et al., 2008; Hanson et al., 2008; Pires et al., 2009). In vitro or when overexpressed in vivo, these appear in different forms as rings, straight filaments, or spirals of varying diameter. Although Snf7 oligomers are the best characterized, filaments of Vps24 and the Vps24-Vps2 subcomplex have also been observed (Ghazi-Tabatabai et al., 2008; Lata et al., 2008). Hanson et al. (2008) observed striking membrane deformations associated with oligomerized Snf7, leading to the hypothesis that the assembly of Snf7 oligomers promotes membrane bending and potentially vesicle formation.

A model explaining how Snf7 oligomers mediate ILV budding must address the observation that ESCRT-II is Y-shaped and has two Vps25 "arms" that can nucleate two Snf7 filaments. In one model, each Vps25 arm binds a Vps20 subunit, which in turn nucleates two Snf7 homo-oligomers that extend out to "encircle" cargo in a ring or spiral (Figure 3iv). Strikingly, mutation of the Vps25-binding sites of Vps36 or Vps22 to generate "one-armed" ESCRT-II does not inhibit Snf7 filament formation but does halt ILV formation (Teis et al., 2010). Thus, a "two-armed" ESCRT-II is necessary for vesicle formation in vivo, and points to a carefully regulated ESCRT-III superstructure of defined size and architecture. This is consistent with the fact that ILVs exhibit a consistent diameter (~25 nm in yeast) and that ESCRT-III is sized as an oligomer of ~450 kDa in yeast extracts. Interestingly, overexpression of Snf7 leads to hyperpolymerization of ESCRT-III into a >600 kDa structure, and the formation of ILVs with varied but significantly larger diameter (up to 360 nm) (Teis et al., 2010).

Because hyperoligomeric ESCRT-III is associated with irregular ILV diameter, how is ESCRT-III oligomer size regulated? The Snf7 oligomer is "capped" by Vps24 and Vps2, which are recruited after Snf7 oligomerization and function to: (1) halt Snf7 filament progression, and (2) recruit Vps4-Vta1 for ESCRT-III complex disassembly. Snf7 oligomerization can also be manipulated by modulating the expression of the ESCRT-III protein Bro1/Alix. Bro1 overexpression enhances Snf7 oligomer stability and leads to ILV invaginations within the endosome that do not undergo scission, indicating that ESCRT-III disassembly is required prior to ILV formation (Wemmer et al., 2011). Notably, this is in contrast to in vitro studies, where vesicle release occurred prior to Vps4-mediated ESCRT-III disassembly (Wollert et al., 2009).

Thus, regulated Snf7 oligomerization represents one model of ESCRT-III-mediated ILV budding. In it, membrane deformation is directly coupled to Snf7 filament assembly and "capping." Snf7 oligomerization also encircles cargo already clustered by the "upstream" ESCRT complexes so it can be deubiquitinated and corralled into the maturing vesicle. Here, the formation of a Snf7 ring acts as a physical barrier preventing transmembrane cargo from escaping the vesicle. This is evidenced by the fact that loss of Snf7 promotes retrograde trafficking of cargo back to the plasma membrane, but loss of Vps24 does not (Teis et al., 2008). However, this model fails to completely provide a mechanism for membrane scission.

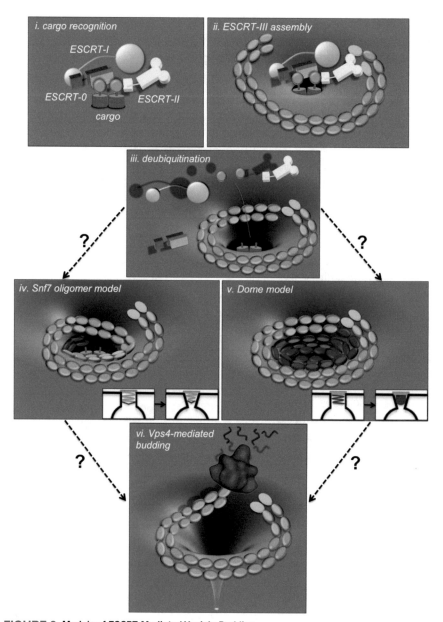

FIGURE 3 Models of ESCRT-Mediated Vesicle Budding

The first three panels represent "upstream" ESCRT complex recognition of ubiquitinated cargo (i), the initiation of ESCRT-III assembly by ESCRT-II (ii), and deubiquitination of the cargo (iii). From there, two different models of ESCRT-III-mediated vesicle maturation are presented. The Snf7 oligomerization model (iv) indicates that vesicle formation is a consequence of controlled Snf7 homo-oligomerization (green) initiated by Vps20 (light green) and "capping" by Vps24 (purple) and Vps2 (dark blue). In the Dome model (v), Snf7 filaments define a "zone" where rings of Vps24 and Vps2 form into a protein hemisphere, which drives vesicle budding. Both models may be dependent on Vps4 for neck constriction (vi), which drives vesicle budding.

Snf7 oligomers observed in vivo appear as flat spirals and require Snf7 truncation or the coexpression of dominant-negative Vps4 to induce membrane curvature (Hanson et al., 2008). Furthermore, the spirals are ~100–300 nm in diameter, and it is unclear how they would allow a narrow ILV neck to form. One possibility is that Snf7 oligomers naturally curve to favor an ideal diameter, and, as oligomerization progresses, filaments narrow to an optimal diameter that is terminated by the "cap" Vps24-Vps2 subcomplex. This narrowing may constrict the ILV neck enough to promote scission.

Snf7 Oligomerization and Membrane Buckling

Another ESCRT-III oligomeric model attempts to explain how ESCRT-III can generate membrane buckling and subsequent fission. In this "buckling ring" model originally proposed by Lenz et al. (2009), ESCRT-III rings have an intrinsic optimal radius but accumulate around one another in varying radii due to the high affinity ESCRT-III subunits have for one another and the membrane. The membrane responds by deforming into an inverted tube to relax the tension on the rings so they can readjust to their preferred diameter (Lenz et al., 2009). The result is an elongated tube containing numerous ESCRT-III rings of preferred diameter contacting each other and the membrane.

This provides a mechanism for how ESCRT-III generates membrane curvature but also does not provide an intuitive mechanism for scission. This model places ESCRT-III rings down the length of an elongated ILV tube but fails to explain how they would be removed prior to budding.

The Dome Model

An alternative model for ESCRT-III-mediated vesicle formation implicates the Vps24-Vps2 subcomplex as its central mediator. It is based on studies by Lata et al. (2008) who observed in vitro cylinders composed of stacked Vps24-Vps2 rings or spirals that bind to liposomes, suggesting that their membrane-binding surface faces the outside of the cylinder. These cylinders form tapered ends of decreasing diameter, terminating in a protein hemisphere or "dome" (Figure 3v). This has given rise to the "Dome Model" of ESCRT-III-mediated scission where sequential ESCRT-III rings or spirals of decreasing diameter stack inside the neck of the forming ILV and are capped by a Vps24/Vps2 hemisphere (Fabrikant et al., 2009). Importantly, membrane attachment to the dome mediates constriction of the ILV neck. Constriction is therefore dictated by the protein affinity for the membrane, generating a highly curved membrane dome that buckles to facilitate scission. Although Vps24-Vps2 filaments drive vesicle scission, they are recruited by Vps20 and the Snf7 oligomer that localize to sites of ILV formation prior to dome formation. Thus, the "Dome Model" is consistent with the late recruitment of Vps24 and Vps2 seen in ESCRT-III assembly (Saksena et al., 2009; Teis et al., 2008).

One critique of the "Dome Model" is that it does not address why the Vps24-Vps2 subcomplex was not necessary for ILV budding into GUVs (Wollert and Hurley, 2010). Likewise, overexpression of Snf7 and mutant Vps4 was responsible for the dramatic plasma membrane deformations observed by Hanson et al. (2008), suggesting that the Vps24-Vps2 complex is not required for Snf7-induced membrane deformations in vivo.

Vps4 and Membrane Scission

Although Vps4 is necessary for ESCRT-III disassembly and MVB sorting, it is unclear if it plays a direct role in ILV scission. There are two general hypotheses for its role. We favor the idea that the Vps4-Vta1 complex engages ESCRT-III following its assembly and, like a mechanical motor, helps generate the constrictive force necessary for scission at the ILV neck. In this model, Vps4-Vta1 oligomers bind ESCRT-III subunits and mediate the stepwise disassembly of the ESCRT-III complex (Figure 3vi). This stepwise disassembly could generate force along the ESCRT-III polymer and mediate neck constriction as the polymer shrinks into smaller and smaller rings, similar to a tightening "purse string" (Saksena et al., 2009). Disassembly could also potentially generate a "sliding" mechanism where Snf7 filaments move past one another, creating constrictive force. Alternatively, the Vps4-Vta1 oligomer could engage multiple ESCRT-III subunits simultaneously and mediate concerted disassembly of the ESCRT-III complex. The drastic removal of the ESCRT-III complex from the neck of the budding ILV may destabilize the highly curved membrane, leading to buckling and scission (similar to the "buckling model" above). A similar model for vesicle scission has been proposed for dynamin, where dynamin oligomers spiral around a membrane tubule but release their hold just prior to scission (Bashkirov et al., 2008). In all models, the late recruitment of Vps4 ensures that ESCRT-III is not prematurely disassembled. The late recruitment of Vps4 is supported by the ordered assembly of ESCRT-III because Vps4 avidity for the ESCRT-III subunits will be greatest only after the complex is fully assembled (Teis et al., 2008).

An alternative model suggests that Vps4 does not mediate ILV budding directly but rather serves only to "recycle" ESCRT-III subunits after scission has occurred (Wollert et al., 2009). This would place the ATP hydrolysis reaction after the scission event, similar to the role of the AAA-ATPase NSF (N-ethylmaleimide sensitive fusion protein) in recycling the SNARE complex after membrane fusion. This is in contrast to other vesicle budding events like dynamin-dependent endocytosis, where the hydrolysis of GTP causes the dynamin oligomer to change conformations and mediate scission. Here,

Vps4 would function primarily to remove ESCRT-III subunits from the membrane and potentially "re-set" them into their autoinhibited conformation. This "closed" conformation places the C terminus in close proximity with the α-helical hairpin formed by the first two helices of the ESCRT-III structure (Bajorek et al., 2009b; Muzioł et al., 2006; Zamborlini et al., 2006). It is intriguing that Vps4 binds the C-terminal MIM motifs, indicating that it could mediate C-terminal conformational changes.

Notably, both the Snf7 oligomeric model and Vps24-Vps2 "Dome" model could be dependent on Vps4 for scission. Vps4-mediated disassembly could provide the constrictive force for Snf7-mediated scission (Saksena et al., 2009). In the "Dome" model, the Vps24-Vps2 hemisphere causes the ILV membrane neck to taper, bringing the membrane into close proximity for spontaneous fission. However, one could speculate that the vesicle neck may be stabilized by its high affinity for Vps24-Vps2 dome. Here, removal of the protein dome by Vps4 could mediate membrane buckling and scission.

Lipid Composition and ILV Budding

Careful consideration should also be placed on the role of lipids in ILV formation. Although in vitro evidence suggests a prominent role for the ESCRTs in vesicle budding, several studies indicate that specific lipids are key mediators of vesicle formation. Sphingomyelin can form liquid ordered domains with cholesterol and, when converted to ceramide by sphingomyelinases, can generate membrane deformations that bud ILVs in vitro (Trajkovic et al., 2008). Lysobisphosphatidic acid (LBPA) has also been shown to generate multivesicular liposomes in a pH-dependent manner, although it is unclear if this lipid exists in yeast (Matsuo et al., 2004). It is highly likely that the clustering of protein complexes like the ESCRTs into small cargo-enriched regions on the endosome drives some lipid ordering. This ordering may reinforce ESCRT-mediated membrane deformation, although the detailed mechanism of this remains to be elucidated.

Summary and Synthesis

Thus, to date, there are two broad models for ESCRT-III-mediated ILV formation, although variants of each of them exist. In one, Snf7 oligomers encircle cargo and recruit "capping" proteins Vps24 and Vps2, which regulate Snf7 oligomer size and recruit Vps4-Vta1. Vesicle formation is thus a consequence of controlled Snf7 oligomerization. In the other model, a Vps24-Vps2 "dome" forms within the neck of a budding ILV, and its electrostatic affinity for membrane drives the constriction of the ILV neck to promote scission. Emphasis is thus placed on the Vps24-Vps2 subcomplex as a membrane-bending and -buckling machine. Notably, either model may be dependent on Vps4, which could provide the energy for budding and scission through sequential or concerted removal of ESCRT-III subunits.

It is notable that these models are not completely mutually exclusive. Snf7 filaments may define a "zone of sequestration" for cargo that is further sculpted by the action of the Vps24-Vps2 subcomplex. This is supported by in vitro reconstitution of ESCRT-III assembly where membrane depressions were observed on liposomes when purified Vps20, Snf7, Vps24, and Vps2 were added in a ratio of ~1:10:5:3 but not seen when individual subunits were left out (Saksena et al., 2009). Both models also satisfy the ordered assembly of the ESCRT-III subunits.

Further study is needed to disentangle the true molecular mechanism of ILV generation. Because crystallography studies indicate that ESCRT-III subunits are structurally similar, how these small coiled-coil proteins achieve apparently diverse functions in the ESCRT pathway must also be investigated. Also, careful consideration must be given to how ESCRT-III mediates viral budding and cytokinesis. Because the topology of membrane deformation is conserved between these processes, it is reasonable to think that ESCRT-III functions in a similar capacity in each of them. Careful thought should also be given to the role of the other ESCRT complexes in vesicle formation. ESCRT-I forms an ~20 nm structure, and this raises the intriguing hypothesis that the elongated shape of ESCRT-I may play a dual role in both cargo binding and setting the size of an ESCRT-III ring. At the same time, the Y shape of ESCRT-II may govern the directions from which ESCRT-III filaments elongate to encircle cargo.

CLOSING

Diversity in ESCRT Function

Thus, from the perspectives of the ESCRT subunits, lipid membrane, and ubiquitinated cargo, the ESCRT pathway can be viewed primarily as a cargo-engaging and lipid bilayer-bending machine conserved from yeast to man. However, recent studies highlight an emerging divergence in ESCRT function. The most striking difference is the differential requirement of some ESCRT proteins in different biological contexts. As mentioned above, this is particularly apparent in retroviral budding. ESCRT-I and ESCRT-III subunits appear essential for viral release because Vps23/Tsg101 and Bro1/Alix bind directly to the L-domain of several viruses. However, ESCRTs -0 and -II and Vps20 appear dispensable for HIV-1 budding (Langelier et al., 2006). Similarly, ESCRT-I and -III are also important for cytokinesis, and their loss impairs abscission (Carlton and Martin-Serrano, 2007). Thus, ESCRT-0 appears negligible in both processes. This is reinforced by the observation that ESCRT-0 only appears present in animals and fungi, posing the question as to how the ESCRT pathway is initiated in other systems (Leung et al., 2008).

Bro1/Alix is an attractive candidate for Snf7 nucleation in processes that do not require ESCRT-II, but interestingly, Bro1 null yeast still contains assembled ESCRT-III complexes, suggesting that in yeast Bro1 is dispensable for Snf7 filament formation (Teis et al., 2008). Bro1 binds to the extreme C terminus of Snf7, allowing it to potentially modulate conformational changes in the C terminus of the protein.

Another significant divergence in ESCRT function is the difference in cargo between processes. MVB sorting requires ubiquitinated cargo, but ubiquitination of the "viral cargo" of an enveloped virus is not essential for its budding. ESCRTs do not appear to utilize ubiquitin for cytokinesis either, where the "cargo" can be viewed as the entire daughter cell interior. This underscores the relative independence of each of the two chief ESCRT functions as a cargo-engaging and membrane-deforming machine.

Remaining Questions in the Field

In closing, since the discovery in 2001 and 2002 of the ESCRT complexes, amazing progress has been made toward the understanding of ESCRT function. Significant achievements have been made in ESCRT protein purification, molecular and genetic characterization, crystallization of intact ESCRT complexes, and reconstitution of ESCRT function in vitro. In parallel with this, the ESCRTs have emerged as key players in diverse cellular and viral processes. This has given insight into the potential ancestral purpose of the ESCRT proteins as membrane-deforming and scission machines necessary for cytokinesis in unicellular organisms—and it is notable that ESCRT-III and Vps4 are involved in cytokinesis in metazoans and *Crenarchaea*. ESCRTs -0, -I, and -II may represent a further specialization on this theme, coupling ubiquitin and PtdIns(3)P binding with scission to form a cargo-sorting pathway conserved from yeast to man. The ability to isolate and study distinct ESCRT complexes also underlines the "division of labor" between complexes necessary to carry out cargo sorting in a tightly regulated manner.

Despite all that we know, however, many questions still remain. In general, a complete elucidation of the ESCRT pathway will require a temporal and spatial understanding of: (1) cargo recognition and ordered recruitment of the ESCRT-0, -I, and -II complexes, (2) ordered assembly (i.e., Vps20, Snf7, Vps24, Vps2) at the membrane surface of ESCRT-III into a sequestering ring/spiral, (3) deubiquitination of cargo, (4) membrane deformation, (5) neck constriction and scission, and (6) Vps4-mediated ESCRT complex disassembly. At present, although stages one, two, and three are more clearly understood, stages four, five, and six require intense study to fully elucidate. To us, the most pressing is the mechanism of ESCRT-III-mediated scission. A deeper analysis of this will potentially aid in the understanding of viral budding, cytokinesis, and MVB sorting. Intertwined with this mystery is the need for an intact ESCRT-III three-dimensional structure, which will provide insight into its mechanism of action. Whether Vps4 helps to mediate scission or only disassemble ESCRT-III also remains a key question in the field.

Also needed is a greater understanding of the inter-ESCRT stoichiometry needed for a single round of MVB sorting. In vitro experiments on GUVs indicate that multiple ESCRT-0, -I, and -II complexes localize to forming ILVs, and further experiments are needed to confirm if this is true in vivo (Wollert et al., 2009). ESCRT-I and ESCRT-II form a supercomplex of 1:1 ratio in vitro (Gill et al., 2007). The ESCRT-0/ESCRT-I stoichiometry is less certain and may not be fixed for MVB sorting. Whether there is an exact stoichiometry to a budding event remains a major challenge to determine. Related to stoichiometry is the issue of how the ESCRTs spatially and temporally interact with ubiquitinated cargo at the endosome. Interestingly, recent studies have shown that mutating ubiquitin-binding moieties of ESCRT-I and -II separately do not result in sorting defects (Shields et al., 2009). This suggests that rather than function in a strictly processive series that can be interrupted if any one step is perturbed, the ESCRT complexes work synergistically to sort cargo.

There are also emerging differences between yeast and mammalian ESCRT pathways. Mammals contain multiple isoforms of ESCRT proteins, many with two or three paralogues. Whether these isoforms are tissue or pathway specific must be further elucidated. Intriguingly, overexpression of CHMP4a, b, and c isoforms elicits dominant-negative effects on viral budding and cytokinesis, but to very different degrees. CHMP4b most potently inhibits viral budding, but CHMP4c most potently inhibits cytokinesis (Carlton et al., 2008).

The role of lipids in ILV formation also remains a key question. Studies have indicated that the MVB sorting requires LBPA in higher eukaryotes, a lipid species that is not detected in yeast (Matsuo et al., 2004). *Crenarchaea* also have very different lipid compositions than mammals, yet both use ESCRTs for cell division. PtdIns(3)P is obviously crucial to the endosomal localization of the ESCRTs, but the impact of lipid composition for the budding and scission events requires further investigation.

The exact mechanism of ESCRT dysfunction in disease remains a key question in medical research. ESCRTs have been associated with numerous pathologies including cancer, AIDS, and neurodegeneration, and an understanding of exactly how perturbations in ESCRT function lead to disease will no doubt aid in the treatment of these diseases.

A final big question is whether additional processes involve ESCRT proteins. Autophagy and mitotic spindle maintenance have recently been implicated as ESCRT-dependent processes, but the mechanics of this need to be further elucidated (Lee et al., 2007; Morita et al., 2010).

In closing, the ESCRT field has made significant progress toward elucidating the mechanism-of-action of these distinctive protein complexes. Although they appear intricate, their function can be distilled down to the basic roles of cargo engagement and membrane deformation, which can be understood from the perspectives of the proteins and lipid bilayers involved. No doubt, future studies will continue to reveal exciting (and potentially unexpected) roles for the ESCRTs and Vps4 AAA ATPase.

ACKNOWLEDGMENTS

We would like to thank members of the Emr laboratory for helpful discussions, especially Jason MacGurn and Chris Stefan for comments on the manuscript. We also thank David Teis and Chris Fromme for stimulating discussion and helpful comments on the manuscript.

REFERENCES

Agromayor, M., Carlton, J.G., Phelan, J.P., Matthews, D.R., Carlin, L.M., Ameer-Beg, S., Bowers, K., and Martin-Serrano, J. (2009). Essential role of hIST1 in cytokinesis. Mol. Biol. Cell 20, 1374–1387.

Alam, S.L., Sun, J., Payne, M., Welch, B.D., Blake, B.K., Davis, D.R., Meyer, H.H., Emr, S.D., and Sundquist, W.I. (2004). Ubiquitin interactions of NZF zinc fingers. EMBO J. 23, 1411–1421.

Amerik, A.Y., Nowak, J., Swaminathan, S., and Hochstrasser, M. (2000). The Doa4 deubiquitinating enzyme is functionally linked to the vacuolar protein-sorting and endocytic pathways. Mol. Biol. Cell 11, 3365–3380.

Amit, I., Yakir, L., Katz, M., Zwang, Y., Marmor, M.D., Citri, A., Shtiegman, K., Alroy, I., Tuvia, S., Reiss, Y., et al. (2004). Tal, a Tsg101-specific E3 ubiquitin ligase, regulates receptor endocytosis and retrovirus budding. Genes Dev. 18, 1737–1752.

Asao, H., Sasaki, Y., Arita, T., Tanaka, N., Endo, K., Kasai, H., Takeshita, T., Endo, Y., Fujita, T., and Sugamura, K. (1997). Hrs is associated with STAM, a signal-transducing adaptor molecule. Its suppressive effect on cytokine-induced cell growth. J. Biol. Chem. 272, 32785–32791.

Azmi, I., Davies, B., Dimaano, C., Payne, J., Eckert, D., Babst, M., and Katzmann, D.J. (2006). Recycling of ESCRTs by the AAA-ATPase Vps4 is regulated by a conserved VSL region in Vta1. J. Cell Biol. 172, 705–717.

Azmi, I.F., Davies, B.A., Xiao, J., Babst, M., Xu, Z., and Katzmann, D.J. (2008). ESCRT-III family members stimulate Vps4 ATPase activity directly or via Vta1. Dev. Cell 14, 50–61.

Babst, M. (2005). A protein's final ESCRT. Traffic 6, 2–9.

Babst, M., Sato, T.K., Banta, L.M., and Emr, S.D. (1997). Endosomal transport function in yeast requires a novel AAA-type ATPase, Vps4p. EMBO J. 16, 1820–1831.

Babst, M., Wendland, B., Estepa, E.J., and Emr, S.D. (1998). The Vps4p AAA ATPase regulates membrane association of a Vps protein complex required for normal endosome function. EMBO J. *17*, 2982–2993.

Babst, M., Katzmann, D.J., Estepa-Sabal, E.J., Meerloo, T., and Emr, S.D. (2002). Escrt-III: an endosome-associated heterooligomeric protein complex required for mvb sorting. Dev. Cell *3*, 271–282.

Babst, M., Katzmann, D.J., Snyder, W.B., Wendland, B., and Emr, S.D. (2002). Endosome-associated complex, ESCRT-II, recruits transport machinery for protein sorting at the multivesicular body. Dev. Cell *3*, 283–289.

Bache, K.G., Raiborg, C., Mehlum, A., and Stenmark, H. (2003). STAM and Hrs are subunits of a multivalent ubiquitin-binding complex on early endosomes. J. Biol. Chem. *278*, 12513–12521.

Bache, K.G., Slagsvold, T., Cabezas, A., Rosendal, K.R., Raiborg, C., and Stenmark, H. (2004). The growth-regulatory protein HCRP1/hVps37A is a subunit of mammalian ESCRT-I and mediates receptor down-regulation. Mol. Biol. Cell *15*, 4337–4346.

Bajorek, M., Morita, E., Skalicky, J.J., Morham, S.G., Babst, M., and Sundquist, W.I. (2009). Biochemical analyses of human IST1 and its function in cytokinesis. Mol. Biol. Cell *20*, 1360–1373.

Bajorek, M., Schubert, H.L., McCullough, J., Langelier, C., Eckert, D.M., Stubblefield, W.M., Uter, N.T., Myszka, D.G., Hill, C.P., and Sundquist, W.I. (2009). Structural basis for ESCRT-III protein autoinhibition. Nat. Struct. Mol. Biol. *16*, 754–762.

Banta, L.M., Robinson, J.S., Klionsky, D.J., and Emr, S.D. (1988). Organelle assembly in yeast: characterization of yeast mutants defective in vacuolar biogenesis and protein sorting. J. Cell Biol. *107*, 1369–1383.

Barends, T.R., Werbeck, N.D., and Reinstein, J. (2010). Disaggregases in 4 dimensions. Curr. Opin. Struct. Biol. *20*, 46–53.

Bashkirov, P.V., Akimov, S.A., Evseev, A.I., Schmid, S.L., Zimmerberg, J., and Frolov, V.A. (2008). GTPase cycle of dynamin is coupled to membrane squeeze and release, leading to spontaneous fission. Cell *135*, 1276–1286.

Bilodeau, P.S., Urbanowski, J.L., Winistorfer, S.C., and Piper, R.C. (2002). The Vps27p Hse1p complex binds ubiquitin and mediates endosomal protein sorting. Nat. Cell Biol. *4*, 534–539.

Bishop, N., and Woodman, P. (2001). TSG101/mammalian VPS23 and mammalian VPS28 interact directly and are recruited to VPS4-induced endosomes. J. Biol. Chem. *276*, 11735–11742.

Bouamr, F., Houck-Loomis, B.R., De Los Santos, M., Casaday, R.J., Johnson, M.C., and Goff, S.P. (2007). The C-terminal portion of the Hrs protein interacts with Tsg101 and interferes with human immunodeficiency virus type 1 Gag particle production. J. Virol. *81*, 2909–2922.

Burton, B.M., Williams, T.L., and Baker, T.A. (2001). ClpX-mediated remodeling of mu transpososomes: selective unfolding of subunits destabilizes the entire complex. Mol. Cell *8*, 449–454.

Carlton, J. (2010). The ESCRT machinery: a cellular apparatus for sorting and scission. Biochem. Soc. Trans. *38*, 1397–1412.

Carlton, J.G., and Martin-Serrano, J. (2007). Parallels between cytokinesis and retroviral budding: a role for the ESCRT machinery. Science *316*, 1908–1912.

Carlton, J.G., Agromayor, M., and Martin-Serrano, J. (2008). Differential requirements for Alix and ESCRT-III in cytokinesis and HIV-1 release. Proc. Natl. Acad. Sci. USA *105*, 10541–10546.

Carpp, L.N., Galler, R., and Bonaldo, M.C. (2011). Interaction between the yellow fever virus nonstructural protein NS3 and the host protein Alix contributes to the release of infectious particles. Microbes Infect. *13*, 85–95.

Chen, B.J., and Lamb, R.A. (2008). Mechanisms for enveloped virus budding: can some viruses do without an ESCRT? Virology *372*, 221–232.

Chu, T., Sun, J., Saksena, S., and Emr, S.D. (2006). New component of ESCRT-I regulates endosomal sorting complex assembly. J. Cell Biol. *175*, 815–823.

Curtiss, M., Jones, C., and Babst, M. (2007). Efficient cargo sorting by ESCRT-I and the subsequent release of ESCRT-I from multivesicular bodies requires the subunit Mvb12. Mol. Biol. Cell *18*, 636–645.

Danilchik, M.V., Funk, W.C., Brown, E.E., and Larkin, K. (1998). Requirement for microtubules in new membrane formation during cytokinesis of Xenopus embryos. Dev. Biol. *194*, 47–60.

Doherty, G.J., and McMahon, H.T. (2009). Mechanisms of endocytosis. Annu. Rev. Biochem. *78*, 857–902.

Dukes, J.D., Richardson, J.D., Simmons, R., and Whitley, P. (2008). A dominant-negative ESCRT-III protein perturbs cytokinesis and trafficking to lysosomes. Biochem. J. *411*, 233–239.

Eastman, S.W., Martin-Serrano, J., Chung, W., Zang, T., and Bieniasz, P.D. (2005). Identification of human VPS37C, a component of endosomal sorting complex required for transport-I important for viral budding. J. Biol. Chem. *280*, 628–636.

Elia, N., Sougrat, R., Spurlin, T.A., Hurley, J.H., and Lippincott-Schwartz, J. (2011). Dynamics of endosomal sorting complex required for transport (ESCRT) machinery during cytokinesis and its role in abscission. Proc. Natl. Acad. Sci. USA *108*, 4846–4851.

Fabrikant, G., Lata, S., Riches, J.D., Briggs, J.A., Weissenhorn, W., and Kozlov, M.M. (2009). Computational model of membrane fission catalyzed by ESCRT-III. PLoS Comput. Biol. *5*, e1000575.

Filimonenko, M., Stuffers, S., Raiborg, C., Yamamoto, A., Malerød, L., Fisher, E.M., Isaacs, A., Brech, A., Stenmark, H., and Simonsen, A. (2007). Functional multivesicular bodies are required for autophagic clearance of protein aggregates associated with neurodegenerative disease. J. Cell Biol. *179*, 485–500.

Fisher, R.D., Wang, B., Alam, S.L., Higginson, D.S., Robinson, H., Sundquist, W.I., and Hill, C.P. (2003). Structure and ubiquitin binding of the ubiquitin-interacting motif. J. Biol. Chem. *278*, 28976–28984.

Gallop, J.L., and McMahon, H.T. (2005). BAR domains and membrane curvature: bringing your curves to the BAR. Biochem. Soc. Symp. *72*, 223–231.

Garrus, J.E., von Schwedler, U.K., Pornillos, O.W., Morham, S.G., Zavitz, K.H., Wang, H.E., Wettstein, D.A., Stray, K.M., Côté, M., Rich, R.L., et al. (2001). Tsg101 and the vacuolar protein sorting pathway are essential for HIV-1 budding. Cell *107*, 55–65.

Gerald, N.J., Damer, C.K., O'Halloran, T.J., and De Lozanne, A. (2001). Cytokinesis failure in clathrin-minus cells is caused by cleavage furrow instability. Cell Motil. Cytoskeleton *48*, 213–223.

Ghazi-Tabatabai, S., Saksena, S., Short, J.M., Pobbati, A.V., Veprintsev, D.B., Crowther, R.A., Emr, S.D., Egelman, E.H., and Williams, R.L. (2008). Structure and disassembly of filaments formed by the ESCRT-III subunit Vps24. Structure *16*, 1345–1356.

Gill, D.J., Teo, H., Sun, J., Perisic, O., Veprintsev, D.B., Emr, S.D., and Williams, R.L. (2007). Structural insight into the ESCRT-I/-II link and its role in MVB trafficking. EMBO J. *26*, 600–612.

Gonciarz, M.D., Whitby, F.G., Eckert, D.M., Kieffer, C., Heroux, A., Sundquist, W.I., and Hill, C.P. (2008). Biochemical and structural studies of yeast Vps4 oligomerization. J. Mol. Biol. *384*, 878–895.

Gromley, A., Yeaman, C., Rosa, J., Redick, S., Chen, C.T., Mirabelle, S., Guha, M., Sillibourne, J., and Doxsey, S.J. (2005). Centriolin anchoring of exocyst and SNARE complexes at the midbody is required for secretory-vesicle-mediated abscission. Cell *123*, 75–87.

Guizetti, J., Schermelleh, L., Mantler, J., Maar, S., Poser, I., Leonhardt, H., Müller-Reichert, T., and Gerlich, D.W. (2011). Cortical constriction during abscission involves helices of ESCRT-III-dependent filaments. Science *331*, 1616–1620.

Haigler, H.T., McKanna, J.A., and Cohen, S. (1979). Direct visualization of the binding and internalization of a ferritin conjugate of epidermal growth factor in human carcinoma cells A-431. J. Cell Biol. *81*, 382–395.

Hanson, P.I., Roth, R., Lin, Y., and Heuser, J.E. (2008). Plasma membrane deformation by circular arrays of ESCRT-III protein filaments. J. Cell Biol. *180*, 389–402.

Henne, W.M., Kent, H.M., Ford, M.G., Hegde, B.G., Daumke, O., Butler, P.J., Mittal, R., Langen, R., Evans, P.R., and McMahon, H.T. (2007). Structure and analysis of FCHo2 F-BAR domain: a dimerizing and membrane recruitment module that effects membrane curvature. Structure *15*, 839–852.

Hierro, A., Sun, J., Rusnak, A.S., Kim, J., Prag, G., Emr, S.D., and Hurley, J.H. (2004). Structure of the ESCRT-II endosomal trafficking complex. Nature *431*, 221–225.

Hinshaw, J.E., and Schmid, S.L. (1995). Dynamin self-assembles into rings suggesting a mechanism for coated vesicle budding. Nature *374*, 190–192.

Hirano, S., Kawasaki, M., Ura, H., Kato, R., Raiborg, C., Stenmark, H., and Wakatsuki, S. (2006). Double-sided ubiquitin binding of Hrs-UIM in endosomal protein sorting. Nat. Struct. Mol. Biol. *13*, 272–277.

Hoeller, D., Crosetto, N., Blagoev, B., Raiborg, C., Tikkanen, R., Wagner, S., Kowanetz, K., Breitling, R., Mann, M., Stenmark, H., and Dikic, I. (2006). Regulation of ubiquitin-binding proteins by monoubiquitination. Nat. Cell Biol. *8*, 163–169.

Hurley, J.H., Boura, E., Carlson, L.A., and Różycki, B. (2010). Membrane budding. Cell *143*, 875–887.

Im, Y.J., and Hurley, J.H. (2008). Integrated structural model and membrane targeting mechanism of the human ESCRT-II complex. Dev. Cell *14*, 902–913.

Jouvenet, N., Zhadina, M., Bieniasz, P.D., and Simon, S.M. (2011). Dynamics of ESCRT protein recruitment during retroviral assembly. Nat. Cell Biol. *13*, 394–401.

Kaksonen, M., Toret, C.P., and Drubin, D.G. (2005). A modular design for the clathrin- and actin-mediated endocytosis machinery. Cell *123*, 305–320.

Katzmann, D.J., Babst, M., and Emr, S.D. (2001). Ubiquitin-dependent sorting into the multivesicular body pathway requires the function of a conserved endosomal protein sorting complex, ESCRT-I. Cell *106*, 145–155.

Katzmann, D.J., Stefan, C.J., Babst, M., and Emr, S.D. (2003). Vps27 recruits ESCRT machinery to endosomes during MVB sorting. J. Cell Biol. *162*, 413–423.

Katzmann, D.J., Sarkar, S., Chu, T., Audhya, A., and Emr, S.D. (2004). Multivesicular body sorting: ubiquitin ligase Rsp5 is required for the modification and sorting of carboxypeptidase S. Mol. Biol. Cell *15*, 468–480.

Kim, B.Y., Olzmann, J.A., Barsh, G.S., Chin, L.S., and Li, L. (2007). Spongiform neurodegeneration-associated E3 ligase Mahogunin ubiquitylates TSG101 and regulates endosomal trafficking. Mol. Biol. Cell *18*, 1129–1142.

Kostelansky, M.S., Sun, J., Lee, S., Kim, J., Ghirlando, R., Hierro, A., Emr, S.D., and Hurley, J.H. (2006). Structural and functional organization of the ESCRT-I trafficking complex. Cell *125*, 113–126.

Kostelansky, M.S., Schluter, C., Tam, Y.Y., Lee, S., Ghirlando, R., Beach, B., Conibear, E., and Hurley, J.H. (2007). Molecular architecture and functional model of the complete yeast ESCRT-I heterotetramer. Cell *129*, 485–498.

Kyuuma, M., Kikuchi, K., Kojima, K., Sugawara, Y., Sato, M., Mano, N., Goto, J., Takeshita, T., Yamamoto, A., Sugamura, K., and Tanaka, N. (2007). AMSH, an ESCRT-III associated enzyme, deubiquitinates cargo on MVB/late endosomes. Cell Struct. Funct. *31*, 159–172.

Langelier, C., von Schwedler, U.K., Fisher, R.D., De Domenico, I., White, P.L., Hill, C.P., Kaplan, J., Ward, D., and Sundquist, W.I. (2006). Human ESCRT-II complex and its role in human immunodeficiency virus type 1 release. J. Virol. *80*, 9465–9480.

Lata, S., Schoehn, G., Jain, A., Pires, R., Piehler, J., Gottlinger, H.G., and Weissenhorn, W. (2008). Helical structures of ESCRT-III are disassembled by VPS4. Science *321*, 1354–1357.

Lee, M.C., Orci, L., Hamamoto, S., Futai, E., Ravazzola, M., and Schekman, R. (2005). Sar1p N-terminal helix initiates membrane curvature and completes the fission of a COPII vesicle. Cell *122*, 605–617.

Lee, J.A., Beigneux, A., Ahmad, S.T., Young, S.G., and Gao, F.B. (2007). ESCRT-III dysfunction causes autophagosome accumulation and neurodegeneration. Curr. Biol. *17*, 1561–1567.

Lenz, M., Crow, D.J., and Joanny, J.F. (2009). Membrane buckling induced by curved filaments. Phys. Rev. Lett. *103*, 038101.

Leung, K.F., Dacks, J.B., and Field, M.C. (2008). Evolution of the multivesicular body ESCRT machinery; retention across the eukaryotic lineage. Traffic *9*, 1698–1716.

Lindås, A.C., Karlsson, E.A., Lindgren, M.T., Ettema, T.J., and Bernander, R. (2008). A unique cell division machinery in the Archaea. Proc. Natl. Acad. Sci. USA *105*, 18942–18946.

Long, K.R., Yamamoto, Y., Baker, A.L., Watkins, S.C., Coyne, C.B., Conway, J.F., and Aridor, M. (2010). Sar1 assembly regulates membrane constriction and ER export. J. Cell Biol. *190*, 115–128.

Lu, Q., Hope, L.W., Brasch, M., Reinhard, C., and Cohen, S.N. (2003). TSG101 interaction with HRS mediates endosomal trafficking and receptor down-regulation. Proc. Natl. Acad. Sci. USA *100*, 7626–7631.

Luhtala, N., and Odorizzi, G. (2004). Bro1 coordinates deubiquitination in the multivesicular body pathway by recruiting Doa4 to endosomes. J. Cell Biol. *166*, 717–729.

Lundmark, R., Doherty, G.J., Vallis, Y., Peter, B.J., and McMahon, H.T. (2008). Arf family GTP loading is activated by, and generates, positive membrane curvature. Biochem. J. *414*, 189–194.

Mao, Y., Nickitenko, A., Duan, X., Lloyd, T.E., Wu, M.N., Bellen, H., and Quiocho, F.A. (2000). Crystal structure of the VHS and FYVE tandem domains of Hrs, a protein involved in membrane trafficking and signal transduction. Cell *100*, 447–456.

Martin-Serrano, J., Zang, T., and Bieniasz, P.D. (2001). HIV-1 and Ebola virus encode small peptide motifs that recruit Tsg101 to sites of particle assembly to facilitate egress. Nat. Med. *7*, 1313–1319.

Matsuo, H., Chevallier, J., Mayran, N., Le Blanc, I., Ferguson, C., Fauré, J., Blanc, N.S., Matile, S., Dubochet, J., Sadoul, R., et al. (2004). Role of LBPA and Alix in multivesicular liposome formation and endosome organization. Science *303*, 531–534.

McCullough, J., Row, P.E., Lorenzo, O., Doherty, M., Beynon, R., Clague, M.J., and Urbé, S. (2006). Activation of the endosome-associated ubiquitin isopeptidase AMSH by STAM, a component of the multivesicular body-sorting machinery. Curr. Biol. *16*, 160–165.

Mizuno, E., Kawahata, K., Kato, M., Kitamura, N., and Komada, M. (2003). STAM proteins bind ubiquitinated proteins on the early endosome via the VHS domain and ubiquitin-interacting motif. Mol. Biol. Cell *14*, 3675–3689.

Morita, E., Sandrin, V., Alam, S.L., Eckert, D.M., Gygi, S.P., and Sundquist, W.I. (2007). Identification of human MVB12 proteins as ESCRT-I subunits that function in HIV budding. Cell Host Microbe *2*, 41–53.

Morita, E., Colf, L.A., Karren, M.A., Sandrin, V., Rodesch, C.K., and Sundquist, W.I. (2010). Human ESCRT-III and VPS4 proteins are required for centrosome and spindle maintenance. Proc. Natl. Acad. Sci. USA *107*, 12889–12894.

Muzioł, T., Pineda-Molina, E., Ravelli, R.B., Zamborlini, A., Usami, Y., Göttlinger, H., and Weissenhorn, W. (2006). Structural basis for budding by the ESCRT-III factor CHMP3. Dev. Cell *10*, 821–830.

Neuwald, A.F., Aravind, L., Spouge, J.L., and Koonin, E.V. (1999). AAA+: a class of chaperone-like ATPases associated with the assembly, operation, and disassembly of protein complexes. Genome Res. *9*, 27–43.

Odorizzi, G., Babst, M., and Emr, S.D. (1998). Fab1p PtdIns(3)P 5-kinase function essential for protein sorting in the multivesicular body. Cell *95*, 847–858.

Odorizzi, G., Katzmann, D.J., Babst, M., Audhya, A., and Emr, S.D. (2003). Bro1 is an endosome-associated protein that functions in the MVB pathway in *Saccharomyces cerevisiae*. J. Cell Sci. *116*, 1893–1903.

Palade, G.E. (1955). Studies on the endoplasmic reticulum. II. Simple dispositions in cells in situ. J. Biophys. Biochem. Cytol. *1*, 567–582.

Peter, B.J., Kent, H.M., Mills, I.G., Vallis, Y., Butler, P.J., Evans, P.R., and McMahon, H.T. (2004). BAR domains as sensors of membrane curvature: the amphiphysin BAR structure. Science *303*, 495–499.

Pires, R., Hartlieb, B., Signor, L., Schoehn, G., Lata, S., Roessle, M., Moriscot, C., Popov, S., Hinz, A., Jamin, M., et al. (2009). A crescent-shaped ALIX dimer targets ESCRT-III CHMP4 filaments. Structure *17*, 843–856.

Polo, S., Sigismund, S., Faretta, M., Guidi, M., Capua, M.R., Bossi, G., Chen, H., De Camilli, P., and Di Fiore, P.P. (2002). A single motif responsible for ubiquitin recognition and monoubiquitination in endocytic proteins. Nature *416*, 451–455.

Pornillos, O., Alam, S.L., Rich, R.L., Myszka, D.G., Davis, D.R., and Sundquist, W.I. (2002). Structure and functional interactions of the Tsg101 UEV domain. EMBO J. *21*, 2397–2406.

Pornillos, O., Higginson, D.S., Stray, K.M., Fisher, R.D., Garrus, J.E., Payne, M., He, G.P., Wang, H.E., Morham, S.G., and Sundquist, W.I. (2003). HIV Gag mimics the Tsg101-recruiting activity of the human Hrs protein. J. Cell Biol. *162*, 425–434.

Prag, G., Watson, H., Kim, Y.C., Beach, B.M., Ghirlando, R., Hummer, G., Bonifacino, J.S., and Hurley, J.H. (2007). The Vps27/Hse1 complex is a GAT domain-based scaffold for ubiquitin-dependent sorting. Dev. Cell *12*, 973–986.

Raiborg, C., and Stenmark, H. (2009). The ESCRT machinery in endosomal sorting of ubiquitylated membrane proteins. Nature *458*, 445–452.

Raiborg, C., Bremnes, B., Mehlum, A., Gillooly, D.J., D'Arrigo, A., Stang, E., and Stenmark, H. (2001). FYVE and coiled-coil domains determine the specific localisation of Hrs to early endosomes. J. Cell Sci. *114*, 2255–2263.

Raiborg, C., Bache, K.G., Gillooly, D.J., Madshus, I.H., Stang, E., and Stenmark, H. (2002). Hrs sorts ubiquitinated proteins into clathrin-coated microdomains of early endosomes. Nat. Cell Biol. *4*, 394–398.

Raiborg, C., Wesche, J., Malerød, L., and Stenmark, H. (2006). Flat clathrin coats on endosomes mediate degradative protein sorting by scaffolding Hrs in dynamic microdomains. J. Cell Sci. *119*, 2414–2424.

Ramachandran, R., Pucadyil, T.J., Liu, Y.W., Acharya, S., Leonard, M., Lukiyanchuk, V., and Schmid, S.L. (2009). Membrane insertion of the pleckstrin homology domain variable loop 1 is critical for dynamin-catalyzed vesicle scission. Mol. Biol. Cell 20, 4630–4639.

Raymond, C.K., Howald-Stevenson, I., Vater, C.A., and Stevens, T.H. (1992). Morphological classification of the yeast vacuolar protein sorting mutants: evidence for a prevacuolar compartment in class E vps mutants. Mol. Biol. Cell 3, 1389–1402.

Reggiori, F., and Pelham, H.R. (2001). Sorting of proteins into multivesicular bodies: ubiquitin-dependent and -independent targeting. EMBO J. 20, 5176–5186.

Reid, B.G., Fenton, W.A., Horwich, A.L., and Weber-Ban, E.U. (2001). ClpA mediates directional translocation of substrate proteins into the ClpP protease. Proc. Natl. Acad. Sci. USA 98, 3768–3772.

Ren, X., and Hurley, J.H. (2010). VHS domains of ESCRT-0 cooperate in high-avidity binding to polyubiquitinated cargo. EMBO J. 29, 1045–1054.

Ren, X., Kloer, D.P., Kim, Y.C., Ghirlando, R., Saidi, L.F., Hummer, G., and Hurley, J.H. (2009). Hybrid structural model of the complete human ESCRT-0 complex. Structure 17, 406–416.

Robinson, J.S., Klionsky, D.J., Banta, L.M., and Emr, S.D. (1988). Protein sorting in Saccharomyces cerevisiae: isolation of mutants defective in the delivery and processing of multiple vacuolar hydrolases. Mol. Cell. Biol. 8, 4936–4948.

Rothman, J.H., and Stevens, T.H. (1986). Protein sorting in yeast: mutants defective in vacuole biogenesis mislocalize vacuolar proteins into the late secretory pathway. Cell 47, 1041–1051.

Rothman, J.E., and Wieland, F.T. (1996). Protein sorting by transport vesicles. Science 272, 227–234.

Rothman, J.H., Howald, I., and Stevens, T.H. (1989). Characterization of genes required for protein sorting and vacuolar function in the yeast Saccharomyces cerevisiae. EMBO J. 8, 2057–2065.

Roux, A., Uyhazi, K., Frost, A., and De Camilli, P. (2006). GTP-dependent twisting of dynamin implicates constriction and tension in membrane fission. Nature 441, 528–531.

Row, P.E., Prior, I.A., McCullough, J., Clague, M.J., and Urbé, S. (2006). The ubiquitin isopeptidase UBPY regulates endosomal ubiquitin dynamics and is essential for receptor down-regulation. J. Biol. Chem. 281, 12618–12624.

Rue, S.M., Mattei, S., Saksena, S., and Emr, S.D. (2008). Novel Ist1-Did2 complex functions at a late step in multivesicular body sorting. Mol. Biol. Cell 19, 475–484.

Rusten, T.E., Vaccari, T., Lindmo, K., Rodahl, L.M., Nezis, I.P., Sem-Jacobsen, C., Wendler, F., Vincent, J.P., Brech, A., Bilder, D., and Stenmark, H. (2007). ESCRTs and Fab1 regulate distinct steps of autophagy. Curr. Biol. 17, 1817–1825.

Sachse, M., Urbé, S., Oorschot, V., Strous, G.J., and Klumperman, J. (2002). Bilayered clathrin coats on endosomal vacuoles are involved in protein sorting toward lysosomes. Mol. Biol. Cell 13, 1313–1328.

Saksena, S., and Emr, S.D. (2009). ESCRTs and human disease. Biochem. Soc. Trans. 37, 167–172.

Saksena, S., Wahlman, J., Teis, D., Johnson, A.E., and Emr, S.D. (2009). Functional reconstitution of ESCRT-III assembly and disassembly. Cell 136, 97–109.

Samson, R.Y., Obita, T., Hodgson, B., Shaw, M.K., Chong, P.L., Williams, R.L., and Bell, S.D. (2011). Molecular and structural basis of ESCRT-III recruitment to membranes during archaeal cell division. Mol. Cell 41, 186–196.

Scott, A., Chung, H.Y., Gonciarz-Swiatek, M., Hill, G.C., Whitby, F.G., Gaspar, J., Holton, J.M., Viswanathan, R., Ghaffarian, S., Hill, C.P., and Sundquist, W.I. (2005). Structural and mechanistic studies of VPS4 proteins. EMBO J. *24*, 3658–3669.

Scott, A., Gaspar, J., Stuchell-Brereton, M.D., Alam, S.L., Skalicky, J.J., and Sundquist, W.I. (2005). Structure and ESCRT-III protein interactions of the MIT domain of human VPS4A. Proc. Natl. Acad. Sci. USA *102*, 13813–13818.

Shestakova, A., Hanono, A., Drosner, S., Curtiss, M., Davies, B.A., Katzmann, D.J., and Babst, M. (2010). Assembly of the AAA ATPase Vps4 on ESCRT-III. Mol. Biol. Cell *21*, 1059–1071.

Shields, S.B., Oestreich, A.J., Winistorfer, S., Nguyen, D., Payne, J.A., Katzmann, D.J., and Piper, R. (2009). ESCRT ubiquitin-binding domains function cooperatively during MVB cargo sorting. J. Cell Biol. *185*, 213–224.

Shiflett, S.L., Ward, D.M., Huynh, D., Vaughn, M.B., Simmons, J.C., and Kaplan, J. (2004). Characterization of Vta1p, a class E Vps protein in *Saccharomyces cerevisiae*. J. Biol. Chem. *279*, 10982–10990.

Shih, S.C., Katzmann, D.J., Schnell, J.D., Sutanto, M., Emr, S.D., and Hicke, L. (2002). Epsins and Vps27p/Hrs contain ubiquitin-binding domains that function in receptor endocytosis. Nat. Cell Biol. *4*, 389–393.

Shim, S., Kimpler, L.A., and Hanson, P.I. (2007). Structure/function analysis of four core ESCRT-III proteins reveals common regulatory role for extreme C-terminal domain. Traffic *8*, 1068–1079.

Shimoni, Y., and Schekman, R. (2002). Vesicle budding from endoplasmic reticulum. Methods Enzymol. *351*, 258–278.

Sierra, M.I., Wright, M.H., and Nash, P.D. (2010). AMSH interacts with ESCRT-0 to regulate the stability and trafficking of CXCR4. J. Biol. Chem. *285*, 13990–14004.

Slagsvold, T., Aasland, R., Hirano, S., Bache, K.G., Raiborg, C., Trambaiolo, D., Wakatsuki, S., and Stenmark, H. (2005). Eap45 in mammalian ESCRT-II binds ubiquitin via a phosphoinositide-interacting GLUE domain. J. Biol. Chem. *280*, 19600–19606.

Sotelo, J.R., and Porter, K.R. (1959). An electron microscope study of the rat ovum. J. Biophys. Biochem. Cytol. *5*, 327–342.

Spitzer, C., Schellmann, S., Sabovljevic, A., Shahriari, M., Keshavaiah, C., Bechtold, N., Herzog, M., Müller, S., Hanisch, F.G., and Hülskamp, M. (2006). The Arabidopsis elch mutant reveals functions of an ESCRT component in cytokinesis. Development *133*, 4679–4689.

Stowell, M.H., Marks, B., Wigge, P., and McMahon, H.T. (1999). Nucleotide-dependent conformational changes in dynamin: evidence for a mechanochemical molecular spring. Nat. Cell Biol. *1*, 27–32.

Strack, B., Calistri, A., Accola, M.A., Palu, G., and Gottlinger, H.G. (2000). A role for ubiquitin ligase recruitment in retrovirus release. Proc. Natl. Acad. Sci. USA *97*, 13063–13068.

Striebel, F., Kress, W., and Weber-Ban, E. (2009). Controlled destruction: AAA+ ATPases in protein degradation from bacteria to eukaryotes. Curr. Opin. Struct. Biol. *19*, 209–217.

Stringer, D.K., and Piper, R.C. (2011). A single ubiquitin is sufficient for cargo protein entry into MVBs in the absence of ESCRT ubiquitination. J. Cell Biol. *192*, 229–242.

Stuchell, M.D., Garrus, J.E., Müller, B., Stray, K.M., Ghaffarian, S., McKinnon, R., Kräusslich, H.G., Morham, S.G., and Sundquist, W.I. (2004). The human endosomal sorting complex required for transport (ESCRT-I) and its role in HIV-1 budding. J. Biol. Chem. *279*, 36059–36071.

Swaminathan, S., Amerik, A.Y., and Hochstrasser, M. (1999). The Doa4 deubiquitinating enzyme is required for ubiquitin homeostasis in yeast. Mol. Biol. Cell *10*, 2583–2594.

Teis, D., Saksena, S., and Emr, S.D. (2008). Ordered assembly of the ESCRT-III complex on endosomes is required to sequester cargo during MVB formation. Dev. Cell *15*, 578–589.

Teis, D., Saksena, S., Judson, B.L., and Emr, S.D. (2010). ESCRT-II coordinates the assembly of ESCRT-III filaments for cargo sorting and multivesicular body vesicle formation. EMBO J. *29*, 871–883.

Teo, H., Perisic, O., González, B., and Williams, R.L. (2004). ESCRT-II, an endosome-associated complex required for protein sorting: crystal structure and interactions with ESCRT-III and membranes. Dev. Cell *7*, 559–569.

Teo, H., Gill, D.J., Sun, J., Perisic, O., Veprintsev, D.B., Vallis, Y., Emr, S.D., and Williams, R.L. (2006). ESCRT-I core and ESCRT-II GLUE domain structures reveal role for GLUE in linking to ESCRT-I and membranes. Cell *125*, 99–111.

Trajkovic, K., Hsu, C., Chiantia, S., Rajendran, L., Wenzel, D., Wieland, F., Schwille, P., Brügger, B., and Simons, M. (2008). Ceramide triggers budding of exosome vesicles into multivesicular endosomes. Science *319*, 1244–1247.

Tsunematsu, T., Yamauchi, E., Shibata, H., Maki, M., Ohta, T., and Konishi, H. (2010). Distinct functions of human MVB12A and MVB12B in the ESCRT-I dependent on their posttranslational modifications. Biochem. Biophys. Res. Commun. *399*, 232–237.

Urata, S., Noda, T., Kawaoka, Y., Morikawa, S., Yokosawa, H., and Yasuda, J. (2007). Interaction of Tsg101 with Marburg virus VP40 depends on the PPPY motif, but not the PT/SAP motif as in the case of Ebola virus, and Tsg101 plays a critical role in the budding of Marburg virus-like particles induced by VP40, NP, and GP. J. Virol. *81*, 4895–4899.

VerPlank, L., Bouamr, F., LaGrassa, T.J., Agresta, B., Kikonyogo, A., Leis, J., and Carter, C.A. (2001). Tsg101, a homologue of ubiquitin-conjugating (E2) enzymes, binds the L domain in HIV type 1 Pr55(Gag). Proc. Natl. Acad. Sci. USA *98*, 7724–7729.

Weber-Ban, E.U., Reid, B.G., Miranker, A.D., and Horwich, A.L. (1999). Global unfolding of a substrate protein by the Hsp100 chaperone ClpA. Nature *401*, 90–93.

Wemmer, M., Azmi, I., West, M., Davies, B., Katzmann, D., and Odorizzi, G. (2011). Bro1 binding to Snf7 regulates ESCRT-III membrane scission activity in yeast. J. Cell Biol. *192*, 295–306.

Williams, R.L., and Urbé, S. (2007). The emerging shape of the ESCRT machinery. Nat. Rev. Mol. Cell Biol. *8*, 355–368.

Wirblich, C., Bhattacharya, B., and Roy, P. (2006). Nonstructural protein 3 of bluetongue virus assists virus release by recruiting ESCRT-I protein Tsg101. J. Virol. *80*, 460–473.

Wollert, T., and Hurley, J.H. (2010). Molecular mechanism of multivesicular body biogenesis by ESCRT complexes. Nature *464*, 864–869.

Wollert, T., Wunder, C., Lippincott-Schwartz, J., and Hurley, J.H. (2009). Membrane scission by the ESCRT-III complex. Nature *458*, 172–177.

Yang, D., Rismanchi, N., Renvoisé, B., Lippincott-Schwartz, J., Blackstone, C., and Hurley, J.H. (2008). Structural basis for midbody targeting of spastin by the ESCRT-III protein CHMP1B. Nat. Struct. Mol. Biol. *15*, 1278–1286.

Yu, Z., Gonciarz, M.D., Sundquist, W.I., Hill, C.P., and Jensen, G.J. (2008). Cryo-EM structure of dodecameric Vps4p and its 2:1 complex with Vta1p. J. Mol. Biol. *377*, 364–377.

Zamborlini, A., Usami, Y., Radoshitzky, S.R., Popova, E., Palu, G., and Göttlinger, H. (2006). Release of autoinhibition converts ESCRT-III components into potent inhibitors of HIV-1 budding. Proc. Natl. Acad. Sci. USA *103*, 19140–19145.

rent Biology

Mechanisms of Autophagosome Biogenesis

David C. Rubinsztein[1,*], Tomer Shpilka[2], Zvulun Elazar[2,*]

[1]Department of Medical Genetics, Cambridge Institute for Medical Research,
Wellcome/MRC Building, Addenbrooke's Hospital, Hills Road, Cambridge CB2 0XY, UK,
[2]Department of Biological Chemistry, The Weizmann Institute of Science,
Rehovot, 76100, Israel
*Correspondence: dcr1000@hermes.cam.ac.uk (D.C.R.), zvulun.elazar@weizmann.ac.il (Z.E.)

Current Biology, Vol. 22, No. 1, R29–R34, January 10, 2012 © 2012 Elsevier Inc.
http://dx.doi.org/10.1016/j.cub.2011.11.034

SUMMARY

Autophagy is a unique membrane trafficking process whereby newly formed membranes, termed phagophores, engulf parts of the cytoplasm leading to the production of double-membraned autophagosomes that get delivered to lysosomes for degradation. This catabolic pathway has been linked to numerous physiological and pathological conditions, such as development, programmed cell death, cancer, pathogen infection, neurodegenerative disorders, and myopathies. In this review, we will focus on recent studies in yeast and mammalian systems that have provided insights into two critical areas of autophagosome biogenesis — the source of the autophagosomal membranes, and the mechanisms regulating the fusion of the edges of the double-membraned phagophores to form autophagosomes.

INTRODUCTION

(Macro)autophagy is a bulk degradation process that mediates the clearance of long-lived proteins and organelles [1]. Initially, double-membraned cup-shaped structures called phagophores engulf portions of cytoplasm. After fusion of the membranes (closing the 'cup'), the autophagosome is formed. These vesicles appear randomly throughout the cytoplasm, then traffic along microtubules with their movement biased towards the microtubule-organising centre, where lysosomes are concentrated [1]. This brings autophagosomes close to lysosomes, enabling fusion and degradation of the contents of the resulting autolysosomes by lysosomal hydrolases.

281

The biology of autophagy was revolutionised by the discovery of so-called Atg (autophagy) genes in yeast, many of which are conserved in humans. The identification of Atg genes allowed for the assessment of the importance of this pathway in various contexts, as well as a detailed dissection of its mechanism of action [2–4]. Autophagy has numerous roles in physiology and disease. Its primordial function from yeast to man is to act as a buffer against starvation by liberating building blocks from macromolecules. It has additional physiological roles, however, including permitting early embryonic development, removal of apoptotic corpses, antigen presentation, protection against cell death insults, and as a degradation route for various aggregate-prone proteins and infectious agents [5]. Abnormalities in autophagy may contribute to pathologies such as tumourigenesis, various neurodegenerative diseases, and certain muscle diseases [5].

The signals that regulate autophagy are diverse. The induction of autophagy in response to starvation is mediated in part via inactivation of the mammalian target of rapamycin (mTOR) and activation of Jun N-terminal kinase (JNK), while energy loss induces autophagy by activation of AMP kinase (AMPK). Other pathways regulating autophagy are regulated by calcium, cyclic AMP, calpains and the inositol trisphosphate (IP_3) receptor [6].

In this review, we will focus on recent studies in yeast and mammalian systems that have provided insights into two critical areas of autophagosome biogenesis. We will discuss the possible sources of the autophagosomal membranes, and the mechanisms regulating the fusion of the edges of the double-membraned phagophores to form autophagosomes.

AUTOPHAGOSOME FORMATION

A conventional view is that autophagosome formation starts at phagophore assembly sites (PAS) [7]. This concept is derived from studies in yeast which observed that a number of the key proteins involved in autophagosome formation colocalise at a single site in the cell [7,8]. This operational definition led to statements in the literature describing autophagosome formation as an event that occurred 'de novo'. In mammalian cells, there are multiple PAS at any one time. The formation of phagophores requires the class III phosphoinositide 3-kinase (PI3K) Vps34, which acts in a large macromolecular complex, along with Beclin-1 (mammalian Atg6), Atg14 and Vps15 (previously known as p150), to form PI 3-phosphate (PI(3)P) [1]. The activity of this complex is dependent on upstream autophagy regulators, including the mammalian Atg1 orthologues ULK1 and ULK2, Atg13 and focal adhesion kinase (FAK)-family interacting protein of 200 kDa (FIP200) [1].

The elongation of membranes that evolve into autophagosomes is regulated by two ubiquitination-like reactions. First, the ubiquitin-like molecule

Atg12 is conjugated to Atg5 by Atg7, which acts like an E1 ubiquitin-activating enzyme, and by Atg10, which is similar to an E2 ubiquitin-conjugating enzyme. The Atg5–Atg12 complex then interacts non-covalently with Atg16L1 and this resulting ternary complex associates with phagophores but dissociates from completed autophagosomes [9].

The second of the ubiquitin-like reactions involves the conjugation of ubiquitin-like molecules of the Atg8 family, which comprises the three subfamilies LC3, GABARAP and GATE-16 [10], to the lipid phosphatidylethanolamine. Microtubule-associated protein 1 light chain 3 (MAP-LC3/Atg8/LC3), the most well-characterised member of the Atg8 protein family, is conjugated to phosphatidylethanolamine by Atg7 (E1-like) and Atg3 (E2-like), resulting in autophagosome-associated LC3-II. The Atg5–Atg12 complex may be able to enhance LC3 conjugation to phosphatidylethanolamine by acting in an E3-like fashion. In this way, the Atg5–Atg12–Atg16L complex may determine the sites of autophagosome synthesis by regulating the targeting of LC3 to Atg5–Atg12-associated membranes.

Although the Atg5–Atg12–Atg16L1 complex localises to phagophores and pre-phagophore structures and dissociates from fully formed (completed) autophagosomes, LC3-II remains associated with autophagosomes until after their fusion with lysosomes. The LC3-II inside the autolysosomes is degraded, while the LC3-II on the cytoplasmic surface can be delipidated and recycled. Thus, Atg5–Atg12–Atg16L1-positive LC3-negative vesicles represent pre-autophagosomal structures (pre-phagophores and possibly early phagophores), Atg5–Atg12–Atg16L1-positive LC3-positive structures can be considered to be phagophores, and Atg5–Atg12–Atg16L1-negative LC3-positive vesicles can be regarded as completed autophagosomes [11]. Note that these definitions do not consider intermediate structures and therefore may end up being imprecise as our understanding of the structure, shape and protein composition of autophagy-related membranes evolves. Also, the PAS, at least in mammalian cells, may include some of these distinct structures.

MEMBRANE SOURCES FOR PHAGOPHORES

For much of the history of the autophagy field, the origins of phagophore membranes have been elusive. However, recent data suggest that membranes from distinct locations may contribute lipid (and possibly key proteins) to evolving autophagosomes (Figure 1).

Endoplasmic Reticulum

The reconsideration of the endoplasmic reticulum (ER) as an autophagosome membrane source was initiated by Ktistakis and colleagues [12], who

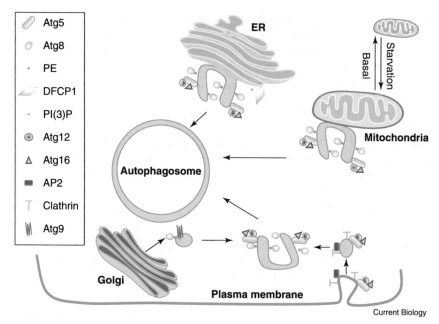

FIGURE 1 Membrane sources for phagophores.
Phagophores require lipids and proteins to mature into autophagosomes. The endoplasmic reticulum is the site of formation of omegasomes, which are essential for phagophore formation and elongation. Mitochondria grow upon starvation and supply lipid vesicles to the phagophore and the Golgi is essential for the trafficking of Atg9-containing vesicles to the phagophore. The plasma membrane contributes membranes to phagophores and autophagosomes under both basal and starvation conditions.

studied a protein called DFCP1, which has two FYVE domains that account for its ability to bind PI(3)P. These authors were intrigued that DFCP1, unlike many other PI(3)P-binding proteins, localises to the ER and Golgi but not to endosomes. Upon starvation, DFCP1 moved to punctuate structures at the ER in a PI(3)P-dependent fashion, and these structures, which they called 'omegasomes', colocalised with LC3 and Atg5 [12].

This study noted that the ER normally contains very little PI(3)P in non-starved cells, and suggested that Vps34 may be delivered to the ER by late endosomes and lysosomes. Subsequent 3D electron tomography studies have revealed interconnections between the ER and forming autophagosomes (Atg16L positive), and have provided further support for a role for the ER by showing that Atg14L targeting to the ER may be important for autophagosome formation [13–15]. However, some questions remain. First, it is not clear whether the ER is an important source of autophagosomal membranes under non-starvation conditions. Second, the topology of the isolation membranes in the 3D electron tomography experiments suggests that there may be ER membrane closely apposed to both sides of the isolation membrane. It would be interesting to understand the next steps of the process that enable the isolation membrane to be freed to engulf cytoplasm and organelles like mitochondria. Alternatively, it is possible that these autophagosomal membranes may show some preference for degradation of ER and its contents, a process known as ER-phagy.

Mitochondria

The Lippincott-Schwartz lab [16] reported that starvation-induced autophagosomes could emerge from sites on mitochondria, via elegant live-cell microscopy studies tracking markers for LC3 and Atg5. Their data suggest that the colocalisation of mitochondria and autophagosomal membranes is not due to mitophagy (autophagy of mitochondria) and that lipids are delivered from mitochondria to autophagosomes that form close to the mitochondria. Importantly, these phenomena appeared to be seen only in glucose-starved cells and were not associated with either basal autophagy or ER-stress-induced autophagy. The idea that mitochondria may be a membrane source for autophagosomes specifically during starvation is interesting to consider in the light of recent data showing that mitochondria are protected from autophagic degradation during starvation because starvation signals cause these organelles to elongate and become refractory to autophagic engulfment [17].

Plasma Membrane

Our own recent data suggest that the plasma membrane contributes to early autophagosomal precursor structures, as we saw the rapid colocalisation of different plasma membrane markers with autophagosomal precursor structures after shifting cells from 4°C to 37°C to enable endocytosis [18]. Importantly, the plasma membrane markers associated with structures that were Atg5–Atg12–Atg16L1-positive but LC3-negative (pre-phagophore structures), structures that contain both Atg5–Atg12–Atg16L1 and LC3 (phagophores), and structures that were LC3-positive but Atg5–Atg12–Atg16L1-negative (completed autophagosomes), suggesting a simple itinerary for plasma membrane from autophagosome precursor to fully-formed autophagosome. The ability of the plasma membrane to contribute to autophagosome formation was associated with the localisation of Atg16L1 at the plasma membrane via Atg16L1–AP-2–clathrin heavy chain interactions. Scission of the Atg16L1-associated clathrin coated pits, leading to the formation of early endosomal-like intermediates, is a crucial step that enables the liberation of these Atg16L1 vesicles and subsequent maturation into autophagosomes. Although they contain Atg16L, Atg5 and Atg12, these autophagosome precursors appear to be membrane structures that precede the phagophore stage [18].

Our data suggest that this process is essential for autophagosome formation under both basal and induced conditions (e.g. starvation), since blocking clathrin-dependent endocytosis attenuated autophagosome biogenesis at the phagophore stage. Although endocytosis appears to be important for autophagosome formation under basal and starvation-induced conditions, it is possible that the ability of the plasma membrane to contribute to autophagosome formation may be particularly important during periods of increased autophagy. At these times, the large surface area of the plasma membrane may serve as an extensive membrane reservoir that allows cells to undergo periods of autophagosome synthesis at much higher rates than under basal conditions, without compromising other processes.

Golgi, Atg9 and Autophagosome Formation in Yeast

Atg9 is the only known multipass-membrane protein that regulates autophagy [19]. Since it cycles between the PAS and different cytoplasmic membranes [20], it has reasonably been assumed to be associated with membrane that contributes to autophagosomes. Understanding its exact intracellular trafficking routes and characterisation of the vehicles that mediate its trafficking is expected to provide valuable information on the cellular mechanism involved in autophagosome biogenesis. Recent studies in yeast suggested that tubulovesicular structures containing Atg9 are delivered to the PAS, a trafficking step essential for autophagosome formation [21]. Klionsky and colleagues [22] identified a set of SNARE molecules that enable the organisation of these tubulovesicular Atg9-containing structures and showed that these SNAREs, in turn, regulated autophagy (Figure 2). This specific function for SNAREs has not yet been studied in mammalian autophagy.

The cycling of yeast Atg9 not only involves specific autophagic factors, but also implicates more general factors involved in intracellular trafficking, including Sec12 and Vps52 [23], Sec7 and Sec2 (guanine nucleotide exchange factors for Arf and Rab GTPases, respectively), Arf1/2 and the Rab Sec4 [24,25], all of which are essential for autophagosome formation via the regulation of Atg9 cycling. Manipulations of the conserved oligomeric Golgi (COG) complex and genes involved in Golgi–endosomal traffic affect autophagosome formation and this is associated with mislocalisation of the autophagy proteins Atg9 and Atg8 [26]. Thus, it is likely that this Golgi route contributes to autophagosome biogenesis in yeast, although it is not clear if this pathway is important in mammalian systems.

Mammalian Atg9 (mAtg9) was more recently implicated in autophagy. Under normal growth conditions, this transmembrane protein, which is synthesised in the ER, localises to the Golgi, the trans-Golgi network, and late endosomes. Following amino-acid starvation, this protein translocates to

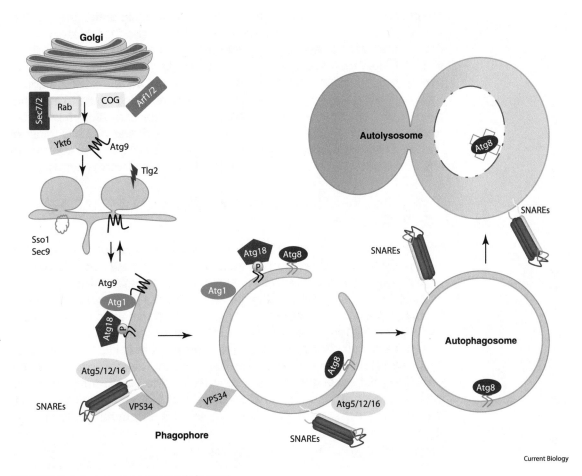

FIGURE 2 Autophagosome biogenesis in yeast.
Atg9-containing vesicles accumulate in tubulovesicular structures and are delivered to the pre-autophagosomal structure along with SNARE proteins to form the phagophore. Atg8 mediates phagophore elongation, and SNARE proteins mediate the fusion of the autophagosome with the lysosome.

LC3-labeled autophagosomes in an ULK1- and PI3K-dependent manner [27]. Cycling of mAtg9 is negatively regulated by p38α MAPK, which competes with mAtg9 for binding to p38IP [28]. In addition, Rab1a and α-synuclein (a protein that causes forms of Parkinson's disease) participate in autophagosome biogenesis, possibly by regulating the localisation of mAtg9 and DFCP1 to omegasomes [29].

When considering the possibility that Atg9 cycling may deliver membrane needed for autophagosome biogenesis, it is important to take into account the fact that Atg9 shuttles back and forth from various cytoplasmic membranes into the pre-autophagosomal membrane (possibly both to phagophores and pre-phagophores). To allow the formation and elongation of the phagophore, the vesicles reaching this membrane should be larger than those carrying Atg9 in the opposite direction. Another non-mutually exclusive possibility is that Atg9 delivers a set of lipids into the autophagosome precursors, while removing other lipids when it departs — this would not require inbound Atg9-containing vesicles to be larger than those returning from the autophagosome precursors. It is also possible that the Atg9-associated membranes are not important for autophagosome formation; perhaps Atg9 has transient structural or catalytic functions at the PAS (or other sites) that regulate autophagy. These are some of the issues that await resolution through further study.

TRANSITION FROM AUTOPHAGOSOME PRECURSOR TO PHAGOPHORE

The data described above reveal that there are autophagosome precursor structures that precede phagophores. It is possible that such structures from various sources coalesce prior to phagophore formation or meet at the phagophore itself, thereby increasing the size of the membranes at the phagophores. In mammalian systems, we have observed that the pre-phagophore Atg5–Atg12–Atg16L1-positive (and LC3-negative) precursors undergo SNARE-mediated homotypic fusion [30]. This process is important for autophagosome formation and autophagy flux, since the homotypic fusion results in vesicles that are larger. Our data suggest that the increase in vesicle size may be a prerequisite for optimal acquisition of LC3 and progression from the autophagosome precursor stage to a nascent phagophore (Figure 3).

TRANSITION FROM PHAGOPHORE TO AUTOPHAGOSOMES — ROLES OF ATG8 PROTEINS

In order for a phagophore to seal its edges and become an autophagosome the membranes probably require elongation and then they need to fuse. In yeast, lipidated Atg8 was found to be essential for elongation of the autophagic membrane [31,32]. This role of Atg8 was further assessed in an *in vitro* liposome-based system, where it was shown to promote membrane fusion [33,34]. Importantly, mutations within Atg8 proteins identified in these systems were found to cause defective autophagosome biogenesis in *Saccharomyces cerevisiae*. However, more direct experiments are needed to determine the mechanism by which Atg8 promotes membrane fusion, including the fusion of Atg9-containing vesicles with the phagophore membrane. For Atg8 (as well as other Atg proteins), it is important to consider other roles in autophagy and cell biology. For instance, Atg8 also binds to other autophagic factors, such as Atg1, which might regulate its recruitment and activity on the autophagic membrane [35]. Other more general

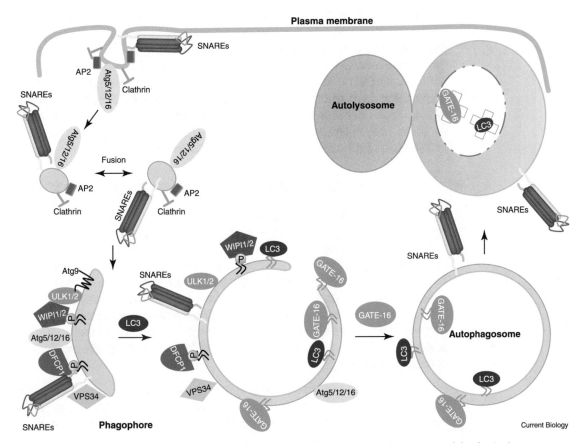

FIGURE 3 Autophagosome biogenesis in mammals, focusing on the plasma membrane as an origin of autophagosome membranes.

Atg16L1-containing vesicles derived from the plasma membrane undergo homotypic fusion, an essential step for phagophore formation. Phagophore elongation is mediated by LC3, and GATE-16 acts downstream in a step coupled to the dissociation of the Atg12–Atg5–Atg16L1 complex. Although SNAREs have been shown to be required for homotypic fusion of Atg12–Atg5–Atg16L1-positive phagophore precursors as well as for autophagosome–lysosome fusion, it is also possible that SNAREs may act at other points of the pathway.

trafficking factors interact with Atg8, including the AAA ATPase Cdc48/p97 and its substrate-recruiting cofactor Shp1. Although these interactions were suggested to control the formation of autophagosomes [36], they may also enable Atg8 to link classical autophagic machinery with general trafficking factors that are crucial for autophagy.

In contrast to yeast, where there is only one Atg8, there are multiple different mammalian Atg8 proteins found on autophagosomes [37–39], all of which undergo post-translational modifications mediated by mammalian orthologs of Atg4, Atg7 and Atg3 [40]. Recently, it has emerged that these are not simply redundant members of a protein family. Our studies indicate that the

LC3 and the GABARAP/GATE-16 subfamilies are essential for autophagy and act at different stages of autophagosome formation: members of the LC3 subfamily are responsible for the elongation of the autophagic membrane, whereas GABARAP/GATE-16 family members act downstream in a step coupled to dissociation of the Atg12–Atg5–Atg16L complex [41]. Both mammalian LC3 and GABARAP/GATE-16 subfamilies assist with membrane fusion after conjugation to phosphatidylethanolamine [34]. This study revealed that the first amino-terminal α helix of LC3 and GABARAP/GATE-16 proteins was both essential and sufficient for this activity.

SYNTHESIS, CHALLENGES AND FUTURE QUESTIONS

Our belief (which will need to be tested) is that there are multiple membrane sources for autophagosomes. Indeed, this may explain why LC3-positive autophagosomes appear to be formed 'de novo' and do not have the same protein compositions as other intracellular membranes, such as the ER. It will be interesting to consider whether different membrane sources have specific contributions in response to different autophagy-inducing stimuli, if and how the different membrane sources may act cooperatively, or whether these sources may have different contributions to specialised forms of autophagy, including autophagy of mitochondria (mitophagy), ER-phagy, and autophagy of bacteria.

Future studies should ideally strive to test the importance of different membrane sources under different conditions and in different cell types. While it is easier to investigate the role of the plasma membrane than other sources by blocking endocytosis, we still need to try to rigorously test all pathways to exclude alternative possibilities, including the possibility that pre-phagophores/phagophores/autophagosomes may 'bump into' other compartments and exchange membrane in a manner that is neither productive nor deleterious, or that autophagosomes/autophagosome precursors may be involved in membrane repair of other organelles.

It is thought that autophagosomes are formed by the delivery of vesicles from different sources leading to the formation of phagophores varying in size between 300 and 900 nm that ultimately are sealed to form double-membraned vesicles. The outer membrane then fuses with the lysosomal membrane, whereas the inner vesicle containing the sequestered cargo is degraded within the lysosomal lumen. Importantly, the two membranes are expected to be asymmetric: the outer membrane is similar in its protein and lipid content to the lysosomal limiting membrane, while the inner membrane contains a small amount of membrane proteins and a lipid composition that renders it amenable to rapid lysosomal degradation. The mechanism responsible for the formation of such asymmetry is poorly understood. Such differential lipid compositions may help explain how phagophore membrane curvature is initiated (given that this is a double membrane). However, at the moment we cannot exclude the possibility that the membrane curvature may be initiated spontaneously and that this serves as a mechanism to enable subsequent inner versus outer membrane differentiation.

Finally, another key issue is to understand how yeast autophagy processes differ from those seen in mammalian cells. Despite the high degree of conservation between mammalian and yeast Atg proteins and conservation of key aspects of the core processes, certain important differences between yeast and mammalian systems are already known, including the fact that yeast has one PAS and one Atg8 whereas there are multiple PAS and a family of Atg8 proteins in mammalian cells. Thus, it would not be too surprising if there were differences in other aspects of the autophagy processes between yeast and mammalian systems. Likewise, various aspects of mammalian autophagy may even vary in different human cell types.

ACKNOWLEDGEMENTS

Z.E. is the incumbent of the Harold Korda Chair of Biology. We are grateful for funding from the Legacy Heritage Fund (Z.E.), the Israeli Science Foundation ISF (Z.E.), the German Israeli Foundation GIF (Z.E.), the Louis Brause Philanthropic Fund (Z.E.), Wellcome Trust Senior Clinical Research Fellowship (D.C.R.), an MRC Programme grant (D.C.R.), Wellcome Trust/MRC Strategic award in Neurodegeneration (D.C.R.), and NIHR Biomendical Research Centre at Addenbrooke's Hospital (D.C.R.).

REFERENCES

1 Weidberg, H., Shvets, E., and Elazar, Z. (2011). Biogenesis and cargo selectivity of autophagosomes. Annu. Rev. Biochem. *80*, 125–156.

2 Tsukada, M., and Ohsumi, Y. (1993). Isolation and characterization of autophagy-defective mutants of Saccharomyces cerevisiae. FEBS Lett. *333*, 169–174.

3 Harding, T.M., Morano, K.A., Scott, S.V., and Klionsky, D.J. (1995). Isolation and characterization of yeast mutants in the cytoplasm to vacuole protein targeting pathway. J. Cell Biol. *131*, 591–602.

4 Thumm, M., Egner, R., Koch, B., Schlumpberger, M., Straub, M., Veenhuis, M., and Wolf, D.H. (1994). Isolation of autophagocytosis mutants of Saccharomyces cerevisiae. FEBS Lett. *349*, 275–280.

5 Ravikumar, B., Sarkar, S., Davies, J.E., Futter, M., Garcia-Arencibia, M., Green-Thompson, Z.W., Jimenez-Sanchez, M., Korolchuk, V.I., Lichtenberg, M., Luo, S., et al. (2010). Regulation of mammalian autophagy in physiology and pathophysiology. Physiol. Rev. *90*, 1383–1435.

6 Metcalf, D.J., Garcia-Arencibia, M., Hochfeld, W.E., and Rubinsztein, D.C. (2010). Autophagy and misfolded proteins in neurodegeneration. Exp. Neurol., epub ahead of print.

7 Suzuki, K., Kirisako, T., Kamada, Y., Mizushima, N., Noda, T., and Ohsumi, Y. (2001). The pre-autophagosomal structure organized by concerted functions of APG genes is essential for autophagosome formation. EMBO J. *20*, 5971–5981.

8 Suzuki, K., Kubota, Y., Sekito, T., and Ohsumi, Y. (2007). Hierarchy of Atg proteins in pre-autophagosomal structure organization. Genes Cells *12*, 209–218.

9 Geng, J., and Klionsky, D.J. (2008). The Atg8 and Atg12 ubiquitin-like conjugation systems in macroautophagy. 'Protein modifications: beyond the usual suspects' review series. EMBO Rep. 9, 859–864.

10 Shpilka, T., Weidberg, H., Pietrokovski, S., and Elazar, Z. (2011). Atg8: an autophagy-related ubiquitin-like protein family. Genome Biol. *12*, 226.

11 Tanida, I. (2011). Autophagy basics. Microbiol. Immunol. *55*, 1–11.

12 Axe, E.L., Walker, S.A., Manifava, M., Chandra, P., Roderick, H.L., Habermann, A., Griffiths, G., and Ktistakis, N.T. (2008). Autophagosome formation from membrane compartments enriched in phosphatidylinositol 3-phosphate and dynamically connected to the endoplasmic reticulum. J. Cell Biol. *182*, 685–701.

13 Hayashi-Nishino, M., Fujita, N., Noda, T., Yamaguchi, A., Yoshimori, T., and Yamamoto, A. (2009). A subdomain of the endoplasmic reticulum forms a cradle for autophagosome formation. Nat. Cell Biol. *11*, 1433–1437.

14 Matsunaga, K., Morita, E., Saitoh, T., Akira, S., Ktistakis, N.T., Izumi, T., Noda, T., and Yoshimori, T. (2010). Autophagy requires endoplasmic reticulum targeting of the PI3-kinase complex via Atg14L. J. Cell Biol. *190*, 511–521.

15 Yla-Anttila, P., Vihinen, H., Jokitalo, E., and Eskelinen, E.L. (2009). 3D tomography reveals connections between the phagophore and endoplasmic reticulum. Autophagy *5*, 1180–1185.

16 Hailey, D.W., Rambold, A.S., Satpute-Krishnan, P., Mitra, K., Sougrat, R., Kim, P.K., and Lippincott-Schwartz, J. (2010). Mitochondria supply membranes for autophagosome biogenesis during starvation. Cell *141*, 656–667.

17 Gomes, L.C., Di Benedetto, G., and Scorrano, L. (2011). During autophagy mitochondria elongate, are spared from degradation and sustain cell viability. Nat. Cell Biol. *13*, 589–598.

18 Ravikumar, B., Moreau, K., Jahreiss, L., Puri, C., and Rubinsztein, D.C. (2010). Plasma membrane contributes to the formation of pre-autophagosomal structures. Nat. Cell Biol. *12*, 747–757.

19 Reggiori, F. (2006). 1. Membrane origin for autophagy. Curr. Top. Dev. Biol. *74*, 1–30.

20 Reggiori, F., Shintani, T., Nair, U., and Klionsky, D.J. (2005). Atg9 cycles between mitochondria and the pre-autophagosomal structure in yeasts. Autophagy *1*, 101–109.

21 Mari, M., Griffith, J., Rieter, E., Krishnappa, L., Klionsky, D.J., and Reggiori, F. (2010). An Atg9-containing compartment that functions in the early steps of autophagosome biogenesis. J. Cell Biol. *190*, 1005–1022.

22 Nair, U., Jotwani, A., Geng, J., Gammoh, N., Richerson, D., Yen, W.L., Griffith, J., Nag, S., Wang, K., Moss, T., *et al.* (2011). SNARE proteins are required for macroautophagy. Cell *146*, 290–302.

23 Reggiori, F., and Klionsky, D.J. (2006). Atg9 sorting from mitochondria is impaired in early secretion and VFT-complex mutants in Saccharomyces cerevisiae. J. Cell Sci. *119*, 2903–2911.

24 Geng, J., Nair, U., Yasumura-Yorimitsu, K., and Klionsky, D.J. (2010). Post-golgi sec proteins are required for autophagy in Saccharomyces cerevisiae. Mol. Biol. Cell *21*, 2257–2269.

25 van der Vaart, A., Griffith, J., and Reggiori, F. (2010). Exit from the golgi is required for the expansion of the autophagosomal phagophore in yeast Saccharomyces cerevisiae. Mol. Biol. Cell *21*, 2270–2284.

26 Ohashi, Y., and Munro, S. (2010). Membrane delivery to the yeast autophagosome from the Golgi-endosomal system. Mol. Biol. Cell *21*, 3998–4008.

27 Young, A.R., Chan, E.Y., Hu, X.W., Kochl, R., Crawshaw, S.G., High, S., Hailey, D.W., Lippincott-Schwartz, J., and Tooze, S.A. (2006). Starvation and ULK1-dependent cycling of mammalian Atg9 between the TGN and endosomes. J. Cell Sci. *119*, 3888–3900.

28 Webber, J.L., and Tooze, S.A. (2010). Coordinated regulation of autophagy by p38alpha MAPK through mAtg9 and p38IP. EMBO J. *29*, 27–40.

29 Winslow, A.R., Chen, C.W., Corrochano, S., Acevedo-Arozena, A., Gordon, D.E., Peden, A.A., Lichtenberg, M., Menzies, F.M., Ravikumar, B., Imarisio, S., et al. (2010). alpha-Synuclein impairs macroautophagy: implications for Parkinson's disease. J. Cell Biol. *190*, 1023–1037.

30 Moreau, K., Ravikumar, B., Renna, M., Puri, C., and Rubinsztein, D.C. (2011). Autophagosome precursor maturation requires homotypic fusion. Cell *146*, 303–317.

31 Abeliovich, H., Dunn, W.A., Jr., Kim, J., and Klionsky, D.J. (2000). Dissection of autophagosome biogenesis into distinct nucleation and expansion steps. J. Cell Biol. *151*, 1025–1034.

32 Xie, Z., Nair, U., and Klionsky, D.J. (2008). Atg8 controls phagophore expansion during autophagosome formation. Mol. Biol. Cell *19*, 3290–3298.

33 Nakatogawa, H., Ichimura, Y., and Ohsumi, Y. (2007). Atg8, a ubiquitin-like protein required for autophagosome formation, mediates membrane tethering and hemifusion. Cell *130*, 165–178.

34 Weidberg, H., Shpilka, T., Shvets, E., Abada, A., Shimron, F., and Elazar, Z. (2011). LC3 and GATE-16 N termini mediate membrane fusion processes required for autophagosome biogenesis. Dev. Cell *20*, 444–454.

35 Kim, J., Huang, W.P., and Klionsky, D.J. (2001). Membrane recruitment of Aut7p in the autophagy and cytoplasm to vacuole targeting pathways requires Aut1p, Aut2p, and the autophagy conjugation complex. J. Cell Biol. *152*, 51–64.

36 Krick, R., Bremer, S., Welter, E., Schlotterhose, P., Muehe, Y., Eskelinen, E.L., and Thumm, M. (2010). Cdc48/p97 and Shp1/p47 regulate autophagosome biogenesis in concert with ubiquitin-like Atg8. J. Cell Biol. *190*, 965–973.

37 Chakrama, F.Z., Seguin-Py, S., Le Grand, J.N., Fraichard, A., Delage-Mourroux, R., Despouy, G., Perez, V., Jouvenot, M., and Boyer-Guittaut, M. (2010). GABARAPL1 (GEC1) associates with autophagic vesicles. Autophagy *6*, 495–505.

38 Kabeya, Y., Mizushima, N., Ueno, T., Yamamoto, A., Kirisako, T., Noda, T., Kominami, E., Ohsumi, Y., and Yoshimori, T. (2000). LC3, a mammalian homologue of yeast Apg8p, is localized in autophagosome membranes after processing. EMBO J. *19*, 5720–5728.

39 Kabeya, Y., Mizushima, N., Yamamoto, A., Oshitani-Okamoto, S., Ohsumi, Y., and Yoshimori, T. (2004). LC3, GABARAP and GATE16 localize to autophagosomal membrane depending on form-II formation. J. Cell Sci. *117*, 2805–2812.

40 Tanida, I., Ueno, T., and Kominami, E. (2004). LC3 conjugation system in mammalian autophagy. Int. J. Biochem. Cell Biol. *36*, 2503–2518.

41 Weidberg, H., Shvets, E., Shpilka, T., Shimron, F., Shinder, V., and Elazar, Z. (2010). LC3 and GATE-16/GABARAP subfamilies are both essential yet act differently in autophagosome biogenesis. EMBO J. *29*, 1792–1802.

rent Biology

Organelle Growth Control through Limiting Pools of Cytoplasmic Components

Nathan W. Goehring*, Anthony A. Hyman

Max Planck Institute of Molecular Cell Biology and Genetics (MPI-CBG), 01307 Dresden, Germany

*Correspondence: goehring@mpi-cbg.de

Current Biology, Vol. 22, No. 9, R330–R339, May 8, 2012 © 2012 Elsevier Inc.
http://dx.doi.org/10.1016/j.cub.2012.03.046

SUMMARY

The critical importance of controlling the size and number of intracellular organelles has led to a variety of mechanisms for regulating the formation and growth of cellular structures. In this review, we explore a class of mechanisms for organelle growth control that rely primarily on the cytoplasm as a 'limiting pool' of available material. These mechanisms are based on the idea that, as organelles grow, they incorporate subunits from the cytoplasm. If this subunit pool is limited, organelle growth will lead to depletion of subunits from the cytoplasm. Free subunit concentration therefore provides a measure of the number of incorporated subunits and thus the current size of the organelle. Because organelle growth rates are typically a function of subunit concentration, cytoplasmic depletion links organelle size, free subunit concentration, and growth rates, ensuring that as the organelle grows, its rate of growth slows. Thus, a limiting cytoplasmic pool provides a powerful mechanism for size-dependent regulation of growth without recourse to active mechanisms to measure size or modulate growth rates. Variations of this general idea allow not only for size control, but also cell-size-dependent scaling of cellular structures, coordination of growth between similar structures within a cell, and the enforcement of singularity in structure formation, when only a single copy of a structure is desired. Here, we review several examples of such mechanisms in cellular processes as diverse as centriole duplication, centrosome and nuclear size control, cell polarity, and growth of flagella.

CellPress

INTRODUCTION

In a series of pioneering experiments, E.G. Conklin [1] provided key insights into how the size of subcellular organelles is set by cytoplasmic determinants. A general correlation was known to exist between the size of a cell and the size of its internal components, including the nucleus, the spindle, and centrosomes (Figure 1A; reviewed in [2]). However, Conklin found that it was not the physical dimensions of the cell, but the amount of associated cytoplasm that was the dominant factor in specifying organelle size. By centrifuging embryos of the sea snail *Crepidula plana*, he was able to induce the separation of yolk from cytoplasm. Because these two phases were often differentially inherited during cleavage, the amount of cytoplasm in the resulting blastomeres was no longer proportional to cell volume (Figure 1B).

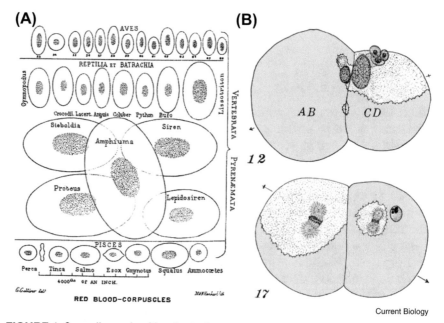

Current Biology

FIGURE 1 Organelles scale with cell cytoplasm.
(A) Sketches of red blood cells taken from various vertebrate species by George Gulliver (1875) highlighting the scaling of nucleus to cell size. Reproduced with permission from [80]. (B) Sketch of two centrifuged *Crepidula* two-cell embryos by E.G. Conklin (1912). Due to centrifugation along the indicated axis (arrows) during cleavage, cell size and the amount of cytoplasm (dotted) and yolk (brown) are not proportional in the two blastomeres. The size of spindles (green), nuclei (purple), and centrosomes or 'spheres' (yellow) depend on the volume of cytoplasm in which they find themselves rather than the physical dimensions of the cell. Adapted with permission from [1]. Note that the organelles do not scale in strict proportion to cytoplasmic volume, possibly reflecting the possibility that other factors may also come into play, e.g. an upper size limit on the mitotic spindle [29].

In these blastomeres, the size of the nucleus, the spindle, and the centro-somes did not scale with the physical dimensions of the cell, but with the amount of cytoplasm in which they were suspended, indicating that the volume of cytoplasm was playing a direct role in setting the size of organelles. But how could cytoplasm regulate the size of cellular structures, let alone provide mechanisms to both measure the size of structures and alter their growth rates accordingly?

Nearly a century after Conklin first posed this question, answers are beginning to emerge. In this review, we explore how these and other examples of organelle growth control can be explained by a consideration of the cytoplasm as a pool of available building blocks. We highlight a general class of mechanisms, which rely on either local or cell-wide depletion of this cytoplasmic pool as a way to regulate the size and number of cellular structures. As long as subunit amounts are limiting, the growth of intracellular structures necessarily leads to depletion of cytoplasmic subunit pools. In turn, reduction in the concentration of free subunits reduces growth rates. Ultimately, this coupling between organelle size, cytoplasmic concentration, and growth rates enables robust mechanisms for size-dependent and number-dependent control of organelle growth in cells. At the same time, because they depend on depletion of diffusible cytoplasmic subunits, these mechanisms require careful consideration of appropriate length and time scales, as well as mechanisms to accurately specify protein concentrations in cells.

HOW CAN THE CYTOPLASM PROVIDE DYNAMIC SIZE CONTROL?

Cytoplasmic Concentrations Govern Assembly Kinetics

The primary role of the cytoplasm in the assembly of cellular structures is to provide a pool of available subunits from which to draw. Typically, the rate of product formation by a chemical reaction is a function of substrate concentration. As a consequence, the growth of a structure such as a microtubule depends on the concentration of its component parts, in this case, free tubulin dimers [3]. Thus, the concentration of subunits in the cytoplasm can directly influence the assembly rate of cellular structures.

Limiting Pools Allow Size-Dependent Growth Rates without Measuring

Just as knowing only the velocity of an object tells one nothing about how far it will travel, in the absence of other constraints, knowledge of the rate of assembly of a structure tells one nothing about its ultimate size. Assembly must therefore be constrained to achieve a target size. Growth can be limited by simple time or physical constraints (Figure 2A,B) or it may reflect a dynamic balance, in which assembly competes with a size-dependent disassembly process (Figure 2C). Recognition that the cytoplasm is finite raises yet another simple mechanism of exerting size-dependent growth control, a so-called 'limiting pool' mechanism (Figure 2D). If the supply of subunits is limited, assembly of a structure will by its very nature deplete the cytoplasmic subunit pool, reducing the cytoplasmic concentration. Consequently, the rate of subunit incorporation will decrease as the structure grows larger. Thus, a limiting pool provides size-dependent control of growth rates without requiring the cell to either measure the current size of a structure or to actively modify assembly rates, for example, through changes in protein activity, protein synthesis, or degradation. Instead, limiting pool models are inherently self-correcting: deviations away from the characteristic size of a structure result in changes in the assembly–disassembly balance that naturally drive the system back towards steady state. For example, if a fully grown structure is severed, subunits are returned to the pool, stimulating growth.

Natural Scaling of a Structure to Cell Size

Another advantage of limiting pools is that they can account for the scaling of structures in cells of different sizes. Consider the case of two cells that possess an identical initial concentration of subunits for a given structure, but differ in size. Because the starting subunit concentration is the same, the initial growth rates of the structures in each cell will also be the same. As the individual structures grow and cytoplasmic pools are depleted, growth rates in both cells will slow. However, because the smaller cell possesses a proportionally smaller pool of subunits (*concentration · volume*) than the larger cell, its subunit concentration will decline faster as the structure grows. Therefore, growth of the structure in the smaller cell will slow more rapidly and plateau at a smaller final size (Figure 2E). Thus, when pools are limiting and the initial subunit concentration is equal across cells, a structure scales to the size of the cell in which it is contained.

Coordination of Growth between Multiple Structures

What about multiple structures growing in a single cell, and therefore sharing a common cytoplasmic subunit pool? If the diffusivity of subunits in the cytoplasm is sufficiently high, local depletion around a growing structure will cause a cell-wide reduction in subunit concentration. In such a regime, the concentration of the cytoplasmic pool provides a measure of the total number of subunits incorporated within individual structures across the cell, thereby providing an indirect measure of the combined total number and size of these structures. In the simplest case,

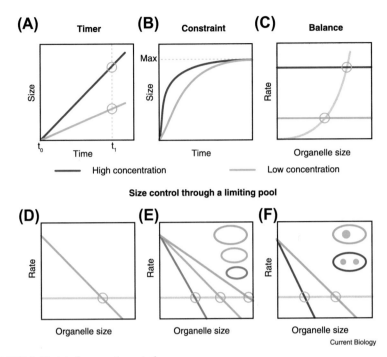

FIGURE 2 Models for growth control.
(A–C) Mechanisms for size control that do not rely on cytoplasmic depletion. In these models, we assume concentration is fixed, resulting in a constant, concentration-dependent assembly rate. (A) Timer: because higher subunit concentrations allow faster assembly, a cell with a higher subunit concentration (red line) will show faster assembly rates compared to a cell with a lower subunit concentration (blue line). Thus, over a given timespan (t_0–t_1, dashed gray line), the cell with a higher concentration can assemble a larger structure. Orange circles indicate the size of each condition at t_1. (B) Physical constraint: a system with a higher concentration will grow faster, but physical constraints limit the maximum achievable size (Max). (C) Size-dependent disassembly: steady-state size is specified by the point at which a size-dependent disassembly rate (gray line) precisely balances the assembly rate (red/blue lines). If the concentration is increased, thereby increasing the rate of assembly, the balance point (shown as an orange circle) shifts to a higher size. See [81] for an example of such a mechanism. (D–F) Size control through depletion of a limiting cytoplasmic pool. As in (C), orange circles indicate the balance point where assembly and disassembly are balanced. (D) Growth of a structure depletes the cytoplasm, thereby reducing the assembly rate (orange). Steady-state size is given by the point at which the rate of assembly exactly matches the rate of disassembly (gray). (E) As in (D), but for three cells containing identical subunit concentrations but differing in volume. Because total pool scales with cell size, for a given increase in size, a smaller cell suffers a proportionally larger depletion of the pool, resulting in a smaller steady-state size. (F) Competition for a limiting pool: if multiple structures compete for subunit from a common pool (red), the pool will be depleted more rapidly relative to a single, isolated structure given an identical cytoplasmic pool (green). Here, total size, independent of number (N), is constant, with mean size scaling as 1/N [63].

if all structures draw from the same cytoplasm, the total combined size of all structures will be limited by the total subunit pool (Figure 2F). Once that pool is depleted, a structure can only grow if another disassembles, ensuring the total combined amount of incorporated subunits remains constant. As we shall see later in other contexts, this notion of competition opens the door to more complex phenomena, such as size equalization between spatially separated structures (flagella) and regulation of structure number through effects on the nucleation of structure assembly (centrioles, cell polarity).

Thus, a simple consideration of the cytoplasm as a finite and potentially limiting pool of subunits for the assembly of cellular structures reveals robust mechanisms for limiting the size of cellular structures, scaling structures to cell size, and coordinating the growth of multiple structures within a single cell. In the following sections, we illustrate the potential impact of limiting pools in several classic examples of size control.

CONTROL OF CENTROSOME SIZE

One clear example of a limiting pool mechanism controlling the size of a cellular organelle is the regulation of centrosome size. The scaling of the centrosome with cell size was recognized by the first decades of the 20th century [1,2]. One purpose of centrosome scaling, at least in *Caenorhabditis elegans*, is to adjust the length of the mitotic spindle to match the rapid decline in cell size that occurs during early embryonic cell divisions. During these divisions, which result in a 558-cell larva that is no bigger than the one-cell zygote [4], metaphase spindle length declines in a manner that correlates with centrosome size [5,6] (Figure 3A). Although the precise mechanism by which centrosomes specify spindle length remains unclear, a direct link was demonstrated by experimental reduction of centrosome size through partial depletion of the centrosome assembly factor SAS-4, which led to corresponding changes in spindle length [5,7]. This observation, however, begs the question of how centrosomes are able to scale with cell size.

Centrosomes are composed of pericentriolar material (PCM), the assembly of which is templated by a pair of centrioles, which specify both the number of centrosomes and where they will form [8]. Centrosome growth occurs through a process of maturation, in which PCM gradually accumulates around centrioles, reaching a peak size during mitosis, when the centrosomes direct organization of the mitotic spindle [9]. The centrosomes then fragment and disassemble during cytokinesis, returning PCM components to the cytoplasm. Presumably, cells must therefore regulate PCM recruitment to ensure formation of centrosomes of the proper size.

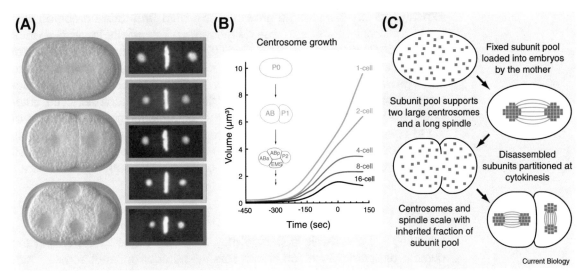

FIGURE 3 Centrosome scaling via a limiting component.
(A) Spindle size and centrosome size scale with cell size during early development of *C. elegans*. (Assembled from images courtesy of G. Greenan.) (B) Centrosome growth is limited by cell volume. Modified with permission from [6]. (C) Schematic of a limiting pool model for centrosome and spindle scaling during development. A fixed precursor pool is loaded into the embryo, which enables growth of two centrosomes in the one-cell embryo. Centrosome disassembly returns precursor to the cytoplasm. Thus, each daughter receives a precursor pool with a size proportional to volume. As a result, the centrosomes in the daughters will scale with cell size. Because centrosome size governs spindle size, the spindles of the two-cell embryo are smaller than in the one-cell embryo.

In a modern re-examination of Conklin's experiments, recent work in *C. elegans* development [6] showed that, as cells become smaller, the growth rate of centrosomes slows, resulting in smaller centrosomes (Figure 3B). Importantly, centrosomes in blastomeres of abnormally small embryos were smaller relative to those in identical blastomeres in normal-sized embryos, consistent with centrosome size being limited not by cell identity but by the volume of cytoplasm. The total volume of all centrosomes in the embryo was also conserved, both through development or in cells manipulated to contain ectopic centrosomes, matching what one would expect if centrosomes are competing for the same pool of PCM components (as in Figure 2F). Thus, a fixed pool of PCM components is loaded into the oocyte, which is then partitioned between cells at each division, with each cell inheriting a pool of PCM components in direct proportion to its volume (Figure 3C). In small cells, the correspondingly smaller PCM pool is depleted more quickly, resulting in smaller centrosomes. Supporting this hypothesis, centrosomes could be induced to grow larger by increasing the pool of available PCM through overexpression of the centriole/centrosome component SPD-2 [6].

Experiments on centrosome growth in syncytial *Drosophila* embryos also found that centrosome size was dependent on the rate of incorporation of the core centrosome component Cnn [10]. Doubling or halving the levels of Cnn yielded corresponding changes in centrosome size. However, centrosome growth did not exhibit any plateau, but rather grew steadily throughout S phase before abruptly ceasing upon mitotic entry. Thus, a cell-cycle timer rather than pool depletion appears to limit centrosome growth. It is tempting to speculate that the extremely large volume of the syncytial embryo in which these nuclear divisions take place provides an effectively unlimited pool, and thus the embryo must rely on other growth control mechanisms.

CENTRIOLE DUPLICATION

Intriguingly, cytoplasmic depletion may also play a role in controlling the number of centrioles, which in turn sets the number of centrosomes. Control of centriole number requires that, during each cell cycle, a single new daughter centriole is nucleated in the immediate vicinity of each mother centriole in a process termed centriole duplication (reviewed in [11]). It is known that cells have sufficient material to form extra centrioles because the removal of existing daughter centrioles from around the mother centriole allows formation of additional daughter centrioles [12]. Thus, depletion of the overall cellular concentration does not limit centriole nucleation. Rather, daughter centrioles locally suppress nucleation events.

One mechanism by which growing daughter centrioles could inhibit nucleation of additional centrioles is through local, as opposed to global, depletion of centriole components around the mother centriole. In contrast to growth rate, nucleation of structures often exhibits strong, non-linear dependence on protein concentrations, leading to effective nucleation thresholds. Thus, one can clearly find concentration regimes where nucleation is effectively suppressed, but growth of existing structures is sustained. One could imagine that the local depletion of centriole components by growing centrioles could be sufficient to push concentrations below the nucleation threshold so that no new centrioles will be nucleated around the mother centriole, while still allowing the daughter centriole to grow. Consistent with this model, overexpression of centriole components, which should overwhelm the effects of local depletion, leads to formation of multiple daughter centrioles [13–15]. Also, as one might expect given the sensitivity of the system to concentration, the cellular concentration of at least one of these components, SAS-6, is tightly regulated [16]. Thus, we speculate that proper spindle morphology relies on a combination of global and local depletion of cytoplasmic protein pools to control, respectively, the size and number of centrosomes.

NUCLEAR SIZE CONTROL — LIMITING COMPONENT OR ACTIVE CONTROL?

As early as the late 19th century, biologists noted a striking correlation between the size of the nucleus and the size of the cell (Figure 1A; reviewed in [2]). The precise scaling relationship, known as the *kern–plasma* (nuclear–cytoplasm) ratio [17], has been the subject of considerable controversy. The ratio appears nearly constant in some systems and more variable in others. This is in part due to limits in obtaining precise measurements of nuclear and cytoplasmic volumes, complicated by the presence of yolk, vacuoles, and other organelles that displace cytoplasm. Moreover, in many systems, the nucleus often continues to grow right up to mitotic entry, meaning that nuclear size also depends on both the duration of interphase and the time at which it is measured [1]. Nonetheless, a general correlation between cell size and nuclear size appears nearly universal [18].

The mechanisms behind nuclear scaling remain largely unknown. One hypothesis, based on a general correlation between DNA amount and nuclear size, is that nuclear size is a function of DNA content [2,19]. However, in yeast, changes in ploidy do not result in direct changes in cell or nuclear size, and no specific increase in nuclear size was seen during S phase, as predicted by a DNA content model [18,20]. Moreover, a direct causal relationship between nuclear size and DNA content makes it difficult to explain why nuclei in different cells of the same organism are of different size [1].

An alternative hypothesis is that the size of the nucleus is a function of cell size. Supporting this view, if a nucleus is transferred from a small donor cell into a larger host cell, the transferred nucleus expands, responding to the size of the host cell (Figure 4A) [21,22]. In both budding and fission yeast, nuclear size increased proportionally with cell growth, maintaining a remarkably constant ratio of nuclear to cell volume over a broad range of cell sizes and growth conditions [18,20]. In fact, a cell-size model likely also explains many cases of coupling between DNA content and nuclear size. Haploid and polyploid cells are typically smaller and larger, respectively, compared with their diploid counterparts [2,23].

How does this work? Similar to the case of centrosomes, in centrifuged *Crepidula* embryos, nuclear size did not correlate with cell size, but rather with the amount of inherited cytoplasm (Figure 1B). This is consistent with a model in which a finite pool of cytoplasmic factors limits nuclear size [1]. Further support for this idea comes from cells containing multiple nuclei. When HeLa nuclei are injected into the cytoplasm of *Xenopus* oocytes, individual or sparsely spaced HeLa nuclei grow much larger than nuclei contained within large clumps of several hundred nuclei (Figure 4B) [21]. A qualitatively similar result was seen in cytokinesis-defective fission yeast, which harbor

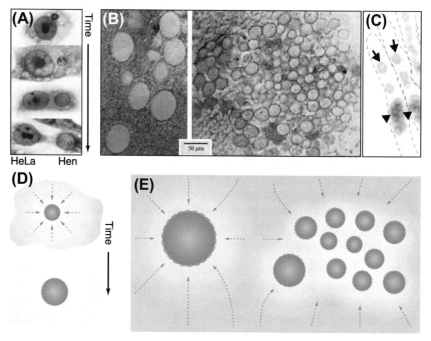

Current Biology

FIGURE 4 The kern-plasma ratio and scaling of nuclei.
(A) Diminutive hen erythrocyte nuclei injected into HeLa cells and fixed at various times post-injection. The hen erythrocyte (right) gradually increases in size until it almost reaches parity with the HeLa nucleus (left). Adapted with permission from [22]. (B) HeLa nuclei injections into *Xenopus* oocytes reveal that nuclei fail to grow substantially when competing with other nuclei in large clusters (right), compared with when nuclei are relatively isolated (left). Adapted with permission from [21]. (C) The same is true in cytokinesis mutants of *Schizosaccharomyces pombe* where nuclei in clusters (arrowheads) are smaller than more isolated nuclei (arrows) in the same cell. Adapted with permission from [18]. © 2007 Rockefeller University Press. Originally published in Journal of Cell Biology. 107:593–600. (D,E) Size control through depletion of a diffusion-limited precursor. Growth of a nucleus will tend to deplete precursor (green) from the local cytoplasm, which is replenished due to diffusion of precursor from elsewhere in the cell (arrows). In a single cell (D), this depletion will eventually reduce concentrations throughout the cytoplasm, limiting growth. (E) If total precursor is not limiting, for example, if cell volume greatly exceeds nuclear volume, or if precursor is constantly synthesized, there will be few limitations on the growth of isolated nuclei, allowing them to grow continuously unless limited by some other factor. However, the effects of depletion may still limit the growth of multiple, clustered nuclei if their combined local uptake sufficiently exceeds the rate of precursor diffusion. This effect will be particularly strong for nuclei at the center of clusters, since nuclei at the edges will tend to take up precursor before it can reach the center. Such a scenario would lead to a situation as in (B) where nuclei in the center of the cluster show almost no growth.

numerous irregularly spaced nuclei (Figure 4C) [18]. Thus, nuclei appear to be competing locally for some diffusion-limited component. Normally, as nuclei take up components, the local pool is replenished by the influx of components from elsewhere in the cell (Figure 4D). However, in nuclear clusters, the nuclei at the edge will take up components as they diffuse in from outside (Figure 4E). Consequently, there is insufficient flux of components into the center of the cluster, keeping the local concentration low and limiting growth. Strikingly, when clusters of HeLa nuclei were injected directly into the nucleus of unfertilized *Xenopus* oocytes, known as the germinal vesicle (GV), the HeLa nuclei expanded uniformly until constrained by the physical limits of the GV membrane [21]. This is presumably because the GV is enriched in whatever components are normally limiting for nuclear growth. Consistent with this interpretation, when the GV was ruptured, HeLa nuclei in the cytoplasm were able to attain a much larger size [21]. The precise nature of these size-limiting components remains unclear, although nuclear lamins are strong candidates due to their known role in nuclear expansion [24–26].

This limiting-pool mechanism for scaling the nucleus to cell size has obvious advantages during early development when cell size often changes rapidly. However, early animal development also presents several situations where a limiting-pool mechanism breaks down. One example is the early *Xenopus laevis* embryo. From the mid-blastula stage onward, the scaling between nuclear size and cell size is typical of what one would expect for a pool-limited model [27,28]. However, nuclear size control is very different within the initial four embryonic cell divisions, where nuclear size remains constant, despite a roughly 16-fold mean reduction in cell volume [27–29]. Why is the typical coupling between cell and nuclear size absent in these early cells? The answer may be a matter of size. *X. laevis* one-cell embryos exceed a diameter of 1.2 mm, ~30-fold larger than the 40–50 μm diameter nucleus typical of cells in these stages, equating to a 27,000-fold difference in volume. Thus, the pool of nuclear components is in such excess that the formation of a nucleus has little effect on the concentration of components in the cytoplasm. Consistent with the pool of components being effectively unlimited, these early nuclei grow unchecked until the nuclear envelope breaks down at mitosis [28]. Contrast this to the mid-blastula where the ratio of nuclear to cell volume is closer to 50. In these smaller cells, nuclear growth reaches steady state during the cell cycle. Moreover, in these later stages, the total volume of all nuclei in the embryo is constant between stages, consistent with a limiting amount of precursor being partitioned into each cell according to volume [27]. Mitotic spindle length in *X. laevis* shows a similar pattern, with scaling to cell size absent in early cell divisions, but evident in embryos entering the mid-blastula stage and beyond [29].

The large size of cells in the early *Xenopus* embryo appears to have prompted development of alternative mechanisms of nuclear size control. Recently, nuclear size was found to be significantly different in both early stage embryos and in oocyte extract from *X. laevis* compared with its smaller relative *X. tropicalis*. These differences were traced to altered rates of import of lamin B [28], an essential architectural component of the nuclear envelope known to be required for nuclear growth [24,25]. The rates of lamin B import were found, in turn, to be set by the levels of importin-α. *X. tropicalis* embryos have lower levels of importin-α, and thus nuclei in *X. tropicalis* grow more slowly than in *X. laevis*. Consequently, *X. tropicalis* nuclei are smaller in size at the onset of mitosis.

Interestingly, this same work also identified lamin B as a potential limiting cytoplasmic component for controlling nuclear size scaling in later stage *Xenopus* embryos. Specifically, increasing lamin B or its import by importin-α in mid-blastula cells allowed nuclei to reach larger sizes before plateauing [28]. Thus, in *Xenopus* there appear to be two size-control regimes, each involving a limitation on the ability of the nucleus to take up lamin B. In large cells, where nuclear growth is not limited by cytoplasmic depletion and nuclei do not scale with cell size, nuclear size is set through a timer mechanism (Figure 2A). Maximum nuclear size attained is a function of the rate of lamin B import and the duration of the interphase period prior to nuclear envelope breakdown. As cell size decreases, cells enter a pool-limiting regime in which depletion of the cytoplasmic pool of lamin B and possibly other components begins to limit growth in a size-dependent fashion, resulting in scaling of nuclei with cell size.

It is tempting to speculate about what would happen to nuclei in the early *Xenopus* embryo if the cell-cycle timer were relaxed and nuclei were allowed to grow over much longer timescales. Intriguingly, the GV (the nucleus of the unfertilized oocyte) reaches a size of nearly 400–500 μm, yielding a nuclear: cytoplasmic ratio much closer to that seen in smaller mid-blastula cells [21,27]. The key difference between the GV and the nucleus of the one-cell embryo may be that GV growth is not time limited. Rather, the GV expands over weeks to months, allowing it to reach sizes where cytoplasmic depletion could begin to limit its growth. Thus, the basic premise of pool limitation may still apply in large cells, so long as there is sufficient time for nuclear growth. However, in the early embryo, the demands for rapid divisions ensure that the duration of interphase is too short for nuclei to reach a size at which cytoplasmic pool limitation comes into play.

CELL POLARITY — GLOBAL CONTROL OF SIZE AND NUMBER

Cell polarity typically involves the formation of a membrane domain — or two opposing membrane domains — that orient the cell along a unique geometric axis. Once a polarity axis is established, the formation of additional polarity domains must be suppressed and the size of existing domains limited to prevent a domain from expanding to occupy the entire cell membrane, thus rendering the cell unpolarized. Although not exactly an assembly process, it raises similar issues of size control: the cell must somehow sense that a domain already exists, determine how big it is, and then adjust the polarization machinery accordingly to prevent both the expansion of existing polarity sites as well as the formation of new sites, all while allowing the existing structure to persist. Although such feedback control of domain size would appear complicated to engineer, a limiting cytoplasmic pool of a critical component provides an elegant and simple solution. Here we focus on the role of a limiting pool in two model systems, Cdc42 polarity in the yeast *Saccharomyces cerevisiae*, and PAR polarity in *C. elegans* (Figure 5), although the general themes we discuss apply to a variety of systems [30].

Cdc42 polarity and PAR polarity share several key features that combine to generate pattern-forming systems. First, the enrichment of a given polarity protein within a domain relies on auto-catalytic feedback that drives local enrichment within domains. Second, the local enrichment involves local conversion of signaling molecules from rapidly diffusing, inactive, cytoplasmic states to more slowly diffusing, active, membrane-associated states. This slow membrane diffusion allows membrane-associated species to be concentrated in space, while rapid cytoplasmic diffusion ensures that local autocatalysis has access to the total cytoplasmic pool of inactive molecules. Finally, there is at least one critical component for which the pool of available protein is limiting. In other words, the total pool is small enough that its recruitment to a polarity domain results in a decline in its cytoplasmic concentration, reducing its ability to be added to the membrane. The importance of feedback and diffusion in pattern-forming systems was first noted by Turing [31] (for an accessible discussion for the general reader see [32]).

Polarization of budding yeast in the absence of pre-existing landmarks depends on a positive feedback loop involving Cdc42 [33,34]. Currently, it is thought that active Cdc42 recruits its own activator, the GTP exchange factor Cdc24, via the scaffold Bem1, resulting in further local recruitment of active Cdc42 [35] (Figure 5B,C). This feedback allows small, local fluctuations of Cdc42 to be rapidly amplified, resulting in the accumulation of active Cdc42 to high levels within a local patch [36] (Figure 5A,B). However, because Cdc42 fluctuations can occur at any time and place within the cell, in the absence of a limiting factor the number of patches would simply increase over time, violating the requirement for a single polarity axis. The limiting factor appears to be the scaffold protein Bem1, which is recruited to the Cdc42 patch as it forms [37]. The eventual depletion of cytoplasmic Bem1 reduces the strength of the positive feedback as the patch expands, limiting further accumulation of Cdc42 to the patch [36]. In addition, because

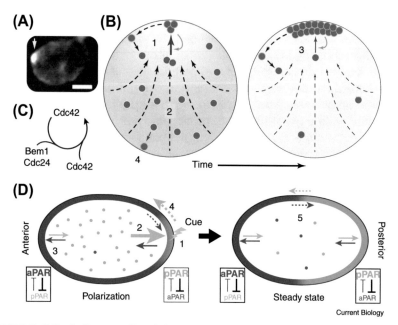

FIGURE 5 Cell polarity: a question of size and number.

(A) Polar Cdc42 cap in yeast. Modified from [38]. (B) Generic scheme for cell polarization of a single component based on Cdc42 polarization: (1) Active signaling molecules at the membrane recruit additional molecules from the cytoplasm. These molecules will diffuse away from the domain and eventually return to the cytoplasm due to spontaneous dissociation or internalization. (2) Local cytoplasmic depletion by the growing domain induces diffusion of molecules from elsewhere in the cell, leading to global depletion of the pool. (3) As the pool is depleted, recruitment by the existing domain slows, leading to stalling of domain growth. (4) At initial time points, high cytoplasmic concentrations may permit nucleation of additional polarity domains. However, the large domain can outcompete the newly formed domain for subunits in a 'winner takes all' situation. As the cytoplasm is depleted, nucleation events become increasingly difficult. (C) Self-amplifying feedback loop: active Cdc42 recruits Bem1 which in turn stimulates further recruitment and activation of Cdc42. (D) Schematic of PAR polarity establishment in *C. elegans*. Polarization involves the formation of two opposing polarity domains, an anterior PAR (aPAR)-dominant domain (red) characterized by net aPAR association and net dissociation of posterior PARs (pPAR), and the converse pPAR-dominant domain characterized by net pPAR association and net aPAR dissociation (blue). Polarization is triggered through a local cue (1) that induces a small pPAR domain. Because cytoplasmic pPAR concentration is high, further addition of pPAR is strongly favored (2). By contrast, a substantial fraction of aPAR is already on the membrane, reducing the pool of available aPAR in the cytoplasm. Thus, accumulation of additional aPAR is less favored (3). Thus, at the onset of polarization the system strongly favors recruitment of pPAR and expansion of the pPAR domain (4). As the pPAR domain grows, the pool of cytoplasmic pPAR is depleted. Simultaneously, the aPAR domain shrinks and aPARs are returned to the cytoplasm. This process eventually reduces the rate of pPAR association and increases the rates of aPAR association until the tendency of the two domains to expand is equalized (5).

depletion reduces cytoplasmic Bem1 concentrations throughout the cell, the probability of stochastic Cdc42 fluctuations being sufficiently large to form a stable second patch drops dramatically. If a second domain is able to form, competition between the two patches for available Bem1 leads to a 'winner takes all' scenario in which only one domain can persist [36,37]. Consistent with this model, overexpression of Bem1, which would reduce the ability of a single focus to deplete the cytoplasmic pool, increased the frequency of multiple Cdc42 patches. (For a recent review of these basic features of the Cdc42 polarity network, see [38].)

In PAR polarity, domain formation is driven by reciprocal negative feedback between two antagonistic groups of PAR proteins (anterior PARs and posterior PARs). Each is enriched on the membrane in the absence of the other, and is capable of displacing the opposing group from the membrane [39,40]. Consequently, a local advantage of one PAR species over the other will tend to be amplified: by displacing its antagonist, a given PAR species enhances its own enrichment at the membrane. Thus, reciprocal negative feedback functions similarly to the single positive feedback loop described for Cdc42. Because of this feedback, the membrane will tend towards one of two states — an anterior-like, high-anterior PAR/low-posterior PAR state, or the reciprocal posterior-like, low-anterior PAR/high-posterior PAR state (Figure 5D). Such a model of reciprocal negative feedback permits segregation of the membrane into domains [41–44], but there is no a priori reason why one domain would be favored over the other, let alone ensure that the cell is divided into two, roughly equal-sized domains. Domain size control appears to rely on limiting pools of PAR protein. Cytoplasmic depletion of the posterior PAR protein PAR-2 was found to be coupled to domain expansion, and overexpression or underexpression of PAR-2 led to corresponding changes in the relative size of the two polarity domains [43]. Thus, a limiting pool of PAR protein ensures that, as a PAR domain expands, it depletes the supply of components required for its further growth. It is currently unclear whether this pool is also necessary to enforce singularity in the system, which may be less of a problem in the C. elegans embryo, given that polarization relies on a polarity cue acting at a single site [45,46].

A survey of current cell polarity models reveals that depletion of a cytoplasmic pool is frequently invoked to limit expansion of polarity domains [30,36,42,47–49]. However, there are relatively few cases beyond Bem1-mediated polarity in yeast and PAR polarity in C. elegans where the role for cytoplasmic depletion has been demonstrated experimentally. Both the robustness and flexibility of limiting pool mechanisms in regulating the size and number of polarity domains suggest a broad role for limiting cytoplasmic pools in polarizing systems.

FLAGELLAR LENGTH CONTROL — COMPETITION FOR LIMITING COMPONENTS

The function of eukaryotic flagella depends on their length [50]. In *Chlamydomonas*, efficient swimming requires not only that the flagella be of proper length to enable an optimal swim stroke but also that the two flagella be of equal length to promote forward-directed (rather than circular) motion [51]. Thus, length control mechanisms must exist both to specify an optimal length for an individual flagellum and to coordinate length between flagella.

The mechanisms governing flagellar length control have been the subject of several excellent reviews but are still the subject of some controversy [50,52]. Assembly and disassembly of subunits into the flagella occur continuously at the tip [53], with the net balance of assembly and disassembly setting the flagellar growth rate [54]. Measurements suggest that, while the rate of subunit dissociation is independent of length [54,55], the rate of incorporation is reduced as flagella lengthen [55,56]. As a consequence, assembly and disassembly will balance at precisely one length. This is the so-called balance point at which flagellar length is stable [55]. But how are assembly rates controlled in a length-dependent fashion? The answer does not appear to be depletion of the cellular pool of flagellar components during flagellar growth. Rather, the rate of subunit transport to the growing tip appears to slow as the flagellum lengthens [50,55,56]. How this change in transport rates occurs remains unclear. Some form of length-dependent feedback signal could be involved [52,57] and the recent discovery of length-dependent phosphorylation of an aurora-like kinase is tantalizing [58]. However, clear support for signaling-based feedback control of flagellar length remains lacking.

One factor that has not been sufficiently explored is the control of assembly rates by a limiting pool, because this does not appear to be the mechanism for specifying length [59]. However, the limiting nature of flagellar components can be revealed in flagella severing experiments. Normally, if both flagella are severed, they regrow in unison to their original lengths [60] (Figure 6A). However, if protein synthesis is blocked, flagella regrow to only half their original length [60,61] (Figure 6B). The flagella will even regrow after a second round of severing, but now to only a quarter of their original length. Thus, synthesis of flagella represents a significant drain on the cytoplasmic precursor pool. Without new synthesis, depletion of this pool eventually reduces assembly rates sufficiently to limit growth. It is important to note that flagella do not stop growing because they run out of precursor. In each case, a significant pool remains in the cell to support regrowth [60,62,63]. Rather, as previously suggested [54], in the absence of new synthesis, regrowth reduces the precursor concentration of the cytoplasmic pool, thus lowering assembly rates, leading to a lower steady-state size. Consistent with a

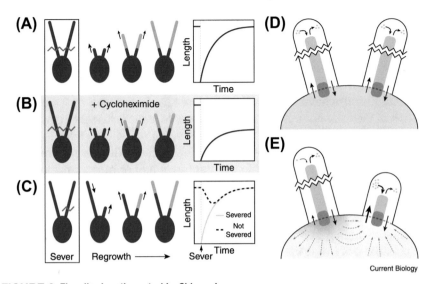

FIGURE 6 Flagellar length control in *Chlamydomonas*.
(A–C) Schematic summary of flagellar-severing experiments. Orange indicates new growth following severing. Graphs show flagellar length over time in each severing experiment. Time of severing is indicated by the dashed vertical line. (A) Normal regrowth of flagella following loss of both flagella. (B) If protein synthesis is blocked, flagella can regrow using the remaining cytoplasmic precursor pool from the cell body. Due to lack of new synthesis, this pool will be depleted, leading to stalling of growth at a shorter length. (C) If only one flagellum is severed, the uncut flagellum (dashed black line) initially shrinks, while the severed flagellum regrows (gray line). Once the two flagella reach equal length, their growth curves converge and they grow out together. (D) At steady state, assembly and disassembly are balanced, the length of the two flagella remains constant, and there are no concentration gradients of precursor since uptake and loss of components presumably remain balanced (black arrows). (E) After severing of one flagellum, the rapid regrowth of the shorter flagellum results in a dramatic increase in precursor uptake (large arrow). This increased uptake locally depletes subunits from the cytoplasm, creating a concentration gradient, which will cause precursor to diffuse toward the site of the regrowing flagellum (dashed arrows), eventually leading to global reductions in precursor concentration. As reduced cytoplasmic concentration begins to limit assembly rates, the balance between assembly and disassembly in the longer flagellum tips to favor disassembly. Consequently, the longer flagellum will shrink, resulting in a net donation of precursor to the cytoplasmic pool, where it can help fuel growth of the shorter flagellum.

general role for limiting pools, the length of the primary cilium in mammalian cells was strongly affected by changes in the availability of free tubulin [64]. However, at least in *Chlamydomonas*, the size-limiting effect of depletion is prevented due to tight regulatory coupling between flagellar growth and precursor synthesis, which ensures sufficiently large pools of precursor [65–67].

Evidence does suggest, however, that competition for a limiting pool underlies equalization of flagellar length. The equalization process can be observed if only one of the two flagella is sheared off at its base. The severed flagellum does not simply regrow to match the remaining one. Rather, as the severed flagellum begins to regrow, the longer flagellum shrinks until the two flagella are equal in size, after which both grow out in unison [54,60] (Figure 6C). How can we explain this rather striking result? From the above experiments, we know that the severing of a flagellum results in the loss of a significant fraction of the total pool of flagellar components, which, at least in the period before flagellar component synthesis is upregulated, will be limiting for flagellar growth. Also, because assembly rates decline with length, the newly shortened flagellum will take up precursor at a higher rate than the longer flagellum. As a consequence, if pools are limiting, the shorter flagellum can outcompete the longer flagellum, ensuring that the short flagellum will grow at the expense of the longer until size is equalized or the pool is sufficiently replenished such that precursor is no longer limiting [54] (Figure 6D,E). If the cytoplasmic pool of flagellar components were simply maintained at high levels such that precursor concentrations were not limiting, such a model would not be possible. Perhaps the selective advantage of rapid flagellar equalization has led to maintenance of flagellar component concentrations within a narrow window.

CONCLUDING REMARKS

The notion of a finite, limiting pool seems intuitive. With more building blocks available, more and larger structures can be assembled. This provides a simple mechanism for both competition between growing structures in the same cell, as well as cell-size-dependent scaling: a larger cell will typically contain more building blocks than a smaller cell. Importantly, subunits need not literally 'run out'. Rather, the simple reduction in cytoplasmic concentration that accompanies structure growth provides size-dependent regulation of assembly rates, enabling the size control, scaling, and competition mechanisms discussed here. As we continue to analyze the assembly of cellular structures, it is worth keeping such mechanisms in mind as the simplicity and adaptability of such mechanisms suggests that they will be ubiquitous.

Such mechanisms are very attractive in early embryonic development, which is typically characterized by rapid, abbreviated cell cycles in which cell divisions are not accompanied by cell growth. This allows for rapid increase in cell numbers, at the expense of an equally rapid decrease in cell size. In each of these cells, the size of organelles must be adjusted accordingly. Typically, the rapid pace of development necessitates that the embryo begins its life with a fixed pool of protein, loaded into the oocyte by the mother. New protein is often not synthesized until the embryonic development is well underway. Under the simplest scenario, cells simply receive a protein pool at cytokinesis that is proportional to cell size. For a limiting pool mechanism, the size of the assembled organelles is then simply a function of this pool. Obviously more elaborate mechanisms may, and likely do, exist. For example, we have seen how the extreme size of cells in the early *Xenopus* embryo appear to have necessitated the evolution of alternative size control strategies. Nonetheless, the simplicity of a limiting cytoplasmic pool as a mechanism for scaling and size control suggest that such mechanisms will be widespread in embryonic systems. Whether such mechanisms are dominant in somatic systems remains to be seen. Here, longer cell cycles would in theory allow more freedom to actively adjust cytoplasmic protein concentration, thus allowing greater regulatory control. Testing such ideas will require a greater understanding of how cells modulate protein concentrations.

Limiting pool models require that cells regulate protein synthesis and/or degradation to achieve precise protein levels. In embryonic systems, there is evidence that protein concentration is tightly controlled for key proteins. For instance, regulation of the centrosome-size determinant SPD-2 and the polarity protein PAR-2 appear to exhibit dosage regulation in *C. elegans* ([6] and our unpublished results). Quantitative analysis in yeast indicates that some level of gene dosage compensation exists for up to ~15% of genes, although <5% showed complete compensation [68]. Thus, the number of pathways involving active regulation of protein amounts through some form of feedback control is not negligible.

Indeed, autoregulatory feedback control of protein amounts is ubiquitous, particularly in the biogenesis of multicomponent complexes, where subunits are required in stoichiometric amounts. The coordination of ribosomal protein expression provided the paradigm [69–71], which has subsequently been implicated in regulating the amounts of proteins as diverse as splicing components [72], tubulin [73], and even an E2 ubiquitin ligase [74,75]. A key feature of all these regulatory systems is that the accumulation of excess, unincorporated molecules feeds back to either inhibit the synthesis of additional molecules and/or promote their own degradation, typically through post-transcriptional mechanisms.

While these examples provide mechanisms to coordinate the level of a protein with cellular demands, how cells precisely regulate protein concentrations remains unclear. For example, we know almost nothing about how the protein production machinery in the *C. elegans* gonad ensures that embryos are loaded with the correct concentrations of proteins, although techniques such as varying codon usage, which has strong effects on expression levels, may provide ways to distinguish among mechanisms [6,43,76]. Combining the tools of control theory and dynamical systems with the analysis of synthetic gene networks has provided substantial insight into the properties of regulatory networks that give rise to stability and gene dosage invariance of network output [77–79]. Such approaches will undoubtedly help provide a way forward. Yet, bridging the gap between idealized, simplified model systems to the mechanisms that underlie dosage control in developmental systems will be an important and non-trivial process.

ACKNOWLEDGEMENTS

We thank S. Reber, C. Brangwynne, O. Wüseke, S. Grill, and several anonymous reviewers for constructive comments on the manuscript. N.W.G. was supported by the Alexander von Humboldt Foundation and a Marie Curie Grant (219286) from the European Commission.

REFERENCES

1 Conklin, E. (1912). Cell size and nuclear size. J. Exp. Embryol. *12*, 1–98.

2 Wilson, E.B. (1925). The Cell in Development and Heredity, Third Edition. (New York: Macmillan).

3 Borisy, G.G., and Olmsted, J.B. (1972). Nucleated assembly of microtubules in porcine brain extracts. Science *177*, 1196–1197.

4 Wood, W.B. (1988). Chapter 8: Embryology. In The Nematode *Caenorhabditis Elegans*, W.B. Wood, ed. (Cold Spring Harbor: Cold Spring Harbor Laboratory).

5 Greenan, G., Brangwynne, C.P., Jaensch, S., Gharakhani, J., Jülicher, F., and Hyman, A.A. (2010). Centrosome size sets mitotic spindle length in Caenorhabditis elegans embryos. Curr. Biol. *20*, 353–358.

6 Decker, M., Jaensch, S., Pozniakovsky, A., Zinke, A., O'Connell, K.F., Zachariae, W., Myers, E., and Hyman, A.A. (2011). Limiting amounts of centrosome material set centrosome size in C. elegans embryos. Curr. Biol. *21*, 1259–1267.

7 Kirkham, M., Müller-Reichert, T., Oegema, K., Grill, S.W., and Hyman, A.A. (2003). SAS-4 is a C. elegans centriolar protein that controls centrosome size. Cell *112*, 575–587.

8 Oegema, K. and Hyman, A.A. Cell division (January 19, 2006), *WormBook*, ed. The *C. elegans* Research Community, WormBook, doi/10.1895/wormbook.1.72.1, http://www.wormbook.org.

9 Hannak, E., Kirkham, M., Hyman, A.A., and Oegema, K. (2001). Aurora-A kinase is required for centrosome maturation in Caenorhabditis elegans. J. Cell Biol. *155*, 1109–1116.

10 Conduit, P.T., Brunk, K., Dobbelaere, J., Dix, C.I., Lucas, E.P., and Raff, J.W. (2010). Centrioles regulate centrosome size by controlling the rate of Cnn incorporation into the PCM. Curr. Biol. *20*, 2178–2186.

11 Hatch, E., and Stearns, T. (2011). The life cycle of centrioles. Cold Spring Harb. Symp. Quant. Biol. *75*, 425–431.

12 Loncarek, J., Hergert, P., Magidson, V., and Khodjakov, A. (2008). Control of daughter centriole formation by the pericentriolar material. Nat. Cell Biol. *10*, 322–328.

13 Duensing, A., Liu, Y., Perdreau, S.A., Kleylein-Sohn, J., Nigg, E.A., and Duensing, S. (2007). Centriole overduplication through the concurrent formation of multiple daughter centrioles at single maternal templates. Oncogene *26*, 6280–6288.

14 Peel, N., Stevens, N.R., Basto, R., and Raff, J.W. (2007). Overexpressing centriole-replication proteins in vivo induces centriole overduplication and de novo formation. Curr. Biol. *17*, 834–843.

15 Strnad, P., Leidel, S., Vinogradova, T., Euteneuer, U., Khodjakov, A., and Gönczy, P. (2007). Regulated HsSAS-6 levels ensure formation of a single procentriole per centriole during the centrosome duplication cycle. Dev. Cell *13*, 203–213.

16 Puklowski, A., Homsi, Y., Keller, D., May, M., Chauhan, S., Kossatz, U., Grünwald, V., Kubicka, S., Pich, A., Manns, M.P., *et al.* (2011). The SCF-FBXW5 E3-ubiquitin ligase is regulated by PLK4 and targets HsSAS-6 to control centrosome duplication. Nat. Cell Biol. *13*, 1004–1009.

17 Hertwig, R. (1908). Ueber neue Probleme der Zellenlehre. Arch. Zellf. *1*, 1–32.

18 Neumann, F.R., and Nurse, P. (2007). Nuclear size control in fission yeast. J. Cell Biol. *179*, 593–600.

19 Cavalier-Smith, T. (1982). Skeletal DNA and the evolution of genome size. Annu. Rev. Biophys. Bioeng. *11*, 273–302.

20 Jorgensen, P., Edgington, N.P., Schneider, B.L., Rupes, I., Tyers, M., and Futcher, B. (2007). The size of the nucleus increases as yeast cells grow. Mol. Biol. Cell *18*, 3523–3532.

21 Gurdon, J.B. (1976). Injected nuclei in frog oocytes: fate, enlargement, and chromatin dispersal. J. Embryol. Exp. Morphol. *36*, 523–540.

22 Harris, H. (1967). The reactivation of the red cell nucleus. J. Cell Sci. *2*, 23–32.

23 Henery, C.C., and Kaufman, M.H. (1992). Relationship between cell size and nuclear volume in nucleated red blood cells of developmentally matched diploid and tetraploid mouse embryos. J. Exp. Zool. *261*, 472–478.

24 Benavente, R., and Krohne, G. (1986). Involvement of nuclear lamins in postmitotic reorganization of chromatin as demonstrated by microinjection of lamin antibodies. J. Cell Biol. *103*, 1847–1854.

25 Newport, J.W., Wilson, K.L., and Dunphy, W.G. (1990). A lamin-independent pathway for nuclear envelope assembly. J. Cell Biol. *111*, 2247–2259.

26 Dechat, T., Adam, S.A., Taimen, P., Shimi, T., and Goldman, R.D. (2010). Nuclear lamins. Cold Spring Harb. Perspect. Biol. *2*, a000547.

27 Gerhart, J.C. (1980). Mechanisms regulating pattern formation in the amphibian egg and early embryo. In Biological Regulation and Development: Molecular Organization and Cell Function, R.F. Goldberger, ed. (New York: Plenum Press), pp. 133–316.

28 Levy, D.L., and Heald, R. (2010). Nuclear size is regulated by importin α and Ntf2 in Xenopus. Cell *143*, 288–298.

29 Wühr, M., Chen, Y., Dumont, S., Groen, A.C., Needleman, D.J., Salic, A., and Mitchison, T.J. (2008). Evidence for an upper limit to mitotic spindle length. Curr. Biol. *18*, 1256–1261.

30 Jilkine, A., and Edelstein-Keshet, L. (2011). A comparison of mathematical models for polarization of single eukaryotic cells in response to guided cues. PLoS Comput. Biol. *7*, e1001121.

31 Turing, A. (1952). The chemical basis of morphogenesis. Philos. Trans. R. Soc. Lond. B. Biol. Sci. *237*, 37–72.

32 Meinhardt, H. (1982). Models of Biological Pattern Formation (London: Academic Press).

33 Wedlich-Soldner, R., Altschuler, S., Wu, L., and Li, R. (2003). Spontaneous cell polarization through actomyosin-based delivery of the Cdc42 GTPase. Science *299*, 1231–1235.

34 Irazoqui, J.E., Gladfelter, A.S., and Lew, D.J. (2003). Scaffold-mediated symmetry breaking by Cdc42p. Nat. Cell Biol. *5*, 1062–1070.

35 Kozubowski, L., Saito, K., Johnson, J.M., Howell, A.S., Zyla, T.R., and Lew, D.J. (2008). Symmetry-breaking polarization driven by a Cdc42p GEF-PAK complex. Curr. Biol. *18*, 1719–1726.

36   Goryachev, A.B., and Pokhilko, A.V. (2008). Dynamics of Cdc42 network embodies a Turing-type mechanism of yeast cell polarity. FEBS Lett. *582*, 1437–1443.

37   Howell, A.S., Savage, N.S., Johnson, S.A., Bose, I., Wagner, A.W., Zyla, T.R., Nijhout, H.F., Reed, M.C., Goryachev, A.B., and Lew, D.J. (2009). Singularity in polarization: rewiring yeast cells to make two buds. Cell *139*, 731–743.

38   Johnson, J.M., Jin, M., and Lew, D.J. (2011). Symmetry breaking and the establishment of cell polarity in budding yeast. Curr. Opin. Genet. Dev. *21*, 740–746.

39   Goldstein, B., and Macara, I.G. (2007). The PAR proteins: fundamental players in animal cell polarization. Dev. Cell *13*, 609–622.

40   Bastock, R., and St Johnston, D. (2011). Going with the flow: an elegant model for symmetry breaking. Dev. Cell *21*, 981–982.

41   Jilkine, A., Marée, A.F.M., and Edelstein-Keshet, L. (2007). Mathematical model for spatial segregation of the Rho-family GTPases based on inhibitory crosstalk. Bull. Math. Biol. *69*, 1943–1978.

42   Mori, Y., Jilkine, A., and Edelstein-Keshet, L. (2008). Wave-pinning and cell polarity from a bistable reaction-diffusion system. Biophys. J. *94*, 3684–3697.

43   Goehring, N.W., Trong, P.K., Bois, J.S., Chowdhury, D., Nicola, E.M., Hyman, A.A., and Grill, S.W. (2011). Polarization of PAR proteins by advective triggering of a pattern-forming system. Science *334*, 1137–1141.

44   Dawes, A.T., and Munro, E.M. (2011). PAR-3 oligomerization may provide an actin-independent mechanism to maintain distinct Par protein domains in the early Caenorhabditis elegans embryo. Biophys. J. *101*, 1412–1422.

45   Cowan, C.R., and Hyman, A.A. (2004). Centrosomes direct cell polarity independently of microtubule assembly in C. elegans embryos. Nature *431*, 92–96.

46   Wallenfang, M.R., and Seydoux, G. (2000). Polarization of the anterior-posterior axis of C. elegans is a microtubule-directed process. Nature *408*, 89–92.

47   Gierer, A., and Meinhardt, H. (1972). A theory of biological pattern formation. Kybernetik *12*, 30–39.

48   Gamba, A., de Candia, A., Di Talia, S., Coniglio, A., Bussolino, F., and Serini, G. (2005). Diffusion-limited phase separation in eukaryotic chemotaxis. Proc. Natl. Acad. Sci. USA *102*, 16927–16932.

49   Arai, Y., Shibata, T., Matsuoka, S., Sato, M.J., Yanagida, T., and Ueda, M. (2010). Self-organization of the phosphatidylinositol lipids signaling system for random cell migration. Proc. Natl. Acad. Sci. USA *107*, 12399–12404.

50   Ishikawa, H., and Marshall, W.F. (2011). Ciliogenesis: building the cell's antenna. Nat. Rev. Mol. Cell Biol. *12*, 222–234.

51   Tam, L.W., and Lefebvre, P.A. (1993). Cloning of flagellar genes in Chlamydomonas reinhardtii by DNA insertional mutagenesis. Genetics *135*, 375–384.

52   Wilson, N.F., Iyer, J.K., Buchheim, J.A., and Meek, W. (2008). Regulation of flagellar length in Chlamydomonas. Semin. Cell Dev. Biol. *19*, 494–501.

53   Johnson, K.A., and Rosenbaum, J.L. (1992). Polarity of flagellar assembly in Chlamydomonas. J. Cell Biol. *119*, 1605–1611.

54   Marshall, W.F., and Rosenbaum, J.L. (2001). Intraflagellar transport balances continuous turnover of outer doublet microtubules: implications for flagellar length control. J. Cell Biol. *155*, 405–414.

55   Marshall, W.F., Qin, H., Rodrigo Brenni, M., and Rosenbaum, J.L. (2005). Flagellar length control system: testing a simple model based on intraflagellar transport and turnover. Mol. Biol. Cell *16*, 270–278.

56 Engel, B.D., Ludington, W.B., and Marshall, W.F. (2009). Intraflagellar transport particle size scales inversely with flagellar length: revisiting the balance-point length control model. J. Cell Biol. *187*, 81–89.

57 Howard, J., Grill, S.W., and Bois, J.S. (2011). Turing's next steps: the mechanochemical basis of morphogenesis. Nat. Rev. Mol. Cell Biol. *12*, 392–398.

58 Luo, M., Cao, M., Kan, Y., Li, G., Snell, W., and Pan, J. (2011). The phosphorylation state of an aurora-like kinase marks the length of growing flagella in Chlamydomonas. Curr. Biol. *21*, 586–591.

59 Sloboda, R.D. (2009). Flagella and cilia: the long and the short of it. Curr. Biol. *19*, R1084–R1087.

60 Rosenbaum, J.L. (1969). Flagellar elongation and shortening in Chlamydomonas: The use of cycloheximide and colchicine to study the synthesis and assembly of flagellar proteins. J. Cell Biol. *41*, 600–619.

61 Coyne, B., and Rosenbaum, J.L. (1970). Flagellar elongation and shortening in chlamydomonas. II. Re-utilization of flagellar proteins. J. Cell Biol. *47*, 777–781.

62 Kuchka, M.R., and Jarvik, J.W. (1982). Analysis of flagellar size control using a mutant of Chlamydomonas reinhardtii with a variable number of flagella. J. Cell Biol. *92*, 170–175.

63 Marshall, W.F. (2011). Centrosome size: scaling without measuring. Curr. Biol. *21*, R594–R596.

64 Sharma, N., Kosan, Z.A., Stallworth, J.E., Berbari, N.F., and Yoder, B.K. (2011). Soluble levels of cytosolic tubulin regulate ciliary length control. Mol. Biol. Cell *22*, 806–816.

65 Weeks, D.P., and Collis, P.S. (1976). Induction of microtubule protein synthesis in Chlamydomonas reinhardi during flagellar regeneration. Cell *9*, 15–27.

66 Lefebvre, P.A., Nordstrom, S.A., Moulder, J.E., and Rosenbaum, J.L. (1978). Flagellar elongation and shortening in Chlamydomonas. IV. Effects of flagellar detachment, regeneration, and resorption on the induction of flagellar protein synthesis. J. Cell Biol. *78*, 8–27.

67 Lefebvre, P.A., and Rosenbaum, J.L. (1986). Regulation of the synthesis and assembly of ciliary and flagellar proteins during regeneration. Annu. Rev. Cell. Biol. *2*, 517–546.

68 Springer, M., Weissman, J.S., and Kirschner, M.W. (2010). A general lack of compensation for gene dosage in yeast. Mol. Syst. Biol. *6*, 368.

69 Dennis, P.P. (1974). In vivo stability, maturation and relative differential synthesis rates of individual ribosomal proteins in Escherichia coli B/r. J. Mol. Biol. *88*, 25–41.

70 Fallon, A.M., Jinks, C.S., Strycharz, G.D., and Nomura, M. (1979). Regulation of ribosomal protein synthesis in Escherichia coli by selective mRNA inactivation. Proc. Natl. Acad. Sci. USA *76*, 3411–3415.

71 Nomura, M., Gourse, R., and Baughman, G. (1984). Regulation of the synthesis of ribosomes and ribosomal components. Annu. Rev. Biochem. *53*, 75–117.

72 Preker, P.J., Kim, K.S., and Guthrie, C. (2002). Expression of the essential mRNA export factor Yra1p is autoregulated by a splicing-dependent mechanism. RNA *8*, 969–980.

73 Yen, T.J., Machlin, P.S., and Cleveland, D.W. (1988). Autoregulated instability of beta-tubulin mRNAs by recognition of the nascent amino terminus of beta-tubulin. Nature *334*, 580–585.

74 Biederer, T., Volkwein, C., and Sommer, T. (1997). Role of Cue1p in ubiquitination and degradation at the ER surface. Science *278*, 1806–1809.

75 Ravid, T., and Hochstrasser, M. (2007). Autoregulation of an E2 enzyme by ubiquitin-chain assembly on its catalytic residue. Nat. Cell Biol. *9*, 422–427.

76 Redemann, S., Schloissnig, S., Ernst, S., Pozniakowsky, A., Ayloo, S., Hyman, A.A., and Bringmann, H. (2011). Codon adaptation-based control of protein expression in C. elegans. Nat. Methods *8*, 250–252.

77 Becskei, A., and Serrano, L. (2000). Engineering stability in gene networks by autoregulation. Nature *405*, 590–593.

78 Acar, M., Pando, B.F., Arnold, F.H., Elowitz, M.B., and van Oudenaarden, A. (2010). A general mechanism for network-dosage compensation in gene circuits. Science *329*, 1656–1660.

79 Bleris, L., Xie, Z., Glass, D., Adadey, A., Sontag, E., and Benenson, Y. (2011). Synthetic incoherent feedforward circuits show adaptation to the amount of their genetic template. Mol. Syst. Biol. *7*, 519.

80 Gulliver, G. (1875). On the size and shape of red corpuscles of the blood of vertebrates. Proc. Zool. Soc. Lond. *1875*, 474–495.

81 Varga, V., Leduc, C., Bormuth, V., Diez, S., and Howard, J. (2009). Kinesin-8 motors act cooperatively to mediate length-dependent microtubule depolymerization. Cell *138*, 1174–1183.

CHAPTER 15

Developmental Cell

Curvature, Lipid Packing, and Electrostatics of Membrane Organelles: Defining Cellular Territories in Determining Specificity

Joëlle Bigay[1], Bruno Antonny[1,*]

[1]Institut de Pharmacologie Moléculaire et Cellulaire, Université de Nice Sophia Antipolis et CNRS, 06560 Valbonne, France

*Correspondence: antonny@ipmc.cnrs.fr

Developmental Cell, Vol. 23, No. 5, November 13, 2012 © 2012 Elsevier Inc.
http://dx.doi.org/10.1016/j.devcel.2012.10.009

SUMMARY

Whereas some rare lipids contribute to the identity of cell organelles, we focus on the abundant lipids that form the matrix of organelle membranes. Observations using bioprobes and peripheral proteins, notably sensors of membrane curvature, support the prediction that the cell contains two broad membrane territories: the territory of loose lipid packing, where cytosolic proteins take advantage of membrane defects, and the territory of electrostatics, where proteins are attracted by negatively charged lipids. The contrasting features of these territories provide specificity for reactions occurring along the secretory pathway, on the plasma membrane, and also on lipid droplets and autophagosomes.

INTRODUCTION

Numerous molecular processes occur at the surface of organelles through the reversible association of proteins from the cytosol. These include signal transduction cascades, nucleation of cytoskeleton structures, formation of transport vesicles and lipid metabolism pathways. In the latter three cases, the organelle membrane is remodeled either physically or chemically. Not surprisingly, cells have built detectors to monitor these changes, allowing the establishment of feedback loops to amplify or arrest the remodeling processes. Herein, we discuss recent examples of such detectors and molecular

319

CellPress

circuits. Some of these examples seem unrelated: the budding of transport vesicles, the phagocytosis of bacteria, and the growth of lipid droplets are obviously very different events. Yet despite their heterogeneity, what unifies these processes is the fact that deceptively simple physical parameters such as membrane charge density, membrane curvature, or lipid packing can be informative inputs to create sharp temporal and spatial responses.

BIOCHEMISTRY ON LIPID MEMBRANE SURFACES

Let's begin with some hallmarks of biochemistry on membrane surfaces. First, lipids are small compared to proteins. In a bilayer, a lipid exposes a surface of ≈ 0.7 nm^2. Therefore, when sitting on a membrane, a protein generally covers several lipids: a dozen in the case of a ≈ 20-kDa small G protein (Liu et al., 2010); a hundred in the case of some elongated proteins (e.g., a BAR domain) (Peter et al., 2004). Second, although some proteins carry a domain specific for a lipid polar head group, peripheral proteins frequently use a combination of weak binding motifs such as lipid modifications or unfolded sequences (McLaughlin and Aderem, 1995). Thus, the adsorption of a protein to a membrane can seldom be reduced to a bimolecular scheme, and binding constants are generally apparent, reflecting multiple interactions. Third, lipids are not evenly accessible. Crossing the layers that separate the polar head group from the acyl chains is not trivial and in some cases the main issue is not lipid recognition per se, but the detection of defects in the geometrical arrangements of lipids (Attard et al., 2000; Davies et al., 2001). Altogether, these considerations suggest that biochemistry on membrane surfaces is a branch of soft matter physics: collective effects arising from multiple low energy interactions have at least the same importance as bimolecular stereospecific interactions.

RECOGNITION OF ORGANELLES BY PERIPHERAL PROTEINS

Cytosolic proteins are surrounded by membrane surfaces equivalent to tens of millimolar lipids. This value implies that strong affinity is not needed; instead the main issue is specificity: how to distinguish one membrane-bound compartment from others. For this aim, a few general strategies have been uncovered. The most straightforward is the specific recognition of rare lipids, most prominently phosphoinositides, which display a decorated polar head group and are restricted to specific compartments (Di Paolo and De Camilli, 2006). The second strategy is kinetic trapping. The protein visits all membranes in a reversible manner through a weak anchor. On the proper compartment, a molecular event occurs that dramatically decreases the protein desorption rate. Two well-known examples are the addition of a

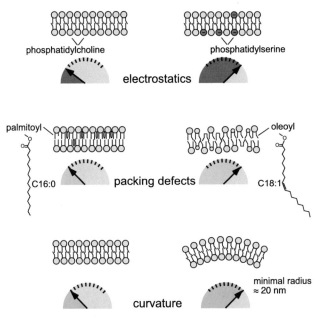

FIGURE 1 Three Physicochemical Parameters for Lipid Membranes
Membrane electrostatics depends on the fraction of negatively charged lipids such as phosphatidylserine (PS) and phosphoinositides. Packing defects are promoted by lipids with unsaturated acyl chains and/or small head group. Membrane curvature, which also results in lipid-packing defects, can reach values of 20 nm (radius) compared to 4 nm for membrane thickness.

second lipid modification (Rocks et al., 2005; Shahinian and Silvius, 1995) and the conformational changes undergone by Arf family G proteins (Liu et al., 2010). The third strategy, which is the topic of this review, is the recognition of a physicochemical parameter of the membrane such as its curvature, its electrostatics, or lipid packing (Figure 1).

MEMBRANE ELECTROSTATICS

In eukaryotic cells, membrane electrostatics largely depends on phosphatidylserine (PS), a negatively charged lipid, the amount of which ranges from a few percent in the endoplasmic reticulum (ER) to more than 10% at the plasma membrane (PM) (Holthuis and Levine, 2005). Remarkably, results of recent experiments suggest that PS is mostly present on the luminal side of the ER, an asymmetric distribution opposite to that of the PM where PS faces the cytosol (Fairn et al., 2011; Kay et al., 2012). Consequently, the gradient of accessible PS seen by cytosolic proteins along membranes of the secretory pathway is probably sharper than what is suggested by bulk measurements (Figure 2A). The highly charged phosphoinositide phosphatidylinositol(4,5)bisphosphate (PIP_2) accentuates this trend because it is

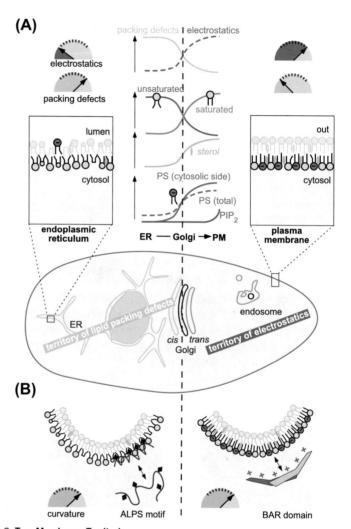

FIGURE 2 Two Membrane Territories

(A) Lipidomic analysis and bioprobe distribution suggest two main membrane territories in the cell. ER and *cis*-Golgi membranes are poorly charged on their cytosolic leaflet but display lipid-packing defects owing to the presence of lipids with monounsaturated chains. Membranes from the *trans*-Golgi to the plasma membrane (PM) harbor negative lipids on their cytosolic face but are tightly packed as their lipids are more saturated.

(B) Curved membranes are present everywhere in the cell, but curvature sensors are adapted to each territory: ALPS motifs to early membranes as they insert hydrophobic residues into lipid-packing defects; BAR domains to late membranes as their positive concave face fits with negatively charged membranes.

restricted to the inner leaflet of the PM (Di Paolo and De Camilli, 2006). The localization of PS and phosphoinositides probably explains the remarkable localization of simple cytosolic bioprobes carrying a defined number of positively charged residues: they label only membranes of the late secretory pathway and in a manner that correlates with their charge density (Yeung et al., 2006).

As rudimentary as it may seem, electrostatics stringently governs the localization of peripheral proteins. Variants of G proteins are good examples of the potency of electrostatics for precise localization. Rac1 and Rac2 are attached to the PM and to endosomes, respectively, owing to the different number of basic residues in their C-terminal region (Magalhaes and Glogauer, 2010; Ueyama et al., 2005; Yeung et al., 2006). K-Ras, the sole Ras variant to harbor a polybasic tail, is restricted to the PM, whereas H-Ras and N-Ras also explore membranes of the early secretory pathway owing to the addition of a second lipid modification (Ahearn et al., 2012). Bacterial toxins targeting these small G proteins also use electrostatics (Mesmin et al., 2004).

In liposome reconstitution assays, the effect of electrostatics can be very steep. WASP, a nucleator of actin filaments, promotes actin polymerization in a narrow range of PIP_2 (Papayannopoulos et al., 2005) because it does not probe PIP_2 through a specific site but via a cryptic polybasic region, which, like Velcro, binds firmly only above a threshold of negative charges. Finally, and this is the hallmark of nonspecific electrostatic interactions, binding depends on the number of positive residues, but is poorly sensitive to the exact amino acid sequence.

In conclusion, electrostatics seems to define two territories (Figure 2A): membranes of the early secretory pathway whose cytosolic leaflet is weakly charged (ER, cis-Golgi), and membranes of the late secretory pathway whose cytosolic leaflet is highly charged (endosomes, PM).

LIPID-PACKING DEFECTS

The concept of lipid-packing defects pertains to the idea that biological membranes display imperfections in the geometrical arrangement of their lipids because they contain substantial levels of lipids whose shape departs from the canonical cylindrical shape (Janmey and Kinnunen, 2006; van den Brink-van der Laan et al., 2004). For example, phosphatidylethanolamine (PE) and diacylglycerol are defined as conical since their polar heads are smaller than that of phosphatidylcholine (PC). Lipid geometry depends also on the acyl chains. An oleyl chain (C18:1) occupies a larger volume than a palmitoyl chain (C16:0) because the double bond induces a "kink" in the middle of the chain (Figure 1). Therefore, lipid packing depends on two ratios: the ratio between small and large polar heads and the ratio between unsaturated and saturated acyl chains. In addition, sterols, which pack preferentially along lipids with saturated chains, notably sphingolipids, improve lipid packing. Under extreme conditions, tight lipid packing leads to the formation of a liquid-ordered phase. However, even within the classical liquid disordered phase, variations in the packing of lipids can be large enough to dramatically influence the binding of peripheral proteins (Antonny et al., 1997; Attard et al., 2000; Davies et al., 2001; Matsuoka et al., 1998).

Our knowledge of the lipidome of most cells remains rudimentary. Nevertheless, several important features are emerging thanks to recent progress in lipid mass spectrometry. First, when a single phospholipid type is considered, its acyl chain composition can display significant differences depending on the organelle. In yeast, PS and PE are more saturated at the PM than at the ER (Schneiter et al., 1999; Tuller et al., 1999). In neurons, polyunsaturated lipids are more abundant in the axon than in the cell body (Yang et al., 2012). Second, manipulations aimed at altering the ratio between conical and nonconical lipids suggest that this ratio is tightly regulated through compensatory mechanisms (Boumann et al., 2006). Overall lipid packing probably increases along the secretory pathway (Figure 2A) (Brügger et al., 2000; Holthuis and Levine, 2005; Klemm et al., 2009). At one extreme, the ER is characterized by loose lipid packing owing to the abundance of unsaturated phospholipids and to the scarcity of cholesterol, which is tightly regulated (Bretscher and Munro, 1993; Radhakrishnan et al., 2008). At the other extreme, the PM is characterized by tight lipid packing due to the presence of saturated lipid species and a high sterol level.

The importance of loose lipid packing at the ER is beginning to emerge. In an extensive screening, 65 fatty acids of varying length and saturation have been tested for their ability to alleviate ER stress in a yeast strain deficient in unsaturated phospholipid synthesis (Deguil et al., 2011). All fatty acids could be incorporated into phospholipids, but only unsaturated fatty acids restored growth. Interestingly, oleate, which bears a single *cis* double bound, was more beneficial than fatty acids bearing multiple *cis* unsaturations or a single *trans* unsaturation. Because the central kink in the oleate chain creates more distortion than any other unsaturations, deviation from the straight conformation and consequently defects in lipid packing seem critical for some functions of the ER.

The small G protein Sar1 is a nice example of the adaptation of a peripheral protein to loose lipid packing. Sar1 is the housekeeping G protein of the ER. Among all small G proteins, Sar1 displays the longest and most hydrophobic sequence for membrane attachment (Huang et al., 2001; Lee et al., 2005). Sar1 therefore strongly contrasts with Rac1 or K-Ras, whose binding is governed by electrostatics. Sar1 binds better to C18:1-C18:1 than to C16:0-C18:1 phospholipid membranes, suggesting that its hydrophobic amino terminal residues insert preferentially into a bilayer with packing defects (Matsuoka et al., 1998).

In conclusion, the territories governed by lipid-packing defects might mirror those defined by electrostatics: membranes of the early secretory pathway seem to combine loose lipid packing and low electrostatic, whereas membranes of the late secretory pathway seem to combine tight lipid packing and high electrostatics (Figure 2A).

MEMBRANE CURVATURE

The ER is composed of a network of tubules and sheets. The Golgi apparatus combines flat cisternae, fenestrations, tubules, and vesicles. In endocytic organelles, outward tubulations permit cargo protein recycling whereas inward invaginations engage cargo proteins in a degradation pathway. At the PM, flat regions coexist with invaginations and protrusions of different sizes, shapes, and dynamics (Shibata et al., 2009). Considering the two broad territories defined above, the question then arises as to whether membrane curvature should be considered independently from membrane electrostatics and lipid-packing defects or whether these parameters combine, at least to some extent. The two major classes of membrane curvature sensors, the BAR domains and the ALPS motifs, suggest that the division of territories between early and late membranes also applies to membrane curvature detection (Figure 2B).

All BAR domains are built on the same fold resulting in a crescent shape (Frost et al., 2009; Peter et al., 2004). BAR domains sense, stabilize, or induce membrane curvature in a manner that depends on protein concentration, protein self-assembly, and additional membrane-interacting regions such as amphipathic helices (Frost et al., 2009; Galic et al., 2012; Peter et al., 2004). Nevertheless, what unifies all BAR domains is their association with late membranes, such as the PM or endosomes. Put differently, a recurrent observation is that BAR domains, even when overexpressed, do not associate with the ER (Peter et al., 2004). This negative observation is informative since ER tubules are abundant and have a diameter of about 50 nm that should fit well with the concave face of many BAR domains (Shibata et al., 2009). However as pointed out above, the electrostatic of the ER is probably kept at minimum due to PS orientation toward the lumen (Fairn et al., 2011; Kay et al., 2012). Because BAR domains interact with membranes through a basic surface, the early secretory pathway is probably not adapted to this family of peripheral proteins.

ALPS motifs form a family of membrane-associated amphipathic helices that are defined by the abundance of serine, glycine, and threonine residues in their polar face (Antonny, 2011; Bigay et al., 2005; Drin et al., 2007). Three salient features characterize the binding of ALPS motifs to lipid membranes: a sharp dependency on membrane curvature, a high sensitivity to lipid shape (C16:0-C16:0 PC << C18:1-C18:1 PC), and no sensitivity to lipid charge (PC = PS). Mutagenesis studies suggest that ALPS motifs use their bulky hydrophobic residues to detect large lipid-packing defects that arise from the conjunction of positive curvature and the presence of lipids with conical shape but essentially ignore membrane surface charge (Drin et al., 2007). Thus, ALPS motifs do not sense membrane geometry per se but the stress corresponding to the mismatch between the actual curvature of the membrane and the spontaneous curvature of its cytosolic leaflet (the curvature that this leaflet would adopt at equilibrium according to its composition and without the constraint of being associated with another leaflet; for discussion on the links between spontaneous curvature, curvature stress, lateral pressure profile, and lipid-packing defects see Attard et al., 2000; Davies et al., 2001; Janmey and Kinnunen, 2006; van den Brink-van der Laan et al., 2004; Antonny, 2011). ALPS motifs have been found in proteins associated with the nuclear envelope (Nup133), the ER (Atg14L/BARKOR), and the *cis*-Golgi (ArfGAP1, GMAP-210), and thus seem adapted to membranes of the early secretory pathway (Cardenas et al., 2009; Doucet et al., 2010; Fan et al., 2011; Levi et al., 2008).

In conclusion, examination of the general features of the two main classes of membrane curvature sensors, the BAR domains and the ALPS motifs, further underlines the general division of territories between membranes of the early secretory pathway and membranes of the late secretory pathway. The code used in early membranes seems to be curvature and lipid-packing defects whereas the code used in late membranes is rather curvature and electrostatics.

CHANGING MEMBRANE PROPERTIES

So far, we have considered three parameters and argued that they are tuned in such a way that they contribute to the identity of cellular organelles. However, cellular membranes are not fixed entities because they exchange materials and can maturate (Bonifacino and Glick, 2004). Necessarily, these processes are accompanied by changes in the bulk properties of the lipid bilayer. On the one hand, these changes can be interpreted from the point of view of homeostasis: a way to maintain compartment identity, which otherwise would vanish due to membrane budding and fusion events. On the other hand, the examples presented below suggest that membrane electrostatics, lipid packing, and membrane curvature are active parameters in the sense that they contribute to the self-organization of reactions at the surface of organelles.

PHAGOCYTOSIS: DECREASE IN ELECTROSTATICS DRIVES MEMBRANE MATURATION

Phagocytosis is the way specialized cells capture pathogens to eliminate them. The membrane surrounding the pathogen derives from the PM and initially displays its bulk features, with charged lipids (PS and PIP_2) enriched in the cytoplasmic leaflet promoting actin polymerization for pathogen engulfment (Yeung et al., 2009). Then the phagosome undergoes a maturation process that leads to its fusion with endosomal and lysosomal compartments and finally pathogen degradation.

Owing to their large size, phagosomes are ideal structures for following membrane maturation under a light microscope. With the parallel development of lipid probes (some specific and some adapted to membrane electrostatics), a clear picture is now emerging (Roy et al., 2000; Yeung et al., 2006, 2009). Once the phagosome membrane is detached from the PM, an abrupt decrease in PIP_2 is observed, allowing actin depolymerization (Figure 3A). Yet, the phagosome remains partially charged due to the persistence of PS. This sequential decrease in electrostatics favors a well-ordered change in the repertoire of associated small G proteins and correct delivery to the lysosome (Magalhaes and Glogauer, 2010; Ueyama et al., 2005; Yeung et al., 2009). Not surprisingly, pathogens have developed survival strategies to resist degradation. *Salmonella typhimurium* injects a phosphoinositide phosphatase (SopB) whose action reduces both the PIP_2 and PS levels. Consequently, the proper sequence of small G protein recruitment cannot occur and the *Salmonella*-containing vacuole escapes lysosomal degradation (Bakowski et al., 2010).

FEEDBACK LOOPS LINKING MEMBRANE CURVATURE AND VESICLE BIOGENESIS

The budding of coated vesicles is one of the best-studied processes of membrane traffic (Bonifacino and Glick, 2004). In contrast to phagocytosis, it involves huge changes in curvature because the membrane ultimately forms a vesicle of ≈30 nm in radius compared to μm for phagosomes. This shape is the result of the mechanical actions of cytosolic proteins that together form a polymerized spherical coat. After vesicle formation, the coat starts to detach, at least partially, and subsequent steps proceed up to membrane fusion.

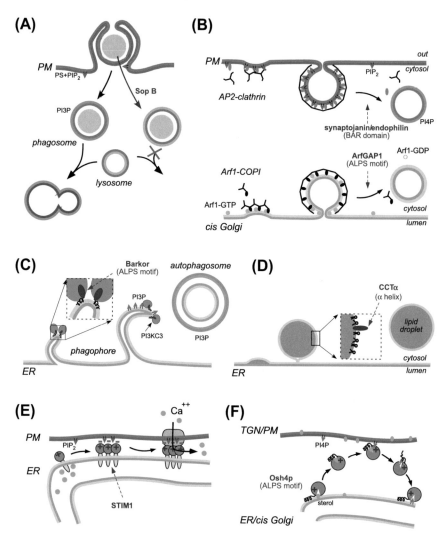

FIGURE 3 Cellular Events Controlled by Changes in the Physical Chemistry of Membranes
(A) During phagocytosis, charged lipids are initially mobilized, but once the phagosome detaches, membrane charge steadily decreases up to degradation by lysosomes. The pathogen effector SopB eliminates negatively charged lipids, preventing phagosome maturation.
(B) Clathrin and COPI coated vesicles display high curvature, which favors the recruitment of synaptojanin via endophilin, which contains a BAR domain, or of ArfGAP1, which contains ALPS motifs. Synaptojanin and ArfGAP1 eliminate PIP_2 and Arf1-GTP, respectively, contributing to coat destabilization.
(C) Autophagosome formation depends on the recruitment of a PI3-kinase. Via an ALPS C-ter motif, Barkor/Atg14(L) specifically targets the PI3 kinase complex to curved ER, promoting elongation of the phagophore and accumulation of PI3P.
(D) During lipid droplet expansion, $CCT\alpha$, a key enzyme for phosphatidylcholine (PC) synthesis, detects the deficit in phospholipid in the droplet monolayer.

The COPI coat is attached to Golgi membranes by Arf1-GTP (Yu et al., 2012); the AP2-clathrin coat is attached to the PM by PIP_2 (Jackson et al., 2010). Although a small G protein and a phosphoinositide are anything but similar, a common basis unites these modes of attachment. The Arf1-binding sites on the COPI coat are spatially related to the PIP_2-binding sites on the AP2 complex (Jackson et al., 2010; Yu et al., 2012). Furthermore, the parallel between the two systems extends to their regulation (Figure 3B). Elimination of Arf1-GTP by the GTPase activating protein ArfGAP1 and PIP_2 hydrolysis by the phosphatase synaptojanin are both stimulated by membrane curvature (Bigay et al., 2003; Chang-Ileto et al., 2011). However, the analogy here is purely functional: the response of ArfGAP1 to membrane curvature relies on its ALPS motifs, whereas synaptojanin seems to respond to membrane curvature by interacting with the BAR-containing protein endophilin (Bigay et al., 2005; Chang-Ileto et al., 2011; Milosevic et al., 2011). These different mechanisms reflect the contrasting properties of the membranes on which the two coats act.

Although the reported effect of membrane curvature on the synaptojanin-endophilin tandem is modest compared to that observed on ArfGAP1, recent studies suggest possibilities for sharper regulation. Structural analysis of endophilin bound to membrane tubes indicates that its SH3 domain undergoes a monomer to dimer transition, which depends very precisely on the tube radius (Mim et al., 2012). Because endophilin recruits synaptojanin through its SH3 domain, the accessibility of this domain could serve to communicate the curvature state of the underlying membrane in a precise manner (Mim et al., 2012). Testing this hypothesis seems possible thanks to the development of micromanipulation assays allowing fine adjustment of membrane tube radius (Roux et al., 2010; Zhu et al., 2012).

BARKOR: INTERPLAY BETWEEN CURVATURE AND PI(3)P SYNTHESIS FOR AUTOPHAGOSOME FORMATION

The autophagosome is a cup-shaped membrane compartment that engulfs organelles and part of the cytosol in a nonselective manner, thereby allowing cells to reduce their volume under starvation conditions

---

(E) Ca^{2+} decrease in the ER induces STIM1 oligomerization. The polybasic end of STIM1 interacts with negative charges at the PM promoting ER-PM bridges and activation of a Ca^{2+} channel.

(F) The lipid transfer protein Osh4 might interact alternatively with loosely packed membranes to extract sterols, and with negatively charged membranes, to deliver sterol and extract PI(4)P.

Red and blue lines represent inner leaflets rich in negative lipids or in lipid-packing defects, respectively.

(Mizushima et al., 2011). The recent identification of an ALPS motif in a protein complex involved at early stages of autophagy suggests an interesting feedback loop to initiate autophagosome formation (Fan et al., 2011). The mammalian protein Atg14L/Barkor forms a complex with three other proteins including a PI3-kinase (Matsunaga et al., 2010). Through its C-terminal ALPS motif, Barkor seems to recognize curved regions of the ER, hence triggering the unusual synthesis of PI(3)P in this compartment (Fan et al., 2011). Because Barkor also binds PI(3)P, the dual detection of positive curvature and PI(3)P may create a positive feedback loop for PI(3)P synthesis and contribute to the emergence of an atypical compartment, both in shape and in lipid composition, from the ER (Figure 3C). Note that here electrostatics increases and thus follows the opposite trend as in the case of phagocytosis. These findings should help to clarify the origin of the autophagosome membrane, which has been heavily discussed (Mizushima et al., 2011).

LIPID-PACKING DEFECTS AS AN INDEX OF LIPID DROPLET EXPANSION

To store carbon sources in the densest way, cells use lipid droplets: a core of triglycerides and sterol esters surrounded by a monolayer of phospholipids and specific peripheral proteins (Wolins et al., 2006). One interesting feature of the lipid droplet monolayer is its composition: it contains mostly PC and PE, whereas PS is barely detectable (Bartz et al., 2007). This composition fits with the hypothesis that lipid droplets emerge from the ER. In this scenario, a lens of triglycerides and sterol esters forms within the ER bilayer and bulges toward the cytoplasm (Figure 3D). Consequently, the droplet monolayer arises from the cytoplasmic leaflet of the ER, which is poor in PS (Fairn et al., 2011; Fan et al., 2011).

Lipid droplets change their volume depending on triglyceride synthesis or consumption, hence requiring adjustment of the monolayer surface. A recent study suggests an elegant feedback mechanism for such an adjustment (Krahmer et al., 2011). Phosphocholine synthetase (CCTα), the rate-limiting enzyme of phosphatidylcholine synthesis, is recruited to lipid droplets in the growing phase, thus allowing the production of more PC molecules to surround the oil core. CCTα contains a C-terminal amphipathic helix, which seems to sense the decrease in packing in the phospholipid monolayer that accompanies droplet expansion. Because this helix was shown to bind preferentially to lipid bilayers displaying a high PE/PC ratio (Attard et al., 2000; Davies et al., 2001), CCTα seems to act as a general sensor of PC deficiency, either in a droplet monolayer or in a lipid bilayer. However, the conformation of phospholipids above an oil core may be quite different from that of phospholipids in a bilayer and the mechanisms by which peripheral proteins selectively bind to lipid droplets remain largely mysterious.

MEMBRANE CONTACT SITES: WHEN LIPID TERRITORIES MEET

Besides exchanging material through tubules and vesicles, cellular compartments make specific contacts by tightly apposing their membranes (Carrasco and Meyer, 2011; Lev, 2010). From the physicochemical perspective, contact sites are interesting as they join membranes displaying different bulk properties. The ER is almost systematically involved, contacting organelles as different as mitochondria, vacuoles, the *trans*-Golgi and the PM.

The yeast transmembrane protein Ist2p provides a straightforward mechanism for membrane contact site formation (Ercan et al., 2009; Lavieu et al., 2010). Although biochemical fractionation indicates that this protein is retained at the ER, it also decorates the PM when observed with light microscopy. This deceptive localization results from the engagement of its cytosolic basic domain with anionic lipids, notably PIP_2, of the PM, which leads to the formation of bridges between the cortical ER and the PM (Ercan et al., 2009). Another related example is the STIM1-Orai1 tandem. STIM1 is an ER transmembrane protein that senses depletion of calcium in the lumen. Upon calcium drop, STIM1 oligomerizes, invades the cortical ER, and activates the PM calcium channel Orai1 in a process referred to as store-operated calcium entry (Carrasco and Meyer, 2011). The formation of the STIM1-Orai1 complex seems to takes advantage of the negative charge of the PM: oligomerization of STIM1 leads to the formation of a cytosolic patch very rich in positive charges and sufficient to surpass the threshold for efficient contact with PIP_2 at the PM (Figure 3E).

Membrane contact sites are also important for lipid transport between organelles. Proteins such as CERT (ceramide trafficking protein), FAPP (four-phosphate adaptor protein), and ORPs (OSBP related proteins) contain a domain that can extract a specific lipid and additional domains or motifs to tether the ER to the *trans*-Golgi (D'Angelo et al., 2007; Hanada et al., 2003; Im et al., 2005; Lev, 2010). Although in most cases the details of the lipid exchange reaction remain to be investigated and the function of these proteins may go beyond exchanging lipids (Mousley et al., 2012), a few reconstitution experiments suggest an exquisite adaptation to membrane interfaces. In vitro, the yeast protein Osh4 exchanges sterol for PI(4)P between liposomes (de Saint-Jean et al., 2011). The rate of lipid exchange is optimal when the sterol donor liposomes are poorly packed and the acceptor liposomes are charged and more strongly packed (de Saint-Jean et al., 2011).

The alternative use of an ALPS motif (Drin et al., 2007) and basic surfaces (Im et al., 2005) might allow Osh4 to rapidly land and take off from each membrane type, hence optimizing lipid transport. Along the same line, CERT extracts preferentially ceramide from poorly packed neutral liposomes (Tuuf et al., 2011).

The fine balance between electrostatics, lipid-packing defects, and curvature might also control other complex reactions involving two different membranes (Drin et al., 2008; Kunding et al., 2011; Park et al., 2012).

DANGEROUS COMBINATIONS

The aforementioned examples illustrate the power of combining surface charge, curvature, and defects in lipid packing to control biochemical reactions on membrane organelles. Yet, certain combinations seem to predominate. Let's examine other formal cases and discuss their relevance in a cellular context (Figure 4).

The first extreme combination is when a membrane displays a high density of anionic lipids together with large lipid-packing defects. In the test tube, this situation is mimicked by pure C18:1-C18:1 PS (DOPS) liposomes where every lipid carries a negative charge and two kinked acyl chains. Experimentally, many peripheral proteins bind avidly to such liposomes. For example, the seemingly innocent replacement of POPS by DOPS causes Osh4 to remain associated to the liposomes after sterol extraction, thereby preventing fast sterol transport (de Saint-Jean et al., 2011), and can even cause

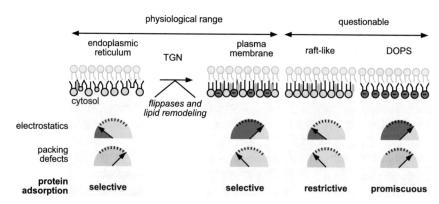

FIGURE 4 Membrane Electrostatics or Lipid-Packing Defects
The inner leaflet of cellular organelles can be rich in negatively charged lipids or in lipid-packing defects. Cells may avoid mixing these two features as this would lead to promiscuous binding of many peripheral proteins. At the other extreme, a highly packed and poorly charged leaflet seems not adapted to most cytosolic proteins.

liposome aggregation when all Osh4 membrane determinants are simultaneously engaged with this too-accommodating membrane surface (Schulz et al., 2009).

The opposite case is when the bilayer contains no anionic lipids and is tightly packed. In the presence of high amounts of cholesterol, lipids with saturated chains form a liquid-ordered ("raft") phase. Although this phase is assumed to favor protein-signaling platforms (Lingwood and Simons, 2010), the majority of intracellular peripheral proteins tested to date do not readily partition into such domains in vitro (Silvius, 2005). This is evidently the case for proteins that adsorb to membranes through electrostatics, but this exclusion also applies to other peripheral proteins. The small G protein Arf partitions exclusively to the disordered phase when incubated with giant liposomes exhibiting phase separation, suggesting that tight packing in the liquid-ordered phase prohibits amphipathic helix insertion (Manneville et al., 2008). Neutral liquid-ordered domains seem adapted to the penetration of dual lipid modifications such as those of GPI-anchored proteins (Silvius, 2005), but this situation applies only to the external leaflet of the PM.

Altogether, and as far as the cytosolic leaflet of membranes is concerned, the division of territories between loose lipid packing and electrostatics seem advantageous because it provides both a broad level of specificity and a means of regulation (Figure 4).

LIMITS OF THE TWO TERRITORIES MODEL AND PERSPECTIVES

We have put the emphasis on the separation between early and late membranes for the association of peripheral proteins from the cytosol. This separation is also important when considering the distribution of transmembrane proteins, as well as other general aspects of membrane traffic (Lippincott-Schwartz and Phair, 2010; Saraste and Goud, 2007; Sharpe et al., 2010). However, the role of membrane asymmetry in mechanisms such as organelle shaping and sorting of luminal proteins remains in most cases to be explored. In this last paragraph we will evoke the merits and limitations of the two territories model, and suggest some general lines for future works.

The development of bioprobes with a defined number of positive charges gives an illuminating picture of the electrostatics of cell membranes (Yeung et al., 2006). For lipid-packing defects, we are far from having a similar level of accuracy. First, lipid packing is a more elusive concept and only recent atomic simulations start depicting lipid-packing defects in a quantitative manner (Cui et al., 2011; van den Brink-van der Laan et al., 2004). Second, the influence of these defects on the functioning of machineries acting on early membranes, notably the ER, in most cases remains to be studied (however, see Fu et al., 2011; Matsuoka et al., 1998; Nilsson et al., 2001). Lastly, cellular approaches will require the development of bioprobes of varying hydrophobicity to map the distribution of membranes with loose versus tight lipid packing. For this aim, amphipathic helices seem promising because their elongated structure is adapted to probe the membrane interface in a repetitive manner (Antonny, 2011).

In a recent study, two amphipathic helices with the most contrasting hydrophobic and polar faces have been compared for their localization in yeast cells (Pranke et al., 2011). The helix with bulky hydrophobic residues (an ALPS motif) decorated small vesicles of the early secretory pathway, whereas the helix with small hydrophobic residues but a highly charged polar face (from α-synuclein) decorated small endocytic vesicles. This distribution seems driven by protein-lipid interactions because the two helices are heterologously expressed and the ALPS localization remains the same after sequence inversion. In the future, other amphipathic helices with peculiar features should help to evaluate the balance between surface charge and lipid-packing defects on various organelles. Among the most interesting helices are those of perilipins, which decorate lipid droplets (Bulankina et al., 2009); synapsin, which binds to synaptic vesicles (Krabben et al., 2011); Hsp12, which binds to the PM under stress conditions (Welker et al., 2010); and the yeast lipin, Pah1p, which associates with the nuclear/ER membrane (Karanasios et al., 2010).

The contrasting physical chemistry of early versus late membranes has an obvious corollary: the key role of lipid metabolism at the Golgi apparatus (Bankaitis et al., 2012; Lippincott-Schwartz and Phair, 2010). Two reactions are particularly interesting: phospholipid flip-flop, to generate asymmetry, and phospholipid remodeling, to change acyl chain composition (Figure 4). Several P4-type ATPases have now been identified that promote PS and other phospholipid translocation from the lumen to the cytoplasmic leaflet of the TGN or the PM (Alder-Baerens et al., 2006; Hua et al., 2002). Their activities help to explain why some mechanisms of membrane shaping at the *trans*-Golgi are reminiscent of those occurring at the PM, involving not only Arf and coats but also actin, Rac, and charged lipids (Anitei et al., 2010; Koronakis et al., 2011; Wang et al., 2003). Phospholipid remodeling is a long known reaction in which a phospholipase and an acyltransferase act sequentially to replace the esterified acyl chains on the phospholipid glycerol backbone; for example, to convert a C18:1-C18:1 into a C16:0-C18:1 lipid. The recent cloning of many acyl transferases and the realization that some members reside at the Golgi suggest interesting possibilities for their role in changing membrane properties across this organelle (Schmidt and Brown, 2009; Shindou et al., 2009; Yang et al., 2011).

So far, we have avoided discussing two lipids whose subcellular distribution is difficult to integrate in a general scheme. These are phosphatidic acid (PA) and phosphatidylinositol (PI). Both are anionic and are present at the ER, thus posing the question of their contribution to the electrostatics of this organelle. PA is a key intermediate in the synthesis of most lipids but its steady state level at the ER seems very low, suggesting no major contribution to general electrostatics. Instead, the level of PA serves as an index to control many lipid metabolism pathways (Loewen, 2012). In contrast to PA, PI is an abundant lipid. However, recent studies suggest that PI is not evenly present in the ER but, instead concentrates at specific subregions (Kim et al., 2011). Together with the luminal distribution of PS (Fairn et al., 2011; Kay et al., 2012), this finding reinforces the idea that the surface of most of the ER is in fact quite neutral and, as such, strongly contrasts with the PM inner leaflet.

Reducing the membrane complexity to two main membrane territories is of course an oversimplification. It does not account for the heterogeneities that exist within a continuous membrane and which involve specific protein-lipid interactions, lipid domain formation, restricted membrane diffusion by fences, and localized enzymatic activities. However, the two territories model is flexible enough to incorporate such variations.

ACKNOWLEDGMENTS

We thank S. Vanni and A. Copic for comments on the manuscript. Work in our laboratory is supported by the ERC, the CNRS, and the ANR.

REFERENCES

Ahearn, I.M., Haigis, K., Bar-Sagi, D., and Philips, M.R. (2012). Regulating the regulator: post-translational modification of RAS. Nat. Rev. Mol. Cell Biol. *13*, 39–51.

Alder-Baerens, N., Lisman, Q., Luong, L., Pomorski, T., and Holthuis, J.C. (2006). Loss of P4 ATPases Drs2p and Dnf3p disrupts aminophospholipid transport and asymmetry in yeast post-Golgi secretory vesicles. Mol. Biol. Cell *17*, 1632–1642.

Anitei, M., Stange, C., Parshina, I., Baust, T., Schenck, A., Raposo, G., Kirchhausen, T., and Hoflack, B. (2010). Protein complexes containing CYFIP/Sra/PIR121 coordinate Arf1 and Rac1 signalling during clathrin-AP-1-coated carrier biogenesis at the TGN. Nat. Cell Biol. *12*, 330–340.

Antonny, B. (2011). Mechanisms of membrane curvature sensing. Annu. Rev. Biochem. *80*, 101–123.

Antonny, B., Huber, I., Paris, S., Chabre, M., and Cassel, D. (1997). Activation of ADP-ribosylation factor 1 GTPase-activating protein by phosphatidylcholine-derived diacylglycerols. J. Biol. Chem. *272*, 30848–30851.

Attard, G.S., Templer, R.H., Smith, W.S., Hunt, A.N., and Jackowski, S. (2000). Modulation of CTP:phosphocholine cytidylyltransferase by membrane curvature elastic stress. Proc. Natl. Acad. Sci. USA 97, 9032–9036.

Bakowski, M.A., Braun, V., Lam, G.Y., Yeung, T., Heo, W.D., Meyer, T., Finlay, B.B., Grinstein, S., and Brumell, J.H. (2010). The phosphoinositide phosphatase SopB manipulates membrane surface charge and trafficking of the Salmonella-containing vacuole. Cell Host Microbe 7, 453–462.

Bankaitis, V.A., Garcia-Mata, R., and Mousley, C.J. (2012). Golgi membrane dynamics and lipid metabolism. Curr. Biol. 22, R414–R424.

Bartz, R., Li, W.H., Venables, B., Zehmer, J.K., Roth, M.R., Welti, R., Anderson, R.G., Liu, P., and Chapman, K.D. (2007). Lipidomics reveals that adiposomes store ether lipids and mediate phospholipid traffic. J. Lipid Res. 48, 837–847.

Bigay, J., Gounon, P., Robineau, S., and Antonny, B. (2003). Lipid packing sensed by ArfGAP1 couples COPI coat disassembly to membrane bilayer curvature. Nature 426, 563–566.

Bigay, J., Casella, J.F., Drin, G., Mesmin, B., and Antonny, B. (2005). ArfGAP1 responds to membrane curvature through the folding of a lipid packing sensor motif. EMBO J. 24, 2244–2253.

Bonifacino, J.S., and Glick, B.S. (2004). The mechanisms of vesicle budding and fusion. Cell 116, 153–166.

Boumann, H.A., Gubbens, J., Koorengevel, M.C., Oh, C.S., Martin, C.E., Heck, A.J., Patton-Vogt, J., Henry, S.A., de Kruijff, B., and de Kroon, A.I. (2006). Depletion of phosphatidylcholine in yeast induces shortening and increased saturation of the lipid acyl chains: evidence for regulation of intrinsic membrane curvature in a eukaryote. Mol. Biol. Cell 17, 1006–1017.

Bretscher, M.S., and Munro, S. (1993). Cholesterol and the Golgi apparatus. Science 261, 1280–1281.

Brügger, B., Sandhoff, R., Wegehingel, S., Gorgas, K., Malsam, J., Helms, J.B., Lehmann, W.D., Nickel, W., and Wieland, F.T. (2000). Evidence for segregation of sphingomyelin and cholesterol during formation of COPI-coated vesicles. J. Cell Biol. 151, 507–518.

Bulankina, A.V., Deggerich, A., Wenzel, D., Mutenda, K., Wittmann, J.G., Rudolph, M.G., Burger, K.N., and Höning, S. (2009). TIP47 functions in the biogenesis of lipid droplets. J. Cell Biol. 185, 641–655.

Cardenas, J., Rivero, S., Goud, B., Bornens, M., and Rios, R.M. (2009). Golgi localisation of GMAP210 requires two distinct cis-membrane binding mechanisms. BMC Biol. 7, 56.

Carrasco, S., and Meyer, T. (2011). STIM proteins and the endoplasmic reticulum-plasma membrane junctions. Annu. Rev. Biochem. 80, 973–1000.

Chang-Ileto, B., Frere, S.G., Chan, R.B., Voronov, S.V., Roux, A., and Di Paolo, G. (2011). Synaptojanin 1-mediated PI(4,5)P2 hydrolysis is modulated by membrane curvature and facilitates membrane fission. Dev. Cell 20, 206–218.

Cui, H., Lyman, E., and Voth, G.A. (2011). Mechanism of membrane curvature sensing by amphipathic helix containing proteins. Biophys. J. 100, 1271–1279.

D'Angelo, G., Polishchuk, E., Di Tullio, G., Santoro, M., Di Campli, A., Godi, A., West, G., Bielawski, J., Chuang, C.C., van der Spoel, A.C., et al. (2007). Glycosphingolipid synthesis requires FAPP2 transfer of glucosylceramide. Nature 449, 62–67.

Davies, S.M., Epand, R.M., Kraayenhof, R., and Cornell, R.B. (2001). Regulation of CTP: phosphocholine cytidylyltransferase activity by the physical properties of lipid membranes: an important role for stored curvature strain energy. Biochemistry 40, 10522–10531.

de Saint-Jean, M., Delfosse, V., Douguet, D., Chicanne, G., Payrastre, B., Bourguet, W., Antonny, B., and Drin, G. (2011). Osh4p exchanges sterols for phosphatidylinositol 4-phosphate between lipid bilayers. J. Cell Biol. *195*, 965–978.

Deguil, J., Pineau, L., Rowland Snyder, E.C., Dupont, S., Beney, L., Gil, A., Frapper, G., and Ferreira, T. (2011). Modulation of lipid-induced ER stress by fatty acid shape. Traffic *12*, 349–362.

Di Paolo, G., and De Camilli, P. (2006). Phosphoinositides in cell regulation and membrane dynamics. Nature *443*, 651–657.

Doucet, C.M., Talamas, J.A., and Hetzer, M.W. (2010). Cell cycle-dependent differences in nuclear pore complex assembly in metazoa. Cell *141*, 1030–1041.

Drin, G., Casella, J.F., Gautier, R., Boehmer, T., Schwartz, T.U., and Antonny, B. (2007). A general amphipathic α-helical motif for sensing membrane curvature. Nat. Struct. Mol. Biol. *14*, 138–146.

Drin, G., Morello, V., Casella, J.F., Gounon, P., and Antonny, B. (2008). Asymmetric tethering of flat and curved lipid membranes by a golgin. Science *320*, 670–673.

Ercan, E., Momburg, F., Engel, U., Temmerman, K., Nickel, W., and Seedorf, M. (2009). A conserved, lipid-mediated sorting mechanism of yeast Ist2 and mammalian STIM proteins to the peripheral ER. Traffic *10*, 1802–1818.

Fairn, G.D., Schieber, N.L., Ariotti, N., Murphy, S., Kuerschner, L., Webb, R.I., Grinstein, S., and Parton, R.G. (2011). High-resolution mapping reveals topologically distinct cellular pools of phosphatidylserine. J. Cell Biol. *194*, 257–275.

Fan, W., Nassiri, A., and Zhong, Q. (2011). Autophagosome targeting and membrane curvature sensing by Barkor/Atg14(L). Proc. Natl. Acad. Sci. USA *108*, 7769–7774.

Frost, A., Unger, V.M., and De Camilli, P. (2009). The BAR domain superfamily: membrane-molding macromolecules. Cell *137*, 191–196.

Fu, S., Yang, L., Li, P., Hofmann, O., Dicker, L., Hide, W., Lin, X., Watkins, S.M., Ivanov, A.R., and Hotamisligil, G.S. (2011). Aberrant lipid metabolism disrupts calcium homeostasis causing liver endoplasmic reticulum stress in obesity. Nature *473*, 528–531.

Galic, M., Jeong, S., Tsai, F.C., Joubert, L.M., Wu, Y.I., Hahn, K.M., Cui, Y., and Meyer, T. (2012). External push and internal pull forces recruit curvature-sensing N-BAR domain proteins to the plasma membrane. Nat. Cell Biol. *14*, 874–881.

Hanada, K., Kumagai, K., Yasuda, S., Miura, Y., Kawano, M., Fukasawa, M., and Nishijima, M. (2003). Molecular machinery for non-vesicular trafficking of ceramide. Nature *426*, 803–809.

Holthuis, J.C., and Levine, T.P. (2005). Lipid traffic: floppy drives and a superhighway. Nat. Rev. Mol. Cell Biol. *6*, 209–220.

Hua, Z., Fatheddin, P., and Graham, T.R. (2002). An essential subfamily of Drs2p-related P-type ATPases is required for protein trafficking between Golgi complex and endosomal/vacuolar system. Mol. Biol. Cell *13*, 3162–3177.

Huang, M., Weissman, J.T., Beraud-Dufour, S., Luan, P., Wang, C., Chen, W., Aridor, M., Wilson, I.A., and Balch, W.E. (2001). Crystal structure of Sar1-GDP at 1.7 A resolution and the role of the NH2 terminus in ER export. J. Cell Biol. *155*, 937–948.

Im, Y.J., Raychaudhuri, S., Prinz, W.A., and Hurley, J.H. (2005). Structural mechanism for sterol sensing and transport by OSBP-related proteins. Nature *437*, 154–158.

Jackson, L.P., Kelly, B.T., McCoy, A.J., Gaffry, T., James, L.C., Collins, B.M., Höning, S., Evans, P.R., and Owen, D.J. (2010). A large-scale conformational change couples membrane recruitment to cargo binding in the AP2 clathrin adaptor complex. Cell *141*, 1220–1229.

Janmey, P.A., and Kinnunen, P.K. (2006). Biophysical properties of lipids and dynamic membranes. Trends Cell Biol. *16*, 538–546.

Karanasios, E., Han, G.S., Xu, Z., Carman, G.M., and Siniossoglou, S. (2010). A phosphorylation-regulated amphipathic helix controls the membrane translocation and function of the yeast phosphatidate phosphatase. Proc. Natl. Acad. Sci. USA *107*, 17539–17544.

Kay, J.G., Koivusalo, M., Ma, X., Wohland, T., and Grinstein, S. (2012). Phosphatidylserine dynamics in cellular membranes. Mol. Biol. Cell *23*, 2198–2212.

Kim, Y.J., Guzman-Hernandez, M.L., and Balla, T. (2011). A highly dynamic ER-derived phosphatidylinositol-synthesizing organelle supplies phosphoinositides to cellular membranes. Dev. Cell *21*, 813–824.

Klemm, R.W., Ejsing, C.S., Surma, M.A., Kaiser, H.J., Gerl, M.J., Sampaio, J.L., de Robillard, Q., Ferguson, C., Proszynski, T.J., Shevchenko, A., and Simons, K. (2009). Segregation of sphingolipids and sterols during formation of secretory vesicles at the trans-Golgi network. J. Cell Biol. *185*, 601–612.

Koronakis, V., Hume, P.J., Humphreys, D., Liu, T., Hørning, O., Jensen, O.N., and McGhie, E.J. (2011). WAVE regulatory complex activation by cooperating GTPases Arf and Rac1. Proc. Natl. Acad. Sci. USA *108*, 14449–14454.

Krabben, L., Fassio, A., Bhatia, V.K., Pechstein, A., Onofri, F., Fadda, M., Messa, M., Rao, Y., Shupliakov, O., Stamou, D., et al. (2011). Synapsin I senses membrane curvature by an amphipathic lipid packing sensor motif. J. Neurosci. *31*, 18149–18154.

Krahmer, N., Guo, Y., Wilfling, F., Hilger, M., Lingrell, S., Heger, K., Newman, H.W., Schmidt-Supprian, M., Vance, D.E., Mann, M., et al. (2011). Phosphatidylcholine synthesis for lipid droplet expansion is mediated by localized activation of CTP:phosphocholine cytidylyltransferase. Cell Metab. *14*, 504–515.

Kunding, A.H., Mortensen, M.W., Christensen, S.M., Bhatia, V.K., Makarov, I., Metzler, R., and Stamou, D. (2011). Intermembrane docking reactions are regulated by membrane curvature. Biophys. J. *101*, 2693–2703.

Lavieu, G., Orci, L., Shi, L., Geiling, M., Ravazzola, M., Wieland, F., Cosson, P., and Rothman, J.E. (2010). Induction of cortical endoplasmic reticulum by dimerization of a coatomer-binding peptide anchored to endoplasmic reticulum membranes. Proc. Natl. Acad. Sci. USA *107*, 6876–6881.

Lee, M.C., Orci, L., Hamamoto, S., Futai, E., Ravazzola, M., and Schekman, R. (2005). Sar1p N-terminal helix initiates membrane curvature and completes the fission of a COPII vesicle. Cell *122*, 605–617.

Lev, S. (2010). Non-vesicular lipid transport by lipid-transfer proteins and beyond. Nat. Rev. Mol. Cell Biol. *11*, 739–750.

Levi, S., Rawet, M., Kliouchnikov, L., Parnis, A., and Cassel, D. (2008). Topology of amphipathic motifs mediating Golgi localization in ArfGAP1 and its splice isoforms. J. Biol. Chem. *283*, 8564–8572.

Lingwood, D., and Simons, K. (2010). Lipid rafts as a membrane-organizing principle. Science *327*, 46–50.

Lippincott-Schwartz, J., and Phair, R.D. (2010). Lipids and cholesterol as regulators of traffic in the endomembrane system. Annu. Rev. Biophys. *39*, 559–578.

Liu, Y., Kahn, R.A., and Prestegard, J.H. (2010). Dynamic structure of membrane-anchored Arf·GTP. Nat. Struct. Mol. Biol. *17*, 876–881.

Loewen, C.J. (2012). Lipids as conductors in the orchestra of life. F1000 Biol. Rep. *4*, 4.

Magalhaes, M.A., and Glogauer, M. (2010). Pivotal advance: phospholipids determine net membrane surface charge resulting in differential localization of active Rac1 and Rac2. J. Leukoc. Biol. *87*, 545–555.

Manneville, J.B., Casella, J.F., Ambroggio, E., Gounon, P., Bertherat, J., Bassereau, P., Cartaud, J., Antonny, B., and Goud, B. (2008). COPI coat assembly occurs on liquid-disordered domains and the associated membrane deformations are limited by membrane tension. Proc. Natl. Acad. Sci. USA *105*, 16946–16951.

Matsunaga, K., Morita, E., Saitoh, T., Akira, S., Ktistakis, N.T., Izumi, T., Noda, T., and Yoshimori, T. (2010). Autophagy requires endoplasmic reticulum targeting of the PI3-kinase complex via Atg14L. J. Cell Biol. *190*, 511–521.

Matsuoka, K., Orci, L., Amherdt, M., Bednarek, S.Y., Hamamoto, S., Schekman, R., and Yeung, T. (1998). COPII-coated vesicle formation reconstituted with purified coat proteins and chemically defined liposomes. Cell *93*, 263–275.

McLaughlin, S., and Aderem, A. (1995). The myristoyl-electrostatic switch: a modulator of reversible protein-membrane interactions. Trends Biochem. Sci. *20*, 272–276.

Mesmin, B., Robbe, K., Geny, B., Luton, F., Brandolin, G., Popoff, M.R., and Antonny, B. (2004). A phosphatidylserine-binding site in the cytosolic fragment of Clostridium sordellii lethal toxin facilitates glucosylation of membrane-bound Rac and is required for cytotoxicity. J. Biol. Chem. *279*, 49876–49882.

Milosevic, I., Giovedi, S., Lou, X., Raimondi, A., Collesi, C., Shen, H., Paradise, S., O'Toole, E., Ferguson, S., Cremona, O., and De Camilli, P. (2011). Recruitment of endophilin to clathrin-coated pit necks is required for efficient vesicle uncoating after fission. Neuron *72*, 587–601.

Mim, C., Cui, H., Gawronski-Salerno, J.A., Frost, A., Lyman, E., Voth, G.A., and Unger, V.M. (2012). Structural basis of membrane bending by the N-BAR protein endophilin. Cell *149*, 137–145.

Mizushima, N., Yoshimori, T., and Ohsumi, Y. (2011). The role of Atg proteins in autophagosome formation. Annu. Rev. Cell Dev. Biol. *27*, 107–132.

Mousley, C.J., Yuan, P., Gaur, N.A., Trettin, K.D., Nile, A.H., Deminoff, S.J., Dewar, B.J., Wolpert, M., Macdonald, J.M., Herman, P.K., et al. (2012). A sterol-binding protein integrates endo-somal lipid metabolism with TOR signaling and nitrogen sensing. Cell *148*, 702–715.

Nilsson, I., Ohvo-Rekilä, H., Slotte, J.P., Johnson, A.E., and von Heijne, G. (2001). Inhibition of protein translocation across the endoplasmic reticulum membrane by sterols. J. Biol. Chem. *276*, 41748–41754.

Papayannopoulos, V., Co, C., Prehoda, K.E., Snapper, S., Taunton, J., and Lim, W.A. (2005). A polybasic motif allows N-WASP to act as a sensor of PIP(2) density. Mol. Cell *17*, 181–191.

Park, Y., Hernandez, J.M., van den Bogaart, G., Ahmed, S., Holt, M., Riedel, D., and Jahn, R. (2012). Controlling synaptotagmin activity by electrostatic screening. Nat. Struct. Mol. Biol. *19*, 991–997.

Peter, B.J., Kent, H.M., Mills, I.G., Vallis, Y., Butler, P.J., Evans, P.R., and McMahon, H.T. (2004). BAR domains as sensors of membrane curvature: the amphiphysin BAR structure. Science *303*, 495–499.

Pranke, I.M., Morello, V., Bigay, J., Gibson, K., Verbavatz, J.-M., Antonny, B., and Jackson, C.L. (2011). α-Synuclein and ALPS motifs are membrane curvature sensors whose contrasting chemistry mediates selective vesicle binding. J. Cell Biol. *194*, 89–103.

Radhakrishnan, A., Goldstein, J.L., McDonald, J.G., and Brown, M.S. (2008). Switch-like control of SREBP-2 transport triggered by small changes in ER cholesterol: a delicate balance. Cell Metab. *8*, 512–521.

Rocks, O., Peyker, A., Kahms, M., Verveer, P.J., Koerner, C., Lumbierres, M., Kuhlmann, J., Waldmann, H., Wittinghofer, A., and Bastiaens, P.I. (2005). An acylation cycle regulates localization and activity of palmitoylated Ras isoforms. Science *307*, 1746–1752.

Roux, A., Koster, G., Lenz, M., Sorre, B., Manneville, J.B., Nassoy, P., and Bassereau, P. (2010). Membrane curvature controls dynamin polymerization. Proc. Natl. Acad. Sci. USA *107*, 4141–4146.

Roy, M.O., Leventis, R., and Silvius, J.R. (2000). Mutational and biochemical analysis of plasma membrane targeting mediated by the farnesylated, polybasic carboxy terminus of K-ras4B. Biochemistry *39*, 8298–8307.

Saraste, J., and Goud, B. (2007). Functional symmetry of endomembranes. Mol. Biol. Cell *18*, 1430–1436.

Schmidt, J.A., and Brown, W.J. (2009). Lysophosphatidic acid acyltransferase 3 regulates Golgi complex structure and function. J. Cell Biol. *186*, 211–218.

Schneiter, R., Brügger, B., Sandhoff, R., Zellnig, G., Leber, A., Lampl, M., Athenstaedt, K., Hrastnik, C., Eder, S., Daum, G., et al. (1999). Electrospray ionization tandem mass spectrometry (ESI-MS/MS) analysis of the lipid molecular species composition of yeast subcellular membranes reveals acyl chain-based sorting/remodeling of distinct molecular species en route to the plasma membrane. J. Cell Biol. *146*, 741–754.

Schulz, T.A., Choi, M.G., Raychaudhuri, S., Mears, J.A., Ghirlando, R., Hinshaw, J.E., and Prinz, W.A. (2009). Lipid-regulated sterol transfer between closely apposed membranes by oxysterol-binding protein homologues. J. Cell Biol. *187*, 889–903.

Shahinian, S., and Silvius, J.R. (1995). Doubly-lipid-modified protein sequence motifs exhibit long-lived anchorage to lipid bilayer membranes. Biochemistry *34*, 3813–3822.

Sharpe, H.J., Stevens, T.J., and Munro, S. (2010). A comprehensive comparison of transmembrane domains reveals organelle-specific properties. Cell *142*, 158–169.

Shibata, Y., Hu, J., Kozlov, M.M., and Rapoport, T.A. (2009). Mechanisms shaping the membranes of cellular organelles. Annu. Rev. Cell Dev. Biol. *25*, 329–354.

Shindou, H., Hishikawa, D., Harayama, T., Yuki, K., and Shimizu, T. (2009). Recent progress on acyl CoA: lysophospholipid acyltransferase research. J. Lipid Res. Suppl. *50*, S46–S51.

Silvius, J.R. (2005). Partitioning of membrane molecules between raft and non-raft domains: insights from model-membrane studies. Biochim. Biophys. Acta *1746*, 193–202.

Tuller, G., Nemec, T., Hrastnik, C., and Daum, G. (1999). Lipid composition of subcellular membranes of an FY1679-derived haploid yeast wild-type strain grown on different carbon sources. Yeast *15*, 1555–1564.

Tuuf, J., Kjellberg, M.A., Molotkovsky, J.G., Hanada, K., and Mattjus, P. (2011). The intermembrane ceramide transport catalyzed by CERT is sensitive to the lipid environment. Biochim. Biophys. Acta *1808*, 229–235.

Ueyama, T., Eto, M., Kami, K., Tatsuno, T., Kobayashi, T., Shirai, Y., Lennartz, M.R., Takeya, R., Sumimoto, H., and Saito, N. (2005). Isoform-specific membrane targeting mechanism of Rac during Fc gamma R-mediated phagocytosis: positive charge-dependent and independent targeting mechanism of Rac to the phagosome. J. Immunol. *175*, 2381–2390.

van den Brink-van der Laan, E., Killian, J.A., and de Kruijff, B. (2004). Nonbilayer lipids affect peripheral and integral membrane proteins via changes in the lateral pressure profile. Biochim. Biophys. Acta *1666*, 275–288.

Wang, Y.J., Wang, J., Sun, H.Q., Martinez, M., Sun, Y.X., Macia, E., Kirchhausen, T., Albanesi, J.P., Roth, M.G., and Yin, H.L. (2003). Phosphatidylinositol 4 phosphate regulates targeting of clathrin adaptor AP-1 complexes to the Golgi. Cell *114*, 299–310.

Welker, S., Rudolph, B., Frenzel, E., Hagn, F., Liebisch, G., Schmitz, G., Scheuring, J., Kerth, A., Blume, A., Weinkauf, S., et al. (2010). Hsp12 is an intrinsically unstructured stress protein that folds upon membrane association and modulates membrane function. Mol. Cell *39*, 507–520.

Wolins, N.E., Brasaemle, D.L., and Bickel, P.E. (2006). A proposed model of fat packaging by exchangeable lipid droplet proteins. FEBS Lett. *580*, 5484–5491.

Yang, H.J., Sugiura, Y., Ikegami, K., Konishi, Y., and Setou, M. (2012). Axonal gradient of arachidonic acid-containing phosphatidylcholine and its dependence on actin dynamics. J. Biol. Chem. *287*, 5290–5300.

Yang, J.S., Valente, C., Polishchuk, R.S., Turacchio, G., Layre, E., Moody, D.B., Leslie, C.C., Gelb, M.H., Brown, W.J., Corda, D., et al. (2011). COPI acts in both vesicular and tubular transport. Nat. Cell Biol. *13*, 996–1003.

Yeung, T., Terebiznik, M., Yu, L., Silvius, J., Abidi, W.M., Philips, M., Levine, T., Kapus, A., and Grinstein, S. (2006). Receptor activation alters inner surface potential during phagocytosis. Science *313*, 347–351.

Yeung, T., Heit, B., Dubuisson, J.F., Fairn, G.D., Chiu, B., Inman, R., Kapus, A., Swanson, M., and Grinstein, S. (2009). Contribution of phosphatidylserine to membrane surface charge and protein targeting during phagosome maturation. J. Cell Biol. *185*, 917–928.

Yu, X., Breitman, M., and Goldberg, J. (2012). A structure-based mechanism for Arf1-dependent recruitment of coatomer to membranes. Cell *148*, 530–542.

Zhu, C., Das, S.L., and Baumgart, T. (2012). Nonlinear sorting, curvature generation, and crowding of endophilin N-BAR on tubular membranes. Biophys. J. *102*, 1837–1845.

nds in Cell Biology

Clathrin-Mediated Endocytosis in Budding Yeast

Jasper Weinberg, David G. Drubin*

Department of Molecular and Cell Biology, University of California, Berkeley, CA 94720-3202, USA

Correspondence: drubin@berkeley.edu

Trends in Cell Biology, Vol. 22, No. 1, January 2012 © 2012 Elsevier Inc.
http://dx.doi.org/10.1016/j.tcb.2011.09.001

SUMMARY

Clathrin-mediated endocytosis in the budding yeast *Saccharomyces cerevisiae* involves the ordered recruitment, activity and disassembly of nearly 60 proteins at distinct sites on the plasma membrane. Two-color live-cell fluorescence microscopy has proven to be invaluable for *in vivo* analysis of endocytic proteins: identifying new components, determining the order of protein arrival and dissociation, and revealing even very subtle mutant phenotypes. Yeast genetics and functional genomics facilitate identification of complex interaction networks between endocytic proteins and their regulators. Quantitative datasets produced by these various analyses have made theoretical modeling possible. Here, we discuss recent findings on budding yeast endocytosis that have advanced our knowledge of how ~60 endocytic proteins are recruited, perform their functions, are regulated by lipid and protein modifications, and are disassembled, all with remarkable regularity.

ENDOCYTOSIS IN YEAST: PAST, PRESENT AND FUTURE

The yeast endocytosis field stretches back decades (Box 1). Budding yeast appears to have only a clathrin-mediated endocytosis pathway, and not the additional clathrin-independent endocytosis pathways found in mammalian cells. Striking parallels between yeast and mammalian endocytosis allow discoveries in one system to be applied to the other. Yeast endocytosis occurs at sites on the plasma membrane marked by the ordered recruitment of endocytic proteins, culminating in the rapid accumulation of actin directly

343

BOX 1 HISTORICAL PERSPECTIVES

Yeast endocytosis occurs at plasma membrane subdomains, called actin patches, that were first identified in the mid 1980s by immunofluorescence and rhodamine-phalloidin staining as polarized, plasma membrane-localized actin punctae of unknown function [1,2]. The intersection of actin and endocytosis in yeast was discovered in 1993 [3]. Using immuno-EM, plasma membrane invaginations were observed at sites of actin density, suggesting that these could be sites of endocytosis [4]. Evidence was growing that actin-regulating proteins affected endocytosis, but direct evidence that actin patches are sites of endocytosis was not presented until the process was analyzed by two-color, real-time fluorescence microscopy in the early 2000s [5].

Alpha factor pheromone and its receptor were cargos of choice for early endocytic studies [90]. Radiolabeled alpha factor was the first cargo for which entry into the cell could be tracked during receptor-mediated endocytosis, and is still used today for quantitative assays. The growing popularity of fluorescent imaging popularized the use of commerical dyes, like Lucifer Yellow [91] as a marker of fluid-phase endocytosis and FM4-64 [92] as a marker for the plasma membrane and internalized lipids. Fluorescently tagged alpha factor was introduced subsequently [7]. Many permeases can be

followed as endocytic cargos because they are expressed highly when their substrates are removed from the media, and then are rapidly internalized when their substrates are added back in excess, dramatically increasing endocytic activity [93]. Ste6p, the alpha factor exporter, and Pdr5p, a multi-drug resistance pump, represent another class of cargos, which are rapidly and constitutively endocytosed under normal conditions [94–96].

Live-cell imaging showed that internalizing fluorescent alpha factor and FM4-64 co-localize with actin patches, confirming that patches are sites of endocytosis [7,8]. Live cell, two-color imaging revealed the highly regular, dynamic nature of endocytosis and the recruitment of actin directly before scission and internalization [5,6,31]. The highly regular ordered recruitment of fluorescently tagged proteins facilitated determination of the order of arrival of proteins at endocytic sites and the detection of subtle changes in dynamics, protein localization, scission and initial vesicle movement caused by mutations or chemical inhibitors [6,12]. Although most of the early work was done in budding yeast, work in fission yeast makes unique contributions and allows comparisons across the vast evolutionary distance separating these yeasts [39,97].

before the scission event. These cortical actin patches were first visualized by immunofluorescence and rhodamine-phalloidin staining [1,2] in budding yeast, where they are polarized to the bud (Figure 1). A growing body of work suggested the involvement of actin in endocytosis [3,4].

The spread of simultaneous two-color imaging and total internal reflection microscopy resulted in rapid advances in the study of endocytosis. These new techniques demonstrated that the previously observed partial colocalization of actin and endocytic proteins was a result of taking snapshots of a highly dynamic process [5]. Upon full spatiotemporal visualization, endocytosis clearly involved the ordered arrival and dissociation of many proteins. The excellent genetics of *Saccharomyces cerevisiae* and the existence of tools such as green fluorescent protein (GFP)-tagged libraries, combined with modern microscopy, enabled researchers to rapidly screen for proteins that localize to endocytic sites and to observe their dynamics in mutant yeast strains [5,6]. The first definitive demonstration that actin patches are sites of endocytosis was attained when two-color live-cell fluorescence

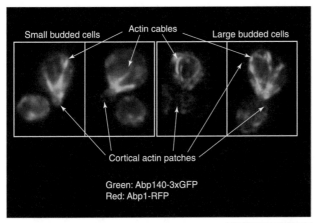

TRENDS in Cell Biology

FIGURE 1 Actin structures in budding yeast.
Actin forms three structures in budding yeast: the cytokinetic ring (not shown) responsible for separating the mother and daughter cell cytoplasm; actin cables, which are parallel bundles of short actin filaments running the length of the cell, nucleated by formins and serving as tracks for myosin-based motility (green); and cortical actin patches, composed of branched filaments nucleated by the Arp2/3 complex, and which are the sites of endocytosis (red). Cables and patches are visualized by imaging Abp140-3xGFP and Abp1-RFP, respectively.

microscopy demonstrated colocalization of actin patches with internalizing alpha-factor, a historically important model cargo, and FM4-64, a fluorescent lipid marker [7,8].

Studies of endocytosis in yeast continue to make important contributions even after several decades of intense research. Newly developed tools and techniques continue to drive the field forward. Mass spectrometry enables sensitive detection of post-translational modifications and low-affinity binding partners [9]. Synthetic genetic arrays and other genomic approaches expand our knowledge of the complex network of interactions and identify novel interacting proteins [10,11]. Studies of post-translational modifications have benefited from the development of analog-sensitive kinases, which allow for a rapid, isothermic, reversible and precise inhibition [12]. These approaches all augment the techniques used to launch the field, including two-color fluorescence microscopy, co-immunoprecipitation and double-mutant analysis. These technological advances allow researchers to identify potential endocytic proteins via interactions revealed by genomic screens, and to then rapidly identify binding partners and potential regulatory modifications by mass spectrometry, and to perform sensitive localization assays using fluorescently tagged proteins under normal conditions and in mutants of interacting proteins.

While the field has advanced in great leaps and bounds there are still many questions to be answered in the coming years. Active fields of study include determining the detailed mechanisms behind endocytic site selection, cargo recruitment, actin assembly, scission and uncoating, and the role of lipid and protein modifications in regulating the timing of recruitment, activation and dissociation of the endocytic machinery.

This review will cover recent advances in the study of budding yeast endocytosis, starting from the early stages of coat protein and cargo recruitment, through the maturation of the patch to the polymerization of actin and scission. The advances provided through ultrastructural analysis and mathematical modeling will be discussed along with modes of regulation and the relationship between yeast and mammalian endocytosis.

EARLY STAGES OF ENDOCYTOSIS, INCLUDING CARGO SELECTION

How endocytic sites are initiated is a mystery (Box 2). Mutations that partially disturb cellular actin affect the polarization of endocytic patches in daughter cells, but this observation is a bit mysterious as actin does not appear at the endocytic site until late in the pathway, after the endocytic site has been selected [13]. Perhaps the polarization of endocytic sites to daughter cells is facilitated by actin cables and their associated secretory cargo (Figure 1),

---

BOX 2 ARE MCC/EISOSOMES SITES OF ENDOCYTOSIS?

Eisosomes, also called membrane compartments of Can1p (MCCs), are static plasma membrane structures associated with long furrows in EM [98]. These furrows were first described in 1963 [99] but they were not associated with a specific set of proteins or a function until much later [98]. Whether eisosomes play a direct role in endocytosis is controversial. Mutations in proteins important for eisosome structure were reported to cause endocytic phenotypes [100] but it is not clear whether these are direct effects due to eisosomes being sites of endocytosis or indirect effects due to changes in membrane composition or cargo sorting and modification. Eisosomes do not colocalize with known endocytic proteins as analyzed by fluorescence microscopy, and cargo localized to the MCC may relocalize out of the eisosome and into the rest of the plasma membrane before endocytosis [101,102]. Some cargos localize to the MCC but their stimulated internalization is not affected by loss of a structural MCC protein [102]. Endocytosis is a very dynamic process, and some cargos can undergo massive, rapid, increases in rates of endocytosis, whereas eisosomes are very static structures, both in location and protein composition [100]. While immuno-EM has been used to localize MCC and eisosome components to long furrows [98], the same techniques reveal that endocytic proteins localize to tubular invaginations [19]. Correlative EM has demonstrated that actin and endocytic proteins localize to finger-like invaginations [81]. Eisosomes are most often found on mother cells or in cells that are in the stationary phase of growth [99,103], whereas endocytosis is thought to be involved in membrane homeostasis and to occur predominantly in locations of high membrane turnover, such as growing buds. Endocytosis can be blocked by inhibiting actin assembly in patches or by mutations in actin patch proteins [104].

or perhaps actin is present at an undetectable level at patches before the rapid polymerization late in the pathway (discussed below).

Among the earliest arriving proteins detected at yeast endocytic sites are the ubiquitin-binding protein Ede1p (Eps15), which is important for the recruitment of many later arriving proteins, and the F-BAR protein Syp1p (FCHo1/SGIP1) (Figure 2) [14,15]. (Throughout this review, the first mention of a yeast protein will be followed in parentheses by the mammalian homolog, when known. For a complete listing, see Table 1.) Syp1p is known to promote the formation of endocytic sites in a polarized fashion at the bud neck, but deletion of *SYP1* does not prevent patch formation in buds, whereas *EDE1* deletion reduces the number of endocytic sites but does not change their polarization [15]. Clearly, some additional factors are involved in the selection and initiation of endocytic sites.

The ability of both Syp1p and Ede1p to bind to membranes, Syp1p via its F-BAR domain and Ede1p in a ubiquitin-dependent manner [16], suggests that patch initiation may occur via a lipid signal, although no such signal

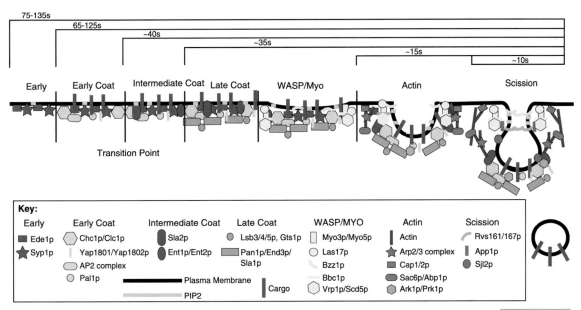

TRENDS in Cell Biology

FIGURE 2 Timeline for endocytic vesicle formation and modular organization of proteins.
Endocytic proteins are dynamically recruited in a highly predictable order. The 'early' and 'early coat' proteins are present at the cell surface for variable lengths of time. After a transition point, possibly defined by cargo recruitment, the lifetimes of endocytic proteins are quite regular. Actin polymerization and BAR domain proteins bend the membrane into an extended tubule, which is pinched off by the combined actions of BAR domain proteins, actin polymerization and possibly by a lipid phase separation. The newly formed vesicle is uncoated by the combined actions of the Ark1p/Prk1p kinases, the Sjl2p lipid-phosphatase and the actions of Arf3p/Gts1p/Lsb5p.

Table 1 *S. cerevisiae* endocytic proteins and their mammalian homologs

Yeast protein	Mammalian protein
Early	
Ede1p	Eps15
Syp1p	FCho1/2
Early coat	
Chc1p	Clathrin heavy chain
Clc1p	Clathrin light chain
Yap1801/2p	AP180
Pal1p	Fungi only
Apl1p	AP2 complex beta subunit
Apl3p	AP2 complex alpha subunit
Apm4p	AP2 complex mu subunit
Aps2p	AP2 complex sigma subunit
Intermediate coat	
Sla2p	Hip1R, Hip1
Ent1/2p	Epsin
Late coat	
Pan1p	Intersectin
Sla1p	Intersectin/CIN85
End3p	Eps15[a]
Lsb3p	SH3YL1[a]
Lsb4p	SH3YL1[a]
Lsb5p	Fungi Only
Gts1p	SMAP2 (small ArfGAP2)[a]
WASP/Myo	
Las17p	WASP/N-WASP
Vrp1p	WIP/WIRE
Bzz1p	syndapin
Scd5p	Saccharomycetales only
Myo3p	myosin-1E (Type 1 Myosin)
Myo5p	myosin-1E (Type 1 Myosin)
Bbc1p	Fungi only
Aim21p	Saccharomycetales only
Actin	
Act1p	Actin
Arc15/18/19/35/40p and Arp2/3p	Arp2/3 complex
Abp1p	ABP1
Cap1p	Capping protein alpha
Cap2p	Capping protein beta
Sac6p	Fimbrin
Scp1p	Transgelin

Table 1 *S. cerevisiae* endocytic proteins and their mammalian homologs—*cont'd*

Yeast protein	Mammalian protein
Twf1p	Twinfilin
Crn1p	Coronin
Ark1p/Prk1p/Akl1p	BMP2-inducible kinase/AP2-associated kinase 1/AAK1
Cof1p	Cofilin
Aip1p	Aip1
Bsp1p	No related protein identified
Pfy1p	Profilin
Aim3p	Saccharomycetales only
Scission	
Rvs161p	Amphiphysin
Rvs167p	Amphiphysin/endophilin
App1p	Fungi only
Sjl2p	Synaptojanin-1

aPhysical homology, but no functional homology identified.

has been identified in yeast. PIP2 may be important for initiation in mammalian endocytosis via FCHo proteins [17] but in yeast, temperature-sensitive mutations of the phosphatidylinositol-4-phosphate 5-kinase Mss4p do not inhibit actin patch initiation [18]. While the F-BAR domain of Syp1p can bind and tubulate membranes [14], electron microscopy (EM) suggests that the membrane remains relatively flat during the initial stages of the endocytic pathway (C. Buser and D. Drubin, unpublished) [19], indicating that Syp1p might bind without tubulation.

The arrival of the early proteins is quickly followed by the accumulation of cargo molecules at the nascent patch (Figure 2). Historically, the difficulties of imaging cargo accumulating in patches prevented solving the long-standing chicken and egg question of the endocytosis field: does the clustering of cargo recruit the endocytic machinery, or is cargo concentrated at a preformed endocytic site? Studies using a fluorescent derivative of alpha factor demonstrated that cargo arrives after the appearance of Ede1p but before the appearance of Sla1p (intersectin/CIN85) (Figure 2) [7]. This timing and apparent differences in bud versus mother cell endocytic patch dynamics [20] support the hypothesis that the variable lifetimes of the early proteins [Ede1p, Syp1p, Clc1p (clathrin light chain), Chc1p (clathrin heavy chain), Yap1801/2p (AP180), Pal1p and the AP2 complex] compared to the regular lifetimes of later proteins like Sla1p might be due to a cargo checkpoint that pauses the pathway after the arrival of early proteins until cargo loading is complete, and that the very regular timing of the later events of the pathway is the result of an endocytic site being fully loaded with cargo (Figure 2). Such a checkpoint has been postulated for mammalian cells [21]. Whereas alpha factor arrives after the early proteins, it is possible that other cargos are recruited later, after the arrival of intermediate or late coat, or may cluster before being recognized by the endocytic machinery [22].

In mammalian cells, knockdown of the AP2 complex, an important cargo adaptor, impairs internalization of several cargos, including the well studied transferrin receptor [23]. In yeast, no endocytic phenotype had been observed in AP2 subunit knockouts until studies of the yeast killer toxin K28 demonstrated that AP2 knockouts are resistant to the toxin, and thus AP2 is likely to be involved in K28 internalization [24]. Other reported cargo-specific adaptors include Yap1801/2p as an adaptor for Snc1p [25] and Syp1p as an adaptor for Mid2p [14].

Ubiquitin has long been recognized as a signal for internalization of membrane proteins, and several early and coat proteins, including Ede1p, Sla1p and Ent1/2p (epsin), have ubiquitin-binding domains (Table 2) [16,26]. It has become clear, however, that although these ubiquitin-binding domains may be important for cargo binding, they also are likely to be important for regulating protein–protein interactions within the endocytic network. Extensive analysis of mutations in the ubiquitin-binding domains of Ent1/2p and Ede1p demonstrate that in the absence of these domains both ubiquitinated and non-ubiquitinated cargos are internalized equally, though at a reduced rate compared to wild type, suggesting that the ubiquitin interaction motifs (UIM; see Glossary) regulate a general, rather than a ubiquitinated receptor-specific, step in endocytosis [27]. Because many endocytic proteins are likely to be ubiquitinated (Table 2), the ubiquitin-binding domains, along with the EH–NPF interactions, are likely to stabilize interactions between some of the ~60 proteins that localize to endocytic patches. Similar mutational analysis defines a role for the Yap1801/2p endocytic adaptors as functioning redundantly with Ent1/2p, likely by stabilizing the endocytic protein network via NPF–EH and UIM–ubiquitin interactions [28].

Early arriving endocytic proteins are candidates to both regulate and facilitate the early steps of patch initiation, patch polarization and cargo recruitment, however, they may also regulate later steps. In addition to its adaptor and endocytic site-specification roles [14,15], Syp1p is implicated as a regulator of Arp2/3 activity late in the endocytic pathway, just before Syp1p leaves the patch [29]. Many endocytic proteins are large, multi-domain proteins with multiple separable functions and may, therefore, act at multiple steps during endocytosis.

The early arriving endocytic proteins are integral for defining how many endocytic sites are formed and where they form, but they are not the only

Table 2 Ubiquitin binding and ubiquitinated endocytic proteins

	Ubiquitin binding [16,27,65,108]	Ubiquitinated by Rsp5p [27,77,106]	Ubiquitinated *in vivo* [107]
Early			
Ede1p	+	+	+
Early coat			
Clc1p	n.d.	–	+
Intermediate coat			
Ent1p	+	–	+
Ent2p	+	+	+
Sla2p	n.d.	–	+
Late coat			
Lsb5p	+	–	–
Sla1p	+	–	+
WASP/Myo			
Myo3p	n.d.	–	+
Actin			
Abp1p	n.d.	–	+
Arc18p	n.d.	–	+
Arc35p	n.d.	–	+
Arc40p	n.d.	–	+
Arp2p	n.d.	–	+
Arp3p	n.d.	–	+
Sac6p	n.d.	–	+
Scission			
Rvs167p	n.d.	+	+

The in vivo *ubiquitination data comes from genomic screens and may not be relevant for endocytosis. Minus sign in column ubiquitinated by Rsp5p: not detected, n.d.: not done.*

players involved. These early proteins bind to cargo and to each other, forming the nascent endocytic protein network. These are the proteins that act in an early temporally variable phase before an apparent cargo-triggered transition point that marks the beginning of the extremely temporally regular later stages of endocytosis (Figure 2).

COAT MATURATION

After the proposed cargo-triggered transition point, other coat proteins join the endocytic site in regular temporal fashion. The next set of proteins to be recruited to the endocytic site, Sla2p (Hip1R) and Ent1/2p, have domains that specifically bind to phosphatidylinositol 4,5-bisphosphate (PIP2)

[16,30]. The UIMs of Ent1/2p were thought to bind ubiquitinated cargo, but recent work suggests they are more important for promoting protein–protein interactions within the endocytic machinery, as discussed above [27,28].

The Pan1p (intersectin)/End3p (Eps15)/Sla1p complex is related to, and may serve a similar function as, the mammalian endocytic protein intersectin [31]. Pan1p has actin nucleation-promoting factor (NPF) activity but its most important role may be to serve as a scaffold to recruit other proteins to endocytic sites. The stability of the Pan1p/End3p/Sla1p complex is regulated by the Ark1p/Prk1p kinases (AAK1), which phosphorylate the complex, promoting disassembly, as discussed below [32,33].

The other proteins in the late coat module arrive along with members of the WASP/Myo module, but are grouped as coat proteins because, in contrast to the WASP/Myo proteins, they are internalized with the newly formed vesicle (Figure 2). Lsb5p and Gts1p (small ArfGAP2) have roles in uncoating [33], whereas Lsb3p (SH3YL1) and Lsb4p/Ysc84p (SH3YL1) promote actin polymerization and bundling [34].

The later arriving components of the endocytic coat are likely to be involved in linking together the early coat, which recruits cargo and begins to shape the membrane, with the actin and scission machinery.

WASP/MYOSIN RECRUITMENT

Las17p (WASP/N-WASP) and Myo3/5p (myosin-1E) are the most important NPFs *in vivo*, based on patch internalization movements in mutants; specific mutation of their NPF domains results in patches that fail to internalize [35], similar to what is observed in latrunculin A-treated cells [5]. Abp1p (ABP1) and Pan1p can stimulate Arp2/3 activity *in vitro* but *in vivo* removal of their NPF domains causes less severe phenotypes than similar mutations in either Las17p or Myo3/5p, demonstrating that they play lesser roles in Arp2/3-mediated actin polymerization during endocytosis [35,36].

The yeast Arp2/3 complex has a higher basal activity than its mammalian counterpart, although NPFs still strongly stimulate its nucleation activity. The myosin activator Vrp1p (WIP/WIRE) arrives after Las17p and is followed by Myo3/5p, which arrives just before actin is detected (Figure 2) [35]. The regulation of Las17p NPF activity is not fully understood. Las17p arrives ~20s before actin is detectable (Figure 2) [6]. During this time, the Las17p might be held in an inactive state by the earlier arriving proteins Syp1p and Sla1p [29,37]. At the onset of internalization, Syp1p departs and Sla1p moves inward with the coat, while Las17p remains closer to the plasma membrane, which may relieve the inhibition of Las17p [35]. As Sla1p internalizes, the Las17p inhibitor Bbc1p localizes to the patch [6,37]. How Bbc1p contributes to Las17p regulation is not understood. Combining null mutants of the inhibitors Sla1p and Bbc1p results in long actin tails associated with deep membrane invaginations (Figure 3A) [6]. The F-BAR protein Bzz1p (syndapin) can stimulate activity of Las17p/Sla1p and it arrives slightly before actin [35]. It is possible that the Bzz1p F-BAR domain might recognize curvature induced by Syp1p, also an F-BAR protein, and thus be recruited to help relieve Sla1p inhibition at endocytic sites. Las17p has many interacting partners that bind to its proline-rich regions via SH3 domains, but the exact roles of many of these proteins are poorly understood [38]. The role of Ysc84p as a Las17p-activating protein was reported recently but its homolog Lsb3p remains relatively uncharacterized [34], so keys to WASP/Myo activation and repression may reside in interactions of NPFs with relatively unstudied proteins.

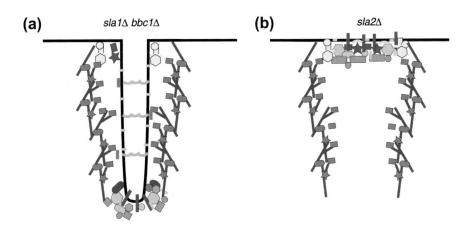

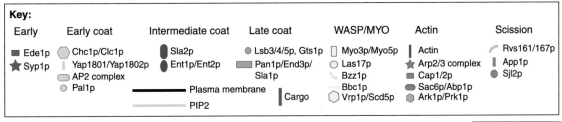

TRENDS in Cell Biology

FIGURE 3 Endocytic mutants affecting actin polymerization.
(a) In *sla1Δ bbc1Δ* mutants, Las17p inhibition is greatly reduced, resulting in excessive actin polymerization. The connection between actin and the endocytic coat is intact, so deep invaginations are formed. (b) In *sla2Δ* mutants, the connection between actin and the endocytic coat is missing, resulting in long, treadmilling actin tails that continuously assemble proximal to the plasma membrane, and flat membranes. Coat proteins and NPFs remain at the plasma membrane whereas all the actin-associated proteins normally associated with endocytosis localize to the comet tails.

Particle-tracking of fluorescently tagged members of the WASP/Myo module reveal that these proteins do not internalize with the newly formed vesicle but, rather, stay at the plasma membrane (Figure 2) [6]. Immuno-EM studies revealed a second population of Myo5p at the invagination tip, as well as Las17p localized on the sides of tubules [19]. Fluorescence recovery after photobleaching (FRAP) analysis of actin tails in *sla2Δ* cells (Figure 3b) suggests that actin is polymerized at the plasma membrane and moves inwards [5], so what is the role of Myo5p at the tip of an invagination? The myosin could be acting as a bridge by simultaneously binding to actin filaments and coat proteins, it could be nucleating filaments for a reason that is presently obscure, or it could be acting as a motor generating force through contact with actin filaments. The latter possibility is supported by the observation that motor domain mutants prevent inward movement, but do not prevent actin polymerization, which suggests that the NPF and motor functions are separable, but that both are important [35].

Despite the identification of a large number of associated players, how the Arp2/3 complex is regulated during endocytosis remains incompletely understood. Additional factors, perhaps unidentified proteins or post-translational and lipid modifications, are likely to act as switches for activating the NPFs in order to promote actin polymerization, membrane invagination and ultimately vesicle scission. Interestingly, it was recently suggested that in fission yeast, actin filaments from one endocytic structure can activate actin assembly at another endocytic site by contacting it [39]. The authors speculate that pre-existing actin filaments in the first structure trigger autocatalytic Arp2/3 activation. This phenomenon also has been observed *in vitro* on NPF-coated beads and rods, where actin polymerization does not occur until a short F-actin primer makes contact with the NFP-coated surface [40]. These results suggest the possibility that endocytic sites are primed to assemble actin and await initiation of the autocatalytic assembly process.

ACTIN POLYMERIZATION

Actin assembly is required for yeast endocytosis. This requirement can be partially overcome by providing osmotic support, which suggests that actin polymerization is needed to counter the turgor pressure at the plasma membrane [41]. A similar phenomenon in mammalian cells may explain discrepancies in the reported dependence on actin polymerization of mammalian clathrin-dependent endocytosis [42]. Severing the attachment between the actin network and the endocytic membrane in *sla2Δ* cells or inhibiting polymerization by latrunculin A sequestration of monomers, stops endocytosis [5]. The organization and physical properties of the actin network are also important for force generation; deletion of

bundling proteins Sac6p (fimbrin) and Scp1p (transgelin) results in nonpro-
ductive endocytic sites, although actin can still polymerize at the plasma
membrane [6,43].

In order to provide force to pull a vesicle out from the planar plasma membrane, the expanding actin network must be tightly connected to the internalizing invagination. In *sla2Δ* cells, which may lack this connection, actin
polymerizes into long 'comet tails' that continue to undergo flux while endocytosis is unproductive (Figure 3b) [5]. Fluorescence microscopy reveals that
while coat proteins remain at the plasma membrane and are not internalized, actin-associated proteins are recruited to the long tails in *sla2Δ* cells
[5]. Electron microscopy confirmed this conclusion as these mutant cells
have very few invaginations (C. Buser and D. Drubin, unpublished). Together,
these data suggest that Sla2p may be integral in connecting the coat to
the polymerizing actin. Sla2p can bind to PIP2 in membranes via its ANTH
domain and can interact with other coat proteins, including Clc1p, Ede1p
and Sla1p, as well as with actin, Las17p and the Arp2/3 complex [30,44–49].
These many interactions suggest that Sla2p may act as an adaptor, connecting the clathrin coat and plasma membrane to actin filaments, which
lets the growing network exert force on the membrane, deforming the membrane and ultimately helping to pinch off a vesicle.

The roles of the Arp2/3 complex and its NPF, Las17p, have been investigated using Las17p-coated microbeads added to yeast extract [50]. These
beads stimulate actin polymerization and recruit a large number of endocytic
proteins. A cloud of polymerized actin forms and a symmetry-breaking event
often follows, resulting in bead motility. Mass spectrometry of the bead-as-
sociated actin networks identified actin-regulating proteins [e.g. Cap1/2p
(capping protein alpha/beta), Sac6p, the Arp2/3 complex] and endocytic
specific proteins (e.g. Sla1p and Syp1p). These results demonstrate that an
NPF (Las17p) is sufficient to form an actin network of biologically relevant
composition. The exact mechanism for creating a branched dendritic network, like that found at endocytic patches [51,52], versus a parallel bundled
network, as exists in cables, is not known. Several recent reports in *Schizo-*
saccharomyces pombe implicate fimbrin (Sac6p) as an important factor
for excluding tropomyosin from endocytic actin filaments via fimbrin actin
binding activity, yet also confirm that its crosslinking activity is important
for endocytic function; a fimbrin truncation mutant that lacks actin bundling
activity yet retains actin binding activity, localizes to endocytic sites and
excludes tropomyosin, but otherwise phenocopies a fimbrin null [53,54].

As polymerization is turned off, disassembly of the actin network ensues
and the action of cofilin/Aip1/coronin (Cof1p/Aip1p/Crn1p) becomes dominant. Cof1p binds to older, ADP-bound actin filaments with the aid of

Crn1p while being inhibited from binding to younger ATP/ADP+Pi-actin by Crn1p [55]. Cof1p induces a twist, causing filaments to break apart [55]. Aip1p is important in breaking down the short actin filaments produced by Cof1p into monomers [56]. Intriguingly, actin oligomer-based polymerization can occur, and this oligomer assembly pathway is enhanced by the loss of Aip1p, suggesting that oligomer annealing occurs *in vivo* at actin patches [56].

Roles for actin after uncoating have been suggested. There are reports that vesicles appear to move along actin cables [7,8]. Some imaging suggests that endosomes move into close proximity with internalizing vesicles, facilitating fusion [7]. Observing actin cables by EM is very difficult, but immuno-EM results support the existence of a link between cables and endocytic sites [4].

The importance of actin in yeast endocytosis is unquestionable. Actin assembly provides force necessary for membrane deformation and scission and actin cables probably play a role in moving the newly formed vesicles away from the plasma membrane. Open questions remain about the regulation of actin assembly, the signals that control the NPFs and initiate polymerization as well as the signals that turn off polymerization to allow for disassembly and the eventual uncoating of the vesicle. However, many of the key players seem to have been identified.

SCISSION AND UNCOATING

In the final steps of endocytosis, the vesicle pinches off from the plasma membrane, moves inward toward the cell center and the endocytic coat disassembles from the newly formed vesicle. While the GTPase dynamin is essential for scission of clathrin-coated vesicles in mammalian cells, the role of dynamin in yeast endocytosis is uncertain. While at least one study suggested that dynamin does not participate, recent reports have suggested that the dynamin Vps1p might be involved in endocytosis (Box 3) [57–59]. However, the involvement of Vps1p in multiple important intracellular trafficking events has prevented definitive conclusions from being drawn.

In order for scission to occur, the two membrane bilayers must be brought into close proximity, at a distance of ~10nm or less, at which point they can fuse and a vesicle can form [60]. To perform this action, yeast take advantage of the membrane tubulating activity of the BAR and F-BAR proteins. While the Rvs161/167p (amphiphysin) heterodimeric BAR protein complex is capable of deforming membranes, this activity, combined with the force derived from actin polymerization, may not be sufficient to pinch off the membrane [60]. Where does the rest of the force come

BOX 3 A ROLE FOR DYNAMIN IN YEAST ENDOCYTOSIS?

Dynamin plays a crucial role in vesicle scission for clathrin-dependent endocytosis in metazoans, but whether it has a role in yeast endocytosis is controversial. The yeast genome encodes three dynamin-like proteins; Dnm1p and Mgm1p are involved in mitochondrial fusion and fission, whereas Vps1p is involved in fusion and fission reactions at other organelles, including the Golgi, vacuole, endosomes and peroxisomes. Some recent reports provide evidence for a role for Vps1p in endocytosis [57,58,105] but other work found no such connection and, unlike the situation in metazoan cells, Vps1p does not have an obligate endocytic function and is detected, at most, at only a subset of endocytic sites. The roles of Vps1p in other trafficking steps may explain some of the apparent endocytic phenotypes; for example, recycling of the endocytic machinery may be impaired in trafficking mutants. It has been noted that Vps1p lacks the PH domain and the C-terminal proline-rich domain found in conventional endocytic dynamins [57]. Similarly, the fission yeast dynamin Vps1p does not appear to be localized to actin patches [97].

from? One possible source is a line tension generated by lipid phase separation, as discussed below. The combination of the line tension, tubulation from BAR proteins and actin polymerization pushing on coat proteins at the bud and perhaps squeezing the tubule are proposed to provide the scission force [61,62].

This model provides a plausible explanation for why the lipid phosphatase synaptojanin is important for scission [63]. Along with creating a proposed lipid phase separation by having higher activity on the bud then on the tubule, Sjl2p (synaptojanin) has a role in vesicle uncoating. This is not surprising because a number of endocytic coat proteins that are internalized with the forming vesicle have PIP2-binding domains (Sla2p, Ent1/2p) and thus the action of Sjl2p reduces their affinity for the vesicle as it destroys the PIP2 [33]. At the same time, the regulatory kinases Ark1p and Prk1p are responsible for phosphorylating a variety of coat proteins, including Pan1p, Sla1p and Ent1/2p, and enabling their disassembly (Figure 4). Arf3p/Lsb5p/Gts1p also promote disassembly of the Pan1p/Sla1p/End3p complex but at the same time *arf3* mutants provide evidence for an Arf3p role in modulation of PIP2 levels [64,65]. These results suggest that a wide variety of factors contribute to the dissociation of coat proteins in a coordinated fashion in which multiple pathways converge to dissociate certain key proteins leading to the disassembly of the entire coat. Weakening of the coat–actin connection is expected to briefly relax tension on the membrane, which may be required for the final vesicle scission event [61].

While a number of uncoating factors are known, proteins specific for removing different endocytic proteins or shutting off actin polymerization may exist. The signals that turn on or recruit uncoating factors are likewise unclear but membrane curvature, actin filaments and lipid composition are likely candidates.

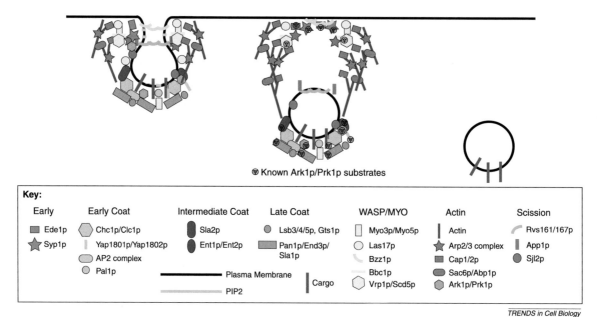

® Known Ark1p/Prk1p substrates

Key:

Early	Early Coat	Intermediate Coat	Late Coat	WASP/MYO	Actin	Scission
▪ Ede1p	⬡ Chc1p/Clc1p	⬭ Sla2p	◦ Lsb3/4/5p, Gts1p	▫ Myo3p/Myo5p	▮ Actin	╱ Rvs161/167p
★ Syp1p	┃ Yap1801p/Yap1802p	⬮ Ent1p/Ent2p	▭ Pan1p/End3p/ Sla1p	○ Las17p	★ Arp2/3 complex	▮ App1p
	▭ AP2 complex			╲ Bzz1p	▪ Cap1/2p	◦ Sjl2p
	○ Pal1p			── Bbc1p	⬭ Sac6p/Abp1p	
		──── Plasma Membrane	┃ Cargo	⬡ Vrp1p/Scd5p	⬡ Ark1p/Prk1p	
		──── PIP2				

TRENDS in Cell Biology

FIGURE 4 Scission and uncoating of endocytic vesicles.

Scission is accomplished by constriction of the bud neck, driven by BAR protein-driven membrane deformation, actin-generated force and the proposed Sjl2p-imposed line tension created by lipid phase separation. After scission, the Pan1 complex proteins Sla1p, Pan1p and End3p are phosphorylated by Ark1p/Prk1p, resulting in their dissociation. Ark1p/Prk1p are also responsible for turning off actin polymerization, allowing Cof1p/Aip1p/Crn1p to disassemble the actin filaments. Sjl2p is responsible for dephosphorylating PIP2, reducing the affinity of Sla2p and Ent1/2p for the vesicles. Arf3p, Gts1p and Lsb5p together are involved in the dissociation of the Pan1 complex.

PHOSPHORYLATION, UBIQUITINATION AND LIPID MODIFICATION

Endocytosis is regulated by a variety of signals, including phosphorylation, ubiquitination and lipid modification. The plasma membrane is a complex sea of proteins and lipids of various types, including phosphatidylcholine, phosphatidylethanolamine, phosphatidylinositols, phosphatidylserine, sphingolipids and ergosterol [66]. These lipids play a poorly understood role in regulating endocytosis. Adding another layer of complexity, the inner and outer leaflets of the plasma membrane are asymmetric, with the inner leaflet being enriched for phosphatidylserine and PIP2. Fluorescent probes combining GFP and lipid-binding protein domains, mutants of enzymes responsible for producing or modifying lipids, and the ability to grow yeast directly in or on media containing exogenous lipids has facilitated studies of how lipids contribute to endocytic regulation.

The importance of PIP2 in yeast endocytosis has been demonstrated genetically; mutants of Mss4p, which produces PIP2 from PIP, and of Sjl1p and

Sjl2p, the PIP2 phosphatases, have endocytic defects [63,67]. Using the PIP2-binding ANTH domain of Sla2p, it was demonstrated that PIP2 is concentrated at endocytic sites as the endocytic coat assembles, and disappears upon scission, soon after Sjl2p is recruited [63]. It should be noted that while PIP2 appears to be concentrated at endocytic sites, Mss4p is more broadly distributed on the plasma membrane, suggesting that PIP2 may be concentrated by PIP2-binding coat proteins such as Sla2p, Ent1/2p and the scission proteins Rvs161/167p.

The phenotypes of synaptojanin mutants suggest multiple roles for PIP2 in endocytosis. Disassembly of PIP2-binding proteins upon vesicle formation is delayed in these mutants, suggesting Sjl2p hydrolysis of PIP2 is important for the uncoating of Sla2p and Ent1/2p [33,63]. Secondly, long extended invaginations with coat proteins at the tips form in $sjl1\Delta$ $sjl2\Delta$ cells, suggesting a role for PIP2 regulation in scission [63]. Thirdly, these mutant cells inappropriately initiate invaginations on invaginations, suggesting a role for PIP2 as a signal for formation of endocytic sites [63]. Mathematical modeling led to the proposal that the role of PIP2 in scission might be in development of a line tension between an area of low PIP2 concentration, the tip of the invagination where Sjl2p is active, and the extended tubule where PIP2 is protected from Sjl2p by the BAR domain proteins [61,63]. The observed phenotype fits this hypothesis; when Sjl2p is missing, a concentration gradient cannot form, and so scission does not occur. Experimental support for this model has been provided by studies of dynamin-independent scission in mammalian cells [68].

While the role of PIP2 in endocytosis may be best studied among the lipids, genetic analyses suggest that sterols and sphingolipids may also be important for endocytosis [69–71]. Deletions of ergosterol biosynthetic enzymes *ERG2* or *ERG6* result in endocytic defects that are exacerbated when combined with each other or with deletions of *ERG3* [71,72]. Temperature-sensitive mutants of *LCB1/END8*, whose product catalyzes the first step in sphingolipid synthesis, are deficient in endocytosis and this block can be overcome by addition of exogenous sphingoid bases [69]. As new tools for the detection and modulation of other specific lipids are developed, our understanding of their contributions to endocytosis will be expanded.

Ubiquitination and phosphorylation have long-studied roles in regulating the internalization of transmembrane cargo molecules. In the canonical model, a receptor gets phosphorylated upon ligand binding, the phosphorylation signals for ubiquitination – in the form of either mono-, multiple mono-, or K63-linked polyubiquitinations – and this leads to internalization mediated by ubiquitin-binding coat molecules [26]. Ubiquitination of cargo molecules is carried out by the WW domain HECT-type ubiquitin ligase Rsp5p assisted by a number of adaptor proteins, including the recently described family of arrestins [73]. Mutations in individual WW domains of Rsp5p, which mediate protein–protein interactions, affect the internalization of specific cargos, whereas temperature-sensitive mutations in this protein generally reduce fluid phase endocytosis [74].

It is possible that Rsp5p also regulates the endocytic machinery directly. Eps15 and epsin, the mammalian homologs of Ede1p and Ent1/2p, respectively, are ubiquitinated *in vivo* and are thought to be regulated by binding to their own modifications [75]. Rsp5p also may regulate endocytosis via effects on the actin cytoskeleton [76]. The BAR domain protein Rvs167p has been reported to be ubiquitinated by Rsp5p, but no phenotype was observed in the K→R mutant, in which the lysine proposed to be ubiquitinated is mutated to arginine (Table 2) [77].

Phosphorylation of Pan1p by Prk1p prevents binding of Pan1p to F-actin, a prerequisite for its activation of the Arp2/3 complex [78]. A role for Ark1p and Prk1p as negative regulators of endocytic actin assembly is supported by the observation that *ark1Δ prk1Δ* cells have large clumps of actin, presumably due to uncontrolled actin polymerization [12,79]. These data suggest that Prk1p activity is important both for uncoating of internalized vesicles and cessation of actin polymerization. Prk1p autophosphorylation provides a mechanism for regulation of its kinase activity. Prk1p phosphorylates multiple targets, including Sla1p. These phosphoproteins then stimulate Prk1p autophosphorylation, which reduces its kinase activity [80]. This mechanism may both prevent hyperphosphorylation of targets and turn off Prk1p activity.

Despite recent advances, many questions about how yeast endocytosis is regulated still need to be addressed: Do lipids in addition to PIP2 play important regulatory roles? What is the nature of the signal that specifies endocytic site selection? Is the endocytic machinery ubiquitinated as a regulatory mechanism? Does phosphorylation only regulate the endocytic machinery during late steps in endocytosis? Dissection of regulatory mechanisms is an exciting and fertile area of current research.

ULTRASTRUCTURAL ANALYSIS

Fluorescence microscopy has proven to be an invaluable tool for studying endocytosis in yeast. With a practical resolution limit of ~200nm, conventional fluorescence microscopy cannot resolve important details, such as the shape of the invagination, where proteins localize along the invagination, or the orientation and organization of the actin network. These details can be resolved only by EM and possibly one day by super-resolution fluorescence microscopy. Unfortunately, there are several complications to studying endocytosis via EM. Firstly, as endocytosis is a highly dynamic process and EM is performed on fixed cells, one captures only snapshots and it may be unclear exactly what step in the process an image reveals. Secondly, the more highly curved the cell, the harder it is to section perpendicular to the plasma membrane so that the invagination is in the section and is oriented properly. Sites of endocytosis are concentrated in the more highly curved small bud (Figure 1). Thirdly, the cytoplasm of yeast is especially dense and filled with ribosomes, making it suboptimal for viewing actin filaments, which are notoriously hard to view by EM in the first place.

Immuno-EM was used to demonstrate precise localization of various endocytic proteins [19]. These results validated many of the proposed localizations of proteins based on fluorescence microscopy studies, such as the BAR domain protein Rvs167p localizing to the sides of tubules. However, the appearance of Myo5p at the base and at the tips of invaginations was an unexpected result, as was the localization of Las17p at both the base of the invagination and partway up the tubule [5,19,35]. These observations demonstrate the importance of ultrastructural analysis in providing critical information on the localization of proteins that could not be acquired by fluorescence microscopy alone [4,19].

Recent innovative use of correlative light and electron microscopy (CLEM) has allowed researchers to tackle the problem of determining which transient stage of endocytosis has been captured in an electron micrograph. An elegant correlative technique was developed in which fixed samples are imaged first by fluorescence microscopy and then by EM [81]. Using fluorescent microbeads as fiduciary marks, the authors could image the same endocytic sites using both techniques. By using two fluorescent markers that localize to endocytic sites at different times, the authors classified the sites into two groups; those with Abp1-RFP only and those with both Abp1-RFP and Rvs167-GFP, a scission protein. Approximately half of the doubly labeled sites had vesicles and half had intact invaginations so the authors concluded that scission must occur halfway through the ~10s period during which the two proteins colocalize. This technique holds promise for applications such as determining exactly when invaginations begin to form relative to recruitment of specific proteins and which proteins are first involved in constricting the neck of an invagination [81].

For this CLEM study, high-pressure freezing with freeze-substitution was used, which is best able to preserve the fine structure of dynamic cellular features such as endocytic invaginations. In the future, combining high-pressure freezing/freeze-substitution with immunostaining promises to both faithfully preserve structural features and provide precise localization of specific proteins. Modulating the stains used in the freeze-substitution steps can enhance the contrast between membranes and ribosome-rich yeast cytoplasm creating even clearer images [82].

One crucial challenge for ultrastructural analysis is to elucidate the exact geometry of the actin network that provides much of the force that drives invagination and scission. Important unanswered questions that could potentially be addressed by ultrastructural analysis include: Where do the filaments attach to the invagination and with what orientation? Does the network enclose the entire invagination or is the tip left exposed? Where exactly are the assembly nucleators? Are they all at the base of the invagination or are some at the tip or on the sides? Further technical innovation will be required to tackle the crucial but very challenging problem of reliably visualizing actin filaments in yeast by EM. Analysis of unroofed yeast cells is one promising avenue to overcoming the challenge presented by the dense cytoplasm of yeast [51].

THEORETICAL MODELING

Mathematical modeling is an important tool for exploring the mechanical and physical aspects of biological processes. If experimentalists can provide theoreticians with sufficient high-quality quantitative data, theoretical models can be developed. Such models allow researchers to understand how the process can and cannot work, they provide testable hypotheses, and they identify areas in which our understanding is incomplete, stimulating further experimentation, which in turn inspires new iterations of the models.

In an attempt to develop a comprehensive model for the endocytic pathway, more than 20 measured parameters from a variety of studies were used to model the recruitment of endocytic proteins, membrane deformation and scission [61]. The most important notion introduced in this model is that membrane curvature is a signal source for biochemical reactions involved in membrane scission. The model accurately predicted phenotypes of some mutations, such as the retraction seen in BAR domain knockouts, validating the predictive powers of the model [61].

The model also established that much of the scission force could be generated by a lipid phase separation. The calculated force of actin polymerization was found to be too small to drive membrane deformation and scission to completion. The authors speculated that scission might be driven by an interfacial force from lipid phase separation created when PIP2 is concentrated in the tubule and depleted in the bud. The model proposed that a pair of feedback loops drive this process (Figure 5). First, binding of BAR domain proteins on the tubule deforms the membrane, thus promoting recruitment of additional BAR domain proteins, which were proposed to protect PIP2

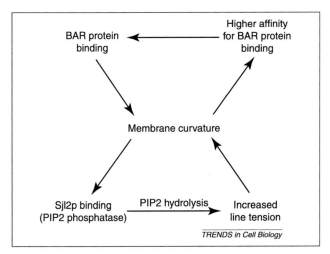

FIGURE 5 Positive feedback loops involved in scission.
During scission, two positive feedback loops are proposed to drive the increase in membrane curvature that leads to pinching off the vesicle. Firstly, BAR domain proteins bind to areas of high membrane curvature generated by actin polymerization forces and, through their binding, enhance the membrane curvature, recruiting more BAR domain proteins. Secondly, Sjl2p, the PIP2 phosphatase, is more active on curved membranes [83]. The activity of Sjl2p on the curved neck decreases levels of PIP2 on the unprotected bud while the PIP2 of the neck is shielded by the BAR domain proteins. The lipid phase separation produces a line tension that further increases membrane curvature, enhancing the activity of Sjl2p.

from dephosphorylation by the synaptojanin Sjl2p, the phosphatase responsible for PIP2 dephosphorylation. Second, BAR protein-induced curvature along the tubule was predicted to drive recruitment and activation of Sjl2p at the unprotected tubule–bud interface. As PIP2 is hydrolyzed, the phase boundary becomes more pronounced, and a line tension at the interface increases curvature, further increasing recruitment and activity of Sjl2p. The model proposes that these two positive feedback loops rapidly generate the force necessary to drive scission [61].

The notion that Sjl2p activity might be stimulated by increasing membrane curvature recently gained experimental support when it was shown in a cell-free system that synaptojanin-1 is more efficient at dephosphorylating PIP2 in more highly curved membranes, especially when a BAR domain protein is added to create very highly curved membranes [83]. This observation validates a major requirement of the model.

This model highlights one of the main differences between mammalian and yeast clathrin-mediated endocytosis: the lack of a requirement in yeast for dynamin participation in scission (Box 3). It was long assumed that the requirement for actin in yeast endocytosis indicated that the primary scission force was provided by actin polymerization, but this model suggests that the phase separation of PIP2 also contributes necessary force.

Another recent mathematical model focused on force generation by actin polymerization during fission yeast endocytosis [84]. Using constants derived from *in vitro* biochemical studies and quantification of local protein concentrations by fluorescence microscopy *in vivo*, the authors developed a model focused on actin polymerization and depolymerization at endocytic sites. Among the novel conclusions drawn from this study were: (i) *in vivo* binding of capping protein and Arp2/3 complex are faster than *in vitro* data suggest; and (ii) disassembly happens faster than can be explained solely by pointed-end depolymerization, which suggests that severing proteins may produce small fragments of actin that diffuse away and depolymerize away from the patch. *In vitro,* Aip1p enhances the depolymerization of short fragments produced by cofilin severing [56].

RELATIONSHIP TO MAMMALIAN ENDOCYTOSIS

Of the ~60 yeast proteins known to be localized to sites of endocytosis, ~85% have homology to mammalian proteins involved in endocytosis (Table 1). There are several striking examples where a single, modular mammalian protein has homology to several yeast proteins that each contain a subset of the modules. For example, mammalian intersectin has two Eps15 homology domains and five SH3 domains and is believed to play a role similar to that of the yeast proteins Pan1p and Sla1p. Together, these proteins have two Eps15 homology domains (Pan1p) and three SH3 domains (Sla1p) and are known to function in a complex together with End3p, which has another two Eps15 homology domains. The functional and domain homology suggests that while individual proteins may not be perfectly conserved, the network of interactions that hold the endocytic machinery together is conserved [85].

The dynamic, ordered recruitment of endocytic proteins is similar between yeast and mammals [85]. Clathrin and adaptor proteins arrive early, followed by additional coat proteins, actin and scission factors, and ending with uncoating proteins. Although several publications have emphasized differences in the dynamics of endocytosis in yeast and mammals, to what extent apparent differences were due to how the process was being studied was not clear. For yeast, but not mammalian cells, fluorescent endocytic proteins are typically expressed from their native promoters without overexpression. Using zinc finger nucleases to edit the genomes of mammalian cell lines, tagged clathrin light chain and dynamin were expressed at endogenous levels and the observed dynamics of mammalian endocytosis were described to be more similar to those of yeast than had been appreciated [86].

9   Toshima, J. *et al.* (2007) Negative regulation of yeast Eps15-like Arp2/3 complex activator, Pan1p, by the Hip1R-related protein, Sla2p, during endocytosis. *Mol. Biol. Cell* 18, 658–668

10 Friesen, H. *et al.* (2006) Characterization of the yeast amphiphysins Rvs161p and Rvs167p reveals roles for the Rvs heterodimer *in vivo*. *Mol. Biol. Cell* 17, 1306–1321

11 Tonikian, R. *et al.* (2009) Bayesian modeling of the yeast SH3 domain interactome predicts spatiotemporal dynamics of endocytosis proteins. *PLoS Biol.* 7, e1000218

12 Sekiya-Kawasaki, M. *et al.* (2003) Dynamic phosphoregulation of the cortical actin cytoskeleton and endocytic machinery revealed by real-time chemical genetic analysis. *J. Cell Biol.* 162, 765–772

13 Wong, M.H. *et al.* (2010) Vrp1p-Las17p interaction is critical for actin patch polarization but is not essential for growth or fluid phase endocytosis *in S. cerevisiae*. *Biochim. Biophys. Acta* 1803, 1332–1346

14 Reider, A. *et al.* (2009) Syp1 is a conserved endocytic adaptor that contains domains involved in cargo selection and membrane tubulation. *EMBO J.* 28, 3103–3116

15 Stimpson, H.E. *et al.* (2009) Early-arriving Syp1p and Ede1p function in endocytic site placement and formation in budding yeast. *Mol. Biol. Cell* 20, 4640–4651

16 Aguilar, R.C. *et al.* (2003) The yeast Epsin Ent1 is recruited to membranes through multiple independent interactions. *J. Biol. Chem.* 278, 10737–10743

17 Henne, W.M. *et al.* (2010) FCHo proteins are nucleators of clathrin-mediated endocytosis. *Science* 328, 1281–1284

18 Homma, K. *et al.* (1998) Phosphatidylinositol-4-phosphate 5-kinase localized on the plasma membrane is essential for yeast cell morphogenesis. *J. Biol. Chem.* 273, 15779–15786

19 Idrissi, F.Z. *et al.* (2008) Distinct acto/myosin-I structures associate with endocytic profiles at the plasma membrane. *J. Cell Biol.* 180, 1219–1232

20 Layton, A.T. *et al.* (2011) Modeling vesicle traffic reveals unexpected consequences for Cdc42p-mediated polarity establishment. *Curr. Biol.* 21, 184–194

21 Loerke, D. *et al.* (2009) Cargo and dynamin regulate clathrin-coated pit maturation. *PLoS Biol.* 7, e57

22 Di Pietro, S.M. *et al.* (2010) Regulation of clathrin adaptor function in endocytosis: novel role for the SAM domain. *EMBO J.* 29, 1033–1044

23 Motley, A. *et al.* (2003) Clathrin-mediated endocytosis in AP-2-depleted cells. *J. Cell Biol.* 162, 909–918

24 Carroll, S.Y. *et al.* (2009) A yeast killer toxin screen provides insights into a/b toxin entry, trafficking, and killing mechanisms. *Dev. Cell* 17, 552–560

25 Burston, H.E. *et al.* (2009) Regulators of yeast endocytosis identified by systematic quantitative analysis. *J. Cell Biol.* 185, 1097–1110

26 Hicke, L. and Riezman, H. (1996) Ubiquitination of a yeast plasma membrane receptor signals its ligand-stimulated endocytosis. *Cell* 84, 277–287

27 Dores, M.R. *et al.* (2010) The function of yeast epsin and Ede1 ubiquitin-binding domains during receptor internalization. *Traffic* 11, 151–160

28 Maldonado-Baez, L. *et al.* (2008) Interaction between Epsin/Yap180 adaptors and the scaffolds Ede1/Pan1 is required for endocytosis. *Mol. Biol. Cell* 19, 2936–2948

29 Boettner, D.R. *et al.* (2009) The F-BAR protein Syp1 negatively regulates WASp-Arp2/3 complex activity during endocytic patch formation. *Curr. Biol.* 19, 1979–1987

30 Sun, Y. *et al.* (2005) Interaction of Sla2p's ANTH domain with PtdIns(4,5)P2 is important for actin-dependent endocytic internalization. *Mol. Biol. Cell* 16, 717–730

31   Kaksonen, M. *et al*. (2006) Harnessing actin dynamics for clathrin-mediated endocytosis. *Nat. Rev. Mol. Cell Biol.* 7, 404–414

32   Zeng, G. *et al*. (2001) Regulation of yeast actin cytoskeleton-regulatory complex Pan1p/Sla1p/End3p by serine/threonine kinase Prk1p. *Mol. Biol. Cell* 12, 3759–3772

33   Toret, C.P. *et al*. (2008) Multiple pathways regulate endocytic coat disassembly in *Saccharomyces cerevisiae* for optimal downstream trafficking. *Traffic* 9, 848–859

34   Robertson, A.S. *et al*. (2009) The WASP homologue Las17 activates the novel actin-regulatory activity of Ysc84 to promote endocytosis in yeast. *Mol. Biol. Cell* 20, 1618–1628

35   Sun, Y. *et al*. (2006) Endocytic internalization in budding yeast requires coordinated actin nucleation and myosin motor activity. *Dev. Cell* 11, 33–46

36   Galletta, B.J. *et al*. (2008) Distinct roles for Arp2/3 regulators in actin assembly and endocytosis. *PLoS Biol.* 6, pe1

37   Rodal, A.A. *et al*. (2003) Negative regulation of yeast WASp by two SH3 domain-containing proteins. *Curr. Biol.* 13, 1000–1008

38   Madania, A. *et al*. (1999) The *Saccharomyces cerevisiae* homologue of human Wiskott-Aldrich syndrome protein Las17p interacts with the Arp2/3 complex. *Mol. Biol. Cell* 10, 3521–3538

39   Basu, R. and Chang, F. (2011) Characterization of dip1p reveals a switch in arp2/3-dependent actin assembly for fission yeast endocytosis. *Curr. Biol.* 21, 905–916

40   Achard, V. *et al*. (2010) A "primer"-based mechanism underlies branched actin filament network formation and motility. *Curr. Biol.* 20, 423–428

41   Aghamohammadzadeh, S. and Ayscough, K.R. (2009) Differential requirements for actin during yeast and mammalian endocytosis. *Nat. Cell Biol.* 11, 1039–1042

42   Boulant, S. *et al*. (2011) Actin dynamics counteract membrane tension during clathrin-mediated endocytosis. *Nat. Cell Biol.* 13, 1124–1131

43   Goodman, A. *et al*. (2003) The *Saccharomyces cerevisiae* calponin/transgelin homolog Scp1 functions with fimbrin to regulate stability and organization of the actin cytoskeleton. *Mol. Biol. Cell* 14, 2617–2629

44   Gourlay, C.W. *et al*. (2003) An interaction between Sla1p and Sla2p plays a role in regulating actin dynamics and endocytosis in budding yeast. *J. Cell Sci.* 116, 2551–2564

45   McCann, R.O. and Craig, S.W. (1999) Functional genomic analysis reveals the utility of the I/LWEQ module as a predictor of protein:actin interaction. *Biochem. Biophys. Res. Commun.* 266, 135–140

46   Gavin, A.C. *et al*. (2002) Functional organization of the yeast proteome by systematic analysis of protein complexes. *Nature* 415, 141–147

47   Newpher, T.M. *et al*. (2006) Novel function of clathrin light chain in promoting endocytic vesicle formation. *Mol. Biol. Cell* 17, 4343–4352

48   Newpher, T.M. and Lemmon, S.K. (2006) Clathrin is important for normal actin dynamics and progression of Sla2p-containing patches during endocytosis in yeast. *Traffic* 7, 574–588

49   Boettner, D.R. *et al*. (2011) Clathrin light chain directs endocytosis by influencing the binding of the yeast Hip1R homologue, Sla2, to F-actin. *Mol. Biol. Cell* 22, 3699–3714

50   Michelot, A. *et al*. (2010) Reconstitution and protein composition analysis of endocytic actin patches. *Curr. Biol.* 20, 1890–1899

51   Rodal, A.A. *et al*. (2005) Actin and septin ultrastructures at the budding yeast cell cortex. *Mol. Biol. Cell* 16, 372–384

52 Young, M.E. *et al.* (2004) Yeast actin patches are networks of branched actin filaments. *J. Cell Biol.* 166, 629–635

53 Skau, C.T. and Kovar, D.R. (2010) Fimbrin and tropomyosin competition regulates endocytosis and cytokinesis kinetics in fission yeast. *Curr. Biol.* 20, 1415–1422

54 Skau, C.T. *et al.* (2011) Actin filament bundling by fimbrin is important for endocytosis, cytokinesis and polarization in fission yeast. *J. Biol. Chem.* 286, 26964–26977

55 Gandhi, M. *et al.* (2009) Coronin switches roles in actin disassembly depending on the nucleotide state of actin. *Mol. Cell* 34, 364–374

56 Okreglak, V. and Drubin, D.G. (2010) Loss of Aip1 reveals a role in maintaining the actin monomer pool and an in vivo oligomer assembly pathway. *J. Cell Biol.* 188, 769–777

57 Smaczynska-de, R., II *et al.* (2010) A role for the dynamin-like protein Vps1 during endocytosis in yeast. *J. Cell Sci.* 123, 3496–3506

58 Nannapaneni, S. *et al.* (2010) The yeast dynamin-like protein Vps1:vps1 mutations perturb the internalization and the motility of endocytic vesicles and endosomes via disorganization of the actin cytoskeleton. *Eur. J. Cell Biol.* 89, 499–508

59 Yu, X. and Cai, M. (2004) The yeast dynamin-related GTPase Vps1p functions in the organization of the actin cytoskeleton via interaction with Sla1p. *J. Cell Sci.* 117, 3839–3853

60 Liu, J. *et al.* (2006) Endocytic vesicle scission by lipid phase boundary forces. *Proc. Natl. Acad. Sci. U.S.A.* 103, 10277–10282

61 Liu, J. *et al.* (2009) The mechanochemistry of endocytosis. *PLoS Biol.* 7, e1000204

62 Youn, J.Y. *et al.* (2010) Dissecting BAR domain function in the yeast Amphiphysins Rvs161 and Rvs167 during endocytosis. *Mol. Biol. Cell* 21, 3054–3069

63 Sun, Y. *et al.* (2007) PtdIns(4,5)P2 turnover is required for multiple stages during clathrin- and actin-dependent endocytic internalization. *J. Cell Biol.* 177, 355–367

64 Smaczynska-de, R., II *et al.* (2008) Yeast Arf3p modulates plasma membrane PtdIns(4,5)P2 levels to facilitate endocytosis. *Traffic* 9, 559–573

65 Costa, R. *et al.* (2005) Lsb5p interacts with actin regulators Sla1p and Las17p, ubiquitin and Arf3p to couple actin dynamics to membrane trafficking processes. *Biochem. J.* 387, 649–658

66 van der Rest, M.E. *et al.* (1995) The plasma membrane of *Saccharomyces cerevisiae*: structure, function, and biogenesis. *Microbiol. Rev.* 59, 304–322

67 Singer-Kruger, B. *et al.* (1998) Synaptojanin family members are implicated in endocytic membrane traffic in yeast. *J. Cell Sci.* 111, 3347–3356

68 Romer, W. *et al.* (2010) Actin dynamics drive membrane reorganization and scission in clathrin-independent endocytosis. *Cell* 140, 540–553

69 Zanolari, B. *et al.* (2000) Sphingoid base synthesis requirement for endocytosis in *Saccharomyces cerevisiae*. *EMBO J.* 19, 2824–2833

70 Friant, S. *et al.* (2001) Sphingoid base signaling via Pkh kinases is required for endocytosis in yeast. *EMBO J.* 20, 6783–6792

71 Munn, A.L. *et al.* (1999) Specific sterols required for the internalization step of endocytosis in yeast. *Mol. Biol. Cell* 10, 3943–3957

72 Heese-Peck, A. *et al.* (2002) Multiple functions of sterols in yeast endocytosis. *Mol. Biol. Cell* 13, 2664–2680

73 Lin, C.H. *et al.* (2008) Arrestin-related ubiquitin-ligase adaptors regulate endocytosis and protein turnover at the cell surface. *Cell* 135, 714–725

74 Gajewska, B. *et al.* (2001) WW domains of Rsp5p define different functions: determination of roles in fluid phase and uracil permease endocytosis in *Saccharomyces cerevisiae*. *Genetics* 157, 91–101

75 Oldham, C.E. *et al.* (2002) The ubiquitin-interacting motifs target the endocytic adaptor protein epsin for ubiquitination. *Curr. Biol.* 12, 1112–1116

76 Kaminska, J. *et al.* (2002) Rsp5p, a new link between the actin cytoskeleton and endocytosis in the yeast *Saccharomyces cerevisiae*. *Mol. Cell. Biol.* 22, 6946–6948

77 Stamenova, S.D. *et al.* (2004) The Rsp5 ubiquitin ligase binds to and ubiquitinates members of the yeast CIN85-endophilin complex, Sla1-Rvs167. *J. Biol. Chem.* 279, 16017–16025

78 Toshima, J. *et al.* (2005) Phosphoregulation of Arp2/3-dependent actin assembly during receptor-mediated endocytosis. *Nat. Cell Biol.* 7, 246–254

79 Cope, M.J. *et al.* (1999) Novel protein kinases Ark1p and Prk1p associate with and regulate the cortical actin cytoskeleton in budding yeast. *J. Cell Biol.* 144, 1203–1218

80 Huang, B. *et al.* (2009) Negative regulation of the actin-regulating kinase Prk1p by patch localization-induced autophosphorylation. *Traffic* 10, 35–41

81 Kukulski, W. *et al.* (2011) Correlated fluorescence and 3D electron microscopy with high sensitivity and spatial precision. *J. Cell Biol.* 192, 111–119

82 Mobius, W. (2009) Cryopreparation of biological specimens for immunoelectron microscopy. *Ann. Anat.* 191, 231–247

83 Chang-Ileto, B. *et al.* (2011) Synaptojanin 1-mediated PI(4,5)P2 hydrolysis is modulated by membrane curvature and facilitates membrane fission. *Dev. Cell* 20, 206–218

84 Berro, J. *et al.* (2010) Mathematical modeling of endocytic actin patch kinetics in fission yeast: disassembly requires release of actin filament fragments. *Mol. Biol. Cell* 21, 2905–2915

85 Taylor, M.J. *et al.* (2011) A high precision survey of the molecular dynamics of mammalian clathrin-mediated endocytosis. *PLoS Biol.* 9, e1000604

86 Doyon, J.B. *et al.* (2011) Rapid and efficient clathrin-mediated endocytosis revealed in genome-edited mammalian cells. *Nat. Cell Biol.* 13, 331–337

87 Newpher, T.M. *et al.* (2005) *In vivo* dynamics of clathrin and its adaptor-dependent recruitment to the actin-based endocytic machinery in yeast. *Dev. Cell* 9, 87–98

88 Tan, P.K. *et al.* (1993) Clathrin facilitates the internalization of seven transmembrane segment receptors for mating pheromones in yeast. *J. Cell Biol.* 123, 1707–1716

89 Ferguson, S.M. *et al.* (2009) Coordinated actions of actin and BAR proteins upstream of dynamin at endocytic clathrin-coated pits. *Dev. Cell* 17, 811–822

90 Jenness, D.D. and Spatrick, P. (1986) Down regulation of the alpha-factor pheromone receptor in *S. cerevisiae*. *Cell* 46, 345–353

91 Riezman, H. (1985) Endocytosis in yeast: several of the yeast secretory mutants are defective in endocytosis. *Cell* 40, 1001–1009

92 Vida, T.A. and Emr, S.D. (1995) A new vital stain for visualizing vacuolar membrane dynamics and endocytosis in yeast. *J. Cell Biol.* 128, 779–792

93 Volland, C. *et al.* (1994) Endocytose and degradation of the uracil permease of *S. cerevisiae* under stress conditions: possible role of ubiquitin. *Folia Microbiol. (Praha)* 39, 554–557

94 Berkower, C. *et al.* (1994) Metabolic instability and constitutive endocytosis of STE6, the a-factor transporter of *Saccharomyces cerevisiae*. *Mol. Biol. Cell* 5, 1185–1198

95 Kolling, R. and Hollenberg, C.P. (1994) The ABC-transporter Ste6 accumulates in the plasma membrane in a ubiquitinated form in endocytosis mutants. *EMBO J.* 13, 3261–3271

96 Egner, R. and Kuchler, K. (1996) The yeast multidrug transporter Pdr5 of the plasma membrane is ubiquitinated prior to endocytosis and degradation *in the vacuole*. *FEBS Lett.* 378, 177–181

97 Sirotkin, V. *et al.* (2010) Quantitative analysis of the mechanism of endocytic actin patch assembly and disassembly in fission yeast. *Mol. Biol. Cell* 21, 2894–2904

98 Stradalova, V. *et al*. (2009) Furrow-like invaginations of the yeast plasma membrane correspond to membrane compartment of Can1. *J. Cell Sci.* 122, 2887–2894

99 Moor, H. and Muhlethaler, K. (1963) Fine structure in frozen-etched yeast cells. *J. Cell Biol.* 17, 609–628

100 Walther, T.C. *et al*. (2006) Eisosomes mark static sites of endocytosis. *Nature* 439, 998–1003

101 Grossmann, G. *et al*. (2008) Plasma membrane microdomains regulate turnover of transport proteins in yeast. *J. Cell Biol.* 183, 1075–1088

102 Brach, T. *et al*. (2011) Reassessment of the role of plasma membrane domains in the regulation of vesicular traffic in yeast. *J. Cell Sci.* 124, 328–337

103 Moreira, K.E. *et al*. (2009) Pil1 controls eisosome biogenesis. *Mol. Biol. Cell* 20, 809–818

104 Ayscough, K.R. *et al*. (1997) High rates of actin filament turnover in budding yeast and roles for actin in establishment and maintenance of cell polarity revealed using the actin inhibitor latrunculin-A. *J. Cell Biol.* 137, 399–416

105 Mishra, R. *et al*. (2011) Expression of Vps1 I649K a self-assembly defective yeast dynamin, leads to formation of extended endocytic invaginations. *Commun. Integr. Biol.* 4, 115–117

106 Gupta, R. *et al*. (2007) Ubiquitination screen using protein microarrays for comprehensive identification of Rsp5 substrates in yeast. *Mol. Syst. Biol.* 3, 116

107 Peng, J. *et al*. (2003) A proteomics approach to understanding protein ubiquitination. *Nat. Biotechnol.* 21, 921–926

108 Stamenova, S.D. *et al*. (2007) Ubiquitin binds to and regulates a subset of SH3 domains. *Mol. Cell* 25, 273–284

GLOSSARY

ANTH domain AP180 N-terminal homology domain

EH-NPF interaction between Eps15 homology domain and asparagine-proline-phenylalanine motif (not to be confused with a nucleation-promoting factor).

HECT ligase E3 ubiquitin ligase with a HECT (homologous to the E6-AP carboxyl terminus) domain. These ligases form a covalent bond with the activated ubiquitin before transferring it directly to the substrate.

Line tension the interfacial energy, or sum of the free energy, between two lipid phases.

UIM ubiquitin-interacting motif.

CHAPTER 17

Life at the Leading Edge

Anne J. Ridley[1,*]

[1]Randall Division of Cell and Molecular Biophysics, King's College London,
New Hunt's House, Guy's Campus, London SE1 1UL, United Kingdom
*Correspondence: anne.ridley@kcl.ac.uk

Cell, Vol. 145, No. 7, June 24, 2011 © 2011 Elsevier Inc.
http://dx.doi.org/10.1016/j.cell.2011.06.010

SUMMARY

Cell migration requires sustained forward movement of the plasma membrane at the cell's front or "leading edge." To date, researchers have uncovered four distinct ways of extending the membrane at the leading edge. In lamellipodia and filopodia, actin polymerization directly pushes the plasma membrane forward, whereas in invadopodia, actin polymerization couples with the extracellular delivery of matrix-degrading metalloproteases to clear a path for cells through the extracellular matrix. Membrane blebs drive the plasma membrane forward using a combination of actomyosin-based contractility and reversible detachment of the membrane from the cortical actin cytoskeleton. Each protrusion type requires the coordination of a wide spectrum of signaling molecules and regulators of cytoskeletal dynamics. In addition, these different protrusion methods likely act in concert to move cells through complex environments in vivo.

INTRODUCTION

To reach their site of action, cells in multicellular animals not only move through the extracellular matrix but also on top of each other, between each other, and even through each other. For example, leukocytes attach to and migrate on endothelial cells lining the bloodstream before crossing the endothelium either between two endothelial cells or by inducing the formation of a membrane channel through a single endothelial cell (Carman, 2009). Cell migration has been studied at many different stages of animal development in vivo, as well as in numerous types of cells cultured in vitro.

373

CellPress

In order to move, cells must extend their plasma membrane forward at the front, or leading edge, of the cell. This is closely coordinated with movement of the cell body (Ridley et al., 2003).

Cells extend four different plasma membrane protrusions at the leading edge: lamellipodia, filopodia, blebs, and invadopodia. Each of these structures uniquely contributes to migration depending on the specific circumstances. For example, lamellipodia can extend long distances through the extracellular matrix in vivo, pulling cells through the tissues (Friedl and Gilmour, 2009). Filopodia explore the cell's surroundings and are particularly important for guidance of neuronal growth cones and angiogenic blood vessels (Eilken and Adams, 2010; Gupton and Gertler, 2007). Membrane blebbing has been described to drive directional cell migration during development (Charras and Paluch, 2008), and invadopodia are protrusions that allow focal degradation of the extracellular matrix, probably to facilitate invasion through the tissues (Buccione et al., 2009). These different types of protrusion can coexist at the leading edge; for example, lamellipodia, filopodia, and blebs have all been observed at the front of migrating zebrafish cells during gastrulation (Diz-Munoz et al., 2010).

Many different molecules and signaling pathways coordinate cell migration, but the actin cytoskeleton and regulators of actin dynamics are involved in all protrusions. Each actin regulator, in turn, is controlled by several signaling molecules, usually including a Rho GTPase, membrane phospholipids, and protein phosphorylation.

Rho GTPases are critical signal transducers that transmit signals from membrane receptors to the cytoskeleton and cell adhesions. Most Rho GTPases switch between an active GTP-bound conformation, which interacts with downstream effectors, and an inactive GDP-bound conformation. GTP hydrolysis converts Rho GTPases from the active to inactive form. Although they have intrinsic GTPase activity, their hydrolysis rates are normally slow and are accelerated in cells by GTPase-activating proteins (GAPs). Exchange of GDP for GTP induces activation, and this is catalyzed by guanine nucleotide exchange factors (GEFs). There are two families of GEFs: DH-PH domain-containing Dbl-related GEFs and DHR2 domain-containing DOCK family GEFs (Buchsbaum, 2007). Rho GTPases interact with membranes through, at least in part, lipid groups covalently attached posttranslationally, including farnesyl and geranylgeranyl isoprenoids. Some Rho GTPases are also regulated by Rho GDP-dissociation inhibitors (RhoGDIs); these proteins bind to isoprenoids and, hence, solubilize and extract the Rho GTPases from membranes (Buchsbaum, 2007).

This Review discusses how actin regulators contribute to the formation of the four protrusion types currently known to occur at the leading edge of migrating cells: lamellipodia, filopodia, blebs, and invadopodia. The Review describes the signaling molecules that activate these actin regulators and thus allow cells to respond dynamically to their extracellular environment with the most appropriate type of protrusion.

LAMELLIPODIA

Actin Regulators

The thin sheet-like region at the leading edge of migrating fibroblasts in culture was first named a "lamellipodium" by Michael Abercrombie in 1970 (Abercrombie et al., 1970). In elegant electron microscopy studies, he and colleagues showed that lamellipodia contain microfilaments (i.e., actin filaments) but not microtubules (Abercrombie et al., 1971). We now know that actin polymerization drives forward protrusion of the plasma membrane in lamellipodia (Ridley et al., 2003). Behind the highly dynamic lamellipodium is a more stable region, called the lamella, which contributes to cell migration by coupling the actin network to myosin II-mediated contractility and substrate adhesion (Ponti et al., 2004). Lamellipodia are observed in many different cell types moving in vivo, such as muscle precursors in chick embryos, epithelial and follicular epithelium border cells in *Drosophila*, and neural crest cells in *Xenopus* and zebrafish (Friedl and Gilmour, 2009; Weijer, 2009).

For many years, the Arp2/3 complex was thought to be the primary mediator of actin polymerization in lamellipodia. First discovered to nucleate actin polymerization in 1998 (Mullins et al., 1998), the Arp2/3 complex binds to the sides of actin filaments and stimulates the formation of branched "dendritic" actin filament networks (Campellone and Welch, 2010) (Figure 1). The Arp2/3 complex remains associated with filament pointed ends, and it is distributed throughout the lamellipodium, but it is incorporated only into the network at the front of the lamellipodium (Lai et al., 2008). In vitro, the nucleation-promoting factors of the WASP (Wiskott–Aldrich syndrome protein) family stimulate the ability of the Arp2/3 complex to induce actin polymerization. These factors, which include WASP itself, N-WASP, WAVE1–3, WASH, and WHAMM proteins, all bind to the Arp2/3 complex through a C-terminal acidic domain. WAVE proteins are known to localize to the leading edge and contribute to lamellipodium extension, whereas N-WASP may localize to and affect lamellipodia in some cell types or indirectly through its role in endocytosis (Campellone and Welch, 2010). WAVE1–3 exist in stable pentameric complexes with Abelson Interacting Protein (Abi), PIR121 (also known as Sra or CYFIP), Nck Associated Protein 1 (Nap1), and HSPC300 (Derivery and Gautreau, 2010). WASH is part of a similar complex that regulates actin polymerization on endosomes (Rottner et al., 2010), and N-WASP

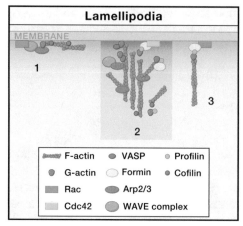

FIGURE 1 Lamellipodia

A model for lamellipodium formation is as follows: (1) Severing of actin filaments by cofilin provides free actin filament barbed ends, which act as sites for actin polymerization and subsequently Arp2/3-mediated nucleation of new filaments. (2) In conditions of steady-state lamellipodial extension, actin polymerization in lamellipodia is nucleated by the Arp2/3 complex, generating a branched actin filament network. The Arp2/3 complex is activated by the WAVE complex, which in turn is activated at the membrane by Rac1. Formins extend Arp2/3 complex-generated filaments. Formins are activated by Cdc42, Rac1, and probably other Rho GTPases. Actin monomers (G-actin) are provided to formins by profilin. VASP also contributes to actin filament extension. Cofilin severs and depolymerizes older actin filaments in the network. (3) Formins can also nucleate actin filaments independent of the Arp2/3 complex, generating unbranched filaments.

can also bind to Abi, but in this case the Abi1/N-WASP complex regulates endocytosis (Takenawa and Suetsugu, 2007).

More recently, other actin nucleators have been found to contribute to lamellipodium extension, including several members of the formin and Spire families. Formins protect barbed (+) ends of actin filaments from capping, and they promote filament elongation without branching (Figure 1). The formin mDia1, a Rho target, localizes at the leading edge (Chesarone et al., 2010). Members of the Spire family have multiple WH2 domains, which bind actin monomers and nucleate unbranched actin filaments. One of these proteins, Cordon-Bleu, is localized in lamellipodia and, when overexpressed, it increases the number of cells with lamellipodia and membrane ruffles (Campellone and Welch, 2010). JMY (junction-mediating and regulatory protein) is an unusual actin nucleator in that it has three WH2 domains and thus can nucleate unbranched actin filaments, like Spire, in the absence of the Arp2/3 complex. In addition, JMY has an acidic domain that binds to and stimulates the Arp2/3 complex and hence induces branched filaments (Zuchero et al., 2009). Although JMY is often in the nucleus, it also localizes at the leading edge, particularly in rapidly migrating neutrophils, and it regulates migration into scratch wounds. It would be interesting to determine whether JMY uses both of its actin nucleation activities at the leading edge to contribute to protrusion.

Branched actin filament networks have been found in electron microscopy images of lamellipodia (Svitkina and Borisy, 1999). However, the extent of actin filament branching in lamellipodia may vary depending on the cell type and conditions of fixation because a recent report found only few filament branches near the leading edge of cells (Urban et al., 2010). It could be that the balance of actin nucleation by the Arp2/3 complex relative to formins and Spires is key to how branched the network is at the leading edge. Indeed, the formin mDia2 is involved in generating long actin filaments in lamellipodia (Yang et al., 2007). In addition, different kinds of actin nucleators can work synergistically to promote actin polymerization (Chesarone and Goode, 2009). For example, mDia1, N-WASP, and WAVE2 all contribute to cell protrusion induced by epidermal growth factor (EGF) (Sarmiento et al., 2008). Adenomatosis polyposis coli (APC) has been recently described to have actin-nucleating activity and to act together with mDia1 (Okada et al., 2010). One possibility is that the Arp2/3 complex and/or Spires initiate nucleation in lamellipodia, whereas formins promote elongation.

The balance of other actin-binding proteins also contributes to the length of actin filaments in the lamellipodium. More capping protein activity reduces actin filament length and increases nucleation by the Arp2/3 complex by diverting actin monomers from elongation to nucleation (Akin and Mullins, 2008). On the other hand, more VASP, which promotes filament elongation, generates more long filaments (Bear and Gertler, 2009; Breitsprecher et al., 2011). Cofilin mediates the severing of existing cortical actin filaments, which generates new barbed ends and hence new filaments, to which the Arp2/3 complex can then bind and stimulate branching (van Rheenen et al., 2009) (Figure 1). Cortactin is a scaffolding protein that stabilizes Arp2/3 complex-induced branches and affects lamellipodial persistence (Lai et al., 2009; Ren et al., 2009). Super-resolution imaging (Toomre and Bewersdorf, 2010) might allow effects of each actin nucleator to be determined more precisely because the technique will permit the observation of actin filament arrangements in lamellipodia by optical microscopy, which requires less harsh fixation conditions than those required for electron microscopy.

There is also strong evidence that, in addition to actin polymerization, myosin II activity is required for stable lamellipodial extension, at least in cultured cells. Periodic myosin II-based contractions occur at the back of the lamellipodium (Giannone et al., 2007). These contractions could allow the protrusion to sense the pliability of the extracellular matrix and other cells and to determine the direction of migration. Myosin II activity is also implicated in actin filament disassembly at the back of the lamellipodium (Wilson et al., 2010).

Finally, several actin nucleators interact directly with microtubules, including the mDia proteins, APC, Spire, and WHAMM (WAS protein homolog associated with actin, golgi membranes, and microtubules). WHAMM's ability to bind microtubules probably relates to its functions in Golgi transport. Spire was reported to localize with Rab11 on endosomes and the Golgi, but it is not known whether it functions to nucleate actin filaments with formins at these sites or interacts with microtubules on a trafficking route to the plasma membrane (Campellone and Welch, 2010). The mDia proteins stabilize microtubules (Chesarone et al., 2010), whereas APC contributes to cell migration by capturing and stabilizing microtubule tips in the lamellipodium (Etienne-Manneville, 2009). It therefore seems likely that mDia and APC could coordinately regulate actin and microtubule cytoskeletons at the leading edge.

Signaling Molecules

Many extracellular stimuli induce the formation of lamellipodia, including growth factors, cytokines, and cell adhesion receptors; a myriad of signaling and structural proteins have been implicated in this process over the past 20 years. Rho GTPases act coordinately with other signals to activate actin regulators in lamellipodia (Figure 1).

Using biosensors, active Rac1, RhoA, and Cdc42 have been shown to localize in lamellipodia during protrusion (Machacek et al., 2009). Activation of Rac1 by itself, using a photoactivatable Rac1, is sufficient to induce lamellipodium extension (Wu et al., 2009), and it would be interesting to know whether this involves RhoA and Cdc42. Rho GTPases can be activated by multiple different GEFs at the leading edge, depending on the cell type and extracellular stimulus (Buchsbaum, 2007). More complex signaling is achieved through activation of GEFs by other Rho GTPases. For example, RhoG activates Rac/Cdc42 through its target protein ELMO (EnguLfment and cell MOtility) and DOCK family GEFs (Cote and Vuori, 2007). RhoG can also induce lamellipodia through an unknown Rac-independent pathway (Meller et al., 2008). Regulated localization of Rho GTPases is also important for their function: Rac is known to be recruited to the plasma membrane at the leading edge through vesicle trafficking (Donaldson et al., 2009), and multiple phosphorylations alter RhoGDI binding to Rho GTPases (Harding and Theodorescu, 2010).

Rac activates the pentameric WAVE complex, but it is currently unknown whether there is any difference in the ability of the three Rac isoforms, Rac1,

Rac2, and Rac3, to interact with the complex. Rac binds to PIR121 in the WAVE complex (Takenawa and Suetsugu, 2007). Structural analysis of the WAVE complex indicates that the C-terminal WCA domain of WAVE, which activates the Arp2/3 complex, is normally sequestered within the complex. It is predicted that Rac binding would induce structural rearrangements to allow the WCA domain to become accessible on the surface (Chen et al., 2010). The Rac target IRSp53 (insulin receptor tyrosine kinase substrate p53) contributes to lamellipodium extension by binding to Rac and WAVE2. Interestingly, the role of IRSp53 in lamellipodia can be selectively inhibited by Kank, an ankyrin repeat-containing protein that inhibits the binding of Rac but not Cdc42 to IRSp53 (Roy et al., 2009). WAVEs are also activated by tyrosine and serine/threonine phosphorylation, and again these phosphorylations are predicted to alter WAVE complex structure. Indeed, phosphomimicking mutations activate actin polymerization and lamellipodium formation (Chen et al., 2010; Sossey-Alaoui et al., 2007). Phosphorylation of Arp2, a component of the Arp2/3 complex, is important for its association with the pointed ends of actin filaments, which is required for it to induce effective filament branching and hence contribute to lamellipodium formation (LeClaire et al., 2008). Finally, the Rac target PAK (p21-activated protein kinase) may be involved in regulating the delivery of WAVE2 to the plasma membrane (Takahashi and Suzuki, 2009).

Cofilin/ADF is inhibited by phosphorylation, by binding to phosphatidylinositol 4,5-bisphosphate, and by increased pH (van Rheenen et al., 2009). Cdc42 and Rac act through their targets PAK and LIMK to phosphorylate and decrease the activity of cofilin (Bernard, 2007), which is probably important to allow cofilin recycling back to the membrane to generate new barbed ends for actin polymerization (van Rheenen et al., 2009) and to regulate the width of the lamellipodium (Delorme et al., 2007). This function of Rac could explain why Rac is most active slightly further back in the lamellipodium than RhoA (Machacek et al., 2009), although RhoA/ROCK can also phosphorylate and inhibit cofilin/ADF (Bernard, 2007). Indeed, both PAK and ROCK appear to regulate cofilin phosphorylation at the leading edge (Delorme et al., 2007). The Rac target NADPH oxidase, which generates reactive oxygen species (ROS), has also been implicated in lamellipodia (Nimnual et al., 2003). One possible mechanism whereby ROS could contribute to lamellipodia is through cofilin; ROS lead to cofilin dephosphorylation through activation of the cofilin phosphatase Slingshot (Kim et al., 2009).

RhoA has predominantly been implicated in tail retraction of migration cells (Ridley et al., 2003), but it is clearly active at the front of lamellipodia where it might act to stimulate mDia1-mediated actin polymerization and/or myosin II-mediated retraction events (Pertz, 2010). Notably, RhoA is also highly active in membrane ruffles that retract backward from lamellipodia (Pertz et al., 2006). The RhoGEF GEF-H1 could be important for activating RhoA in lamellipodia because GEF-H1 knockdown decreases RhoA activation at the leading edge and reduces forward protrusion (Nalbant et al., 2009). Although RhoA depletion induces loss of lamellipodia (Heasman et al., 2010), too much RhoA activity at the front inhibits lamellipodial extension. RhoA levels have been reported to be regulated locally in lamellipodia by Smurf1- (SMAD-specific E3 ubiquitin protein ligase) and ubiquitin-mediated degradation. Moreover, decreasing the expression of Smurf1 reduces lamellipodia and increases membrane blebbing (Sahai et al., 2007; Wang et al., 2003). However, interpretation of these results is complicated by the fact that Smurf1 has multiple targets, several of which affect cell migration (Huang, 2010).

Lamellipodial extension is abruptly terminated by contact inhibition, when two cells of the same type touch. Michael Abercrombie first described contact inhibition between migrating fibroblasts (Abercrombie and Heaysman, 1954), and studies more recently have explored the mechanisms underlying contact inhibition. For example, ephrin (Eph) receptors inhibit lamellipodial extension by activating myosin II-mediated retraction through a combination of Cdc42/MRCK and RhoA/ROCK signaling (Astin et al., 2010; Groeger and Nobes, 2007). In vivo, contact inhibition is used during development to guide the migration of neural crest cells. Noncanonical Wnt signaling leads to activation of RhoA at sites of contact between neural crest cells, which then represses lamellipodia through its target ROCK (Carmona-Fontaine et al., 2008), presumably through increased actomyosin contractility. It is interesting that RhoA can either contribute to or inhibit lamellipodial extension depending on the circumstances, which might reflect involvement of different RhoGEFs and downstream targets (Heasman et al., 2010).

FILOPODIA

Actin Regulators

Filopodia are exploratory extensions from the plasma membrane that contain parallel bundles of actin filaments (Figure 2). Fascin is the major actin-bundling protein that localizes to filopodia and is important for filopodium stability (Machesky and Li, 2010). One model for filopodia assembly is that they emerge from the lamellipodial F-actin network nucleated by the Arp2/3 complex through the binding of proteins such as fascin and the anti-capping protein VASP (Vasodilator-stimulated phosphoprotein) (Gupton and Gertler, 2007). Indeed, N-WASP is required for filopodium formation in certain situations, and the Arp2/3 complex can be active in filopodia (Johnston et al., 2008; Takenawa and Suetsugu, 2007). However, filopodia can also be observed independent of N-WASP, the Arp2/3 complex, and lamellipodia

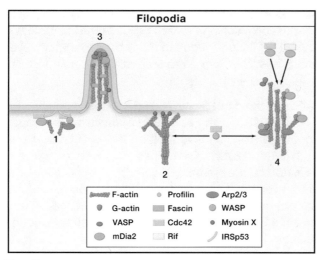

FIGURE 2 Filopodia

A model for filopodium formation is as follows: (1) IRSp53 initiates filopodia by bending the membrane and recruiting Cdc42 and Cdc42 targets, mDia2 and WASP/N-WASP, which then stimulate actin polymerization. (2) Actin filaments could also be provided from lamellipodia, where Myosin X could cluster WASP/Arp2/3-nucleated actin filaments. (3) Filopodia subsequently extend through the addition of actin monomers (G-actin) onto actin filaments (F-actin). VASP, Myosin X, and mDia2 are localized to the tip of filopodia. Myosin X moves dynamically in filopodia and could contribute to delivery of proteins to the filopodial tip. (4) Actin polymerization in filopodia is nucleated by mDia2 in concert with VASP, which delivers actin monomers to the filopodial tip. Profilin binds to and provides actin monomers directly to mDia2. Cdc42 and Rif stimulate mDia2-mediated actin polymerization, and Cdc42 also stimulates WASP/Arp2/3-driven polymerization.

(Takenawa and Suetsugu, 2007), and it is now clear that formins, in particular the mDia proteins, are major contributors to actin polymerization in filopodia (Mellor, 2010; Campellone and Welch, 2010).

VASP and its relatives Mena and Evl (known as Ena/VASP proteins) localize to tips of filopodia, and at least in certain systems, they are essential for filopodium extension (Bear and Gertler, 2009). In vitro, Ena/VASP proteins have an anti-capping protein function, and they promote filament elongation (Hansen and Mullins, 2010). VASP oligomers could stimulate filament elongation in filopodia by delivering actin monomers to the growing tips of filopodia (Applewhite et al., 2007; Breitsprecher et al., 2008) but also by inhibiting filament capping (Bear and Gertler, 2009).

Overexpression of a variety of proteins can increase the number of filopodia on cells. For example, filopodia can be induced by proteins containing I-BAR domains, which bend the plasma membrane outwards (Figure 2). The IRSp53 protein is a multidomain protein that induces filopodia through its I-BAR domain. The I-BAR domain alone induces small dynamic filopodium-like membrane protrusions lacking F-actin. IRSp53 also interacts with N-WASP, which is required for IRSp53-induced filopodium formation, even though in other conditions N-WASP is not required for filopodium formation (Ahmed et al., 2010; Takenawa and Suetsugu, 2007).

Myosin X traffics to the tip of filopodia and can induce filopodium assembly (Sousa and Cheney, 2005). The selectivity of myosin X for filopodia appears to be due to its preferential movement on fascin-actin bundles (Nagy and Rock, 2010). Two mechanisms have been proposed to explain how myosin X stimulates filopodium assembly (Figure 2). First, it could deliver cargo, such as actin monomers, to the growing tips of filopodia and, hence, accelerate filament elongation (Zhuravlev et al., 2010). Second, its motor function could induce actin filament convergence at the leading edge to initiate filopodium extension (Tokuo et al., 2007).

Signaling Molecules

Cdc42 was the first Rho GTPase found to induce filopodia. Cdc42 could bring together three of its targets, IRSp53 (I-BAR protein), mDia2, and N-WASP, all of which can contribute to filopodium initiation and extension (Ahmed et al., 2010) (Figure 2). I-BAR domains may activate membrane protrusion by clustering membrane phosphatidylinositol 4,5-bisphosphate (PIP_2) (Zhao et al., 2011), which could then contribute to activation of PIP_2-binding proteins, such as N-WASP. N-WASP is also activated by tyrosine phosphorylation (Takenawa and Suetsugu, 2007), and thus, it is regulated cooperatively by multiple signals (Figure 2). In contrast, IRSp53 function is inhibited by threonine phosphorylation and subsequent binding to 14-3-3 proteins (Robens et al., 2010). In addition to Cdc42, other Rho GTPases can also induce filopodia (Mellor, 2010). For example, RhoF/Rif induces filopodia via mDia2, and this is important in the early stages of dendritic spine assembly in neurons (Hotulainen et al., 2009). Whether RhoF also recruits I-BAR proteins remains to be determined.

Fascin and Ena/VASP binding to actin filaments is inhibited by phosphorylation (Bear and Gertler, 2009; Machesky and Li, 2010). Protein kinase C (PKC) phosphorylates fascin, and Rac regulates the interaction of fascin with PKC (Parsons and Adams, 2008), and thus, Rac might inhibit filopodium assembly. Alternatively, it is possible that fascin in association with PKC has a separate function independent of its actin-bundling activity (Hashimoto et al., 2007). Interestingly, Rab35 interacts directly with fascin and could be important for its delivery to filopodia (Zhang et al., 2009). Rab35 induces long filopodium-like protrusions and is required in *Drosophila* cells for Cdc42 delivery to the plasma membrane (Chua et al., 2010; Shim et al., 2010). It will be interesting to know whether Cdc42 and fascin delivery is coordinated by Rab35 to ensure that they act together in stimulating filopodium extension.

How is filopodium extension terminated? Formin displacers or inhibitors could be important in this process. In budding yeast, BUD14 (Bud site selection protein 14) directly binds to the FH2 domain of the Bnr formin and displaces it from the barbed end of actin filaments (Chesarone et al., 2009). It is not yet known whether mammalian cells have a protein that acts similarly to BUD14. In mammalian cells, overexpression of Dia-interacting protein (DIP) inhibits mDia2-induced actin assembly in vitro and filopodium formation in vivo (Eisenmann et al., 2007).

INVADOPODIA

Actin Regulators

Invadopodia were first described as actin-rich matrix-degrading protrusions in Rous sarcoma virus-transformed fibroblasts, driven by oncogenic Src tyrosine kinase (Chen, 1989). Invadopodia and related structures known as podosomes are important for degrading the extracellular matrix during cell invasion (Buccione et al., 2009), particularly when cells cross the basement membrane (Schoumacher et al., 2011). Invadopodium extension in three dimensions (3D) requires force driven by actin polymerization. Many studies on invadopodia are carried out on two-dimensional (2D) surfaces coated with extracellular matrix proteins, where they are present on the ventral surface. In 3D, invading cells often extend long protrusions that degrade the matrix (Friedl and Gilmour, 2009; Schoumacher et al., 2010; Wolf and Friedl, 2009).

Although many of the actin-regulatory proteins found in invadopodia are also in filopodia and lamellipodia, the key difference is that invadopodia degrade the extracellular matrix, and thus, they require the delivery of vesicles containing matrix-degrading proteases, particularly membrane type 1 metalloprotease (MT1-MMP). These vesicles are targeted to invadopodia by the vesicle-tethering exocyst complex (Poincloux et al., 2009) (Figure 3). Microtubules are also important for invadopodium extension, probably for the delivery of vesicles, and intermediate filaments might provide mechanical stability (Schoumacher et al., 2010).

The Arp2/3 complex is an essential component of invadopodia, and N-WASP (and WASP in hematopoietic cells) appears to be the major Arp2/3 complex activator (Buccione et al., 2009) (Figure 3). However, the formins mDia1–3 are also required for invadopodium assembly and invasion, implying that Arp2/3 complex and formins cooperate to induce actin polymerization in invadopodia (Lizarraga et al., 2009), as in lamellipodia and filopodia. Several

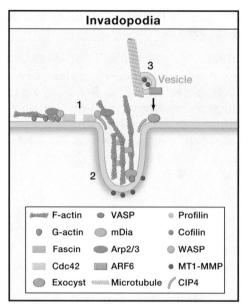

FIGURE 3 Invadopodia

A model for invadopodium assembly is as follows: (1) Actin polymerization in invadopodia is mediated by Cdc42-mediated N-WASP/WASP activation of the Arp2/3 complex. Cofilin severs actin filaments (F-actin) to provide new sites for actin nucleation. CIP4 might bend membranes and also help to recruit Cdc42/N-WASP. (2) Actin filament elongation in the invadopodium requires mDia formins and profilin/actin monomers (G-actin). Actin filaments are bundled by fascin. (3) MT1-MMP and possibly other MMPs are transported to the invadopodial tip by vesicle trafficking, initially on microtubules, and requiring ARF6. The exocyst captures vesicles at the plasma membrane.

actin-binding proteins also contribute to formation of invadopodia. For example, cortactin binds to and buffers cofilin: cortactin phosphorylation releases cofilin so that it can sever filaments to create new barbed ends for actin polymerization; then dephosphorylated cortactin inhibits cofilin's actin-severing activity to promote filament elongation (Oser et al., 2009). In addition, the actin-bundling protein fascin has recently been shown to be critical for invadopodium stability (Machesky and Li, 2010).

Signaling Molecules

Cdc42 is the main Rho GTPase implicated in the formation of invadopodia, and it appears to coordinate actin filament assembly with matrix degradation. Cdc42 is required for N-WASP/WASP targeting; it is not yet known whether Cdc42 also regulates mDia proteins in invadopodia. Cdc42 could be activated to form invadopodia by the RhoGEF Fgd1 (faciogenital dysplasia protein), which is mutated in faciogenital dysplasia (Ayala et al., 2009).

Generation of membrane curvature is also important for invadopodium assembly, as for filopodia. Recently, the F-BAR-containing protein CIP4 has been implicated in invadopodia (Pichot et al., 2010). CIP4 also binds Cdc42 and N-WASP (Figure 3), thereby acting as a membrane-curving scaffolding protein similar to IRSp53 in filopodia.

A role for RhoA in invadopodia has also been suggested, but its contribution is not clear, apart from a possible role in regulating exocyst binding to IQGAP1, which can also be mediated by Cdc42 (Buccione et al., 2009).

Tyrosine kinases of the Src family stimulate invadopodium assembly, and indeed many Src substrates are in invadopodia, including cortactin and WASP/N-WASP, for which tyrosine phosphorylation is important for podosome assembly (Dovas and Cox, 2010). Reactive oxygen species generated by NADPH oxidases are required for assembly of invadopodia (Diaz et al., 2009; Weaver, 2009), and NADPH oxidase components, such as Tks, are Src substrates implicated in formation of invadopodia (Buccione et al., 2009). Abl tyrosine kinases also appear to be important for targeting or retaining MT1-MMP in invadopodia (Smith-Pearson et al., 2010).

In summary, it appears that Cdc42 and tyrosine kinases act coordinately to drive both the actin polymerization required for invadopodium extension and the delivery and retention of MT1-MMP to the surface of invadopodia (Figure 3).

BLEBS

Actin Regulators

Membrane blebbing was first described in migrating amphibian and fish cells in several papers between the 1940s and the 1970s (Charras and Paluch, 2008). Recently, blebbing has received renewed interest in the migration field with the observations that multiple cell types move by blebbing under certain conditions, including cancer cells and *Dictyostelium* cells in vitro and several cell types in vivo (Charras and Paluch, 2008; Fackler and Grosse, 2008). In vitro, blebbing is often observed on or within pliable extracellular matrix environments, in contrast to the predominance of lamellipodia on rigid substrates.

Blebs form when the plasma membrane detaches focally from the underlying actin filament cortex, allowing cytoplasmic flow to push the membrane outwards rapidly due to hydrostatic pressure in the cell interior (Bovellan et al., 2010). Myosin II-induced actomyosin contraction increases hydrostatic pressure locally or globally leading to focal rupture of the actin cortex from the membrane, thereby driving the formation of blebs (Tinevez et al.,

2009). Reduced association between the cortex and the membrane could also drive blebbing, for example in filamin null cells (filamin is an actin filament crosslinking protein critical for stability of the actin cortex) (Charras and Paluch, 2008).

Once blebs have extended, actin filaments reassemble on the bleb membrane to form a new actin cortex (Figure 4). The actin nucleator required to stimulate actin polymerization is likely to vary depending on the cell type. Several formins have been implicated in membrane blebbing. Stabilization of the actin cortex requires ERM (Ezrin, Radixin, and Moesin) proteins, which link actin filaments to the plasma membrane and to membrane receptors. Finally, Myosin II is recruited, and actomyosin contraction can power retraction of membrane blebs. Thus, contractility not only can induce blebbing but also contribute to bleb termination (Charras and Paluch, 2008; Fackler and Grosse, 2008). In migrating cells, however, blebs are not always retracted; instead, new blebs extend out of existing blebs (Kardash et al., 2010).

Leading edge extension through blebbing and lamellipodia is not mutually exclusive. For example, both structures can be observed in different regions or at different times of the extending membrane in zebrafish prechordal plate precursor cells during gastrulation (Diz-Munoz et al., 2010). However, an increase in blebbing leads to a decrease in lamellipodia and vice versa (Derivery et al., 2008), reflecting the very different mechanical processes: blebbing requires loss of actin filament interaction with the membrane,

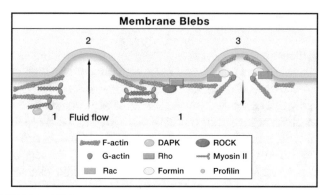

FIGURE 4 Membrane Blebs

A model for membrane blebbing is as follows: (1) Membrane blebs are induced by local weakening of plasma membrane/cortical actin interactions, coupled to actomyosin contractility on the membrane. Actomyosin contractility can be induced by Rho/ROCK and/or DAPK stimulation of myosin light chain phosphorylation on myosin II. (2) This leads to fluid flow pushing the membrane outwards locally. (3) Actin polymerization on the membrane in blebs leads to re-engagement of the plasma membrane with cortical actin filaments and retraction of blebs. This polymerization might be mediated by formins such as mDia proteins or FHOD1 and require activation by Rac1.

whereas lamellipodium extension requires close interaction of actin filaments with the membrane.

Signaling Molecules

Cells bleb during the initial stages of adhesion to extracellular matrix proteins before they adhere firmly (Dubin-Thaler et al., 2008). This appears to correlate with rates of membrane endocytosis and exocytosis, suggesting that blebbing is due to excess membrane (Norman et al., 2010). Rho and its target ROCK induce blebbing through increased actomyosin contractility, but it is not known which of the three Rho isoforms (RhoA, RhoB, or RhoC) is actually responsible for blebbing during migration. Microtubule depolymerization activates RhoA/ROCK, which then leads to membrane blebbing and bleb-based migration (Pletjushkina et al., 2001; Takesono et al., 2010). Germ cells in zebrafish extend membrane blebs at the leading edge during migration. Interestingly, both RhoA and Rac1 are active in this region (Kardash et al., 2010). Inhibition of Rho isoforms prevents membrane blebbing, whereas Rac1 is required for actin polymerization at the front. Rac1-induced actin polymerization, in turn, is dependent on E-cadherin engagement, which is well known to stimulate Rac1 (Yap and Kovacs, 2003). This suggests a model in which Rho/ROCK-induced bleb extension is followed by adhesion-stimulated Rac-driven actin polymerization, both of which are required for sustained membrane extension during directional migration in vivo.

Apart from ROCK, other kinases can also stimulate myosin light chain (MLC) phosphorylation in membrane blebbing, including the Death-associated protein kinase (DAPK) (Bovellan et al., 2010). DAPK can act redundantly with ROCK to regulate phosphorylated MLC levels (Neubueser and Hipfner, 2010), and hence ROCK-independent blebbing could involve DAPK.

Several formins are linked to membrane blebbing (Figure 4), but whether they contribute to physiological blebbing during migration is not yet known. Diaphanous-interacting protein (DIP) is a scaffold protein that induces membrane blebbing when overexpressed, presumably because it binds to and inhibits the activity of the formin mDia2 (Eisenmann et al., 2007). This suggests that mDia2 could be involved in stabilizing the actin cortex, thereby inhibiting detachment of the membrane from the cortex. Overexpression of the formin FHOD1 (FH2 domain-containing protein 1) reduces the size of ROCK1-induced membrane blebs but increases their number (Hannemann et al., 2008). This suggests that FHOD1 too might stabilize the actin cortex and/or stimulate actin polymerization in blebs to promote bleb retraction. On the other hand, overexpression of a constitutively activated form of the formin FMNL1 (Formin-like protein 1) alone induces membrane blebbing, independently of ROCK (Han et al., 2009), and thus this response might require DAPK. It will be interesting to determine whether FMNL1 stimulates blebbing by thickening the actin cortex and hence increasing cortical tension.

COLLECTIVE CELL MIGRATION

Collective migration is the simultaneous movement of multiple cells attached to each other through cell-cell adhesion, which occurs reiteratively during development and wound healing (Weijer, 2009). Live-cell imaging during development has shown that the leading cells of collectively migrating groups selectively extend lamellipodia, filopodia, and/or blebs, whereas cells behind rarely extend protrusions (Friedl and Gilmour, 2009; Diz-Munoz et al., 2010).

Recent data have indicated the involvement of Rho GTPases and other signaling pathways in collective cell migration. It is clear that Rac-driven lamellipodial extension is important in the leader cells of collective groups in a number of models. For example, during angiogenesis, the tip cells at the front of sprouting blood vessels extend long protrusions, which are presumed to be required for navigation sensing, such as guidance toward VEGF (vascular endothelial growth factor) (Eilken and Adams, 2010) . Rac1 is required for VEGF-induced endothelial sprouting, and local downregulation of myosin II on the cortex is important for tip branching in 3D (Eilken and Adams, 2010). Similarly, duct initiation in mammary epithelial morphogenesis requires Rac (Ewald et al., 2008), and *Drosophila* epithelial border cells extend long Rac-driven protrusions in order to migrate as a group between egg chamber cells toward the oocyte (Friedl and Gilmour, 2009). Rac activation in the leading cell, using a photoactivatable Rac1, is sufficient to drive polarization of the border cell cluster (Wang et al., 2010). Consistent with its role in lamellipodia (see above), the balance of Rac-regulated cofilin phosphorylation is also critical for collective border cell migration. It is possible that high levels of phospho-cofilin could suppress protrusion in all cells of the group, except the leading cell (Zhang et al., 2011). Alternatively, cofilin and its regulator LIMK could be critical for matrix degradation, which is particularly important in the leading cell of collectively migrating cancer cells (Scott et al., 2010).

Maintaining intact cell-cell adhesions is essential for collective migration. In A431 cancer cells, this maintenance has been shown to require the transmembrane receptor DDR1 (Discoidin domain receptor tyrosine kinase 1), which recruits the Par polarity complex to cell-cell junctions and reduces actomyosin contractility (Hidalgo-Carcedo et al., 2011). The Par complex, in turn, is known to be important for assembly of cell-cell junctions and is often regulated by Cdc42 (Goldstein and Macara, 2007). However, whether Cdc42 contributes to collective migration of cancer cells has not been investigated.

Suppression of actomyosin contractility at cell-cell junctions also enhances endothelial vessel sprouting (Abraham et al., 2009), indicating that the balance of forces acting on cell-cell interactions is critical for collective movement of cells.

CONCLUSIONS AND PERSPECTIVES

Since the "textbook" model for actin dynamics in lamellipodia was first described by Pollard and Borisy (2003), our understanding of how protrusions are initiated, extended, and retracted has increased on numerous fronts. It is now clear that multiple actin nucleators are involved in each type of protrusion and that some formins and Spire family proteins have important roles at the leading edge. Different protrusions can exist together, such as filopodia and lamellipodia at the leading edge of some cell types, whereas in other cases they act independently. For example, filopodia are initiated in the absence of lamellipodia as the starting point for dendritic spines on neurons (Yoshihara et al., 2009). Cells can also switch rapidly between different types of protrusion. For example, in *Dictyostelium*, blebs interchange rapidly with filopodia and lamellipodia at the leading edge (Yoshida and Soldati, 2006).

Multiple signals regulate protrusions, but how they act together to coordinate protrusion extension and retraction is currently not yet clear. For example, although we now know that the Rho GTPases RhoA, Rac1, and Cdc42 are all active in lamellipodia, we still do not know precisely where and when they interact with each of their downstream targets. RhoA is active in areas of membrane blebbing, but where exactly it stimulates actomyosin contraction with respect to where the bleb extends is not clear. We know even less about where Rho GTPases are active in filopodia or invadopodia. In many cases, Rho GTPases need to act synergistically with other signals, and thus new methods are needed to follow Rho GTPase activity simultaneously with other signaling molecules in cells.

Even though there are 20 Rho GTPases and most of them affect the cytoskeleton in some way, our understanding of how they regulate protrusions is based primarily on Cdc42, Rac1/2, and RhoA. Perhaps this is because these are the mostly highly conserved GTPases in eukaryotes (Boureux et al., 2007), and hence, they are the ones that are actually central to protrusion, whereas the other ones serve more specialized functions in specific cell types.

New microscopy techniques should allow us to visualize in ever-greater detail the interactions and localization of proteins in protrusions. Super-resolution microscopy provides important in-depth snapshots of protein localization, although so far it is not possible to use these methods for rapid live-cell imaging (Toomre and Bewersdorf, 2010). Rapid imaging techniques now being used to visualize cells in vivo will provide us with better insight into how cells coordinate membrane protrusions at the leading edge in physiological and pathological environments.

Cell migration is central to many chronic human diseases, including cancer, cardiovascular disease, and chronic inflammation. Therefore, new insights into the crucial molecules required for cell protrusion will be important in designing therapies to counter these diseases.

ACKNOWLEDGMENTS

I am grateful to Cancer Research UK, Bettencourt-Schueller Foundation, and King's College London BHF Centre of Excellence for grant support.

REFERENCES

Abercrombie, M., and Heaysman, J.E. (1954). Observations on the social behaviour of cells in tissue culture. II. Monolayering of fibroblasts. Exp. Cell Res. 6, 293–306.

Abercrombie, M., Heaysman, J.E., and Pegrum, S.M. (1970). The locomotion of fibroblasts in culture. II. "Ruffling". Exp. Cell Res. 60, 437–444.

Abercrombie, M., Heaysman, J.E., and Pegrum, S.M. (1971). The locomotion of fibroblasts in culture. IV. Electron microscopy of the leading lamella. Exp. Cell Res. 67, 359–367.

Abraham, S., Yeo, M., Montero-Balaguer, M., Paterson, H., Dejana, E., Marshall, C.J., and Mavria, G. (2009). VE-Cadherin-mediated cell-cell interaction suppresses sprouting via signaling to MLC2 phosphorylation. Curr. Biol. 19, 668–674.

Ahmed, S., Goh, W.I., and Bu, W. (2010). I-BAR domains, IRSp53 and filopodium formation. Semin. Cell Dev. Biol. 21, 350–356.

Akin, O., and Mullins, R.D. (2008). Capping protein increases the rate of actin-based motility by promoting filament nucleation by the Arp2/3 complex. Cell 133, 841–851.

Applewhite, D.A., Barzik, M., Kojima, S., Svitkina, T.M., Gertler, F.B., and Borisy, G.G. (2007). Ena/VASP proteins have an anti-capping independent function in filopodia formation. Mol. Biol. Cell 18, 2579–2591.

Astin, J.W., Batson, J., Kadir, S., Charlet, J., Persad, R.A., Gillatt, D., Oxley, J.D., and Nobes, C.D. (2010). Competition amongst Eph receptors regulates contact inhibition of locomotion and invasiveness in prostate cancer cells. Nat. Cell Biol. 12, 1194–1204.

Ayala, I., Giacchetti, G., Caldieri, G., Attanasio, F., Mariggio, S., Tete, S., Polishchuk, R., Castronovo, V., and Buccione, R. (2009). Faciogenital dysplasia protein Fgd1 regulates invadopodia biogenesis and extracellular matrix degradation and is up-regulated in prostate and breast cancer. Cancer Res. 69, 747–752.

Bear, J.E., and Gertler, F.B. (2009). Ena/VASP: towards resolving a pointed controversy at the barbed end. J. Cell Sci. 122, 1947–1953.

Bernard, O. (2007). Lim kinases, regulators of actin dynamics. Int. J. Biochem. Cell Biol. 39, 1071–1076.

Boureux, A., Vignal, E., Faure, S., and Fort, P. (2007). Evolution of the Rho family of ras-like GTPases in eukaryotes. Mol. Biol. Evol. 24, 203–216.

Bovellan, M., Fritzsche, M., Stevens, C., and Charras, G. (2010). Death-associated protein kinase (DAPK) and signal transduction: blebbing in programmed cell death. FEBS J. *277*, 58–65.

Breitsprecher, D., Kiesewetter, A.K., Linkner, J., Urbanke, C., Resch, G.P., Small, J.V., and Faix, J. (2008). Clustering of VASP actively drives processive, WH2 domain-mediated actin filament elongation. EMBO J. *27*, 2943–2954.

Breitsprecher, D., Kiesewetter, A.K., Linkner, J., Vinzenz, M., Stradal, T.E., Small, J.V., Curth, U., Dickinson, R.B., and Faix, J. (2011). Molecular mechanism of Ena/VASP-mediated actin-filament elongation. EMBO J. *30*, 456–467.

Buccione, R., Caldieri, G., and Ayala, I. (2009). Invadopodia: specialized tumor cell structures for the focal degradation of the extracellular matrix. Cancer Metastasis Rev. *28*, 137–149.

Buchsbaum, R.J. (2007). Rho activation at a glance. J. Cell Sci. *120*, 1149–1152.

Campellone, K.G., and Welch, M.D. (2010). A nucleator arms race: cellular control of actin assembly. Nat. Rev. Mol. Cell Biol. *11*, 237–251.

Carman, C.V. (2009). Mechanisms for transcellular diapedesis: probing and pathfinding by 'invadosome-like protrusions'. J. Cell Sci. *122*, 3025–3035.

Carmona-Fontaine, C., Matthews, H.K., Kuriyama, S., Moreno, M., Dunn, G.A., Parsons, M., Stern, C.D., and Mayor, R. (2008). Contact inhibition of locomotion in vivo controls neural crest directional migration. Nature *456*, 957–961.

Charras, G., and Paluch, E. (2008). Blebs lead the way: how to migrate without lamellipodia. Nat. Rev. Mol. Cell Biol. *9*, 730–736.

Chen, W.T. (1989). Proteolytic activity of specialized surface protrusions formed at rosette contact sites of transformed cells. J. Exp. Zool. *251*, 167–185.

Chen, Z., Borek, D., Padrick, S.B., Gomez, T.S., Metlagel, Z., Ismail, A.M., Umetani, J., Billadeau, D.D., Otwinowski, Z., and Rosen, M.K. (2010). Structure and control of the actin regulatory WAVE complex. Nature *468*, 533–538.

Chesarone, M.A., and Goode, B.L. (2009). Actin nucleation and elongation factors: mechanisms and interplay. Curr. Opin. Cell Biol. *21*, 28–37.

Chesarone, M., Gould, C.J., Moseley, J.B., and Goode, B.L. (2009). Displacement of formins from growing barbed ends by bud14 is critical for actin cable architecture and function. Dev. Cell *16*, 292–302.

Chesarone, M.A., DuPage, A.G., and Goode, B.L. (2010). Unleashing formins to remodel the actin and microtubule cytoskeletons. Nat. Rev. Mol. Cell Biol. *11*, 62–74.

Chua, C.E., Lim, Y.S., and Tang, B.L. (2010). Rab35–a vesicular traffic-regulating small GTPase with actin modulating roles. FEBS Lett. *584*, 1–6.

Cote, J.F., and Vuori, K. (2007). GEF what? Dock180 and related proteins help Rac to polarize cells in new ways. Trends Cell Biol. *17*, 383–393.

Delorme, V., Machacek, M., DerMardirossian, C., Anderson, K.L., Wittmann, T., Hanein, D., Waterman-Storer, C., Danuser, G., and Bokoch, G.M. (2007). Cofilin activity downstream of Pak1 regulates cell protrusion efficiency by organizing lamellipodium and lamella actin networks. Dev. Cell *13*, 646–662.

Derivery, E., and Gautreau, A. (2010). Generation of branched actin networks: assembly and regulation of the N-WASP and WAVE molecular machines. Bioessays *32*, 119–131.

Derivery, E., Fink, J., Martin, D., Houdusse, A., Piel, M., Stradal, T.E., Louvard, D., and Gautreau, A. (2008). Free Brick1 is a trimeric precursor in the assembly of a functional wave complex. PLoS ONE *3*, e2462.

Diaz, B., Shani, G., Pass, I., Anderson, D., Quintavalle, M., and Courtneidge, S.A. (2009). Tks5-dependent, nox-mediated generation of reactive oxygen species is necessary for invadopodia formation. Sci. Signal. *2*, ra53.

Diz-Munoz, A., Krieg, M., Bergert, M., Ibarlucea-Benitez, I., Muller, D.J., Paluch, E., and Heisenberg, C.P. (2010). Control of directed cell migration in vivo by membrane-to-cortex attachment. PLoS Biol. *8*, e1000544.

Donaldson, J.G., Porat-Shliom, N., and Cohen, L.A. (2009). Clathrin-independent endocytosis: a unique platform for cell signaling and PM remodeling. Cell. Signal. *21*, 1–6.

Dovas, A., and Cox, D. (2010). Regulation of WASp by phosphorylation: Activation or other functions? Commun. Integr. Biol. *3*, 101–105.

Dubin-Thaler, B.J., Hofman, J.M., Cai, Y., Xenias, H., Spielman, I., Shneidman, A.V., David, L.A., Dobereiner, H.G., Wiggins, C.H., and Sheetz, M.P. (2008). Quantification of cell edge velocities and traction forces reveals distinct motility modules during cell spreading. PLoS ONE *3*, e3735.

Eilken, H.M., and Adams, R.H. (2010). Dynamics of endothelial cell behavior in sprouting angiogenesis. Curr. Opin. Cell Biol. *22*, 617–625.

Eisenmann, K.M., Harris, E.S., Kitchen, S.M., Holman, H.A., Higgs, H.N., and Alberts, A.S. (2007). Dia-interacting protein modulates formin-mediated actin assembly at the cell cortex. Curr. Biol. *17*, 579–591.

Etienne-Manneville, S. (2009). APC in cell migration. Adv. Exp. Med. Biol. *656*, 30–40.

Ewald, A.J., Brenot, A., Duong, M., Chan, B.S., and Werb, Z. (2008). Collective epithelial migration and cell rearrangements drive mammary branching morphogenesis. Dev. Cell *14*, 570–581.

Fackler, O.T., and Grosse, R. (2008). Cell motility through plasma membrane blebbing. J. Cell Biol. *181*, 879–884.

Friedl, P., and Gilmour, D. (2009). Collective cell migration in morphogenesis, regeneration and cancer. Nat. Rev. Mol. Cell Biol. *10*, 445–457.

Giannone, G., Dubin-Thaler, B.J., Rossier, O., Cai, Y., Chaga, O., Jiang, G., Beaver, W., Dobereiner, H.G., Freund, Y., Borisy, G., and Sheetz, M.P. (2007). Lamellipodial actin mechanically links myosin activity with adhesion-site formation. Cell *128*, 561–575.

Goldstein, B., and Macara, I.G. (2007). The PAR proteins: fundamental players in animal cell polarization. Dev. Cell *13*, 609–622.

Groeger, G., and Nobes, C.D. (2007). Co-operative Cdc42 and Rho signalling mediates ephrinB-triggered endothelial cell retraction. Biochem. J. *404*, 23–29.

Gupton, S.L., and Gertler, F.B. (2007). Filopodia: the fingers that do the walking. Sci. STKE *2007*, re5.

Han, Y., Eppinger, E., Schuster, I.G., Weigand, L.U., Liang, X., Kremmer, E., Peschel, C., and Krackhardt, A.M. (2009). Formin-like 1 (FMNL1) is regulated by N-terminal myristoylation and induces polarized membrane blebbing. J. Biol. Chem. *284*, 33409–33417.

Hannemann, S., Madrid, R., Stastna, J., Kitzing, T., Gasteier, J., Schonichen, A., Bouchet, J., Jimenez, A., Geyer, M., Grosse, R., et al. (2008). The Diaphanous-related formin FHOD1 associates with ROCK1 and promotes Src-dependent plasma membrane blebbing. J. Biol. Chem. *283*, 27891–27903.

Hansen, S.D., and Mullins, R.D. (2010). VASP is a processive actin polymerase that requires monomeric actin for barbed end association. J. Cell Biol. *191*, 571–584.

Harding, M.A., and Theodorescu, D. (2010). RhoGDI signaling provides targets for cancer therapy. Eur. J. Cancer *46*, 1252–1259.

Hashimoto, Y., Parsons, M., and Adams, J.C. (2007). Dual actin-bundling and protein kinase C-binding activities of fascin regulate carcinoma cell migration downstream of Rac and contribute to metastasis. Mol. Biol. Cell *18*, 4591–4602.

Heasman, S.J., Carlin, L.M., Cox, S., Ng, T., and Ridley, A.J. (2010). Coordinated RhoA signaling at the leading edge and uropod is required for T cell transendothelial migration. J. Cell Biol. *190*, 553–563.

Hidalgo-Carcedo, C., Hooper, S., Chaudhry, S.I., Williamson, P., Harrington, K., Leitinger, B., and Sahai, E. (2011). Collective cell migration requires suppression of actomyosin at cell-cell contacts mediated by DDR1 and the cell polarity regulators Par3 and Par6. Nat. Cell Biol. *13*, 49–58.

Hotulainen, P., Llano, O., Smirnov, S., Tanhuanpaa, K., Faix, J., Rivera, C., and Lappalainen, P. (2009). Defining mechanisms of actin polymerization and depolymerization during dendritic spine morphogenesis. J. Cell Biol. *185*, 323–339.

Huang, C. (2010). Roles of E3 ubiquitin ligases in cell adhesion and migration. Cell. Adh. Migr. *4*, 10–18.

Johnston, S.A., Bramble, J.P., Yeung, C.L., Mendes, P.M., and Machesky, L.M. (2008). Arp2/3 complex activity in filopodia of spreading cells. BMC Cell Biol. *9*, 65.

Kardash, E., Reichman-Fried, M., Maitre, J.L., Boldajipour, B., Papusheva, E., Messerschmidt, E.M., Heisenberg, C.P., and Raz, E. (2010). A role for Rho GTPases and cell-cell adhesion in single-cell motility in vivo. Nat. Cell Biol. *12*, 47–53 supp. 1–11.

Kim, J.S., Huang, T.Y., and Bokoch, G.M. (2009). Reactive oxygen species regulate a slingshot-cofilin activation pathway. Mol. Biol. Cell *20*, 2650–2660.

Lai, F.P., Szczodrak, M., Block, J., Faix, J., Breitsprecher, D., Mannherz, H.G., Stradal, T.E., Dunn, G.A., Small, J.V., and Rottner, K. (2008). Arp2/3 complex interactions and actin network turnover in lamellipodia. EMBO J. *27*, 982–992.

Lai, F.P., Szczodrak, M., Oelkers, J.M., Ladwein, M., Acconcia, F., Benesch, S., Auinger, S., Faix, J., Small, J.V., Polo, S., et al. (2009). Cortactin promotes migration and platelet-derived growth factor-induced actin reorganization by signaling to Rho-GTPases. Mol. Biol. Cell *20*, 3209–3223.

LeClaire, L.L., 3rd, Baumgartner, M., Iwasa, J.H., Mullins, R.D., and Barber, D.L. (2008). Phosphorylation of the Arp2/3 complex is necessary to nucleate actin filaments. J. Cell Biol. *182*, 647–654.

Lizarraga, F., Poincloux, R., Romao, M., Montagnac, G., Le Dez, G., Bonne, I., Rigaill, G., Raposo, G., and Chavrier, P. (2009). Diaphanous-related formins are required for invadopodia formation and invasion of breast tumor cells. Cancer Res. *69*, 2792–2800.

Machacek, M., Hodgson, L., Welch, C., Elliott, H., Pertz, O., Nalbant, P., Abell, A., Johnson, G.L., Hahn, K.M., and Danuser, G. (2009). Coordination of Rho GTPase activities during cell protrusion. Nature *461*, 99–103.

Machesky, L.M., and Li, A. (2010). Fascin: Invasive filopodia promoting metastasis. Commun. Integr. Biol. *3*, 263–270.

Meller, J., Vidali, L., and Schwartz, M.A. (2008). Endogenous RhoG is dispensable for integrin-mediated cell spreading but contributes to Rac-independent migration. J. Cell Sci. *121*, 1981–1989.

Mellor, H. (2010). The role of formins in filopodia formation. Biochim. Biophys. Acta *1803*, 191–200.

Mullins, R.D., Heuser, J.A., and Pollard, T.D. (1998). The interaction of Arp2/3 complex with actin: nucleation, high affinity pointed end capping, and formation of branching networks of filaments. Proc. Natl. Acad. Sci. USA *95*, 6181–6186.

Nagy, S., and Rock, R.S. (2010). Structured post-IQ domain governs selectivity of myosin X for fascin-actin bundles. J. Biol. Chem. *285*, 26608–26617.

Nalbant, P., Chang, Y.C., Birkenfeld, J., Chang, Z.F., and Bokoch, G.M. (2009). Guanine nucleotide exchange factor-H1 regulates cell migration via localized activation of RhoA at the leading edge. Mol. Biol. Cell *20*, 4070–4082.

Neubueser, D., and Hipfner, D.R. (2010). Overlapping roles of Drosophila Drak and Rok kinases in epithelial tissue morphogenesis. Mol. Biol. Cell *21*, 2869–2879.

Nimnual, A.S., Taylor, L.J., and Bar-Sagi, D. (2003). Redox-dependent downregulation of Rho by Rac. Nat. Cell Biol. *5*, 236–241.

Norman, L.L., Bruges, J., Sengupta, K., Sens, P., and Aranda-Espinoza, H. (2010). Cell blebbing and membrane area homeostasis in spreading and retracting cells. Biophys. J. *99*, 1726–1733.

Okada, K., Bartolini, F., Deaconescu, A.M., Moseley, J.B., Dogic, Z., Grigorieff, N., Gundersen, G.G., and Goode, B.L. (2010). Adenomatous polyposis coli protein nucleates actin assembly and synergizes with the formin mDia1. J. Cell Biol. *189*, 1087–1096.

Oser, M., Yamaguchi, H., Mader, C.C., Bravo-Cordero, J.J., Arias, M., Chen, X., Desmarais, V., van Rheenen, J., Koleske, A.J., and Condeelis, J. (2009). Cortactin regulates cofilin and N-WASp activities to control the stages of invadopodium assembly and maturation. J. Cell Biol. *186*, 571–587.

Parsons, M., and Adams, J.C. (2008). Rac regulates the interaction of fascin with protein kinase C in cell migration. J. Cell Sci. *121*, 2805–2813.

Pertz, O. (2010). Spatio-temporal Rho GTPase signaling - where are we now? J. Cell Sci. *123*, 1841–1850.

Pertz, O., Hodgson, L., Klemke, R.L., and Hahn, K.M. (2006). Spatiotemporal dynamics of RhoA activity in migrating cells. Nature *440*, 1069–1072.

Pichot, C.S., Arvanitis, C., Hartig, S.M., Jensen, S.A., Bechill, J., Marzouk, S., Yu, J., Frost, J.A., and Corey, S.J. (2010). Cdc42-interacting protein 4 promotes breast cancer cell invasion and formation of invadopodia through activation of N-WASp. Cancer Res. *70*, 8347–8356.

Pletjushkina, O.J., Rajfur, Z., Pomorski, P., Oliver, T.N., Vasiliev, J.M., and Jacobson, K.A. (2001). Induction of cortical oscillations in spreading cells by depolymerization of microtubules. Cell Motil. Cytoskeleton *48*, 235–244.

Poincloux, R., Lizarraga, F., and Chavrier, P. (2009). Matrix invasion by tumour cells: a focus on MT1-MMP trafficking to invadopodia. J. Cell Sci. *122*, 3015–3024.

Pollard, T.D., and Borisy, G.G. (2003). Cellular motility driven by assembly and disassembly of actin filaments. Cell *112*, 453–465.

Ponti, A., Machacek, M., Gupton, S.L., Waterman-Storer, C.M., and Danuser, G. (2004). Two distinct actin networks drive the protrusion of migrating cells. Science *305*, 1782–1786.

Ren, G., Crampton, M.S., and Yap, A.S. (2009). Cortactin: Coordinating adhesion and the actin cytoskeleton at cellular protrusions. Cell Motil. Cytoskeleton *66*, 865–873.

Ridley, A.J., Schwartz, M.A., Burridge, K., Firtel, R.A., Ginsberg, M.H., Borisy, G., Parsons, J.T., and Horwitz, A.R. (2003). Cell migration: integrating signals from front to back. Science *302*, 1704–1709.

Robens, J.M., Yeow-Fong, L., Ng, E., Hall, C., and Manser, E. (2010). Regulation of IRSp53-dependent filopodial dynamics by antagonism between 14-3-3 binding and SH3-mediated localization. Mol. Cell. Biol. *30*, 829–844.

Rottner, K., Hanisch, J., and Campellone, K.G. (2010). WASH, WHAMM and JMY: regulation of Arp2/3 complex and beyond. Trends Cell Biol. *20*, 650–661.

Roy, B.C., Kakinuma, N., and Kiyama, R. (2009). Kank attenuates actin remodeling by preventing interaction between IRSp53 and Rac1. J. Cell Biol. *184*, 253–267.

Sahai, E., Garcia-Medina, R., Pouyssegur, J., and Vial, E. (2007). Smurf1 regulates tumor cell plasticity and motility through degradation of RhoA leading to localized inhibition of contractility. J. Cell Biol. *176*, 35–42.

Sarmiento, C., Wang, W., Dovas, A., Yamaguchi, H., Sidani, M., El-Sibai, M., Desmarais, V., Holman, H.A., Kitchen, S., Backer, J.M., et al. (2008). WASP family members and formin proteins coordinate regulation of cell protrusions in carcinoma cells. J. Cell Biol. *180*, 1245–1260.

Schoumacher, M., Goldman, R.D., Louvard, D., and Vignjevic, D.M. (2010). Actin, microtubules, and vimentin intermediate filaments cooperate for elongation of invadopodia. J. Cell Biol. *189*, 541–556.

Schoumacher, M., Louvard, D., and Vignjevic, D. (2011). Cytoskeleton networks in basement membrane transmigration. Eur. J. Cell Biol. *90*, 93–99 Published online July 6, 2010.

Scott, R.W., Hooper, S., Crighton, D., Li, A., Konig, I., Munro, J., Trivier, E., Wickman, G., Morin, P., Croft, D.R., et al. (2010). LIM kinases are required for invasive path generation by tumor and tumor-associated stromal cells. J. Cell Biol. *191*, 169–185.

Shim, J., Lee, S.M., Lee, M.S., Yoon, J., Kweon, H.S., and Kim, Y.J. (2010). Rab35 mediates transport of Cdc42 and Rac1 to the plasma membrane during phagocytosis. Mol. Cell. Biol. *30*, 1421–1433.

Smith-Pearson, P.S., Greuber, E.K., Yogalingam, G., and Pendergast, A.M. (2010). Abl kinases are required for invadopodia formation and chemokine-induced invasion. J. Biol. Chem. *285*, 40201–40211.

Sossey-Alaoui, K., Li, X., and Cowell, J.K. (2007). c-Abl-mediated phosphorylation of WAVE3 is required for lamellipodia formation and cell migration. J. Biol. Chem. *282*, 26257–26265.

Sousa, A.D., and Cheney, R.E. (2005). Myosin-X: a molecular motor at the cell's fingertips. Trends Cell Biol. *15*, 533–539.

Svitkina, T.M., and Borisy, G.G. (1999). Arp2/3 complex and actin depolymerizing factor/cofilin in dendritic organization and treadmilling of actin filament array in lamellipodia. J. Cell Biol. *145*, 1009–1026.

Takahashi, K., and Suzuki, K. (2009). Membrane transport of WAVE2 and lamellipodia formation require Pak1 that mediates phosphorylation and recruitment of stathmin/Op18 to Pak1-WAVE2-kinesin complex. Cell. Signal. *21*, 695–703.

Takenawa, T., and Suetsugu, S. (2007). The WASP-WAVE protein network: connecting the membrane to the cytoskeleton. Nat. Rev. Mol. Cell Biol. *8*, 37–48.

Takesono, A., Heasman, S.J., Wojciak-Stothard, B., Garg, R., and Ridley, A.J. (2010). Microtubules regulate migratory polarity through Rho/ROCK signaling in T cells. PLoS ONE *5*, e8774.

Tinevez, J.Y., Schulze, U., Salbreux, G., Roensch, J., Joanny, J.F., and Paluch, E. (2009). Role of cortical tension in bleb growth. Proc. Natl. Acad. Sci. USA *106*, 18581–18586.

Tokuo, H., Mabuchi, K., and Ikebe, M. (2007). The motor activity of myosin-X promotes actin fiber convergence at the cell periphery to initiate filopodia formation. J. Cell Biol. *179*, 229–238.

Toomre, D., and Bewersdorf, J. (2010). A new wave of cellular imaging. Annu. Rev. Cell Dev. Biol. *26*, 285–314.

Urban, E., Jacob, S., Nemethova, M., Resch, G.P., and Small, J.V. (2010). Electron tomography reveals unbranched networks of actin filaments in lamellipodia. Nat. Cell Biol. *12*, 429–435.

van Rheenen, J., Condeelis, J., and Glogauer, M. (2009). A common cofilin activity cycle in invasive tumor cells and inflammatory cells. J. Cell Sci. *122*, 305–311.

Wang, H.R., Zhang, Y., Ozdamar, B., Ogunjimi, A.A., Alexandrova, E., Thomsen, G.H., and Wrana, J.L. (2003). Regulation of cell polarity and protrusion formation by targeting RhoA for degradation. Science *302*, 1775–1779.

Wang, X., He, L., Wu, Y.I., Hahn, K.M., and Montell, D.J. (2010). Light-mediated activation reveals a key role for Rac in collective guidance of cell movement in vivo. Nat. Cell Biol. *12*, 591–597.

Weaver, A.M. (2009). Regulation of cancer invasion by reactive oxygen species and Tks family scaffold proteins. Sci. Signal. *2*, pe56.

Weijer, C.J. (2009). Collective cell migration in development. J. Cell Sci. *122*, 3215–3223.

Wilson, C.A., Tsuchida, M.A., Allen, G.M., Barnhart, E.L., Applegate, K.T., Yam, P.T., Ji, L., Keren, K., Danuser, G., and Theriot, J.A. (2010). Myosin II contributes to cell-scale actin network treadmilling through network disassembly. Nature *465*, 373–377.

Wolf, K., and Friedl, P. (2009). Mapping proteolytic cancer cell-extracellular matrix interfaces. Clin. Exp. Metastasis *26*, 289–298.

Wu, Y.I., Frey, D., Lungu, O.I., Jaehrig, A., Schlichting, I., Kuhlman, B., and Hahn, K.M. (2009). A genetically encoded photoactivatable Rac controls the motility of living cells. Nature *461*, 104–108.

Yang, C., Czech, L., Gerboth, S., Kojima, S., Scita, G., and Svitkina, T. (2007). Novel roles of formin mDia2 in lamellipodia and filopodia formation in motile cells. PLoS Biol. *5*, e317.

Yap, A.S., and Kovacs, E.M. (2003). Direct cadherin-activated cell signaling: a view from the plasma membrane. J. Cell Biol. *160*, 11–16.

Yoshida, K., and Soldati, T. (2006). Dissection of amoeboid movement into two mechanically distinct modes. J. Cell Sci. *119*, 3833–3844.

Yoshihara, Y., De Roo, M., and Muller, D. (2009). Dendritic spine formation and stabilization. Curr. Opin. Neurobiol. *19*, 146–153.

Zhang, J., Fonovic, M., Suyama, K., Bogyo, M., and Scott, M.P. (2009). Rab35 controls actin bundling by recruiting fascin as an effector protein. Science *325*, 1250–1254.

Zhang, L., Luo, J., Wan, P., Wu, J., Laski, F., and Chen, J. (2011). Regulation of cofilin phosphorylation and asymmetry in collective cell migration during morphogenesis. Development *138*, 455–464.

Zhao, H., Pykalainen, A., and Lappalainen, P. (2011). I-BAR domain proteins: linking actin and plasma membrane dynamics. Curr. Opin. Cell Biol. *23*, 14–21. http://dx.doi.org/10.1016/j.ceb.2010.10.005. Published online November 17, 2010.

Zhuravlev, P.I., Der, B.S., and Papoian, G.A. (2010). Design of active transport must be highly intricate: A possible role of myosin and Ena/VASP for G-actin transport in filopodia. Biophys. J. *98*, 1439–1448.

Zuchero, J.B., Coutts, A.S., Quinlan, M.E., Thangue, N.B., and Mullins, R.D. (2009). p53-cofactor JMY is a multifunctional actin nucleation factor. Nat. Cell Biol. *11*, 451–459.

Use the Force: Membrane Tension as an Organizer of Cell Shape and Motility

Alba Diz-Muñoz[1,2], Daniel A. Fletcher[1,3],*, Orion D. Weiner[2,*]

[1]Bioengineering Department and Biophysics Program, University of California, Berkeley, CA 94720, USA, [2]Cardiovascular Research Institute and Department of Biochemistry, University of California San Francisco, San Francisco, CA 94143, USA, [3]Physical Biosciences Division, Lawrence Berkeley National Laboratory, Berkeley, CA 94720, USA

*Correspondence: fletch@berkeley.edu, orion.weiner@ucsf.edu

Trends in Cell Biology, Vol. 23, No. 2, February 2013 © 2013 Elsevier Inc.
http://dx.doi.org/10.1016/j.tcb.2012.09.006

SUMMARY

Many cell phenomena that involve shape changes are affected by the intrinsic deformability of the plasma membrane (PM). Far from being a passive participant, the PM is now known to physically, as well as biochemically, influence cell processes ranging from vesicle trafficking to actin assembly. Here we review current understanding of how changes in PM tension regulate cell shape and movement, as well as how cells sense PM tension.

INTRODUCTION

One way that cells interact with the world around them is biochemically. For example, binding of soluble extracellular ligands to receptors on the cell membrane can trigger intracellular signaling cascades. More recently, it has become clear that physical interactions are also an important currency of information transfer in cells and tissues [1,2] (reviewed in [3]). In particular, tension in the PM has been shown to regulate many cell behaviors, including vesicle trafficking [4] and cell motility [5,6]. The PM is often described by the fluid mosaic model [7], which characterizes it as a 2D continuous fluid bilayer of lipids with freely diffusing embedded proteins. In this paradigm, the bilayer is considered a uniform semipermeable barrier that serves as a passive matrix for membrane proteins. However, this model is incomplete; lipids are now known to have a much more active role

397

in regulating membrane structure and biological function [8–10] (reviewed in [11]) and the mechanical properties of the PM need to be included for a complete picture.

Mechanically, membranes have a low shear modulus (a result of the fluid nature of the lipid bilayer; $4–10 \times 10^{-3}$ N/m [12–14]), a high elastic modulus (due to the small stretch in bilayers; 10^3 N/m^2 [15,16]), a variable viscosity (which depends on membrane composition; $0.36–2.1 \times 10^{-3}$ Pa s for an erythrocyte [16]), and a bending stiffness strongly influenced by membrane proteins and cytoskeletal elements (10^{-19} N m [17–19]). Membrane tension is related to the force needed to deform a membrane. Historically, the term membrane tension has been applied to define different concepts, and this has lead to confusion in the literature. Initially, membrane tension was measured in lipid vesicles (Box 1) in which the force needed to stretch the membrane is the in-plane membrane tension (T_m, N/m). In cells, the force needed to deform the PM is greater than that for a pure lipid vesicle due to contributions from membrane proteins and membrane-to-cortex attachments (MCA) (γ, N/m), which link the membrane to the underlying cortex and also resist membrane deformation. Thus, PM tension, also known as apparent membrane tension or effective membrane tension, is the sum of T_m and γ (Box 2).

Research in recent decades has established the importance of PM tension as a physical regulator of cell motility and morphology [6,20–23], but the mechanism of tension sensation and how membrane tension is integrated in the cell's mechanical properties are unknown. We will focus on how PM tension affects and is affected by other cellular processes and will outline possible mechanisms for membrane tension sensation.

FEEDBACK BETWEEN PM TENSION AND CELLULAR PROCESSES

Some studies point to PM tension being a constant parameter within a given cell type [24]. However, it is unclear whether cells have a preferred 'set point' for PM tension and, if so, how cells measure their PM tension. Moreover, in cells, biological membranes are active in the sense that they are constantly maintained out of equilibrium by cellular processes that contribute to changes in PM area, composition, and MCA protein activity. This adds complexity but also gives the cell multiple routes of adjustment. Several cellular processes affect and are affected by PM tension (Figure 1).

Exocytosis and Clathrin-Mediated Endocytosis
PM tension regulates the balance between exocytosis and endocytosis in numerous systems; exocytosis (which is stimulated by high membrane tension) acts to decrease PM tension, whereas endocytosis (which is stimulated

BOX 1 TECHNIQUES TO MEASURE AND MANIPULATE MEMBRANE TENSION

Measuring membrane tension

Experiments to determine mechanical properties of biological membranes began in the 1930s using sea urchin eggs and red blood cells [60,61]. Since then, new higher-resolution techniques have been developed. Here we list the most commonly used approaches for measuring membrane tension.

Compression of the cell with two plates (Figure Ia) **and micropipette aspiration** (Figure Ib). These techniques have been extensively used for studying the mechanical

Quantifying membrane tension

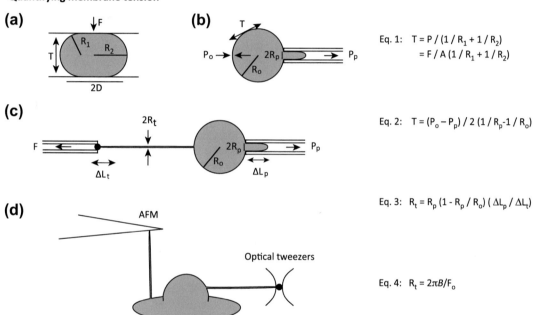

Eq. 1: $T = P / (1 / R_1 + 1 / R_2)$
$\qquad = F / A (1 / R_1 + 1 / R_2)$

Eq. 2: $T = (P_o - P_p) / 2 (1 / R_p - 1 / R_o)$

Eq. 3: $R_t = R_p (1 - R_p / R_o) (\Delta L_p / \Delta L_t)$

Eq. 4: $R_t = 2\pi B / F_o$

TRENDS in Cell Biology

FIGURE I Quantifying membrane tension.
(a) The spherical cell or lipid vesicle is compressed with known force (F) between two parallel plates. R1 and R2 are the radii of the principal curvatures of the surface. The internal pressure P (P = F/A) is the applied force F divided by the contacted area (A = πD^2) between the plate and the cell or vesicle. This pressure is in equilibrium with the surface tension T (see Equation [1] in figure). **(b)** The membrane of a spherical cell or lipid vesicle is deformed by a micropipette. Pp is the pressure in the pipette, and Po is the pressure in the reservoir. Ro and Rp are the radii of the cell or vesicle and the pipette. The resulting isotropic stress in the membrane is the surface tension T and is determined by Equation [2] in figure. **(c)** Initially, tethers were formed using a micropipette to hold samples with suction pressure. A bead in a second pipette was used to extract a membrane tether with force F. Rp and Ro are the radii of the pipette and the cell or vesicle. The tether radius Rt can be calculated from the change in the length of membrane projection in the pipette (ΔLp) caused by the tether length change (ΔLt) as seen in Equation [3] in figure or can be derived from the membrane bending stiffness/static tether force relationship as seen in Equation [4] in figure [17] for cells with simple morphologies, such as red blood cells or lipid vesicles. **(d)** Atomic force microscopy cantilevers and optical tweezers provide higher-resolution measurements of plasma membrane (PM) tension [65,66,79,80]. See Box 2 for an in-depth description of how tethers can be used to measure membrane tension.

Continued

BOX 1 TECHNIQUES TO MEASURE AND MANIPULATE MEMBRANE TENSION—CONT'D

properties of membranes in lipid vesicles, urchin eggs, and red blood cells [15,18,60–63]. However, they are applicable only to lipid vesicles or suspension cells with simple morphologies and cannot be used for cells with complex morphologies such as neurons or neutrophils. Moreover, in cells, isolating the contribution of PM tension to these measurements is complicated by the fact that a significant portion of the measured forces can be due to deformations of the cytoskeleton, in particular the actin cortex that lies immediately under the PM.

Tethers. Tethers (which lack a continuous cytoskeleton) have been studied to measure PM tension. Initially these experiments were performed using a micropipette to hold cells or lipid vesicles and a second pipette to extract a membrane tether (Figure Ic) [64]. More recently atomic force microscope (AFM) cantilevers, optical tweezers, and magnetic tweezers have enabled higher-resolution measurements of PM tension (Figure Id) [65–67]. See Box 2 for an in-depth description of how tethers can be used to measure membrane tension.

Fluorescence resonance energy transfer (FRET)-based biosensors. These were recently developed for assaying tension in the cytoskeleton and at sites of adhesion [68]. Although the field currently lacks comparable imaging-based sensors for membrane tension, such a tool would enable less invasive analysis of the spatial and temporal dynamics of PM tension in living cells. Moreover, intracellular organelles are not accessible for tether experiments, and such technology would allow us to determine whether the PM is the only organelle that can act as a mechanical sensor.

Manipulating membrane tension

Vesicle fusion, lipid addition, and changes in osmolarity. These have been used to manipulate T_m [6,69], but none is quantitative unless combined with simultaneous measurements of tension such as tether pulling. Moreover, how much those techniques affect cytoskeletal components is unknown.

Multiple tethers. Pulling multiple tethers with an AFM [70] is an alternative method that can be used to measure and manipulate membrane tension simultaneously.

by low membrane tension) increases it [25] (reviewed in [4,24,26,27]). These opposing effects of vesicle trafficking could enable cells to keep tension close to a set point [28]. When the PM reservoir is reduced following cell spreading, there is a twofold increase in PM tension followed by activation of exocytosis and myosin-based contraction [20]. The rate of spreading and the time point at which exocytosis and myosin contraction occur are highly dependent on PM tension. These data implicate tension in coordinating membrane trafficking, actomyosin contraction, and PM area change. More recently, MCA has been shown to determine the actin dependence of clathrin coat assembly [21]. Clathrin-mediated endocytosis is independent of actin dynamics in many circumstances but requires actin polymerization in others. On the apical surface of polarized cells where MCA is higher [21,29] or following cell swelling, actin engagement is necessary to convert a coated pit into a vesicle [21].

Caveolae

Caveolae, invaginations of the PM that are formed by caveolins, are physiological membrane reservoirs that have recently been shown to enable cells to accommodate sudden changes in PM tension [30,31]. Increases in

BOX 2 QUANTIFYING MEMBRANE TENSION AND MCAs FROM TETHERS

The mathematical relationship between tether force (F_o) and tension is known for lipid vesicles [15,63,71]. In cells, the tether force is generated by a combination of: (i) T_m of the lipid-bilayer; (ii) membrane bending stiffness (B); and (iii) MCA (γ) [17,72]. The T_m and cytoskeleton adhesion terms are difficult to separate and are therefore combined into a single term (T) that is known by multiple names: PM tension, apparent membrane tension, or effective membrane tension [17]:

$$T = T_m + \gamma = F_o^2/8B\pi^2 \qquad \text{[I]}$$

(i) T_m is the result of the membrane being an inelastic fluid that equilibrates stresses within milliseconds [40,73]:

$$T_m = k(\Delta A/A) \qquad \text{[II]}$$

where k is the elastic area stretch modulus, which depends on lipid composition, and A is the cell surface area.

T_m and k completely characterize the differential equation of state for planar surfaces. However, when the surface is not a plane (e.g., when it is rippled due to thermally driven fluctuations), measurements of these parameters include the entropic elasticity of the membrane.

T_m appears to be uniform throughout the whole cell, even across the junctions of epithelial cells [28], and on cell blebs in which the membrane separates from the acto-myosin cortex [29,74].

(ii) The membrane bending stiffness (B) relates to the force needed to bend the membrane for a given radius of curvature. It has been experimentally measured for tethers in lipid vesicles. It is approximately 10^{-19} N m for a typical lipid bilayer, red cell, or neutrophil membrane [17–19]. It can also be calculated from measurements of tether radius as a function of the static force:

$$B = F_o R_t/2\pi \qquad \text{[III]}$$

(iii) The MCA force can be expressed as adhesion energy per unit area (γ). It results from various MCA proteins that connect the actin cytoskeleton and the plasma membrane [75] and nonspecific binding of the membrane-to-cortex components. It was long believed that MCA was the result of only specific protein–protein interactions, but some experimental findings suggest otherwise. For instance, the fact that tether forces are rapidly reversible with no hysteresis has been used to favor a continuum model with nonspecific binding of the membrane-to-cortex components [17,24].

To separate MCA and T_m, we can measure F_b, the tether force in the absence of MCA contribution. This can be achieved experimentally by performing tether measurements on nascent blebs, which are locally devoid of cytoskeletal support, or on cells in which the cytoskeleton has been depolymerized. Under these conditions, the adhesion term (γ) equals 0 [29], and T_m is given by:

$$T_m = F_b^2/8B\pi^2 \qquad \text{[IV]}$$

The adhesion energy can then be calculated using Equation [I] and [IV] if we assume that T_m is constant over smooth regions of the cell surface:

$$\gamma = T - T_m = (F_o^2/8B\pi^2) - (F_b^2/8B\pi^2) = (F_o^2 - F_b^2)/8B\pi^2 \qquad \text{[V]}$$

A recent model has related pulling force–velocity profiles to the density of crosslinkers and the lipid bilayer viscosity [76], providing a possible means of discriminating the two PM tension components in a wider range of cellular contexts.

Experimentally, the tether force in cytoskeletally unsupported regions is typically less than half of that in regions supported by the cytoskeleton ($F_b < 0.5F_o$). Applying this inequality to Equation [V], we see that over 75% of the PM tension term is the result of MCA. T_m can increase markedly with hypotonic swelling [30,77]. However, under normal conditions, large changes in PM tension are thought to primarily reflect changes in MCA [78].

Feedback between PM tension and cellular processes
T = Tm + γ ; with Tm = k (ΔA/A)

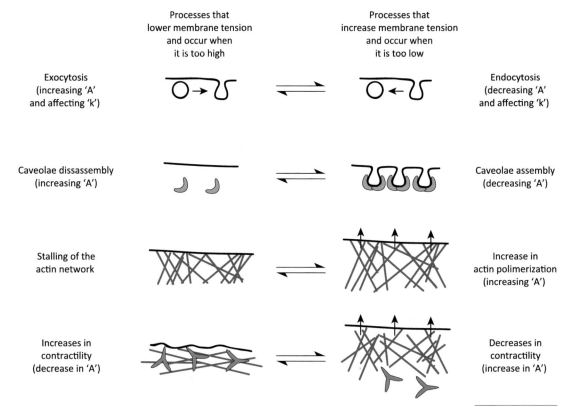

	Processes that lower membrane tension and occur when it is too high		Processes that increase membrane tension and occur when it is too low	
Exocytosis (increasing 'A' and affecting 'k')				Endocytosis (decreasing 'A' and affecting 'k')
Caveolae dissassembly (increasing 'A')				Caveolae assembly (decreasing 'A')
Stalling of the actin network				Increase in actin polimerization (increasing 'A')
Increases in contractility (decrease in 'A')				Decreases in contractility (increase in 'A')

TRENDS in Cell Biology

FIGURE 1 Feedback between plasma membrane (PM) tension and cellular processes.
Examples of cellular processes that occur when PM tension is too high and that lead to its reduction (left) or that occur when PM tension is too low and lead to its increase (right) – vesicle trafficking, caveola formation, actin polymerization, and changes in myosin. In brackets we comment on the parameters of Equations [I] and [II] in Box 2 that are predicted to change in each of these processes.

tension through cell stretching or hypo-osmotic shock induce disassembly of caveolae, whereas recovery of iso-osmolarity leads to complete caveolar reassembly [30]. How caveolae buffer PM tension is not yet fully understood, because the amount of area released on membrane tension surge is very small (approximately 0.3%) [30].

Actin Network Assembly
To generate lamellipodium-like protrusions during cell crawling, growing actin filaments must generate sufficient local force to displace the PM

[5,20,23]. Indeed, actin-based protrusion can lead to an increase in T_m as the force of polymerization unfolds wrinkles in the membrane during cell spreading [20]. Moreover, an increase in PM tension constrains the spread of the existing leading edge and prevents the formation of secondary fronts in chemotactic cells such as neutrophils [6]. In these cells, increasing cell tension by micropipette aspiration is sufficient to act as a long-range inhibitor of the signals that promote actin assembly at the leading edge. Conversely, the reduction of PM tension through hyperosmotic shock produces global activation of leading edge signals [6]. Because the front is the likely source of tension, any fluctuation in front size is immediately balanced by compensatory changes in tension levels, providing a possible mechanism of homeostasis [5,6].

Models of the PM as a Global Mechanical Regulator

Several models suggest a role of PM tension as a global mechanical regulator that coordinates cell protrusion and retraction. PM tension has been suggested to optimize motility by streamlining filament polymerization in the direction of movement [22]. A model of actin network polymerization in an inextensible membrane bag can quantitatively predict both cell shape and speed and recapitulate the natural phenotypic variability in a large population of motile epithelial fish keratocytes [23]. If PM tension is assumed to be spatially homogeneous at all points along the cell boundary, the force per filament is inversely proportional to the local filament density. Therefore, at the center of the leading edge, the membrane resistance per filament is small, allowing filaments to grow rapidly and generate protrusion. As filament density gradually decreases towards the cell sides and the cell rear, the forces per filament caused by PM tension increase until polymerization is stalled and the actin network disassembles [23]. More recently, Ofer *et al.* [32] hypothesized a simple disassembly clock mechanism in which the rear position of a lamellipodium is determined by where the actin network has disassembled enough for membrane tension to crush the actin network and haul it forward. Finally, PM tension could also limit bleb expansion [33], but direct experimental evidence is still missing.

ROLE OF CYTOSKELETAL TENSION VERSUS PM TENSION

Both cytoskeletal tension (also referred to as contractility) and membrane tension are capable of transmitting forces over long range to spatially and temporally regulate cell polarity and cell migration [23,29,34]. Membrane tension antagonizes actin-based protrusion by being the barrier that growing actin filaments fight to protrude the membrane [23] and contractility opposes protrusion by pulling actin filaments away from the membrane [35]. The relative contribution of cytoskeletal versus membrane tension is likely to vary in different cell types.

In *Dictyostelium*, contractility plays an important role in restricting signals to the leading edge. Upon deletion of myosin 2, cytoskeletal tension is reduced dramatically [36] and there is an increase in lateral pseudopod number [37] and in Ras activation [38]. These data support a predominant role of contractility in *Dictyostelium* polarity, but whether PM tension also plays a significant role remains unknown.

In fibroblasts, a combination of cytoskeletal and membrane tension limits cell protrusion. Increasing membrane tension by hypo-osmotic shock halts spreading, whereas decreasing it by adding lipids increases the rate of cell spreading, enhances lamellipodial extension, and transiently causes uniform spreading [5]. Decreasing contractility through myosin inhibition causes faster spreading and a larger final spread area [35], and increasing it with biaxial cellular stretching downregulates Rac activity [39].

In fish keratocytes, decreasing contractility through myosin inhibition does not destroy keratocyte polarity and only slightly reduces migration speed, suggesting a predominant role for PM tension in this system [23,40].

In neutrophils, membrane tension also appears to be the dominant inhibitory mechanism for cell polarization. Membrane tension increases during neutrophil protrusion and decreasing membrane tension through hypo-osmotic shock results in the expansion of leading-edge signals and loss of polarity [6]. Decreasing cytoskeletal tension with myosin inhibition has no effect on leading-edge signals [6].

To what extent myosin inhibition, osmotic shock, or other tension perturbations affect both membrane tension and contractility remains unknown. Moreover, it is likely that cytoskeletal and membrane tension are interdependent; myosin 2 activity is required to reduce PM tension at the end of spreading [20], and its inhibition increases PM tension in resting neutrophils [6].

Finally, it is important to note that these conclusions (along with most other investigations of cytoskeletal tension) rely on myosin inhibition, but it is also possible that filament disassembly-based changes in cytoskeletal tension could contribute to cell polarity and movement in the absence of myosin activity.

SENSING PM TENSION

The molecular mechanisms by which cells sense and respond to mechanical signals are not fully understood. There are several mechanisms by which a cell could read PM tension (Figure 2).

Sensing PM tension

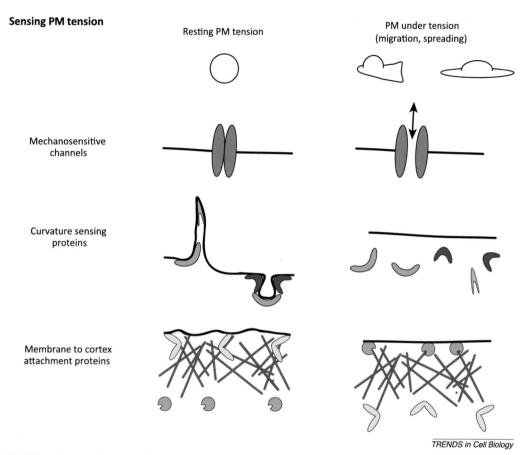

FIGURE 2 Sensing membrane tension.
Plasma membrane (PM) in a resting cell (left) or following an increase in PM tension, as observed during cell protrusion or cell spreading (right). PM tension could be sensed by the opening of stretch-activated ion channels (top), the dissociation of curvature-sensitive membrane-binding proteins (middle), or changes in the activity of membrane-to-cortex attachment (MCA) proteins (bottom).

Mechanosensitive Channels

Stretch-activated ion channels are the best understood sensors of PM tension. For these channels, changes in PM tension affect the probability of channel opening. Some examples are found in prokaryotes (the ion channel MscL) [41], primary osteoblasts [42], and specialized sensory cells [43]. Mechanosensitive channels can sense membrane tension over a wide dynamic range. The magnitude of tension sensing varies from signals barely above the thermal noise in hair cells [43] to a set point for activation near the lytic tension of the bilayer for MscL [41].

Perturbation of the ion gradients across the PM and the influx/efflux of water can also dramatically increase/decrease PM tension. Osmotic changes have been used to manipulate PM tension [6,20,21,30], but whether cells use this mechanism to change membrane tension remains to be seen.

Curvature-Sensing Proteins

Numerous proteins have domains (like BAR or ALPS domains) that associate with curved membranes, either because they are sensitive to curvature or because they induce curvature (or both) (reviewed in [44]). High PM tension could reduce the binding of I-BAR proteins by limiting the membrane bending that is necessary for their binding to the membrane [45]. In this manner, the many GEFs and GAPs with curvature-sensing domains could regulate GTPases in a tension-dependent manner [46]. Indeed, ArfGAP1, which contains an ALPS domain, has a preference for positively curved membranes (like those generated during vesicle formation) or areas with disrupted packing of lipids. ArfGAP1 preferentially induces hydrolysis of the GTP of Arf in these regions [47,48]. High curvature could be both sensed and generated at the leading edge through the action of individual proteins such as amphiphysin I (BAR domain-containing effector in clathrin-mediated endocytosis), whose density on the membrane determines whether it senses or induces curvature [49]. Additionally, the activation of N-WASP-mediated actin polymerization by proteins containing an F-BAR domain depends on membrane curvature [50]. This suggests the possibility of feedback between curvature and actin dynamics [51]: curvature-sensing/inducing proteins could stimulate actin polymerization in a curvature-dependent manner, and actin polymerization could decrease curvature to maintain homeostasis.

MCA Proteins

MCA proteins, which provide links between the PM and the actin cytoskeleton, could also sense PM tension. External forces have been found to modulate the activity of some MCA proteins. Candidates include filamin [52] and the myosin 1 family of single-headed and membrane-associated myosins [53], both of which can interact simultaneously with the cytoskeleton and the PM. Filamin A is a central mechanotransduction element of the cytoskeleton that interacts with FilGAP, a GTPase-activating protein specific for Rac; the loss of this interaction due to high stresses increases Rac activation and actin polymerization [54]. Myosin 1c is an MCA protein that dynamically provides tension to sensitize mechanosensitive ion channels responsible for hearing [55]. Myosin 1b dramatically alters its motile properties in response to external force; the rate of myosin 1b detachment from actin decreases 75-fold under forces of 2 pN or less [56]). This suggests a potential mechanism that remains to be tested: if sensation of PM tension decreases MCA protein activity, it could also generate a homeostatic feedback loop.

GLOBAL VERSUS LOCAL MEMBRANE TENSION

Asymmetries in contractility are sufficient to polarize both protrusion and adhesion [57]. Are there also inhomogeneities in membrane tension, and if so are they functionally relevant? Membrane lipids flow like a liquid and can almost instantaneously equilibrate T_m across the cell [40]. The lack of large-scale flows has been interpreted as indicating uniform PM tension in several cell lines, including keratocytes [23]. However, lack of flow could also be achieved by the presence of local barriers that limit lipid movement, which are known to exist, at least over short timescales [58]. Interestingly, PM tension is inhomogeneous in epithelia and neurons [21,24,30,31] (reviewed in [4]). For these cells, it was observed that T_m is homogeneous across a cell, and only the MCA component differs between different membrane compartments. Is this always the case? A septin ring has been observed in T lymphocyte migration [59]; if the ring provides a lipid diffusion barrier, it could enable transient differences in PM tension in immune cells during movement. MCA and T_m can be distinguished by different means (Box 1), and future studies should assess the sources and prevalence of PM tension inhomogeneity. It is important to note that even uniform membrane tension could orchestrate the initiation and maintenance of cell polarity if it opposes cytoskeletal protrusions that are locally regulated (in which protrusions grow until they generate enough tension that enables some protrusions to survive and all others to be extinguished).

FUTURE DIRECTIONS

There are many open questions regarding membrane tension. Do cells have a set point for PM tension? If so, how is it maintained? Does membrane tension regulation differ for isolated cells versus cells in a tissue? How do motile cells interpret changes in PM tension? How do cytoskeletal tension and PM tension interrelate? Further research will be necessary to determine which signaling currencies are altered by changes in PM tension and to clarify how membrane tension contributes to and is affected by endocytosis, exocytosis, actin dynamics, and myosin activity.

ACKNOWLEDGMENTS

We apologize for not being able to cite all contributions because of space restrictions and acknowledge the many scientists who have contributed to the field of PM tension. We thank Ewa Paluch, Christer S. Ejsing, Martin Bergert, and Patricia Bassereau for critical reading of the manuscript. We would like to also thank Oliver Hoeller for the beautiful cover. This work was supported by NIH GM084040 and NIH GM074751.

REFERENCES

1 Weber, G.F. *et al*. (2012) A mechanoresponsive cadherin-keratin complex directs polarized protrusive behavior and collective cell migration. *Dev. Cell* 22, 104–115

2 Yonemura, S. *et al*. (2010) alpha-Catenin as a tension transducer that induces adherens junction development. *Nat. Cell Biol.* 12, 533–542

3 Yu, H. *et al*. (2011) Forcing form and function: biomechanical regulation of tumor evolution. *Trends Cell Biol.* 21, 47–56

4 Apodaca, G. (2002) Modulation of membrane traffic by mechanical stimuli. *Am. J. Physiol. Renal Physiol.* 282, F179–F190

5 Raucher, D. and Sheetz, M.P. (2000) Cell spreading and lamellipodial extension rate is regulated by membrane tension. *J. Cell Biol.* 148, 127–136

6 Houk, A.R. *et al*. (2012) Membrane tension maintains cell polarity by confining signals to the leading edge during neutrophil migration. *Cell* 148, 175–188

7 Singer, S. and Nicolson, G. (1972) The fluid mosaic model of the structure of cell membranes. *Science* 175, 720–731

8 Hamill, O.P. and Martinac, B. (2001) Molecular basis of mechanotransduction in living cells. *Physiol. Rev.* 81, 685–740

9 Helms, J. and Zurzolo, C. (2004) Lipids as targeting signals: lipid rafts and intracellular trafficking. *Traffic* 5, 247–254

10 van Meer, G. and Sprong, H. (2004) Membrane lipids and vesicular traffic. *Curr. Opin. Cell Biol.* 16, 373–378

11 Cao, X. *et al*. (2012) Polarized sorting and trafficking in epithelial cells. *Cell Res.* 22, 793–805

12 Evans, E. (1973) New membrane concept applied to the analysis of fluid shear- and micropipette-deformed red blood cells. *Biophys. J.* 13, 941–954

13 Engelhardt, H. and Sackmann, E. (1988) On the measurement of shear elastic moduli and viscosities of erythrocyte plasma membranes by transient deformation in high frequency electric fields. *Biophys. J.* 54, 495–508

14 Hénon, S. *et al*. (1999) A new determination of the shear modulus of the human erythrocyte membrane using optical tweezers. *Biophys. J.* 76, 1145–1151

15 Hochmuth, R. *et al*. (1973) Measurement of the elastic modulus for red cell membrane using a fluid mechanical technique. *Biophys. J.* 13, 747–762

16 Hochmuth, R. and Waugh, R. (1987) Erythrocyte membrane elasticity and viscosity. *Annu. Rev. Physiol.* 49, 209–219

17 Hochmuth, F.M. *et al*. (1996) Deformation and flow of membrane into tethers extracted from neuronal growth cones. *Biophys. J.* 70, 358–369

18 Bo, L. and Waugh, R.E. (1989) Determination of bilayer membrane bending stiffness by tether formation from giant, thin-walled vesicles. *Biophys. J.* 55, 509–517

19 Shao, J-Y. and Xu, J. (2002) A modified micropipette aspiration technique and its application to tether formation from human neutrophils. *J. Biomech. Eng.* 124, 388–396

20 Gauthier, N.C. *et al*. (2011) Temporary increase in plasma membrane tension coordinates the activation of exocytosis and contraction during cell spreading. *Proc. Natl. Acad. Sci. U.S.A.* 108, 11467–11472

21 Boulant, S. *et al.* (2011) Actin dynamics counteract membrane tension during clathrin-mediated endocytosis. *Nat. Cell Biol.* 13, 1124–1131

22 Batchelder, E.L. *et al.* (2011) Membrane tension regulates motility by controlling lamellipodium organization. *Proc. Natl. Acad. Sci. U.S.A.* 108, 114429–114434

23 Keren, K. *et al.* (2008) Mechanism of shape determination in motile cells. *Nature* 453, 475–480

24 Dai, J. and Sheetz, M.P. (1995) Regulation of endocytosis, exocytosis, and shape by membrane tension. *Cold Spring Harb. Symp. Quant. Biol.* 60, 567–571

25 Gauthier, N.C. *et al.* (2009) Plasma membrane area increases with spread area by exocytosis of a GPI-anchored protein compartment. *Mol. Biol. Cell* 20, 3261–3272

26 Sheetz, M.P. (1993) Glycoprotein motility and dynamic domains in fluid plasma membranes. *Annu. Rev. Biophys. Biomol. Struct.* 22, 417–431

27 Gauthier, N.C. *et al.* (2012) Mechanical feedback between membrane tension and dynamics. *Trends Cell Biol.* 22, 527–535

28 Sheetz, M. (2001) Cell control by membrane-cytoskeleton adhesion. *Nat. Rev. Mol. Cell Biol.* 2, 392–396

29 Dai, J. and Sheetz, M.P. (1999) Membrane tether formation from blebbing cells. *Biophys. J.* 77, 3363–3370

30 Sinha, B. *et al.* (2011) Cells respond to mechanical stress by rapid disassembly of caveolae. *Cell* 144, 402–413

31 Gervásio, O.L. *et al.* (2011) Caveolae respond to cell stretch and contribute to stretch-induced signaling. *J. Cell Sci.* 124, 3581–3590

32 Ofer, N. *et al.* (2011) Actin disassembly clock determines shape and speed of lamellipodial fragments. *Proc. Natl. Acad. Sci. U.S.A.* 108, 20394–20399

33 Tinevez, J-Y. *et al.* (2009) Role of cortical tension in bleb growth. *Proc. Natl. Acad. Sci. U.S.A.* 106, 18581–18586

34 Mayer, M. *et al.* (2010) Anisotropies in cortical tension reveal the physical basis of polarizing cortical flows. *Nature* 467, 617–621

35 Cai, Y. *et al.* (2010) Cytoskeletal coherence requires myosin-IIA contractility. *J. Cell Sci.* 123, 413–423

36 Pasternak, C. *et al.* (1989) Capping of surface receptors and concomitant cortical tension are generated by conventional myosin. *Nature* 341, 549–551

37 Wessels, D. *et al.* (1988) Cell motility and chemotaxis in *Dictyostelium* amebae lacking myosin heavy chain. *Dev. Biol.* 128, 164–177

38 Lee, S. *et al.* (2010) Involvement of the cytoskeleton in controlling leading-edge function during chemotaxis. *Mol. Biol. Cell* 21, 1810–1824

39 Katsumi, A. *et al.* (2002) Effects of cell tension on the small GTPase Rac. *J. Cell Biol.* 158, 153–164

40 Kozlov, M.M. and Mogilner, A. (2007) Model of polarization and bistability of cell fragments. *Biophys. J.* 93, 3811–3819

41 Sukharev, S. (1999) Mechanosensitive channels in bacteria as membrane tension reporters. *FASEB J.* 13, 55–61

42 Charras, G.T. *et al.* (2004) Estimating the sensitivity of mechanosensitive ion channels to membrane strain and tension. *Biophys. J.* 87, 2870–2884

43 Denk, W. and Webb, W.W. (1992) Forward and reverse transduction at the limit of sensitivity studied by correlating electrical and mechanical fluctuations in frog saccular hair cells. *Hear. Res.* 60, 89–102

44 Zimmerberg, J. and Kozlov, M.M. (2006) How proteins produce cellular membrane curvature. *Nat. Rev. Mol. Cell Biol.* 7, 9–19

45 Zhao, H. *et al.* (2011) I-BAR domain proteins: linking actin and plasma membrane dynamics. *Curr. Opin. Cell Biol.* 23, 14–21

46 de Kreuk, B.J. *et al.* (2011) The F-BAR domain protein PACSIN2 associates with Rac1 and regulates cell spreading and migration. *J. Cell Sci.* 124, 2375–2388

47 Bigay, J. *et al.* (2003) Lipid packing sensed by ArfGAP1 couples COPI coat disassembly to membrane bilayer curvature. *Nature* 426, 563–566

48 Ambroggio, E. *et al.* (2010) ArfGAP1 generates an Arf1 gradient on continuous lipid membranes displaying flat and curved regions. *EMBO J.* 29, 292–303

49 Sorre, B. *et al.* (2012) Nature of curvature coupling of amphiphysin with membranes depends on its bound density. *Proc. Natl. Acad. Sci. U.S.A.* 109, 173–178

50 Takano, K. *et al.* (2008) EFC/F-BAR proteins and the N-WASP-WIP complex induce membrane curvature-dependent actin polymerization. *EMBO J.* 27, 2817–2828

51 Peleg, B. *et al.* (2011) Propagating cell-membrane waves driven by curved activators of actin polymerization. *PLoS ONE* 6, e18635

52 Tempel, M. *et al.* (1994) Insertion of filamin into lipid membranes examined by calorimetry, the film balance technique, and lipid photolabeling. *Biochemistry* 33, 12565–12572

53 Nambiar, R. *et al.* (2009) Control of cell membrane tension by myosin-I. *Proc. Natl. Acad. Sci. U. S. A.* 106, 11972–11977

54 Ehrlicher, A.J. *et al.* (2011) Mechanical strain in actin networks regulates FilGAP and integrin binding to filamin A. *Nature* 478, 260–263

55 Holt, J. *et al.* (2002) A chemical-genetic strategy implicates myosin-1c in adaptation by hair cells. *Cell* 108, 371–381

56 Laakso, J.M. *et al.* (2008) Myosin I can act as a molecular force sensor. *Science* 321, 133–136

57 Wittmann, T. and Waterman-Storer, C.M. (2001) Cell motility: can Rho GTPases and microtubules point the way? *J. Cell Sci.* 114, 3795–3803

58 Kusumi, A. *et al.* (2005) Paradigm shift of the plasma membrane concept from the two-dimensional continuum fluid to the partitioned fluid: high-speed single-molecule tracking of membrane molecules. *Annu. Rev. Biophys. Biomol. Struct.* 34, 351–378

59 Tooley, A.J. *et al.* (2009) Amoeboid T lymphocytes require the septin cytoskeleton for cortical integrity and persistent motility. *Nat. Cell Biol.* 11, 17–26

60 Cole, K.S. (1932) Surface forces of the *Arbacia* egg. *J. Cell. Comp. Physiol.* 1, 1–9

61 Norris, C.H. (1939) The tension at the surface, and other physical properties of the nucleated erythrocyte. *J. Cell. Comp. Physiol.* 4, 117–128

62 Evans, E. (1980) Minimum energy analysis of membrane deformation applied to pipet aspiration and surface adhesion of red blood cells. *Biophys. J.* 30, 265–284

63 Evans, E. and Yeung, A. (1994) Hidden dynamics in rapid changes of bilayer shape. *Chem. Phys. Lipids* 73, 39–56

64 Hochmuth, R.M. *et al.* (1982) Extensional flow of erythrocyte membrane from cell body to elastic tether. II. Experiment. *Biophys. J.* 39, 83–89

65 Dai, J. and Sheetz, M.P. (1998) Cell membrane mechanics. *Methods Cell Biol.* 55, 157–171

66 Diz-Muñoz, A. *et al.* (2010) Control of directed cell migration in vivo by membrane-to-cortex attachment. *PLoS Biol.* 8, e1000544

67 Heinrich, V. and Waugh, R.E. (1996) A piconewton force transducer and its application to measurement of the bending stiffness of phospholipid membranes. *Ann. Biomed. Eng.* 24, 595–605

68 Grashoff, C. *et al*. (2010) Measuring mechanical tension across vinculin reveals regulation of focal adhesion dynamics. *Nature* 466, 263–266

69 Stewart, M.P. *et al*. (2011) Hydrostatic pressure and the actomyosin cortex drive mitotic cell rounding. *Nature* 469, 226–230

70 Sun, M. *et al*. (2005) Multiple membrane tethers probed by atomic force microscopy. *Biophys. J.* 89, 4320–4329

71 Waugh, R. *et al*. (1992) Local and nonlocal curvature elasticity in bilayer membranes by tether formation from lecithin vesicles. *Biophys. J.* 61, 974–982

72 Sheetz, M.P. and Dai, J. (1996) Modulation of membrane dynamics and cell motility by membrane tension. *Trends Cell Biol.* 6, 85–89

73 Needham, D. and Evans, E. (1988) Structure and mechanical properties of giant lipid (DMPC) vesicle bilayers from 20 degrees C below to 10 degrees C above the liquid crystal-crystalline phase transition at 24 degrees C. *Biochemistry* 27, 8261–8269

74 Charras, G. and Paluch, E. (2008) Blebs lead the way: how to migrate without lamellipodia. *Nat. Rev. Mol. Cell Biol.* 9, 730–736

75 Sheetz, M. *et al*. (2006) Continuous membrane-cytoskeleton adhesion requires continuous accommodation to lipid and cytoskeleton dynamics. *Annu. Rev. Biophys. Biomol. Struct.* 35, 417–434

76 Brochard-Wyart, F. *et al*. (2006) Hydrodynamic narrowing of tubes extruded from cells. *Proc. Natl. Acad. Sci. U.S.A.* 103, 7660–7663

77 Dai, J. *et al*. (1998) Membrane tension in swelling and shrinking molluscan neurons. *J. Neurosci.* 18, 6681–6692

78 Raucher, D. *et al*. (2000) Phosphatidylinositol 4,5-bisphosphate functions as a second messenger that regulates cytoskeleton-plasma membrane adhesion. *Cell* 100, 221–228

79 Dai, J. and Sheetz, M.P. (1995) Mechanical properties of neuronal growth cone membranes studied by tether formation with laser optical tweezers. *Biophys. J.* 68, 988–996

80 Krieg, M. *et al*. (2008) A bond for a lifetime: employing membrane nanotubes from living cells to determine receptor-ligand kinetics. *Angew. Chem. Int. Ed. Engl.* 47, 9775–9777

nds in Cell Biology

Thinking Outside the Cell: How Cadherins Drive Adhesion

Julia Brasch[1], Oliver J. Harrison[1,2], Barry Honig[1,2,3,*], Lawrence Shapiro[1,*]

[1]Department of Biochemistry and Molecular Biophysics, Columbia University, 1150 Saint Nicholas Avenue, New York, NY 10032, USA, [2]Howard Hughes Medical Institute, Columbia University, 1130 Saint Nicholas Avenue, New York, NY 10032, USA, [3]Center for Computational Biology and Bioinformatics, Columbia University, 1130 Saint Nicholas Avenue, New York, NY 10032, USA

*Correspondence: BH6@columbia.edu, LSS8@columbia.edu

Trends in Cell Biology, Vol. 22, No. 6, June 2012 © 2012 Elsevier Inc.
http://dx.doi.org/10.1016/j.tcb.2012.03.004

SUMMARY

Cadherins are a superfamily of cell surface glycoproteins whose ectodomains contain multiple repeats of β-sandwich extracellular cadherin (EC) domains that adopt a similar fold to immunoglobulin domains. The best characterized cadherins are the vertebrate 'classical' cadherins, which mediate adhesion via *trans* homodimerization between their membrane-distal EC1 domains that extend from apposed cells, and assemble intercellular adherens junctions through *cis* clustering. To form mature *trans* adhesive dimers, cadherin domains from apposed cells dimerize in a 'strand-swapped' conformation. This occurs in a two-step binding process involving a fast-binding intermediate called the 'X-dimer'. *Trans* dimers are less flexible than cadherin monomers, a factor that drives junction assembly following cell–cell contact by reducing the entropic cost associated with the formation of lateral *cis* oligomers. Cadherins outside the classical subfamily appear to have evolved distinct adhesive mechanisms that are only now beginning to be understood.

THE CLASSICAL CADHERIN FAMILY

Cadherins constitute a large superfamily of cell surface receptors, many of which function in calcium-dependent cell–cell recognition and adhesion. Cadherins are found in a wide array of species ranging from unicellular animals with multicellular life stages [1,2] to mammals, in which they are

413

CellPress

involved in morphogenetic processes such as embryonic cell layer separation, synapse formation and specificity in the central nervous system [3,4], mechanotransduction [5,6], cell signaling [7,8], and physical homeostasis of mature tissues [9,10]. Consistent with these roles, decreased cadherin expression, which may allow cells to escape normal viability requirements for cellular cohesion [11–13], is a common feature of metastasis.

Members of the cadherin superfamily are defined by a common structural component, the EC domain – an approximately 110 residue β-fold domain – and cadherins can be classified into multiple subfamilies based on the number and arrangement of EC domains (Box 1, Figure I). By far the best understood of these subfamilies are the vertebrate classical cadherins, comprising six 'type I' and 13 'type II' cadherins in typical vertebrate genomes,

BOX 1   MEETING THE FAMILY

Cadherins are membrane associated glycoproteins, many of which function in calcium-dependent cell adhesion or recognition processes. Each EC domain comprises a seven stranded β-barrel [23–26,30,51,65,70,71,75,80] with the N and C termini located on opposite sides allowing consecutive domains to be arranged in tandem. Most EC domains contain conserved Ca^{2+}-binding sites that coordinate three Ca^{2+} ions in the linker regions between consecutive domains [24], rigidifying the ectodomain structure [81] and protecting it from proteolysis [82] (Box 2). Less frequently, and mostly in very long cadherins, canonical EC domains can lack Ca^{2+}-binding residues resulting in Ca^{2+}-free linker regions, suggesting flexibility that could result in more globular overall structures [2,75]. The number of EC domains, overall domain organization and other sequence characteristics vary widely between different cadherins, dividing the superfamily into several subfamilies (Figure I) [83,84]. Vertebrate classical type I and type II cadherins are single-pass transmembrane proteins with ectodomains comprising five EC repeats (after removal of the N-terminal prodomain), and a short, highly conserved cytoplasmic domain with binding motifs for the armadillo proteins p120 and β-catenin (reviewed in [8]). Desmosomal cadherins, expressed in all vertebrate animals, have a domain organization similar to that of classical cadherins (reviewed in [60]). However, they are attached via distinct cytoplasmic proteins to intermediate filaments forming specialized cell–cell junctions, referred to as desmosomes, in tissues exposed to high mechanical stress. The largest cadherin subfamily is the protocadherins, divided into the gene-clustered α-, β- and γ- and non-clustered protocadherins (reviewed in [85,86], respectively). They are single-pass transmembrane proteins with six or seven EC domains and distinct cytoplasmic domains, and are expressed primarily in the nervous system of mammals. Clustered protocadherins are thought to play an important role in neural patterning [66,85]. Other subfamilies are more divergent; for example, Flamingo/CELSR cadherins, which mediate planar cell polarity in vertebrates and invertebrates, have nine EC repeats, EGF, laminin-G like and hormone receptor-like domains and, uniquely in the cadherin family, a seven-pass transmembrane structure [87]. Some atypical cadherins, such as FAT and Dachsous, are involved in adhesion-mediated signaling and planar cell polarity [88]. These cadherins have many EC domains, but have a close phylogenetic sequence resemblance to protocadherins in their N-terminal region [2]. Invertebrate 'classical' cadherins, typified by *Drosophila* N- and E-cadherin, are found in adherens junction-like structures but deviate greatly in their domain organization. The heterophilic adhesive pair cadherin-23 and protocadherin-15 appear to form a long braided structure, the 'tip-link', which is involved in auditory mechanotransduction [6,89]. Each vertebrate genome contains a solitary truncated (T-) cadherin [90], which regulates neurite outgrowth and has a similar overall domain organization to the ectodomain of classical cadherins, but the transmembrane and cytoplasmic domain are replaced by a GPI anchor. T-cadherin binds adhesively through the 'X-dimer' interface [51], which functions as a binding intermediate in vertebrate classical cadherins [48].

BOX 1 MEETING THE FAMILY—CONT'D

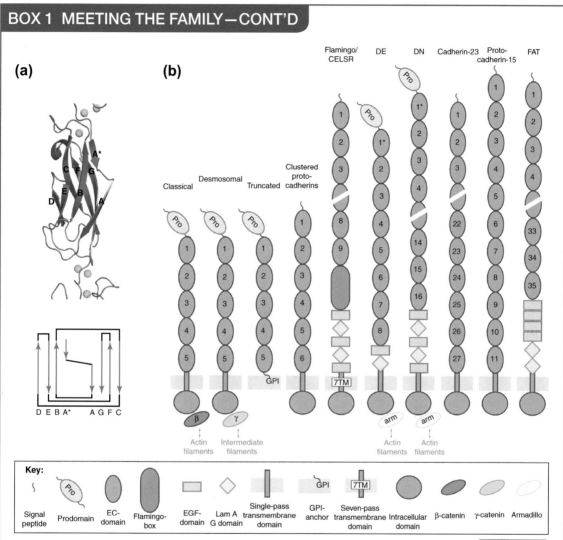

TRENDS in Cell Biology

FIGURE I Schematic representation of members of the cadherin family, which share a common structural motif: the EC domain.

(a) Typical folding of an EC domain shown in ribbon representation (top panel from pdb-ID: 1L3W). Seven antiparallel β-strands (A–G) assemble two β-sheets as shown in the topology diagram (lower panel). Note that the A strand is split into two halves, the A* and A strands. These are connected by a loop, referred to as the 'hinge'. Three Ca^{2+} ions (green spheres) are coordinated between consecutive EC domains. (b) Schematic representation of overall domain organization of selected cadherin family members. All cadherins have two or more EC domains in their extracellular regions (blue ovals, numbered from membrane distal to membrane proximal domain), which can also contain non-EC domains such as EGF-repeats (green rectangles), laminin A G domains (cyan diamonds) and flamingo boxes (pink oval). Some cadherins have, in addition to the signal peptide, a prodomain (grey ovals) that is removed by a furin protease on the cell surface. Hashed domains indicate four or more EC domains omitted for clarity. *The first EC domain of *Drosophila* E- (DE) and N- (DN) cadherin is predicted from sequence analysis [75].

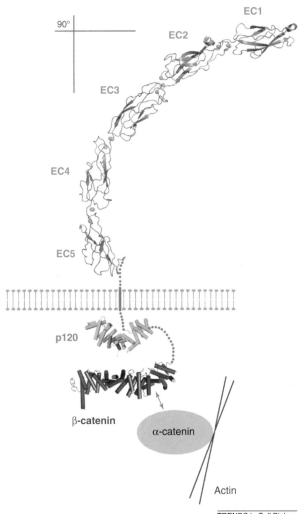

EC1

EC2

90°

EC3

EC4

EC5

p120

β-catenin

α-catenin

Actin

TRENDS in Cell Biology

FIGURE 1 Overall architecture of classical cadherins.

The extracellular domain of C-cadherin (pdb-ID: 1L3W) is depicted as a ribbon diagram (orange). Ca^{2+} ions (green spheres) are coordinated between consecutive domains, stabilizing an overall curved shape of the ectodomain, with an angle of close to 90° between domains EC1 and EC5. The structure of the stalk region, the transmembrane domain and parts of the intracellular domain are unknown and are shown as dotted lines. The cytoplasmic domain of cadherins binds to intracellular binding partners p120 (green barrels representing α-helices; pdb-ID: 3L6X) in the juxta-membrane region and β-catenin (blue barrels representing α-helices; pdb-ID: 1I7X) in the C-terminal region. β-catenin interacts with α-catenin, which in turn binds to actin filaments linking cadherins to the cytoskeleton. The depicted orientation, position and size of the intracellular binding partners relative to each other and to C-cadherin are schematic; the overall structural arrangement of the cytoplasmic side of adherens junctions is unknown.

BOX 2 CALCIUM DEPENDENCE OF CADHERIN ADHESION

Cadherins are named for the dependence of their adhesive function on the presence of extracellular calcium. Before their structures were known, it was speculated that Ca^{2+} ions might bridge the adhesive interface. However, the role of calcium in cadherin function is far more complex. Calcium binds to cadherins at stereotyped binding sites situated between successive EC domains. Each of these sites binds three Ca^{2+} ions in a highly cooperative manner such that each five-domain classical cadherin coordinates twelve Ca^{2+} ions in total [17,24,80]. The binding affinities of the Ca^{2+} sites vary, but all bind with a dissociation constant (K_D) lower than the Ca^{2+} concentration characteristic of the extracellular milieu, approximately 1mM [91,92]. Thus, it is expected that cadherin ectodomains will be fully Ca^{2+}-occupied under physiological conditions.

Three roles are now understood for Ca^{2+} binding in classical cadherins. The first is rigidifying the ectodomain so that it adopts a characteristic crescent shape [81], although this structure retains considerable conformational flexibility [55,70]. The crescent shape is critical to adhesive binding because the axes of the membrane-distal and membrane-proximal EC domains must be approximately 90° apart to satisfy the geometrical requirements of *trans* binding [17,24]. Notably, chelation of Ca^{2+} leads to the loss of *trans* binding and its concomitant replacement by binding to other cadherins on the same cell through the adhesive interface [93]. Thus, Ca^{2+}-mediated rigidification is critical to adhesive *trans* binding.

A second role for Ca^{2+} ions is in defining the structure of the X-dimer interface surfaces. The X-dimer binding intermediate of classical cadherins is centered around the EC1–EC2 Ca^{2+} binding region, which is unstructured in the absence of Ca^{2+} [48,51,80,94,95]. Thus, in the absence of Ca^{2+}, the mature adhesive strand-swap interface is likely to be kinetically unfavorable due to the slow exchange inherent in domain swap binding.

The third role for Ca^{2+} involves direct energetic effects on strand swapping. NMR experiments [46] and molecular simulations [96] reveal that Ca^{2+} ligation favors the opening of the A strand. The underlying molecular mechanism has recently been described [34] and is discussed in the text.

which share a conserved cytoplasmic domain and an ectodomain containing five tandem EC domains (Figure 1). Linkers between successive EC domains are each stabilized by the binding of three Ca^{2+} ions resulting in a characteristic curvature of the ectodomain (Figure 1). The role of Ca^{2+} binding in classical cadherin mediated adhesion is summarized in Box 2.

Classical cadherins provide the prototypical example of calcium-dependent homophilic cell–cell adhesion. They are often concentrated at adherens junctions (reviewed in [14]), specialized cell–cell adhesion structures characterized by parallel apposed plasma membranes with an intermembrane space of approximately 15–30nm. In these junctions, cadherins form *trans* bonds bridging the intermembrane space via their ectodomains, while their cytoplasmic domains bind to the adaptor proteins β-catenin, which links cadherins indirectly to the cytoskeleton (reviewed in [8]), and p120 catenin which regulates cadherin turnover and modulates actin assembly (reviewed [15,16]).

Recent studies suggest that the ectodomains of classical cadherins, in the absence of cytoplasmic regions, are sufficient to drive the initial assembly of adherens junctions [17–19]. This process is mediated by cooperative formation of distinct cadherin–cadherin interfaces in *cis* (on the same cell) and in *trans* (on different cells). These prototypical interfaces of classical cadherins, and their roles in adhesion, are described in detail below. The molecular mechanisms of non-classical cadherins are less clear; recent structural and functional insights into this diverse group of proteins suggest various ecto-domain interactions beyond our current knowledge.

EXTRACELLULAR CADHERIN DOMAINS DRIVE ADHESION FROM OUTSIDE THE CELL

The relative contributions to adhesion of the extracellular and intracellular regions of classical cadherins are only now becoming clear. Early cell adhesion studies using cadherins engineered to lack p120 and β-catenin binding sites in the cytoplasmic domain demonstrated loss of adhesion, initially leading to the conclusion that the cytoplasmic machinery is essential for cadherin clustering and junction formation [20]. However, a recent study using E-cadherin similarly lacking the β-catenin and p120 binding sites but, crucially, with an endocytic clathrin adapter binding motif also deleted, showed effective junction forma-tion [17,18]. In A431 cells, which have background expression of wild-type E-cadherin, these 'tail-less' cadherins are effectively recruited into wild-type adherens junctions [18]. In addition, in cadherin-deficient A431-D cells, the tail-less cadherins form clusters at cell contact sites that closely resemble wild-type adherens junctions observed in the same study [18]. Similarly, in MDCK II cells and other epithelial cell lines, transfected catenin-uncoupled E-, N- and VE-cadherins [19] were also found to form adherens junction-like clusters in the lateral membranes. These results suggest that the cytoplasmic region is dispensable for the initial assembly of adherens junctions, yet to visualize these junctions in cells requires the uncoupling of endocytosis so that cadherin cell surface lifetimes are increased [17,18].

To test the hypothesis that classical cadherin extracellular domains can self-assemble to form junction-like structures, two groups separately devel-oped cell-free liposome systems in which cadherin ectodomains bound to the liposome surface were assessed for adhesion and junction formation [17,21]. Cryo-electron microscopy (cryo-EM) revealed that both liposome-attached E-cadherin [17] and VE-cadherin [21] extracellular domains clustered at sites of adhesive contact between liposomes and formed 'artificial adher-ens junctions' characterized by dense cadherin clustering and flattening of the apposed membranes [17]. Taken together, these cellular and biophys-ical studies demonstrate that vertebrate classical cadherin extracellular domains are sufficient to form initial cell–cell contacts and assemble adher-ens junction-like structures even without contributions from the cytoplasmic machinery. Cytoplasmic interactions in adherens junction formation, such as stabilization of junctions by actin fiber recruitment (reviewed in [14,22]), are likely to function downstream of these initial extracellular events.

MECHANISM OF ADHESIVE BINDING BETWEEN SINGLE CADHERIN MOLECULES FROM APPOSED CELLS

Classical cadherin ectodomains protrude from opposing cell surfaces and form *trans* adhesive homodimers through their membrane-distal EC1 domains, bridging the intermembrane space between neighboring cells (Figure 2a). The interface underlying this interaction has been character-ized in detail from atomic resolution structures [17,23–27], revealing that all classical cadherins share a common binding mechanism in which the most N-terminal portion of the β-A strand, the A* strand, is swapped between EC1 domains of the adhesive partner protomers (Figure 2a). Key to this mech-anism is the docking of conserved hydrophobic anchor residues located on the A* strand – tryptophan at position 2 (Trp2) for type I cadherins and Trp2 and Trp4 in type II cadherins – into a conserved hydrophobic pocket in the body of the partnering EC1 domain. The physiological relevance of this 'strand-swapped' adhesive interface has been confirmed in numerous mutation, electron microscopy, structural and cell studies [8,14,28–30].

The exchange of β-strands observed in classical cadherins is an example of the '3D domain swapping' protein interaction mechanism [31] in which the swapping domain (the A*-strand) can dock into its own pocket to form a 'closed' monomer (Figure 2b, left panel) or can dock into the pocket of the partner EC1 domain to form a swapped dimer (Figure 2b, right panel). A nec-essary step in the transition of the closed monomer to the swapped dimer is an open monomer state in which the swapping domain, the A* strand, is undocked, allowing dimer formation between two open monomers. Notably, the swapping domain is located in a closely similar residue environment in the 'closed' monomer and swapped dimer states. Therefore, the closed monomer form can be thought of as a competitive inhibitor for the swapped dimer. This competition is responsible for the weak binding affinities of clas-sical cadherins [32], and requires that structural differences exist that stabi-lize the dimer and/or destabilize the monomer to drive dimerization.

A comparison of cadherin domains that engage in strand swapping (the EC1 domains of classical cadherins) with non-swapping cadherin domains (EC2–5) identified numerous factors that favor the formation of strand-swapped dimers [33]. Swapping cadherin domains were found to have a shortened β-A strand, in addition to the conserved tryptophan at position 2, which is replaced by a phenylalanine in other EC domains. A glutamic acid residue (Glu11) at the base of the A strand coordinates Ca^{2+} in all classical cadher-ins, and anchoring of the A strand at both ends – at the base by Ca^{2+} binding to Glu11 and at the N terminus by Trp2 docking – induces strain in the short-ened A strand. This in turn destabilizes the closed monomer and thus favors

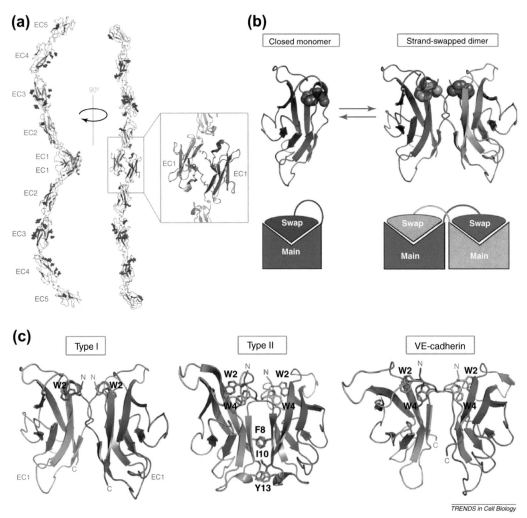

FIGURE 2 Classical cadherins from adhesive dimers by exchange of the N-terminal β-strand.

(a) A classical cadherin *trans* dimer is shown as a ribbon diagram in two orthogonal orientations; one protomer is shown in blue, one in orange (from pdb-ID: 3Q2W). Membrane distal EC1 domains overlap and exchange N-terminal β-strands (expanded view). Note that substantial O- and N-linked glycosylation (magenta and green spheres, respectively) is found on extracellular domains on EC2–4, but not on adhesive EC1 domains. Ca^{2+} ions are shown as green spheres. (b) The adhesive mechanism of classical cadherins is an example of 3D domain swapping. EC1 domains are shown for the monomer and the dimer (ribbon representation). The swapping element, residue Trp2 (side chain depicted as spheres), has an identical residue environment in the monomer (left panel) and the 'strand-swapped' dimer (right panel). Adapted with permission from [8]. (c) Ribbon presentations of strand-swapped EC1 domains of type I E-cadherin (pdb-ID: 2QVF), type II cadherin-11 (pdb-ID: 2A4E) and VE-cadherin (pdb-ID: 3PPE). Residues characteristic of the adhesive interfaces of each subfamily are depicted as sticks. In type I cadherins, residue Trp2 in domain EC1 is swapped between binding partners. In type II cadherins, two Trp residues, Trp2 and Trp4, are exchanged, and, in addition, hydrophobic interactions occur between conserved residues Phe8, Ile10 and Tyr13 giving rise to an extended interface. VE-cadherin exchanges Trp2 and Trp4 like type II cadherins, but the interface is limited to the apex of the domain, as in type I cadherins.

swapped dimer formation, in which this strain is released [34–36]. A naturally monomeric non-swapping EC2 domain was successfully 'converted' into a strand swap-binding EC1-like domain by introducing point mutations, thereby validating this mechanism [34].

Interestingly, although strain in the 'closed' monomer favors swapped dimer formation, selective pressure also appears to have kept adhesive binding weak. Type I classical cadherin EC1 sequences include a conserved Pro5-Pro6 motif that prevents continuous β-sheet hydrogen bonding between cadherin EC1 domains of adhesive dimers. When the diproline motif is mutated to alanine in E- and N-cadherins, dimer affinity is enhanced [34] and, as opposed to their wild-type counterparts [37] (see below), the mutant N- and E-cadherin dimerization affinities become indistinguishable. Crystal structures of these mutants reveal continuous β-strand hydrogen bonds between the A strands of partner EC1 domains, explaining the loss of binding specificity [34]. The diproline motif thus appears to be a required structural element underlying the differential binding affinities of N- and E-cadherin.

All vertebrate classical cadherins utilize a similar strand-swapping mechanism to form adhesive dimers; however, the interfaces found in the crystal structures of type I and type II cadherins are different (Figure 2c). The adhesive interface of type I cadherins is restricted to the pocket region near the apex of EC1 (Figure 2c, left panel) and the partner A* strand region, which includes the anchoring tryptophan residue Trp2. By contrast, in type II cadherins, two tryptophan residues, Trp2 and Trp4, are swapped. Moreover, the dimer interface in type II family members extends along the entire face of the EC1 domain involving conserved hydrophobic residues at position 8, 10 and 13 (Figure 2c, middle panel) [26]. Interestingly, VE-cadherin, a divergent classical cadherin and the crucial adhesion protein of the vascular endothelium [38], blurs the definition between type I and type II cadherin interfaces. In common with type II cadherins, VE-cadherin docks Trp2 and Trp4 into the hydrophobic pocket of its partner, but lacks the hydrophobic interactions along the rest of the EC1 domain (Figure 2c, right panel) and thus has an overall dimer arrangement more similar to that of type I cadherins [27].

Classical cadherin homophilic binding specificity at the cellular level is governed by EC1, as shown in domain shuffling experiments [26,39–42], suggesting that differences in the strand-swapping interface modulate specificity. Type I cadherins in general do not bind to type II cadherins [8,26,37,43] consistent with the substantial differences in the canonical adhesive interface structures of each subfamily. Interestingly, classical cadherins interact promiscuously within subfamilies, consistent with the close similarity in the interface region between individual members [26,37,43,44]. Thus, within subfamilies, cadherins exhibit both homophilic and heterophilic binding properties, which combine to yield the homophilic aggregation behavior of cadherin-expressing cells [37].

SPEED DATING: THE X-DIMER INTERMEDIATE

Formation of strand-swapped dimers requires refolding of each partner protomer to transition from the 'closed' monomer form (Figure 2b, left panel) to the 'open' dimer form (Figure 2b, right panel). This conformational change could render dimerization kinetically unfavorable because, this interconversion can occur over long periods of time in other proteins that engage in 3D domain swapping [31], yet binding is fast for cadherins [35,37,45]. Two alternative mechanisms have been proposed to explain the kinetics of cadherin interaction: 'selected fit', in which cadherin monomers exist in equilibrium between open and closed forms and dimerization results from collision of two open monomers; and 'induced fit', in which cadherin monomers first form a non-swapped intermediate – an 'encounter complex' – that lowers the activation energy required for strand swapping to occur [46]. Recently, single-molecule fluorescence resonance energy transfer (FRET) experiments provided evidence for an encounter complex in E-cadherin, strongly favoring the induced fit pathway for classical cadherin mediated interaction [47]. When strand swapping was ablated by a Trp2 to Ala mutation, dimers still formed between EC1 domains, with slightly altered FRET distances compared with swapped dimers, suggesting the existence of a non-swapped dimer form. Additionally, atomic force microscope (AFM) experiments showed the non-swapped mutant dimers to be weaker than strand-swapped wild-type dimers, energetically consistent with a role as a binding intermediate [47].

Crystallographic studies of strand swap-impaired cadherin mutants revealed the molecular details of this encounter complex [48]. For numerous swapping-incompetent mutants, a dimer with its interface centered around the EC1–EC2 interdomain linker region and the apex of EC2 (Figure 3, middle panel) is observed. This structure is now referred to as the 'X-dimer' because the relative orientation of the interacting protomers is reminiscent of an 'X' shape (Figure 3). The X-dimer requires no refolding for its interaction, enabling fast binding kinetics. Importantly, the X-dimer positions the A strands of each protomer parallel to each other in close proximity as if poised to swap [48]. A similar structure was observed in a strand swap-deficient mutant of type II cadherin-6 [48].

The role of the X-dimer as an encounter complex is confirmed by the observation that mutations designed to prevent X-dimer formation, while leaving strand swapping intact, significantly slowed the binding rate of E-cadherin

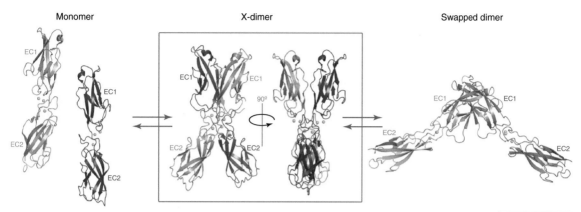

TRENDS in Cell Biology

FIGURE 3 **Strand-swapped adhesive dimers of classical cadherins form through a non-swapped intermediate.**
E-cadherin monomers (orange and blue ribbon diagrams (left panel); only EC1–2 shown for clarity) associate via an 'X-dimer' interface in which N-terminal strands are not swapped but are closely apposed (middle panel). Swapping of strands leads to formation of mature strand-swapped dimers (right panel). Assembly and disassembly of swapped dimers is likely to proceed via the same pathway. Protomers shown as orange and blue ribbon diagrams with only EC1–2 domains shown for clarity.

and cadherin-6. Specifically, no dimerization is observed in short term SPR assays, but there is no loss of affinity in long term analytical ultracentrifugation experiments [48]. Furthermore, unlike wild-type proteins, X-dimer mutant monomers and dimers could be resolved as stable monomer and dimer species in size exclusion chromatography and velocity ultracentrifugation experiments, indicating slow exchange rates between these two forms [48]. In unpublished work, we find a similar structure and binding behavior for the X-dimer of N-cadherin and, in addition, mutation of the predicted X-dimer interface in N-cadherin has been shown to abolish cell–cell aggregation activity [49]. Similar to the encounter complex observed via FRET experiments, X-dimers were found to have weaker binding affinity than wild-type swapped dimers [48]. In transfected epithelial cells, cadherin X-dimer mutants formed extraordinarily stable cell–cell junctions [50], consistent with slowed monomer–dimer exchange rates observed in cell free experiments [48], although effects on dimer dissociation were emphasized by the authors. Taken together, current data favor a mechanism in which the X-dimer functions as an intermediate in the formation and disassembly of the 'mature' adhesive dimer. The structural and functional observation of X-dimers in type I E-cadherin and the relatively distant type II cadherin-6 (34% identity over EC1–EC2), together with sequence conservation patterns of interfacial residues [17], suggests that the X-dimer mechanism may be general among members of the two subfamilies of vertebrate classical cadherins.

Interestingly, T-cadherin, a divergent classical cadherin anchored to the plasma membrane via a glycosylphosphatidyl inositol (GPI) anchor (Box 1), does not strand swap and adopts an X-dimer conformation for its mature adhesive binding interface [51]. Mutations targeting the X-dimer interface in T-cadherin were found to abolish its function in modulation of neurite outgrowth, whereas targeted strand dimer mutations, analogous to those that abolish strand-swap binding in classical cadherins, had no effect on T-cadherin function or homodimerization [51]. The close phylogenetic relation to type I classical cadherins suggests that T-cadherin represents a classical cadherin that has lost its swapping ability. Other roles for X-dimers outside the classical cadherin subfamily remain unknown.

FROM BONDS TO JUNCTIONS

Cell–cell adhesion in mature tissues is mediated in part by adherens junctions where numerous cadherin *trans* dimers assemble. In principle, a passive diffusion trap mechanism, whereby cadherins would become concentrated at cell–cell contact sites through their adhesive interactions, could explain the accumulation of cadherins at sites of intercellular contact [52]. However, mutations in a crucial *cis* interface (described below; Figure 4a) which leave adhesive binding intact show that the diffusion trap mechanism is insufficient to achieve the level of concentration at cell–cell contacts observed for wild-type cadherins [17]. It therefore appears likely that lateral or *cis* interactions could account for the enhanced localization of classical cadherins at cell contact sites.

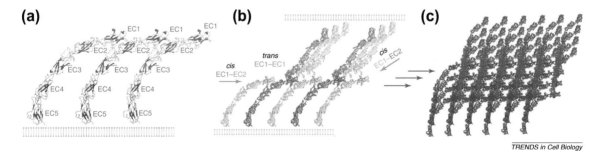

TRENDS in Cell Biology

FIGURE 4 Extracellular structure of adherens junctions formed through *cis* and *trans* ectodomain interactions.
(**a**) Selected region of the N-cadherin EC1–5 crystal lattice (blue ribbon presentation; pdb-ID: 3Q2W) showing an array of N-cadherin molecules oriented as if emanating from the same cell membrane and connected by a *cis* interface formed between the EC1 and EC2 domains of neighboring molecules. (**b**) Strand-swapped *trans* dimers form together with *cis* interactions in the same crystal lattice. *trans* interactions orient opposing *cis* arrays approximately perpendicularly such that each *cis* array (blue) forms *trans* interactions with multiple opposing *cis* arrays (orange). (**c**) The combination of *cis* and *trans* interactions enables cadherin ectodomains to form an ordered network that is thought to be the basis for the extracellular architecture of adherens junction. Adapted from [17].

A potential lateral interaction site, apparently conserved among type I cadherins, has been observed in crystal structures of full-length ectodomains of C- [24], N- and E-cadherins [17]. Despite forming crystals that are unrelated to one another, in addition to the adhesive strand-swap interface, all three structures reveal a lateral interface formed between the base of the EC1 domain of one protomer and a region near the apex of EC2 of a parallel partner (Figure 4a). The combination of *cis* and *trans* interactions engaged by each cadherin molecule (Figure 4b) creates a molecular layer within each crystal that is likely to correspond to the extracellular structure of the adherens junction [17,24]. The region of EC1 involved in this *cis* interface is opposite to the strand-swapping site, so that *cis* and *trans* interactions can form simultaneously resulting in a continuous two-dimensional lattice with dimensions near to that expected for adherens junctions (Figure 4c).

When cadherin ectodomains are bound to the surface of liposomes, in the absence of other proteins, cryo-EM analysis reveals ordered junction-like structures that resemble the molecular layer observed in C- [24], E- and N-cadherin crystals [17]. This system, as well as cell-based experiments, was used to test the idea that the *cis* interface underlies the lateral assembly of cadherins in adherens junctions. Mutations that targeted the *cis* interface of E-cadherin (without interfering with *trans* strand-swapped dimerization) still allowed a reduced level of adhesion between liposomes; however, the ordered structure of the reconstituted junctions was lost [17]. Consistently, incorporation of these mutants into endogenous wild-type cellular junctions caused these junctions to become unstable and transient [17]. In cells lacking endogenous cadherin, *cis* mutant protein localized to sites of cell contact but failed to cluster into junction-like structures [17]. Taken together, these data suggest that the *cis* interface identified in structural studies functions to laterally assemble cadherin *trans* dimers into adherens junctions. *cis* oligomerization of cadherins at adherens junctions might account for previous observations of multiple adhesive states between cadherin monolayers in molecular force experiments [45,53,54] that were initially interpreted as multiple *trans* dimer states, but could be explained by combinations of *cis* and *trans* interactions.

Interestingly, the *cis* interaction is too weak to be detected in solution binding experiments (which are limited to a detection level of approximately 1mM) [17], yet as discussed above it appears to play a crucial biological role. This is not surprising because the strength of interaction between proteins in solution can differ significantly from that of the same interaction in the context of restricted motion when membrane bound [55]. Indeed, *in silico* simulations suggest that when *trans* ectodomain dimers form, flexibility is dramatically reduced because the two interacting protomers are now attached to each other via the adhesive interface, and in addition are tethered to each of the apposed cell membranes [55]. Thus, when *trans* dimers are formed, conformational flexibility is decreased, which lowers the entropic penalty associated with *cis* dimer formation [55,56]. This model, in which *cis* assembly requires *trans* dimerization, would account for observations that cadherins do not cluster in the absence of cognate adhesion to an apposed cadherin-expressing cell [12,13].

Large cellular adherens junctions such as the zonula adherens that circumscribe epithelial cells appear less dense than desmosomes (see below) and it is possible that they are assembled from numerous subdomains, each with the defined layer structure described above. The cadherin extracellular lattice structure is directional such that two such subdomains would have to meet with an appropriate orientation to achieve a continuous merger. Whether zonula adherens are formed from continuous structures or collections of defined puncta is not clear from current data. Maturation of these structures requires cytoskeletal activity, which could play a role in driving their assembly from smaller puncta [57]. Currently, we favor the view that small punctate clusters are likely to auto-assemble via their ectodomains when two cognate cadherin-expressing cells come into contact, and these may later be incorporated into mature large adherens junctions by cytoplasmic processes. Further investigation will be needed to elucidate the interplay between extracellular and cytoplasmic mechanisms in cadherin assembly.

Although type II classical cadherins have the same adhesive mechanism as type I cadherins, the *cis* interface described above has not been found in any of the multidomain crystal structures of type II cadherins [26,27,48]. Nonetheless, there is evidence that at least some members of the type II cadherin family may form adherens junction-like structures. In particular, junctions mediated by the divergent VE-cadherin appear similar to those of type I cadherins seen by EM [58]. Cadherin-11, a type II cadherin, has also been observed to colocalize with p120-catenin, α-catenin and actin filaments at cell–cell contacts [59], but it remains to be determined whether cadherin-11 and other type II cadherins form junctions with ultrastructures similar to those observed for type I cadherins. Thus, VE-cadherin and possibly other type II cadherins may form cell–cell junctions via a different lateral interface yet to be determined.

NON-CLASSICAL CADHERIN SUBFAMILIES SUGGEST DIVERSITY OF ADHESIVE MECHANISM

Desmosomal Cadherins

Sequence conservation analyses suggest that desmosomal cadherins (Box 1, reviewed in [60]), the major component of desmosome junctions, also adhere through a strand-swap binding mechanism. These cadherins have the classical Trp residue conserved at position 2, and hydrophobic residues corresponding to the Trp binding pocket in classical cadherins [33,60]. Moreover, mutation of Trp2 or the hydrophobic pocket abolishes *trans* binding of desmocollin 2 in cross-linking experiments [61]. A nuclear magnetic resonance (NMR) structure of an EC1 fragment of human des-moglein-2 (pdb-ID: 2YQG) shows a domain fold remarkably similar to type I classical cadherins. This structure is monomeric and Trp2 is self-docked, probably owing to the inclusion of ten residues preceding the native N terminus due to cloning artifacts. Similar extensions are known to inhibit strand-swap dimerization in classical cadherins [25,48].

Cryo-electron tomography of vitreous sections of desmosomes from human skin [62] and electron tomography of mouse skin sections embedded in plastic [63] show an extracellular arrangement compatible with *trans* dimerization via EC1 domains. Desmosomes in human skin showed a highly ordered arrangement in the extracellular region, whereas those of mouse skin were relatively disordered, probably due to differences in sample preparation. Fitting of structures of C-cadherin ectodomains [24] into a 34Å resolution cryo-EM map of desmosomes in human skin suggested a molecular array comprising a linear 'zipper' formed from alternating EC1-mediated *cis*- and *trans*-interactions [62] distinct from the two-dimensional array observed for type I classical cadherins [17,24]. Alternatively, the possibility that desmo-somal cadherins form similar assemblies to classical cadherins has also been suggested, based on EM of lanthanide infiltrated desmosomes from guinea pig heart [64]. Because an atomic resolution structure of *cis* and *trans* dimers of desmosomal cadherins is not yet available, further mutational and structural studies are needed to reveal their detailed binding mechanism. Interestingly, binding interactions between desmogleins and desmocollins have also been shown to display a high degree of isoform specificity [61].

Clustered Protocadherins

Clustered protocadherins, named because they are encoded in three novel gene clusters (α, β and γ) are predominantly expressed in vertebrate brain (Box 1) and constitute the largest cadherin subfamily. However, their adhesive properties are poorly understood. Numerous single domain structures have been determined for protocadherins (pdb ID: 2EE0, 2YST, 1WYJ, 1WUZ; [65]), but none appear to include a functional adhesive binding site, which remains elusive. Transfected cell aggregation studies showed strict homo-philic binding specificity for seven members of the protocadherin γ-cluster [66]. In the same system, domain shuffling experiments showed that con-secutive domains EC1 through EC3 are crucial for *trans* adhesion and, interestingly, domains EC2 and EC3 were found to govern protocadherin specificity in cell aggregation assays [66]. Notably, these domains show the highest sequence diversity among individual protocadherin isoforms [66]. Since individual neurons express multiple protocadherin isoforms [67,68], homophilic specificity of this type could give rise to an enormous range of potential cellular affinities. It has further been suggested that multiple isoforms can associate as *cis* tetramers on the same cell surface to mediate combinatorial specificity [66], although this model remains untested.

Large Cadherins with Many EC Domains

Numerous members of the cadherin superfamily – in both vertebrates and invertebrates – are large proteins containing many EC domains (Box 1). Although relatively little is known about their structure/function relations, early studies suggest that some of these proteins adopt extended conformations, whereas others may form structures more like folded globular 'super-domains'. Two atypical members of the superfamily that appear to adopt extended structures, cadherin-23 (27 EC domains) and protocadherin-15 (11 ECs) (Box 1 and Figure I) [69], link stereocilia of hair cells by formation of an extracellular structure, known as the tip-link, assembled by *trans* heterophilic interaction between *cis* homodimers [6]. Recently, the atomic resolution structure of an N-terminal EC1–2 domain fragment of cadherin-23 was determined [70,71] revealing a domain architecture closely similar to that of other known cadherins (Figure 5a), as well as features unique to cadherin-23 including a 3_{10} helix in the A strand, an α-helix between β strands C and D of EC1 and, notably, an additional novel calcium binding site, referred to as site 0, at the apex of EC1 [70,71]. N-terminal fragments of cadherin-23 and protocadherin-15 comprising EC1–3 are sufficient for *trans* heterophilic binding, but not for *cis* homodimerization [6,70,71]. The structures of the heterocomplex of cadherin-23 and protocadherin-15 and of longer ecto-domain fragments are needed to identify the *trans* and *cis* interfaces that form the tip-link.

DN-cadherin and DE-cadherin in *Drosophila melanogaster* are invertebrate classical cadherin orthologs, in that they mediate Ca^{2+}-dependent cell–cell adhesion, have conserved armadillo binding domains in their cytoplasmic region [72], and form adherens junctions with intermembrane distances of 20–30nm, similar to those of mammals [14,73,74]. However, the extracellular domain organization is very different from that of vertebrate classical cadherins: there are 8 and 16 sequence-predicted EC domains arrayed in tandem for DE- and DN-cadherin [75], respectively, followed by epidermal growth factor (EGF)-like and laminin-G domains (Box 1, Figure I). Structures of the N-terminal portion of DN-cadherin were recently determined [75] and revealed that a Ca^{2+}-free linker region between domains EC2 and EC3 results in an acute interdomain angle in all three crystal forms that causes the otherwise linear structure of EC1–4 to 'jackknife' (Figure 5b). Bioinformatics

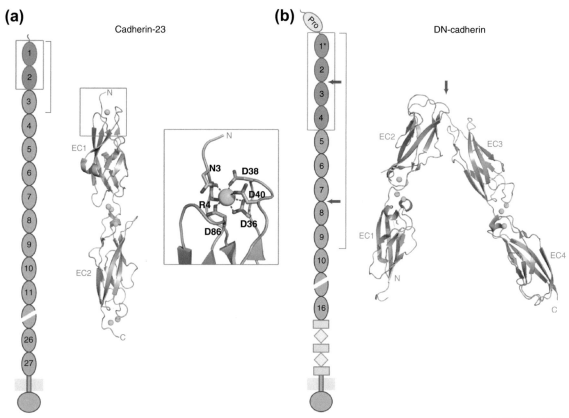

FIGURE 5 Crystal structures of cadherin-23 and *Drosophila* N-cadherin reveal unique features of atypical cadherins.
(a) Structures of mouse cadherin-23 EC1–2, which are involved in adhesive binding to protocadherin-15 (binding domain indicated by brackets in schematic) reveal successive EC domains (ribbon diagram, pdb-ID: 3MVS, 2WHV) with three Ca^{2+} ions (green spheres) coordinated in the linker region. Uniquely, a Ca^{2+} binding site was identified at the apex of EC1 (box), referred to as Ca^{2+} binding site 0. Structural determination of a complex of cadherin-23 with protocadherin-15 will help to identify the heterophilic binding interface. (b) Structures of DN-cadherin EC1–4, which is part of the adhesive interface for homodimerization (EC1–9, bracket in schematic), reveal four consecutive EC domains (ribbon diagram). Interestingly, Ca^{2+} coordination was found only between domains EC1–2 and EC3–4 and not between EC2–3 (pink arrow). This Ca^{2+}-free linker introduces a 'kink' in the otherwise linear structure. Sequence analysis suggests a second occurrence of a Ca^{2+}-free linker between EC7 and EC8 in the ectodomain of DN-cadherin; this may contribute to folding of the 16 EC domains into a compact form within the intermembrane space of *Drosophila* adherens junctions.

analysis finds that other long cadherins such as FAT, FAT-like, Dachsous and CELSR/Flamingo also contain interdomain linker regions that lack some or all of the residues required for Ca^{2+} binding [75]. These findings suggest that long cadherin ectodomains might fold onto themselves, resulting in a more compact arrangement compatible with the relatively narrow intermembrane distance of adherens junctions. It remains to be determined whether these Ca^{2+}-free linker regions allow for flexibility or whether they introduce a fixed bend as observed in the three crystal structures so far determined. Deletion mutagenesis has mapped the minimal adhesive binding site of DN-cadherin to the EC1–9 domain region. The apparent requirement for nine EC domains is remarkably different from the vertebrate classical cadherins, for which all adhesive contacts are formed through EC1–EC1 interactions. The jackknifed bend between domains EC3 and EC4 – and another predicted between EC7 and EC8 (Figure 5b) – is reminiscent of Dscam immunoglobulin superfamily adhesion proteins, which fold into a super-domain platform that positions multiple immunoglobulin domains for engagement in adhesive binding [76].

CONCLUDING REMARKS

The structural basis of the adhesive function of vertebrate classical cadherins is becoming increasingly clear. Adhesive binding between cells uses a *trans* strand-swapping mechanism that is enabled by a fast-binding intermediate, the X-dimer. Vertebrate classical cadherins on isolated cells diffuse freely in the plasma membrane, but when they are bound by cognate cadherins from a contacting apposed cell, *trans* binding lowers the entropic penalty for the formation of *cis* interactions, initiating lateral oligomerization. These early processes depend only on the properties of cadherin ectodomains, yet subsequent events such as junction strengthening clearly involve interactions of the cadherin cytoplasmic region with regulatory and cytoplasmic elements.

The picture is far less clear for other cadherin subfamilies. Only vertebrate desmosomal cadherins – close relatives of the classical subfamilies – contain sequence elements indicative of strand-swap binding. Other members of the cadherin superfamily, including all invertebrate cadherins, seem likely to engage in adhesive binding by other means, and may adopt diverse binding mechanisms. It is remarkable how many different cadherin–cadherin interfaces have already been discovered, revealing a surprising complexity in the interactions of classical cadherins. However, longer cadherins appear to form interfaces through surfaces not yet defined. Moreover, some cadherins are known to bind to other proteins and, although for some of these the structural basis is known (for example E-cadherin binding to NKLRG1 [77] and to internalin [78]), for others, such as integrins [79], we have little structural insight into how such binding occurs. It is clear that, at this stage, our structural understanding is limited to only a small portion of the wider cadherin universe, which appears to exploit the remarkable versatility of the cadherin fold in forming diverse sets of protein–protein interfaces. Progress in our understanding of classical cadherins emphasizes the utility of combining insights from structural, cell biological, biophysical and computational studies. The new mechanistic insights that have been inferred may be applicable to many other classes of adhesion receptors.

ACKNOWLEDGMENTS

This work was supported by grants R01 GM062270 (to L.S.) from the National Institutes of Health and MCB-0918535 (to B.H.) from the National Science Foundation.

REFERENCES

1 Abedin, M. and King, N. (2008) The premetazoan ancestry of cadherins. *Science* 319, 946–948

2 Hulpiau, P. and van Roy, F. (2011) New insights into the evolution of metazoan cadherins. *Mol. Biol. Evol.* 28, 647–657

3 Williams, M.E. *et al.* (2011) Cadherin-9 regulates synapse-specific differentiation in the developing hippocampus. *Neuron* 71, 640–655

4 Osterhout, J.A. *et al.* (2011) Cadherin-6 mediates axon-target matching in a non-image-forming visual circuit. *Neuron* 71, 632–639

5 Leckband, D.E. *et al.* (2011) Mechanotransduction at cadherin-mediated adhesions. *Curr. Opin. Cell Biol.* 23, 523–530

6 Kazmierczak, P. *et al.* (2007) Cadherin 23 and protocadherin 15 interact to form tip-link filaments in sensory hair cells. *Nature* 449, 87–91

7 Watanabe, T. *et al.* (2009) Cadherin-mediated intercellular adhesion and signaling cascades involving small GTPases. *Cold Spring Harb. Perspect. Biol.* 1, a003020

8 Shapiro, L. and Weis, W.I. (2009) Structure and biochemistry of cadherins and catenins. *Cold Spring Harb. Perspect. Biol.* 1, a003053

9 Harris, T.J. and Tepass, U. (2010) Adherens junctions: from molecules to morphogenesis. *Nat. Rev. Mol. Cell Biol.* 11, 502–514

10 Nishimura, T. and Takeichi, M. (2009) Remodeling of the adherens junctions during morphogenesis. *Curr. Top. Dev. Biol.* 89, 33–54

11 Berx, G. and van Roy, F. (2009) Involvement of members of the cadherin superfamily in cancer. *Cold Spring Harb. Perspect. Biol.* 1, a003129

12 Hajra, K.M. and Fearon, E.R. (2002) Cadherin and catenin alterations in human cancer. *Genes Chromosomes Cancer* 34, 255–268

13 Gumbiner, B.M. (2005) Regulation of cadherin-mediated adhesion in morphogenesis. *Nat. Rev. Mol. Cell Biol.* 6, 622–634

14 Meng, W. and Takeichi, M. (2009) Adherens junction: molecular architecture and regulation. *Cold Spring Harb. Perspect. Biol.* 1, a002899

15 Reynolds, A.B. and Carnahan, R.H. (2004) Regulation of cadherin stability and turnover by p120ctn: implications in disease and cancer. *Semin. Cell Dev. Biol.* 15, 657–663

16 Yonemura, S. (2011) Cadherin-actin interactions at adherens junctions. *Curr. Opin. Cell Biol.* 23, 515–522

17 Harrison, O.J. *et al.* (2011) The extracellular architecture of adherens junctions revealed by crystal structures of type I cadherins. *Structure* 19, 244–256

18 Hong, S. *et al*. (2010) Spontaneous assembly and active disassembly balance adherens junction homeostasis. *Proc. Natl. Acad. Sci. U.S.A.* 107, 3528–3533

19 Ozaki, C. *et al*. (2010) The extracellular domains of E- and N-cadherin determine the scattered punctate localization in epithelial cells and the cytoplasmic domains modulate the localization. *J. Biochem.* 147, 415–425

20 Nagafuchi, A. and Takeichi, M. (1988) Cell binding function of E-cadherin is regulated by the cytoplasmic domain. *EMBO J.* 7, 3679–3684

21 Taveau, J.C. *et al*. (2008) Structure of artificial and natural VE-cadherin-based adherens junctions. *Biochem. Soc. Trans.* 36, 189–193

22 Pokutta, S. *et al*. (2008) Biochemical and structural analysis of alpha-catenin in cell–cell contacts. *Biochem. Soc. Trans.* 36, 141–147

23 Shapiro, L. *et al*. (1995) Structural basis of cell–cell adhesion by cadherins. *Nature* 374, 327–337

24 Boggon, T.J. *et al*. (2002) C-cadherin ectodomain structure and implications for cell adhesion mechanisms. *Science* 296, 1308–1313

25 Haussinger, D. *et al*. (2004) Proteolytic E-cadherin activation followed by solution NMR and X-ray crystallography. *EMBO J.* 23, 1699–1708

26 Patel, S.D. *et al*. (2006) Type II cadherin ectodomain structures: implications for classical cadherin specificity. *Cell* 124, 1255–1268

27 Brasch, J. *et al*. (2011) Structure and binding mechanism of vascular endothelial cadherin: a divergent classical cadherin. *J. Mol. Biol.* 408, 57–73

28 Harrison, O. *et al*. (2005) Cadherin adhesion depends on a salt bridge at the N-terminus. *J. Cell Sci.* 118, 4123–4130

29 Troyanovsky, R. *et al*. (2003) Adhesive and lateral E-cadherin dimers are mediated by the same interface. *Mol. Cell. Biol.* 23, 7965–7972

30 Patel, S.D. *et al*. (2003) Cadherin-mediated cell–cell adhesion: sticking together as a family. *Curr. Opin. Struct. Biol.* 13, 690–698

31 Bennett, M.J. *et al*. (1995) 3D domain swapping: a mechanism for oligomer assembly. *Protein Sci.* 4, 2455–2468

32 Chen, C.P. *et al*. (2005) Specificity of cell–cell adhesion by classical cadherins: critical role for low-affinity dimerization through beta-strand swapping. *Proc. Natl. Acad. Sci. U.S.A.* 102, 8531–8536

33 Posy, S. *et al*. (2008) Sequence and structural determinants of strand swapping in cadherin domains: do all cadherins bind through the same adhesive interface? *J. Mol. Biol.* 378, 952–966

34 Vendome, J. *et al*. (2011) Molecular design principles underlying beta-strand swapping in the adhesive dimerization of cadherins. *Nat. Struct. Mol. Biol.* 18, 693–700

35 Vunnam, N. and Pedigo, S. (2011) Prolines in betaA-sheet of neural cadherin act as a switch to control the dynamics of the equilibrium between monomer and dimer. *Biochemistry* 50, 6959–6965

36 Vunnam, N. and Pedigo, S. (2011) Calcium-induced strain in the monomer promotes dimerization in neural cadherin. *Biochemistry* 50, 8437–8444

37 Katsamba, P. *et al*. (2009) Linking molecular affinity and cellular specificity in cadherin-mediated adhesion. *Proc. Natl. Acad. Sci. U.S.A.* 106, 11594–11599

38 Harris, E.S. and Nelson, W.J. (2010) VE-cadherin: at the front, center, and sides of endothelial cell organization and function. *Curr. Opin. Cell Biol.* 22, 651–658

39 Klingelhofer, J. *et al*. (2000) Amino-terminal domain of classic cadherins determines the specificity of the adhesive interactions. *J. Cell Sci.* 113, 2829–2836

40 Nose, A. *et al.* (1990) Localization of specificity determining sites in cadherin cell adhesion molecules. *Cell* 61, 147–155

41 Shan, W. *et al.* (2000) Functional cis-heterodimers of N- and R-cadherins. *J. Cell Biol.* 148, 579–590

42 Shan, W.S. *et al.* (1999) The adhesive binding site of cadherins revisited. *Biophys. Chem.* 82, 157–163

43 Shimoyama, Y. *et al.* (2000) Identification of three human type-II classic cadherins and frequent heterophilic interactions between different subclasses of type-II classic cadherins. *Biochem. J.* 349, 159–167

44 Shimoyama, Y. *et al.* (1999) Biochemical characterization and functional analysis of two type II classic cadherins, cadherin-6 and -14, and comparison with E-cadherin. *J. Biol. Chem.* 274, 11987–11994

45 Bayas, M.V. *et al.* (2006) Lifetime measurements reveal kinetic differences between homophilic cadherin bonds. *Biophys. J.* 90, 1385–1395

46 Miloushev, V.Z. *et al.* (2008) Dynamic properties of a type II cadherin adhesive domain: implications for the mechanism of strand-swapping of classical cadherins. *Structure* 16, 1195–1205

47 Sivasankar, S. *et al.* (2009) Characterizing the initial encounter complex in cadherin adhesion. *Structure* 17, 1075–1081

48 Harrison, O.J. *et al.* (2010) Two-step adhesive binding by classical cadherins. *Nat. Struct. Mol. Biol.* 17, 348–357

49 Emond, M.R. *et al.* (2011) A complex of protocadherin-19 and N-cadherin mediates a novel mechanism of cell adhesion. *J. Cell Biol.* 195, 1115–1121

50 Hong, S. *et al.* (2011) Cadherin exits the junction by switching its adhesive bond. *J. Cell Biol.* 192, 1073–1083

51 Ciatto, C. *et al.* (2010) T-cadherin structures reveal a novel adhesive binding mechanism. *Nat. Struct. Mol. Biol.* 17, 339–347

52 Perez, T.D. *et al.* (2008) Immediate-early signaling induced by E-cadherin engagement and adhesion. *J. Biol. Chem.* 283, 5014–5022

53 Sivasankar, S. *et al.* (2001) Direct measurements of multiple adhesive alignments and unbinding trajectories between cadherin extracellular domains. *Biophys. J.* 80, 1758–1768

54 Zhu, B. *et al.* (2003) Functional analysis of the structural basis of homophilic cadherin adhesion. *Biophys. J.* 84, 4033–4042

55 Wu, Y. *et al.* (2011) Transforming binding affinities from three dimensions to two with application to cadherin clustering. *Nature* 475, 510–513

56 Wu, Y. *et al.* (2010) Cooperativity between trans and cis interactions in cadherin-mediated junction formation. *Proc. Natl. Acad. Sci. U.S.A.* 107, 17592–17597

57 Vasioukhin, V. *et al.* (2000) Directed actin polymerization is the driving force for epithelial cell–cell adhesion. *Cell* 100, 209–219

58 Uehara, K. (2006) Distribution of adherens junction mediated by VE-cadherin complex in rat spleen sinus endothelial cells. *Cell Tissue Res.* 323, 417–424

59 Kiener, H.P. *et al.* (2006) The cadherin-11 cytoplasmic juxtamembrane domain promotes alpha-catenin turnover at adherens junctions and intercellular motility. *Mol. Biol. Cell* 17, 2366–2376

60 Thomason, H.A. *et al.* (2010) Desmosomes: adhesive strength and signalling in health and disease. *Biochem. J.* 429, 419–433

61 Nie, Z. *et al.* (2011) Membrane-impermeable cross-linking provides evidence for homophilic, isoform-specific binding of desmosomal cadherins in epithelial cells. *J. Biol. Chem.* 286, 2143–2154

62 Al-Amoudi, A. *et al*. (2007) The molecular architecture of cadherins in native epidermal desmosomes. *Nature* 450, 832–837

63 He, W. *et al*. (2003) Untangling desmosomal knots with electron tomography. *Science* 302, 109–113

64 Garrod, D.R. *et al*. (2005) Hyper-adhesion in desmosomes: its regulation in wound healing and possible relationship to cadherin crystal structure. *J. Cell Sci.* 118, 5743–5754

65 Morishita, H. *et al*. (2006) Structure of the cadherin-related neuronal receptor/protocadherin-alpha first extracellular cadherin domain reveals diversity across cadherin families. *J. Biol. Chem.* 281, 33650–33663

66 Schreiner, D. and Weiner, J.A. (2010) Combinatorial homophilic interaction between gamma-protocadherin multimers greatly expands the molecular diversity of cell adhesion. *Proc. Natl. Acad. Sci. U.S.A.* 107, 14893–14898

67 Kaneko, R. *et al*. (2006) Allelic gene regulation of Pcdh-alpha and Pcdh-gamma clusters involving both monoallelic and biallelic expression in single Purkinje cells. *J. Biol. Chem.* 281, 30551–30560

68 Esumi, S. *et al*. (2005) Monoallelic yet combinatorial expression of variable exons of the protocadherin-alpha gene cluster in single neurons. *Nat. Genet.* 37, 171–176

69 Vollrath, M.A. *et al*. (2007) The micromachinery of mechanotransduction in hair cells. *Annu. Rev. Neurosci.* 30, 339–365

70 Sotomayor, M. *et al*. (2010) Structural determinants of cadherin-23 function in hearing and deafness. *Neuron* 66, 85–100

71 Elledge, H.M. *et al*. (2010) Structure of the N terminus of cadherin 23 reveals a new adhesion mechanism for a subset of cadherin superfamily members. *Proc. Natl. Acad. Sci. U.S.A.* 107, 10708–10712

72 Laprise, P. and Tepass, U. (2011) Novel insights into epithelial polarity proteins in *Drosophila*. *Trends Cell. Biol.* 21, 401–408

73 Oda, H. and Takeichi, M. (2011) Evolution: structural and functional diversity of cadherin at the adherens junction. *J. Cell Biol.* 193, 1137–1146

74 Tepass, U. and Hartenstein, V. (1994) The development of cellular junctions in the *Drosophila* embryo. *Dev. Biol.* 161, 563–596

75 Jin, X. *et al*. (2011) Crystal structures of *Drosophila* N-cadherin ectodomain regions reveal a widely used class of Ca^{2+}-free interdomain linkers. *Proc. Natl. Acad. Sci. U.S.A.* 109, E127–E134

76 Sawaya, M.R. *et al*. (2008) A double S shape provides the structural basis for the extraordinary binding specificity of Dscam isoforms. *Cell* 134, 1007–1018

77 Li, Y. *et al*. (2009) Structure of natural killer cell receptor KLRG1 bound to E-cadherin reveals basis for MHC-independent missing self recognition. *Immunity* 31, 35–46

78 Schubert, W. *et al*. (2002) Structure of internalin, a major invasion protein of *Listeria monocytogenes*, in complex with its human receptor E-cadherin. *Cell* 111, 825–836

79 Weber, G.F. *et al*. (2011) Integrins and cadherins join forces to form adhesive networks. *J. Cell Sci.* 124, 1183–1193

80 Nagar, B. *et al*. (1996) Structural basis of calcium-induced E-cadherin rigidification and dimerization. *Nature* 380, 360–364

81 Pokutta, S. *et al*. (1994) Conformational changes of the recombinant extracellular domain of E-cadherin upon calcium binding. *Eur. J. Biochem.* 223, 1019–1026

82 Takeichi, M. (1991) Cadherin cell adhesion receptors as a morphogenetic regulator. *Science* 251, 1451–1455

83 Nollet, F. *et al.* (2000) Phylogenetic analysis of the cadherin superfamily allows identification of six major subfamilies besides several solitary members. *J. Mol. Biol.* 299, 551–572

84 Hulpiau, P. and van Roy, F. (2009) Molecular evolution of the cadherin superfamily. *Int. J. Biochem. Cell Biol.* 41, 349–369

85 Yagi, T. (2008) Clustered protocadherin family. *Dev. Growth Differ.* 50, S131–S140

86 Kim, S.Y. *et al.* (2011) Non-clustered protocadherin. *Cell. Adh. Migr.* 5, 97–105

87 Usui, T. *et al.* (1999) Flamingo, a seven-pass transmembrane cadherin, regulates planar cell polarity under the control of Frizzled. *Cell* 98, 585–595

88 Sopko, R. and McNeill, H. (2009) The skinny on Fat: an enormous cadherin that regulates cell adhesion, tissue growth, and planar cell polarity. *Curr. Opin. Cell Biol.* 21, 717–723

89 Ahmed, Z.M. *et al.* (2006) The tip-link antigen, a protein associated with the transduction complex of sensory hair cells, is protocadherin-15. *J. Neurosci.* 26, 7022–7034

90 Ranscht, B. and Dours-Zimmermann, M.T. (1991) T-cadherin, a novel cadherin cell adhesion molecule in the nervous system lacks the conserved cytoplasmic region. *Neuron* 7, 391–402

91 Alattia, J.R. *et al.* (1997) Lateral self-assembly of E-cadherin directed by cooperative calcium binding. *FEBS Lett.* 417, 405–408

92 Koch, A.W. *et al.* (1997) Calcium binding and homoassociation of E-cadherin domains. *Biochemistry* 36, 7697–7705

93 Troyanovsky, R.B. *et al.* (2003) Adhesive and lateral E-cadherin dimers are mediated by the same interface. *Mol. Cell. Biol.* 23, 7965–7972

94 Pertz, O. *et al.* (1999) A new crystal structure, Ca2+ dependence and mutational analysis reveal molecular details of E-cadherin homoassociation. *EMBO J.* 18, 1738–1747

95 Haussinger, D. *et al.* (2002) Calcium-dependent homoassociation of E-cadherin by NMR spectroscopy: changes in mobility, conformation and mapping of contact regions. *J. Mol. Biol.* 324, 823–839

96 Sotomayor, M. and Schulten, K. (2008) The allosteric role of the Ca2+ switch in adhesion and elasticity of C-cadherin. *Biophys. J.* 94, 4621–4633

nds in Cell Biology

Actin Cortex Mechanics and Cellular Morphogenesis

Guillaume Salbreux[1,*], Guillaume Charras[2,3], Ewa Paluch[4,5,*]

[1]Max Planck Institute for the Physics of Complex Systems, Dresden, 01187, Germany,
[2]London Centre for Nanotechnology, London, WC1H 0AH, UK, [3]Department of Cell
and Developmental Biology, University College London, London, WC1E 6BT, UK,
[4]Max Planck Institute of Molecular Cell Biology and Genetics, Dresden, 01307, Germany,
[5]International Institute of Molecular and Cell Biology, Warsaw, 02-109, Poland
*Correspondence: salbreux@pks.mpg.de, paluch@mpi-cbg.de

Trends in Cell Biology, Vol. 22, No. 10, October 2012 © 2012 Elsevier Inc.
http://dx.doi.org/10.1016/j.tcb.2012.07.001

SUMMARY

The cortex is a thin, crosslinked actin network lying immediately beneath the plasma membrane of animal cells. Myosin motors exert contractile forces in the meshwork. Because the cortex is attached to the cell membrane, it plays a central role in cell shape control. The proteic constituents of the cortex undergo rapid turnover, making the cortex both mechanically rigid and highly plastic, two properties essential to its function. The cortex has recently attracted increasing attention and its functions in cellular processes such as cytokinesis, cell migration, and embryogenesis are progressively being dissected. In this review, we summarize current knowledge on the structural organization, composition, and mechanics of the actin cortex, focusing on the link between molecular processes and macroscopic physical properties. We also highlight consequences of cortex dysfunction in disease.

THE ACTOMYOSIN CORTEX CONTROLS ANIMAL CELL SHAPE

The cellular cortex comprises a layer of actin filaments, myosin motors, and actin-binding proteins and lies under the plasma membrane of most eukaryotic cells lacking a cell wall, including animal cells and amoebae. Mechanically, the cortical actin mesh plays a role comparable to the cell wall in bacteria or plant cells (Figure 1); it is the main determinant of

437

CellPress

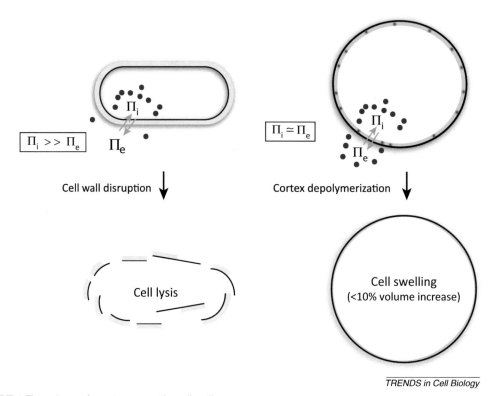

$\Pi_i \gg \Pi_e$ Π_i Π_e

$\Pi_i \simeq \Pi_e$ Π_i Π_e

Cell wall disruption   Cortex depolymerization

Cell lysis

Cell swelling
(<10% volume increase)

TRENDS in Cell Biology

FIGURE 1 The actomyosin cortex versus the cell wall.
The cell wall in plant and prokaryotic cells is a rigid structure outside the plasma membrane, the mechanical function of which is to resist intracellular osmotic pressure. Disruption of the cell wall leads to lysis of the cell resulting from the high difference in osmotic pressure between the interior and exterior of the cell (~10^5 Pa). In animal cells, the osmotic pressure differential is much lower (~10^2–10^3 Pa [34]). As a consequence, depolymerization of the actin cortex leads to only a slight increase in cell volume [2,47].

stiffness of the cell surface, resists external mechanical stresses [1], and is thought to oppose intracellular osmotic pressure [2]. However, in contrast to cell walls, the cortex can undergo dynamic remodeling on timescales of seconds, because of turnover of its protein constituents and network rearrangement through myosin-mediated contractions. This dynamic plasticity is a key feature of animal cell survival in a changing extracellular environment, because it allows cells to rapidly change shape, move, and exert forces.

The idea that a contractile material is present at the periphery of animal cells arose in the late 19th century from observations of surface contractions in migrating and dividing cells. As early as 1876, mechanical models of cytokinesis proposed that furrow ingression results from an equatorial increase in surface tension [3], driven by active forces produced in an elastic gel, termed superficial plasmagel, at the cell surface [4]. Early investigators speculated that the flows of superficial plasmagel observed during cytokinesis, and the contractions seen at the rear of migrating lymphocytes, could be driven by similar mechanisms [5], an idea revisited in a seminal review by Bray and White [1].

In the molecular biology era, the actin cortex has been little investigated compared with other cellular actin networks such as the lamellipodium [6]. Many basic features of the cortex, including its molecular composition, regulation of assembly, and structural organization, are surprisingly poorly understood. However, in the past 10 years, cortex mechanics has attracted increasing attention as a result of growing interest in the control of cell shape and movement during embryonic development and cancer metastasis.

In this review, we summarize our current understanding of cortex proteic composition, physical properties, and function in cell morphogenesis. We first discuss cortex organization and dynamics, emphasizing how, together, they determine cortex mechanics. We then summarize current conceptual physical models describing cortex mechanics. Finally, we discuss potential implications of cortex dysregulation in disease.

STRUCTURE AND COMPOSITION OF THE CORTICAL NETWORK

The mechanical properties of the actin cortex are central to its functions. As with any polymer network, understanding cortex mechanics necessitates knowledge of the mechanics and dynamics of single actin filaments together with their geometrical arrangement. However, the ultrastructural organization of the actin cortex has been relatively little studied. This is partly because of its localization underneath the cell body plasma membrane, which makes obtaining high-resolution electron micrographs more difficult than for flat actin structures immediately adjacent to culture substrates, such as lamellipodia or filopodia [7].

When examined by light microscopy, the cortex is usually identified morphologically as a thin, actin-rich layer directly under the membrane. In suspension cells [8], cells rounded for mitosis [9], or cells undergoing amoeboid-like motion [10], a well-defined, uniformly distributed cortex can easily be observed. Although often more difficult to distinguish, cortical actin is also present under the membrane of cells spread over flat surfaces [11]. The presence and localization of the cortex in cells within tissues have been investigated in only a few specific cases. A uniformly distributed actin cortex extends around the surface of *Xenopus* blastomeres [12]. In *Caenorhabditis elegans* and *Drosophila* epithelia, a contractile actomyosin cortex distinct from junctional actin is present underneath the apical cell membrane [13,14] whereas actin is less prominent at the basal membrane [15,16]. By contrast, in progenitor cells of the zebrafish retinal neuroepithelium, a basally enriched contractile cortex may drive nucleus displacement [17]. Contrary to other cytoskeletal structures such as filopodia, microvilli, and adherens junctions, the cortex lacks well-established proteic markers, such that its presence and specific localization within a cell often remain difficult to pinpoint.

Electron microscopy studies in several cultured cell lines show that the cortex comprises a dense, crosslinked network of actin filaments directly adjacent to the plasma membrane [18]. Filaments appear to form an isotropic network parallel to the membrane with mesh sizes ranging from 20 to 250nm [18,19]. Some filaments oriented perpendicularly to the membrane are also observed [18]. Whether each filament overlap is crosslinked by an actin-binding protein is unknown, and the exact three-dimensional organization and polarity of filaments in the network are not understood. Electron microscopy in cultured mammalian cells and in *Dictyostelium discoideum* suggests that cortex thickness is in the 50–100-nm range [19–21]. However, uncertainties due to sample preparation of actin structures for electron microscopy cloud conclusions [22]. Future studies using subdiffraction microscopy techniques may allow direct measurement of cortex thickness in live cells.

Although its proteic composition has not been exhaustively characterized, numerous actin-binding proteins are seen to localize to the cortex using immunostaining and GFP reporter constructs in dividing cells, in which a prominent actin cortex is clearly visible at the cell membrane [9], and in membrane blebs, cellular protrusions where *de novo* assembly of the actin cortex can be observed [19]. Most of the classical actin-binding proteins localize to the cortex, including actin bundling and crosslinking proteins (α-actinin, filamin, fimbrin), proteins involved in contractility (myosins, tropomyosin, tropomodulin), and linker proteins (proteins of the ezrin–radixin–moesin [ERM] family, filamin).

How cortical components interact to regulate the steady-state properties of the network remains to be thoroughly investigated. In particular, little is known about how cortical actin is assembled. Based on indirect evidence, several nucleators of the formin family (Fhod1, Diaph1, Diaph3, Fmnl1) [23–25] as well as the Arp2/3 complex [14] are suggested to play a role in cortex assembly, but no consensus on cortex nucleation has emerged. One class of proteins, those that tether actin to the plasma membrane, has been more extensively investigated because loss of function results in clear phenotypes affecting not only cell shape control but also the organization of membrane domains [26,27]. Membrane-to-cortex adhesion energy can be assessed experimentally through the extraction of membrane tethers with optical tweezers or atomic force microscopy (AFM) and measurements on various cells point to roles for ERM proteins, myosin 1 motors, and filamin in regulating membrane–cortex attachment [26,28–30].

MECHANICAL PROPERTIES OF THE CORTEX

Because the diameter of actin filaments and the mesh size of the cortical network are small compared with the size of the cell, cellular mechanics is determined by emerging macroscopic properties of the cortex. Understanding how cell shape is controlled at the molecular level requires a combination of quantitative experiments and physical modeling, relating the microscopic organization and dynamics of cortical components to the resulting global mechanical properties. These include viscoelasticity – a central material property of semiflexible polymer networks (Box 1) [31] – and

BOX 1 BASIC PHYSICAL PROPERTIES OF SEMIFLEXIBLE POLYMER NETWORKS

Actin filaments are semiflexible polymers; their persistence length (approximately the average length of straight portions of filaments) is 10–15µm, comparable to the maximal length of actin filaments in the cell. This means that the energy generated by thermal agitation is not high enough to strongly distort filament shape. However, the bending modulus of the filaments is small enough so that bending contributes to the filament mechanical response. As a result, cellular actin filaments display a moderate level of bending and fluctuations, which defines them as semiflexible polymers.

The mechanical properties of passive semiflexible polymer networks have been extensively investigated both theoretically and experimentally [42,108]. Several crucial parameters set the mechanical properties of such networks: the network mesh size (i.e. the average distance between filaments in the network), the distribution of filament lengths, and the mechanical properties and binding and unbinding rates of crosslinking molecules. Networks of polymers exhibit an elastic response at short times due to entanglement and crosslinking of filaments, which hinder relaxation of the mechanical stresses. A further generic property of semiflexible polymer networks is a strongly nonlinear response characterized by an increase in elastic modulus on stretching of the network, a behavior termed strain stiffening. On longer timescales, following either rearrangements of filaments in the network [31] or detachment of crosslinkers allowing network reorganization, polymer networks can become fluid-like and relax mechanical stresses as in a viscous liquid.

Actin networks differ from passive semiflexible filament networks in that they display active properties, depending on ATP consumption; actin filaments undergo dynamic polarization and depolymerization, and myosin motors can exert stresses on the filaments. The generic properties of active semiflexible polymer networks, also called active gels, are only beginning to be addressed [32,42]. In vitro experiments indicate that addition of myosin motors to actin gels significantly modifies the mechanical properties of the network; the viscous relaxation time of a network of entangled filaments can be reduced through myosin-induced filament sliding [109] and myosins can also increase the elasticity of a crosslinked network of filaments, mimicking the effect of an external stress applied to the network [108,110]. Furthermore, turnover of actin filaments contributes to the viscoelastic relaxation of stressed networks. In vivo, the combined effects of myosin contractility, actin and crosslinker turnover, and regulation by a large number of actin-binding proteins all contribute to generation of the mechanical properties of the cortical network. Understanding how these properties depend on cortex spatial organization, dynamics, and regulation is a long-term goal in the field.

cortical tension (see Glossary), resulting from active processes generating contractility in the network [32].

Elastic Properties of the Cortex

Material properties of the cortex determine how the cell deforms in response to external forces. On timescales shorter than its turnover, the cortex essentially behaves like an elastic solid [33]. Its response to perturbation is then characterized by an elastic modulus, which relates the applied stress to network deformation. In the case of bending deformations, the cortex response depends on bending rigidity, which can be related to the elastic modulus [20,34]. Cortex elasticity can be estimated from biophysical experiments involving cell deformation. For example, analysis of the relationship between the shape of blebs induced by laser ablation and cortical tension suggests a cortical elastic modulus of the order of 10^3 Pa in cultured fibroblasts [34], similar to what is found by analyzing the stretching deformation of fibroblasts by optical forces [35]. In AFM indentation experiments, the cellular response is influenced by both cortex elasticity and cortical tension, yielding an effective elastic response designated 'cortex stiffness' (Box 2). When the ultrastructural organization of the actin filaments within the cortex is known, cortical elasticity can also be estimated from the mesh size of the network and the mechanical properties of single actin filaments [20].

Protein Turnover and Cortex Viscoelasticity

The cortex is a dynamic structure in which the protein constituents undergo continuous turnover. As a consequence, on long timescales the cortex behaves like a viscous fluid because actin turnover and dissociation of crosslinkers dissipate stresses and enable the polymer network to remodel [33]. Turnover dynamics of cortical components has been measured in numerous systems using fluorescence recovery after photobleaching (FRAP) (Table 1). The entirety of cortical actin turns over within approximately 1 min, suggesting that, in response to any deformation applied for longer timescales, the cortex behaves mostly like a viscous fluid. Interestingly, actin turnover can be modulated by actin-binding proteins. In the cleavage furrow of dividing mammalian cells, actin turnover is accelerated by myosin activity [36,37], perhaps due to direct disassembly or enhanced breakage of actin filaments by myosin motors as observed *in vitro* [38] and at the rear of lamellipodia in migrating cells [39]. By contrast, cortical actin turnover in dividing cells is slowed by overexpression of the crosslinking protein α-actinin [40], suggesting that crosslinks hinder actin depolymerization. The dynamics of actin crosslinkers and myosin motors is usually 5–10 times faster than actin turnover (Table 1). This suggests that the timescale of viscoelastic relaxation of the cortex is dominated by the turnover of crosslinks rather than of the actin

BOX 2 CORTEX STIFFNESS AND TENSION

The contractile tension of the cortex can generate an apparent elastic response to deformation by indentation, independent of actual cortex elasticity. This apparent elastic response, also termed 'cortex stiffness', is sometimes used to characterize changes in the mechanical properties of the cortex, particularly in AFM indentation experiments (Figure I) [68,72,111]. This quantity is an effective global cellular parameter that could depend simultaneously on cortical tension, the elastic modulus of the cortex, and cell geometry. Assuming that the cortex is a layer under constant tension surrounding a liquid cytoplasm, a simple relationship connects the indentation force to cortex tension [112]. However, this relationship will not be satisfied if cortex elasticity cannot be neglected [113]. In cases where cortex elasticity and cell geometry are unchanged between two experimental conditions, cortex stiffness provides a relative measure of cortical tension.

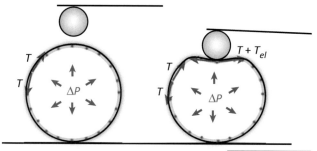

TRENDS in Cell Biology

FIGURE I Schematic of an atomic force microscopy indentation experiment on a round cell with a cortex under tension. A bead is often used instead of an AFM tip to probe global cell mechanics. The force exerted by the cell on the cantilever can depend both on the tension, T, of the cortical network and on the elastic deformation of the network on deformation.

filaments themselves. How the interplay of crosslinker and actin filament dynamics determines the viscoelastic properties of the cortex remains an active research field. The simplest descriptions consider the cortex as a simple viscoelastic material modeled with springs and dashpots [41]. However, experiments examining cortical rheology over a wide range of frequencies reveal more complex behaviors [42], possibly due to uncoordinated turnover of actin crosslinkers [43]. This suggests that subtle changes in the turnover dynamics of crosslinkers may strongly affect cortex rheology. Precise regulation of cortical component turnover appears to be essential during shape changes such as cytokinesis, in which crosslinker dynamics at the cell poles and equator (Table 1) control the speed of furrow ingression [40,44].

Cortical Tension

Cortical tension is the force per unit length locally exerted by the surrounding network on a cortex cross-section (Figure 2a). Cortex tension is the main determinant of cell surface tension and its precise regulation is essential for

Table 1 Experimental measurements of turnover of cortex components

Cell type	Protein	Turnover half-time	Refs
LLCPK1 cells during cytokinesis	Actin	45 s (polar cortex) 26 s (contractile ring)	[37]
NRK cells, anaphase and telophase	Actin	15 s	[36]
Dictyostelium, contractile ring	Myosin II	7s	[104]
Drosophila S2 cells	Myosin II	~6 s (metaphase) ~14 s (anaphase)	[105]
NRK cells	Alpha-actinin (crosslinker)	~8 s (equator in cytokinesis) ~19 s (poles in cytokinesis)	[40]
Dictyostelium	Dynacortin (crosslinker)	0.45 s (interphase) 0.98 s (equator in cytokinesis) 0.51 s (poles in cytokinesis)	[44]
Dictyostelium	Fimbrin (crosslinker)	0.26 s (interphase) 0.58 s (equator in cytokinesis) 0.31 s (poles in cytokinesis)	[44]
Dictyostelium	Cortexillin-I (crosslinker)	3.3 s (interphase) 5.4 s (equator in cytokinesis) 4.5 s (poles in cytokinesis)	[44]

Turnover times of cortical actin, myosin II, and crosslinkers measured by FRAP. Myosin and crosslinkers typically turn over faster than actin filaments. Notably, non-muscle myosin II aggregates in mini-filaments in the cortex and it is unclear whether FRAP experiments measure the timescale of turnover of individual myosins or of entire mini-filaments.

cell shape control, both in single cells and in tissues [45]. Molecularly, tension is mostly governed by myosin-generated contractility, although other contractile processes have been suggested (see below). When the cortex undergoes flow or deformation, its viscoelastic resistance contributes to total tension, because elastic stretching or viscous flows can increase total surface tension [46,47] (Box 2).

Different experimental techniques allow estimation of cortical tension. Assuming that, at the timescales of the experiment, the cytoplasm behaves as a fluid, cortical tension can be obtained from the pressure required to aspirate a cell inside a micropipette [48]. Tension can also be inferred from the force necessary to compress a cell between two plates [49] or from AFM indentation measurements of cellular stiffness [50] (Box 2). Finally, relative values of cortical tension can be estimated by monitoring how much the cortex opens following laser ablation, assuming that cortex retraction velocity is proportional to its tension [51]. Cortical tension varies by two orders of magnitude between cell types, ranging from approximately 30 pN/μm in blood granulocytes [48] and progenitor cells isolated from early zebrafish embryos [50], to several thousand

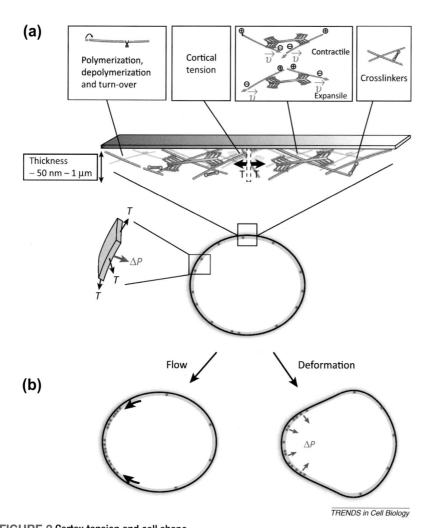

FIGURE 2 Cortex tension and cell shape.
(a) Actin filaments assemble into a thin network connected to the cell membrane and undergoing continuous turnover. Myosin motors exert forces in the network, giving rise to a cortical tension, T. Motors are assembled into mini-filaments, which connect pairs of actin filaments and slide them with respect to each other. This can result in contractile or expansile stresses, depending on the position of the motors on the actin filaments (top right). Why contractile stresses dominate, giving rise to a positive tension in the network, is not fully understood. Crosslinks in the network could contribute to the generation of tension; crosslink turnover is a major determinant of cortex viscoelasticity. (b) Because of the cell curvature, cortical tension gives rise to a hydrostatic pressure in the cytoplasm. Gradients of motor-generated contractility within the cortex can drive tangential flows of cortex in the plane of the membrane (left), whereas normal forces can drive cell deformations with net displacement of the cytoplasm (right).

Table 2 Experimental measurements of cortical tension

Cell type	Technique	Value (pN/µm)	Refs
Blood granulocyte	Micropipette	30–35	[48,106]
Dictyostelium	Micropipette	~1500	[30]
Dictyostelium	Micropipette	4330	[52]
Fibroblasts	Micropipette	400	[34]
Fibroblasts	Micropipette	~300	[107]
Zebrafish progenitor cells	AFM indentation	~50	[50]

pN/µm in *D. discoideum* cells [30,52] (Table 2). It is unclear whether these differences result from subtle variations in myosin activity and organization of the cortical network, or if they reflect more fundamental cortical differences between cell types, related to specific cell functions. Precise control of cortical tension in space and time is important for cellular functions such as migration and division [1] (see below) and differences in cortex tension between cells can contribute to cell sorting in development [45,50].

PHYSICAL DESCRIPTIONS OF THE CORTEX

Understanding how the macroscopic properties of the actin cortex stem from single-filament mechanics, dynamics, and network ultrastructural organization is a challenge that can be addressed with the help of physical models or numerical simulations. Theoretical modeling of the cortex has mostly focused on investigating: (i) how contractility is generated at the microscopic level; and (ii) how the macroscopic properties of the cortex govern whole-cell mechanics.

Generation of Contractile Forces

Non-muscle myosin II isoforms are the main source of cortical contractility, similar to myosin function in muscle contraction. However, whereas muscle fibers are ideally arranged for contraction, with sarcomeres of antiparallel actin filaments connected by myosin filaments [33], the generation of contractility within the isotropic cortical network is far less well understood. In a disordered actin array, myosin II can induce both contractile and expansile stresses (Figure 2a), and why contractility dominates over expansion remains unclear. Possible mechanisms include dependence of the attachment rate of myosins on their position with respect to the plus end of actin filaments [53], a role for nonlinear elasticity of the actin network [54,55] – possibly due to filament buckling [56,57] – and instability of expansile configurations [58].

A rough estimate indicates that myosin contractility could account for most of the tension accumulated in the cortex. With myosins assembling into bipolar filaments with approximately 10 heads on each side [59], with a duty ratio of 0.2–0.4 [60] and a force per head of 3.5 pN [33], a single layer of contractile actomyosin units separated by 100nm – the typical mesh size of the cortex [18,19] – would exert a tension of approximately 100 pN/μm, in the lower range of experimentally measured values (Table 1). Higher values could easily be obtained taking into account that multiple contractile units are likely to be superimposed across the thickness of the cortex and that the duty ratio of myosin is enhanced under load [60].

Interestingly, it is suggested that crosslinking may generate contractile forces within the cortex independent of myosin activity. Crosslinkers can in theory passively induce gelation and contraction in a polymer meshwork [61]. Moreover, end-tracking crosslinkers can in principle generate sustained contractility in a network undergoing continuous turnover [62,63]. Such a mechanism of force generation could be involved in cleavage furrow ingression, consistent with a recent study suggesting that the crosslinking function of myosin, rather than its motor activity, is essential for cytokinesis [64].

Addressing how cortex contractility is controlled will require quantitative experiments correlating contractility with network organization in wild-type cells and on depletion of potential molecular regulators. Molecular simulations, which have been successfully used to model microtubule structures [65], are likely to provide insight into how contractility emerges from molecular processes.

Coarse-Grained Descriptions

Molecular simulations of the cortex require detailed knowledge of the molecular processes underlying cortex mechanics. Because experimental data on cortex organization and dynamics remain scarce, simulations describing global cortex behaviors remain a long-term goal. However, to investigate the mechanics of whole cells, coarse-grained models using macroscopic parameters that retain only the most salient characteristics of cortex mechanics without attempting to incorporate molecular details have been used as an alternative to molecular modeling. The strength of coarse-grained approaches is that they do not require knowledge of the microscopic organization of the cortex and the biochemical processes that underlie its dynamics, both of which are poorly understood. One such model, the active gel theory, describes the cortex as a continuous gel submitted to internal stresses resulting from active, ATP-dependent mechanisms [32]. It has been successfully used to describe cortical flows and contractions in numerous processes, including cell migration [66], *C. elegans* zygote polarization [46], bleb expansion [34], and cell deformations during cytokinesis [67] (discussed below). However, coarse-grained models generally require the introduction of phenomenological coefficients such as contractility or viscoelastic parameters. For our comprehension of cortical mechanics to be complete, the validity of each introduced phenomenological coefficient must be verified experimentally and related to underlying molecular processes. Coarse-grained approaches could help bridge the gap between the dynamics of single filaments and deformations of entire cells.

CORTEX CONTRACTILITY AND CELL SHAPE CONTROL

Cortex mechanics, and in particular cortical tension, is a key determinant of animal cell morphology. Both global and local changes in cortical tension, in response to intrinsic and extrinsic factors, drive cell shape changes. Global control of tension is tightly linked to cell volume, because cortex tension appears to balance intracellular osmotic pressure [2,34]. In mitotic HeLa cells, cortex depolymerization leads to small increases in cell volume, driven by osmotic pressure [2] (Figure 1). During mitosis, a global increase in myosin activity, downstream of the small GTPase RhoA, results in an approximately threefold increase in cortex stiffness and is thought to cause mitotic cell rounding [2,68]. During apoptosis, contractility increases globally, resulting in the formation of numerous blebs and, coupled to lamin proteolysis, nuclear disintegration [69].

When cortical tension is not uniform throughout the cell, contractility gradients can result in local contractions and cell deformations (Figure 2b, right). During lamellipodial cell migration, local contractions at the rear of the cell cause cell body retraction [59]; in amoeboid motion, cortex contractions drive the formation of pressure-driven blebs at the leading edge of the cell [70,71]. In cytokinesis, the cortex accumulates at the cell equator and drives ingression of the cleavage furrow [1]. Cortex stiffness is higher at the equator than at the poles [72], equatorial actin turns over faster [40] (Table 1), and actin filaments often display a certain level of alignment at the equator, which could favor contraction [73]. Nonetheless, a substantial amount of cortical actin and myosin remains at the poles throughout cytokinesis [40,74] and plays a fundamental role in the mechanics of cell division. Polar cortex tension slows cleavage furrow ingression [44,49]. Moreover, polar cortex contraction during cytokinesis can affect the shape of the dividing cell. During neuroblast division in *C. elegans* and *Drosophila*, cortex accumulation at one of the poles appears to directly cause the generation of daughter cells of unequal sizes through unbalanced contraction of the actin-rich pole [75,76] (Figure 3). Unbalanced cortex contractions can also occur in symmetrically dividing cells, where they destabilize the position of the cleavage furrow and result in shape oscillations, with each pole cyclically contracting at the expense of the other [67]. Such

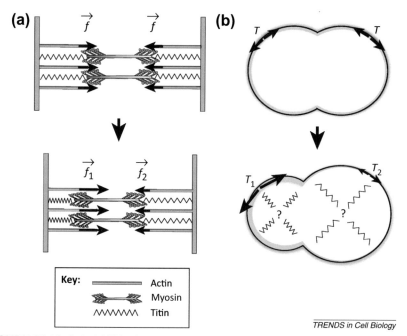

FIGURE 3 Intrinsic instabilities in actomyosin contractile structures.
(a) During contraction of muscle sarcomeres, deviation of the position of the myosin filament from the center of the sarcomere results in more myosin heads engaging with actin filaments on one side of the sarcomere, driving myosin further away from the center. The symmetric position of myosin is therefore potentially unstable. Elastic elements such as the protein titin have been proposed to counteract contraction, thus stabilizing the sarcomere against unbalanced contractions [80]. (b) Unbalanced polar cortex contractions can trigger cell shape instabilities during cytokinesis [67]. As one pole contracts, actin and myosin accumulate at the contracting pole, increasing cortical tension and leading to further contraction. Elastic elements could prevent this instability by counteracting deviations from symmetric cell shape.

oscillations can occur spontaneously or on depletion of various cortical components and usually lead to division failure [77]. Cortex oscillations are also commonly observed during epithelial contractions in numerous developmental systems [78,79]. A physical model suggests that instabilities leading to dramatic contractions are an intrinsic mechanical behavior of the cortex [67] (Figure 3) and that maintaining a stable contractile cortex requires mechanical stabilization. In dividing cells, global cellular elasticity appears to play this stabilizing role [67], although the molecular origin of this elasticity is unclear. A similar contractile instability occurs in muscle cells, where the structural protein titin is thought to provide elastic resistance, stabilizing muscle fibers against unbalanced sarcomere contractions [80] (Figure 3). In dividing cells, contractile instabilities result in cortex oscillations due to dynamic cortex turnover, which provides a restoring force reversing the direction of contractions (Figure 3) [67]. It will be interesting to explore whether similar mechanical principles underlie cortex oscillations in contracting epithelia.

In addition to localized changes in cell shape, gradients of contractility can lead to cortical flows in the plane of the membrane (Figure 2b, left), commonly observed in processes such as amoeboid cell migration, cytokinesis, cell polarization, and epithelial constriction during development [78]. In cell division, flows of actin and myosin from the poles toward the equator contribute to cleavage furrow formation [1,81]. A coarse-grained model of cortical flows suggests that they could contribute to local alignment of actin filaments at the equator [82]. Cortical flows also drive the segregation of polarity determinants during the polarization of *C. elegans* zygotes [83,84]. Future studies will have to investigate how in some instances contractility gradients give rise to cortical flows and in others to contractions with net displacement of cytoplasmic material (Figure 2b).

DYSREGULATION IN DISEASE

Changes in cellular mechanics are emerging as an important factor in numerous diseases [85]. However, linking disease to cortex malfunction or misregulation remains challenging, because few studies have specifically investigated cortical changes, even when actin-associated protein mutations are identified as the cause of disease.

One pathology in which a role for cortical actin is particularly well documented is cancer. The well-studied biochemical changes occurring during the progression of cancer are accompanied by changes in cellular mechanics. Pilot studies suggest that cells become softer as they become more malignant [86–88]. However, further investigations are needed to explore the generality of this feature and clarify the importance of the cortex in cancer cell mechanics. Intriguingly, in epithelial tissues (from which most cancers arise and where cortical actin is particularly prominent), softer transformed cells are extruded from the epithelium [88,89]. Extrusion generally occurs apically, thereby removing transformed cells from the organism, and basal extrusion has been proposed as a first step toward metastasis [90]. Once transformation has occurred, further mutations leading to epithelial–mesenchymal transition contribute to the dissemination of transformed cells [91]. Cells can migrate out of the primary tumor by either collective or individual migration [92]. The latter can occur via mesenchymal motility that utilizes proteases to tunnel through the matrix or via 'amoeboid' motility that is protease-independent but depends strongly on contractility. Cells undergoing amoeboid motion display more rounded shapes and often protrude blebs [93]. Plasticity in migration modes is believed to facilitate the dissemination of cancer, because it allows cells to dynamically adapt their migration to their environment [94]. Although the molecular pathways promoting specific migration modes have been extensively investigated [95], the underlying mechanical processes are less well understood [96,97]. It will be interesting to investigate how the cellular cortex is dynamically remodeled during transitions between mesenchymal and amoeboid migration, and which molecular pathways trigger these changes.

As its prominence during mitosis suggests, proper regulation of the cortex is a prerequisite for correct cell division. *In vitro*, depletion of the ERM protein moesin results in a softer cortex and incorrect spindle alignment during mitosis, whereas silencing of its negative regulator PP1-87B leads to a stiffer cortex [68,98,99]. Similarly, *in vivo*, cells in the epidermis round up during mitosis and display a marked enrichment in cortical actin [100]. In mice whose epidermis was conditionally depleted of serum response factor (SRF), a key transcription factor regulating cellular actin levels [100], mitotic cells within the epidermis presented cortical anomalies similar to those observed in moesin-depleted cultured cells: they were no longer able to round up and displayed a weaker cortex, and ERM and myosin phosphorylation were downregulated. This resulted in defects in spindle orientation [100], which can lead to division defects and aneuploidy.

CONCLUDING REMARKS

The actomyosin cortex of animal cells is an essential, but still poorly understood, cellular actin network. Proper control of cortical mechanics is crucial for cell division and motility, and misregulation is often the cause of disease. Recent studies have begun to unveil the molecular composition, structure, and dynamics of the cortical network. Although our understanding of the biophysics and regulation of cortical mechanics is progressing rapidly, many open questions remain.

It is unclear, for instance, to what extent the cortices observed in various cell types represent a unique and well-defined actin structure. It will be important to investigate how cortex composition and physical properties change during the cell cycle and how they differ between cell types, particularly *in vivo*. Measurements of cortex mechanical properties suggest a wide degree of variability in cortical properties between cell types (Tables 1 and 2). These may reflect qualitative differences between cortices required for specific physiological functions. Importantly, the shape of a cell is the result of global cellular mechanics, and variations in parameters such as tension can be compensated by changes in other mechanical properties [44,67]. How cellular mechanical properties are coregulated to achieve controlled cell shape changes remains an open question.

Another important question is the interplay between the cortex and other cellular components. Both microtubules and intermediate filaments interact with the cortical actin network [101,102]. However, their function in cortex regulation and their contribution to cortical mechanics are poorly understood. The local regulation of cortical actin during endocytosis – where the cortex must be locally remodeled to allow for passage of endocytotic vesicles that are often larger than the typical mesh size of the cortex – is also unclear [18]. Finally, it will be important to investigate how the plasma membrane influences cortex regulation and mechanics. The lipid composition of the plasma membrane influences membrane-to-cortex attachment as well as actin filament dynamics [103]. How the plasma membrane and the actin cortex physically constrain each other and give rise to global effective cell surface mechanics is poorly understood. Experiments assessing the geometry of the plasma membrane during cortex contractions, and modulating membrane mechanical properties, will be required to address the role the membrane plays in cortex-driven cell shape changes.

The actin cortex is attracting increasing attention as a key regulator of animal cell mechanics and shape changes. The development of new biological methods for precise characterization of cortex composition and spatial organization, together with novel theoretical approaches, will be necessary to reach a systems-level understanding of cortex function in cell morphogenesis.

ACKNOWLEDGMENTS

We thank present and past members of our laboratories as well as J. Howard and S.W Grill for discussions and A.G. Clark and K. Dierkes for comments on the manuscript. We acknowledge the financial support of the Polish Ministry of Science and Higher Education (E.P.), the Max Planck Society (E.P. and G.S.), and a Human Frontier Young Investigator Grant to G.C. and E.P. G.C. is a Royal Society University Research Fellow.

REFERENCES

1 Bray, D. and White, J.G. (1988) Cortical flow in animal cells. *Science* 239, 883–888

2 Stewart, M.P. *et al.* (2011) Opposing activities of hydrostatic pressure and the actomyosin cortex drive mitotic cell rounding. *Nature* 469, 226–230

3 Bütschli, O. (1876) Studien über die erste Entwicklungsvorgange der Eizelle, der Zelltheilung und die Conjugation der Infusorien. *Abhandlungen Senckenberg. Naturf. Gesellsch.* 10, 213–464

4 Rappaport, R. (1996) *Cytokinesis in Animal Cells.* Cambridge University Press

5 Lewis, W. (1939) The role of a superficial plasmagel layer in changes of form, locomotion and division of cells in tissue cultures. *Arch. Exp. Zellforsch.* 23, 1–7

6 Pollard, T.D. and Borisy, G.G. (2003) Cellular motility driven by assembly and disassembly of actin filaments. *Cell* 112, 453–465

7 Svitkina, T.M. *et al.* (2003) Mechanism of filopodia initiation by reorganization of a dendritic network. *J. Cell Biol.* 160, 409–421

8 Tooley, A.J. *et al.* (2009) Amoeboid T lymphocytes require the septin cytoskeleton for cortical integrity and persistent motility. *Nat. Cell Biol.* 11, 17–26

9 Kunda, P. and Baum, B. (2009) The actin cytoskeleton in spindle assembly and positioning. *Trends Cell Biol.* 19, 174–179

10 Wolf, K. *et al.* (2003) Compensation mechanism in tumor cell migration: mesenchymal–amoeboid transition after blocking of pericellular proteolysis. *J. Cell Biol.* 160, 267–277

11 Estecha, A. *et al.* (2009) Moesin orchestrates cortical polarity of melanoma tumour cells to initiate 3D invasion. *J. Cell Sci.* 122, 3492–3501

12 Kofron, M. *et al.* (2002) Plakoglobin is required for maintenance of the cortical actin skeleton in early *Xenopus* embryos and for cdc42-mediated wound healing. *J. Cell Biol.* 158, 695–708

13 Blanchard, G.B. *et al.* (2010) Cytoskeletal dynamics and supracellular organisation of cell shape fluctuations during dorsal closure. *Development* 137, 2743–2752

14 Roh-Johnson, M. *et al.* (2012) Triggering a cell shape change by exploiting preexisting actomyosin contractions. *Science* 335, 1232–1235

15 Rauzi, M. *et al.* (2010) Planar polarized actomyosin contractile flows control epithelial junction remodelling. *Nature* 468, 1110–1114

16 Wang, Y.C. *et al.* (2012) Differential positioning of adherens junctions is associated with initiation of epithelial folding. *Nature* 484, 390–393

17 Norden, C. *et al.* (2009) Actomyosin is the main driver of interkinetic nuclear migration in the retina. *Cell* 138, 1195–1208

18 Morone, N. *et al.* (2006) Three-dimensional reconstruction of the membrane skeleton at the plasma membrane interface by electron tomography. *J. Cell Biol.* 174, 851–862

19 Charras, G.T. *et al.* (2006) Reassembly of contractile actin cortex in cell blebs. *J. Cell Biol.* 175, 477–490

20 Charras, G.T. *et al.* (2008) Life and times of a cellular bleb. *Biophys. J.* 94, 1836–1853

21 Hanakam, F. *et al.* (1996) Myristoylated and non-myristoylated forms of the pH sensor protein hisactophilin II: intracellular shuttling to plasma membrane and nucleus monitored in real time by a fusion with green fluorescent protein. *EMBO J.* 15, 2935–2943

22 Small, J.V. *et al.* (2008) Unravelling the structure of the lamellipodium. *J. Microsc.* 231, 479–485

23 Hannemann, S. *et al.* (2008) The diaphanous-related formin FHOD1 associates with ROCK1 and promotes Src-dependent plasma membrane blebbing. *J. Biol. Chem.* 283, 27891–27903

24 Han, Y. *et al.* (2009) Formin-like 1 (FMNL1) is regulated by N-terminal myristoylation and induces polarized membrane blebbing. *J. Biol. Chem.* 284, 33409–33417

25 Eisenmann, K.M. *et al.* (2007) Dia-interacting protein modulates formin-mediated actin assembly at the cell cortex. *Curr. Biol.* 17, 579–591

26 Stossel, T.P. *et al.* (2001) Filamins as integrators of cell mechanics and signalling. *Nat. Rev. Mol. Cell Biol.* 2, 138–145

27 Fehon, R.G. *et al.* (2010) Organizing the cell cortex: the role of ERM proteins. *Nat. Rev. Mol. Cell Biol.* 11, 276–287

28 Diz-Muñoz, A. *et al.* (2010) Control of directed cell migration *in vivo* by membrane-to-cortex attachment. *PLoS Biol.* 8, e1000544

29 Nambiar, R. *et al.* (2009) Control of cell membrane tension by myosin-I. *Proc. Natl. Acad. Sci. U.S.A.* 106, 11972–11977

30 Dai, J. *et al.* (1999) Myosin I contributes to the generation of resting cortical tension. *Biophys. J.* 77, 1168–1176

31 de Gennes, P.-G. (1979) *Scaling Concepts in Polymer Physics*. Cornell University Press

32 Kruse, K. *et al.* (2005) Generic theory of active polar gels: a paradigm for cytoskeletal dynamics. *Eur. Phys. J. E* 16, 5–16

33 Howard, J. (2001) *Mechanics of Motor Proteins and Cytoskeleton*. Sinauer Associates

34 Tinevez, J.Y. *et al.* (2009) Role of cortical tension in bleb growth. *Proc. Natl. Acad. Sci. U.S.A.* 106, 18581–18586

35 Ananthakrishnan, R. *et al.* (2006) Quantifying the contribution of actin networks to the elastic strength of fibroblasts. *J. Theor. Biol.* 242, 502–516

36 Guha, M. *et al.* (2005) Cortical actin turnover during cytokinesis requires myosin II. *Curr. Biol.* 15, 732–736

37 Murthy, K. and Wadsworth, P. (2005) Myosin-II-dependent localization and dynamics of F-actin during cytokinesis. *Curr. Biol.* 15, 724–731

38 Haviv, L. *et al.* (2008) A cytoskeletal demolition worker: myosin II acts as an actin depolymerization agent. *J. Mol. Biol.* 375, 325–330

39 Wilson, C.A. *et al.* (2010) Myosin II contributes to cell-scale actin network treadmilling through network disassembly. *Nature* 465, 373–377

40 Mukhina, S. *et al.* (2007) Alpha-actinin is required for tightly regulated remodeling of the actin cortical network during cytokinesis. *Dev. Cell* 13, 554–565

41 Bausch, A.R. *et al.* (1998) Local measurements of viscoelastic parameters of adherent cell surfaces by magnetic bead microrheometry. *Biophys. J.* 75, 2038–2049

42 Wen, Q. and Janmey, P.A. (2011) Polymer physics of the cytoskeleton. *Curr. Opin. Solid State Matter Sci.* 15, 177–182

43 Broedersz, C.P. *et al.* (2010) Cross-link-governed dynamics of biopolymer networks. *Phys. Rev. Lett.* 105, 238101

44 Reichl, E.M. *et al.* (2008) Interactions between myosin and actin crosslinkers control cytokinesis contractility dynamics and mechanics. *Curr. Biol.* 18, 471–480

45 Lecuit, T. and Lenne, P.F. (2007) Cell surface mechanics and the control of cell shape, tissue patterns and morphogenesis. *Nat. Rev. Mol. Cell Biol.* 8, 633–644

46 Mayer, M. *et al.* (2010) Anisotropies in cortical tension reveal the physical basis of cortical flow in polarising *C. elegans* zygotes. *Nature* 467, 617–621

47 Clark, A.G. and Paluch, E. (2011) Mechanics and regulation of cell shape during the cell cycle. *Cell Cycle Dev.* 53, 31–73

48 Evans, E. and Yeung, A. (1989) Apparent viscosity and cortical tension of blood granulocytes determined by micropipet aspiration. *Biophys. J.* 56, 151–160

49 Yoneda, M. and Dan, K. (1972) Tension at the surface of the dividing sea-urchin egg. *J. Exp. Biol.* 57, 575–587

50 Krieg, M. *et al.* (2008) Tensile forces govern germ-layer organization in zebrafish. *Nat. Cell Biol.* 10, 429–436

51 Mayer, M. *et al.* (2012) Biophysics of cell developmental processes: a lasercutter's perspective. *Comp. Biophys.* 7, 194–207

52 Schwarz, E.C. *et al.* (2000) *Dictyostelium* myosin IK is involved in the maintenance of cortical tension and affects motility and phagocytosis. *J. Cell Sci.* 113, 621–633

53 Kruse, K. and Jülicher, F. (2000) Actively contracting bundles of polar filaments. *Phys. Rev. Lett.* 85, 1778–1781

54 MacKintosh, F.C. and Levine, A.J. (2008) Nonequilibrium mechanics and dynamics of motor-activated gels. *Phys. Rev. Lett.* 100, 018104

55 Liverpool, T. *et al.* (2009) Mechanical response of active gels. *Europhys. Lett.* 85, 18007

56 Soares e Silva, M. *et al.* (2011) Active multistage coarsening of actin networks driven by myosin motors. *Proc. Natl. Acad. Sci. U.S.A.* 108, 9408–9413

57 Lenz, M. *et al.* (2012) Requirements for contractility in disordered cytoskeletal bundles. *New J. Phys.* 14, 033037

58 Dasanayake, N.L. *et al.* (2011) General mechanism of actomyosin contractility. *Phys. Rev. Lett.* 107, 118101

59 Vicente-Manzanares, M. *et al.* (2009) Non-muscle myosin II takes centre stage in cell adhesion and migration. *Nat. Rev. Mol. Cell Biol.* 10, 778–790

60 Kovacs, M. *et al.* (2007) Load-dependent mechanism of nonmuscle myosin 2. *Proc. Natl. Acad. Sci. U.S.A.* 104, 9994–9999

61 Sun, S.X. *et al.* (2010) Cytoskeletal cross-linking and bundling in motor-independent contraction. *Curr. Biol.* 20, R649–R654

62 Dickinson, R.B. *et al.* (2004) Force generation by cytoskeletal filament end-tracking proteins. *Biophys. J.* 87, 2838–2854

63 Zumdieck, A. *et al.* (2007) Stress generation and filament turnover during actin ring constriction. *PLoS ONE* 2, e696

64 Ma, X. *et al.* (2012) Nonmuscle myosin II exerts tension but does not translocate actin in vertebrate cytokinesis. *Proc. Natl. Acad. Sci. U.S.A.* 109, 4509–4514

65 Karsenti, E. *et al.* (2006) Modelling microtubule patterns. *Nat. Cell Biol.* 8, 1204–1211

66 Hawkins, R.J. *et al.* (2011) Spontaneous contractility-mediated cortical flow generates cell migration in three-dimensional environments. *Biophys. J.* 101, 1041–1045

67 Sedzinski, J. *et al.* (2011) Polar acto-myosin contractility destabilises the position of the cytokinetic furrow. *Nature* 476, 462–466

68 Kunda, P. *et al.* (2008) Moesin controls cortical rigidity, cell rounding, and spindle morphogenesis during mitosis. *Curr. Biol.* 18, 91–101

69 Croft, D.R. *et al.* (2005) Actin-myosin-based contraction is responsible for apoptotic nuclear disintegration. *J. Cell Biol.* 168, 245–255

70 Charras, G. and Paluch, E. (2008) Blebs lead the way: how to migrate without lamellipodia. *Nat. Rev. Mol. Cell Biol.* 9, 730–736

71 Fackler, O.T. and Grosse, R. (2008) Cell motility through plasma membrane blebbing. *J. Cell Biol.* 181, 879–884

72 Matzke, R. *et al.* (2001) Direct, high-resolution measurement of furrow stiffening during division of adherent cells. *Nat. Cell Biol.* 3, 607–610

73 Maupin, P. and Pollard, T.D. (1986) Arrangement of actin filaments and myosin-like filaments in the contractile ring and of actin-like filaments in the mitotic spindle of dividing HeLa cells. *J. Ultrastruct. Mol. Struct. Res.* 94, 92–103

74 Robinson, D.N. *et al.* (2002) Quantitation of the distribution and flux of myosin-II during cytokinesis. *BMC Cell Biol.* 3, 4

75 Cabernard, C. *et al.* (2010) A spindle-independent cleavage furrow positioning pathway. *Nature* 467, 91–94

76 Ou, G. *et al.* (2010) Polarized myosin produces unequal-size daughters during asymmetric cell division. *Science* 330, 677–680

77 Green, R. *et al.* Cytokinesis in animal cells. *Annu. Rev. Cell Dev. Biol.* http://dx.doi.org/10.1146/annurev-cellbio-101011-155718 (in press)

78 Levayer, R. and Lecuit, T. (2011) Biomechanical regulation of contractility: spatial control and dynamics. *Trends Cell Biol.* 22, 61–81

79 Martin, A.C. (2010) Pulsation and stabilization: contractile forces that underlie morphogenesis. *Dev. Biol.* 341, 114–125

80 Agarkova, I. *et al*. (2003) M-band: a safeguard for sarcomere stability? *J. Muscle Res. Cell Motil.* 24, 191–203

81 Zhou, M. and Wang, Y.L. (2008) Distinct pathways for the early recruitment of myosin II and actin to the cytokinetic furrow. *Mol. Biol. Cell* 19, 318–326

82 Salbreux, G. *et al*. (2009) Hydrodynamics of cellular cortical flows and the formation of contractile rings. *Phys. Rev. Lett.* 103, 058102

83 Munro, E. *et al*. (2004) Cortical flows powered by asymmetrical contraction transport PAR proteins to establish and maintain anterior-posterior polarity in the early *C. elegans* embryo. *Dev. Cell* 7, 413–424

84 Goehring, N.W. *et al*. (2011) Polarization of PAR proteins by advective triggering of a pattern-forming system. *Science* 334, 1137–1141

85 Kumar, S. and Weaver, V.M. (2009) Mechanics, malignancy, and metastasis: the force journey of a tumor cell. *Cancer Metastasis Rev.* 28, 113–127

86 Cross, S.E. *et al*. (2007) Nanomechanical analysis of cells from cancer patients. *Nat. Nanotechnol.* 2, 780–783

87 Remmerbach, T.W. *et al*. (2009) Oral cancer diagnosis by mechanical phenotyping. *Cancer Res.* 69, 1728–1732

88 Hogan, C. *et al*. (2009) Characterization of the interface between normal and transformed epithelial cells. *Nat. Cell Biol.* 11, 460–467

89 Kajita, M. *et al*. (2010) Interaction with surrounding normal epithelial cells influences signalling pathways and behaviour of Src-transformed cells. *J. Cell Sci.* 123, 171–180

90 Marshall, T.W. *et al*. (2011) The tumor suppressor adenomatous polyposis coli controls the direction in which a cell extrudes from an epithelium. *Mol. Biol. Cell* 22, 3962–3970

91 Yilmaz, M. and Christofori, G. (2009) EMT, the cytoskeleton, and cancer cell invasion. *Cancer Metastasis Rev.* 28, 15–33

92 Friedl, P. and Gilmour, D. (2009) Collective cell migration in morphogenesis, regeneration and cancer. *Nat. Rev. Mol. Cell Biol.* 10, 445–457

93 Friedl, P. and Wolf, K. (2003) Tumor-cell invasion and migration: diversity and escape mechanisms. *Nat. Rev. Cancer* 3, 362–374

94 Friedl, P. and Wolf, K. (2010) Plasticity of cell migration: a multiscale tuning model. *J. Cell Biol.* 188, 11–19

95 Olson, M.F. and Sahai, E. (2009) The actin cytoskeleton in cancer cell motility. *Clin. Exp. Metastasis* 26, 273–287

96 Lammermann, T. and Sixt, M. (2009) Mechanical modes of 'amoeboid' cell migration. *Curr. Opin. Cell Biol.* 21, 636–644

97 Bergert, M. *et al*. (2012) Cell mechanics control rapid transitions between blebs and lamellipodia during migration. *Proc. Natl. Acad. Sci. U. S. A.* 109, 14434–14439

98 Carreno, S. *et al*. (2008) Moesin and its activating kinase Slik are required for cortical stability and microtubule organization in mitotic cells. *J. Cell Biol.* 180, 739–746

99 Roubinet, C. *et al*. (2011) Molecular networks linked by moesin drive remodeling of the cell cortex during mitosis. *J. Cell Biol.* 195, 99–112

100 Luxenburg, C. *et al*. (2011) Developmental roles for Srf, cortical cytoskeleton and cell shape in epidermal spindle orientation. *Nat. Cell Biol.* 13, 203–214

101 Chang, L. and Goldman, R.D. (2004) Intermediate filaments mediate cytoskeletal crosstalk. *Nat. Rev. Mol. Cell Biol.* 5, 601–613

102 Rodriguez, O.C. *et al*. (2003) Conserved microtubule-actin interactions in cell movement and morphogenesis. *Nat. Cell Biol.* 5, 599–609

103 Sheetz, M.P. *et al.* (2006) Continuous membrane–cytoskeleton adhesion requires continuous accomodation to lipid and cytoskeleton dynamics. *Annu. Rev. Biophys. Biomol. Struct.* 35, 417–434

104 Yumura, S. (2001) Myosin II dynamics and cortical flow during contractile ring formation in *Dictyostelium* cells. *J. Cell Biol.* 154, 137–146

105 Uehara, R. *et al.* (2010) Determinants of myosin II cortical localization during cytokinesis. *Curr. Biol.* 20, 1080–1085

106 Hochmuth, R.M. (2000) Micropipette aspiration of living cells. *J. Biomech.* 33, 15–22

107 Thoumine, O. *et al.* (1999) Changes in the mechanical properties of fibroblasts during spreading: a micromanipulation study. *Eur. Biophys. J.* 28, 222–234

108 Bausch, A.R. and Kroy, K. (2006) A bottom-up approach to cell mechanics. *Nat. Phys.* 2, 231–238

109 Humphrey, D. *et al.* (2002) Active fluidization of polymer networks through molecular motors. *Nature* 416, 413–416

110 Koenderink, G.H. *et al.* (2009) An active biopolymer network controlled by molecular motors. *Proc. Natl. Acad. Sci. U.S.A.* 106, 15192–15197

111 Harris, A.R. and Charras, G.T. (2011) Experimental validation of atomic force microscopy-based cell elasticity measurements. *Nanotechnology* 22, 345102

112 Lomakina, E.B. *et al.* (2004) Rheological analysis and measurement of neutrophil indentation. *Biophys. J.* 87, 4246–4258

113 Sen, S. *et al.* (2005) Indentation and adhesive probing of a cell membrane with AFM: theoretical model and experiments. *Biophys. J.* 89, 3203–3213

GLOSSARY

Cortex tension force per unit length exerted on a piece of the cortex by the cortical network around it (Figure 2a). A local increase in cortical tension energetically favors a local decrease in surface area.

Duty ratio the fraction of time spent by a molecular motor attached to its filament during one motor cycle.

Persistence length owing to thermal fluctuations, a polymer in solution can bend and adopt a more or less convoluted shape, depending on its bending rigidity. As a result, the correlation between two unit vectors tangential to the filament at two points on the filament decays as the distance between the two points increases. The characteristic length of the decay is the persistence length of the filament.

Stiffness coefficient characterizing the response of an object to deformation. Stiffness is defined as the ratio between the force exerted on the object and the deformation induced by this force. By contrast to the elastic modulus, which is an intrinsic property of a material, stiffness is an effective property of an object and can depend on various physical parameters, including tension and elasticity (Box 2).

Viscoelasticity property of a material displaying rheological properties characteristic of both elastic solids and liquids on deformation. In Kelvin–Voigt viscoelastic materials (which are viscous at short timescales and elastic at long timescales) or Maxwell viscoelastic materials (which are elastic at short timescales and viscous at long timescales) the transition from one behavior to the other occurs on a characteristic timescale called the viscoelastic relaxation time. Biological networks are usually elastic on short timescales and viscous on long timescales because of turnover and remodeling.

Index

Note: Page numbers with "f" denote figures; "t" tables; "b" boxes.